High Marks:
REGENTS CHEMISTRY
MADE EASY
The Physical Setting

SHARON WELCHER

ADJUNCT INSTRUCTOR
Chemistry and Physics
City University of New York

CHAIRPERSON
Machon Academy High School

Teacher of High School Chemistry,
Physics, Biology,
Earth Science,
REGENTS REVIEW COURSES

See our Website:
http//www.HighMarksinSchool.com

High Marks Made Easy
Forest Hills, NY   *(877) 600-7466*

DEDICATION

I dedicate this book to
my father, Rav Jacob Joseph Mazo, ז"צל,
my mother, Claire Mazo, ע"ה
my husband, Dr. Marvin Welcher,
and my children

HIGH MARKS: REGENTS CHEMISTRY MADE EASY, THE PHYSICAL SETTING

Copyright © 2019 by Sharon Welcher. All rights reserved.
Printed in the United States of America.
No part of this book may be used or reproduced in any manner whatsoever without written permission, except in the case of brief quotations embodied in critical articles or reviews. For information, contact HIGH MARKS MADE EASY, 62-27 108th Street, Forest Hills, NY 11375, (877) 600-7466.
Graphics, editing and typesetting: **S. Malkah Cohen**
ISBN: 0-9714662-4-6
Second Edition:
First Printing: High Marks Made Easy, October, 2001.
Second Printing: High Marks Made Easy, November, 2001.
Third Printing: High Marks Made Easy, January, 2002.
Fourth Printing: High Marks Made Easy, February, 2002.
Fifth Printing: High Marks Made Easy, June, 2002.
Sixth Printing: High Marks Made Easy, August, 2002.
Seventh Printing: High Marks Made Easy, November, 2002.
Eighth Printing: High Marks Made Easy, February, 2003.
Ninth Printing: High Marks Made Easy, August, 2003.
Tenth Printing: High Marks Made Easy, September, 2003.
Eleventh Printing: High Marks Made Easy, March, 2004.
Twelfth Printing: High Marks Made Easy, August, 2004.
Thirteenth Printing: High Marks Made Easy, September, 2004.
Fourteenth Printing: High Marks Made Easy, February, 2005.
Fifteenth Printing: High Marks Made Easy, August, 2005.
Sixteenth Printing: High Marks Made Easy, September, 2005.
Seventeenth Printing: High Marks Made Easy, February, 2006.
Eighteenth Printing: High Marks Made Easy, July, 2006.
Nineteenth Printing: High Marks Made Easy, October, 2006.
Twentieth Printing: High Marks Made Easy, March, 2007.
Twenty First Printing: High Marks Made Easy, July, 2007.
Twenty Second Printing: High Marks Made Easy, October, 2007.
Twenty Third Printing: High Marks Made Easy, July, 2008.
Twenty Fourth Printing: High Marks Made Easy, September, 2008.
Twenty Fifth Printing: High Marks Made Easy, March, 2009.
Twenty Sixth Printing: High Marks Made Easy, August, 2009.
Twenty Seventh Printing: High Marks Made Easy, September, 2009.
Twenty Eighth Printing: High Marks Made Easy, August, 2010.
Twenty Ninth Printing: High Marks Made Easy, March, 2011.
Thirtieth Printing: High Marks Made Easy, August, 2011.
Thirty First Printing: High Marks Made Easy, March, 2012.
Thirty Second Printing: High Marks Made Easy, August, 2012.
Thirty Third Printing: High Marks Made Easy, August, 2013.
Thirty Fourth Printing: High Marks Made Easy, October, 2014.
Thirty Fifth Printing: High Marks Made Easy, July, 2015.
Thirty Sixth Printing: High Marks Made Easy, October, 2016.
Thirty Seventh Printing: High Marks Made Easy, September, 2017.
Thirty Eighth Printing: High Marks Made Easy, May, 2019.
10 9 8 7 6 5 4 3 2 1

All sales through
(718) 271-7466 / (877) 600-7466

INTRODUCTION

WHY IS THIS BOOK SO GOOD AND SO NECESSARY?

1 This book contains **everything you need to know** to get high marks on the New York State Chemistry Regents–The Physical Setting. This book also contains **additional** material, not on the NY State Regents exam, clearly marked with **dashed lines** on the **side** of the **page**.

2 This book is in **simple, clear, easy language**, explaining everything you need to get **high marks** on the Chemistry Regents and all chemistry exams. Chemistry is a very difficult subject, and **this book makes Chemistry easy.**

3 High Marks: Regents Chemistry Made Easy can also be used as an **easy to understand** chemistry textbook because it follows the New York State chemistry - physical setting curriculum and teaches what the students need to know for the New York State Regents.

4 If you **don't understand** a topic in chemistry, **read** the same topic in this **book**, and it will **help** you **understand** it. This book is your **private tutor.**

5 All questions are **Regents** and **Regents-type questions**. Included in the book are **constructed response questions**, a type of question that is in the revised Chemistry Regents. Sample questions are directly from the chapter or applications. These **sample questions** have **solutions** with full explanations.

6 This book **emphasizes** how to use the **Reference Tables**. Questions on the **Regents** will **involve** these **tables**. These are very **easy points** to get and require little studying. Each chapter of the book explains, in simple, clear language, **which table** to use for that chapter and **how to use it.** For emphasis, the **name** of the **Table** is indicated on the **side** of the **pages** where it is taught.

In addition, all the Reference Tables are put together in the **Reference Table section** at the end of the book, with a short **description** under each **Reference Table**. If you need more explanation of how to use a specific Table, go back to the original page in the book where it was taught.

7 High Marks: Regents Chemistry Made Easy is **continuously updated,** based on the latest Regents, enabling the students to get high marks in chemistry.

8 At the beginning of the Exam section is a detailed description of the Chemistry Regents and **test-taking strategies** to get higher marks.

9 The exam section contains **June** and **January Regents**.

10 If you don't understand a chemistry term, go to the Glossary, which has definitions in simple, easy language.

11 At the end of the book is an **index**, which makes it easier for you to find what you are looking for.

12 **Answer keys** to homework questions and Regents are available to the **students** at the **teacher's request.**

This book can be used as a basic chemistry textbook; it teaches what is required for the students to know in a clear, easy, understandable way.

With this clear and simple book, Chemistry is made **EASY**, and you can get **High Marks** on the **Chemistry Regents** and all **Chemistry exams**.

Good Luck!

FOREWORD

This is the second edition of **High Marks: Regents Chemistry Made Easy**, which conforms to the United States' national standards. This book is based on all information currently available from the New York State Education Department. Because this is a new curriculum, it is quite possible that the New York State Education Department may make some modifications to it in the next few years. Teachers should consult the State Education website, www.NYSED.gov, for updates.

The first edition I wrote was very successful. Numerous chairmen, teachers, parents, and students informed me that the book **High Marks: Regents Chemistry Made Easy** was a tremendous help and benefit to the students. I am thanking those parents who specifically called to tell me how this book helped their sons or daughters get a much higher grade on the Chemistry Regents.

My students are the ones who gave me the idea to write a chemistry book. Year after year, my students compared my notes and my review sheets to the other chemistry books in the stores. They realized that my review sheets were in simple, clear, easy language, while the other books were difficult for many of them to understand.

I decided to follow their suggestion and write easy-to-understand chemistry books to help students get high marks on the Regents, other state exams, tests, and quizzes. I have taken all my strategies of teaching over the years and incorporated them into chemistry books to help all students, and not only my own.

Furthermore, this book also will give you a good foundation for college chemistry. When I began teaching college chemistry, I realized that my own students had been given a good background, and college chemistry would be much easier for them. I am including in this book everything you need for the Regents and additional material, not required by the NY State Regents, which is clearly marked by dashed lines along the side of the page, for the student who wants an enriched course in chemistry.

ACKNOWLEDGMENTS

I thank my brilliant father, Rav Jacob Joseph Mazo, זצ"ל, for teaching me how to be an excellent teacher and how to help students get high marks on exams. I thank my dear mother, Claire Mazo, ע"ה, for encouraging me to write this book. I also thank my thoughtful husband, Dr. Marvin Welcher, for typing the manuscript and helping to proofread the book, and my children for being considerate and good.

I express my gratitude to S. Malkah Cohen, who did a professional job in editing, typesetting and computer graphics and in bringing this book to publication. Helpful suggestions from Judith Dinowitz of JM Publishers are appreciated.

I thank my students for giving me the idea to write a chemistry book. They knew my book would be in simple, clear, easy language, would make the difficult subject of chemistry **EASY** to learn, and would help students get high marks on the Regents and all chemistry exams.

I express my thanks to the department chairmen, teachers, parents and students for informing me how **High Marks: Regents Chemistry Made Easy** helped the students tremendously to understand chemistry and get higher marks on their state exams.

Sharon Welcher

TABLE OF CONTENTS

Dashed line on the side of pages in this book means topics or sections that are not part of the NYS Regents Chemistry Curriculum.

ABOUT THE AUTHOR: SHARON WELCHER

Sharon Welcher, an adjunct instructor of **chemistry** and **physics** at **City University of New York**, Science **Chairperson** of Machon Academy High School, and author of many high school science books, is a proven **master teacher**, teaching **chemistry, physics, living environment,** and **earth science** in both **public** and **private** high schools in New York City. **Almost all** her students **pass** the chemistry **Regents** every year, because she explains chemistry in a simple, clear, **easy to understand** manner and **concentrates** on what the New York State **Regents emphasizes**. She has incorporated all these strategies of teaching in her chemistry books.

Mrs. Welcher teaches **Chemistry Regents review courses** before the Regents and, when students are failing, she is called in as an **emergency tutor** to help them pass the Regents. Her ability to focus on what is important for the Regents and make it easy to understand has helped innumerable students get **excellent marks** on the Chemistry **Regents**.

She decided to take the advice of many of her students and wrote the first edition of **High Marks: Regents Chemistry Made Easy.** This book was a great success.

She now wrote the second edition of her book, **High Marks: Regents Chemistry Made Easy, The Physical Setting**. based on the revised New York State Regents curriculum, to help **all** chemistry students pass and get **high marks** on chemistry tests, Regents and other statewide exams. Included in the second edition is additional material not required by the New York State Regents, which is clearly marked by dashed lines along the sides of the page, for the student who wants an enriched course in chemistry.

Sharon Welcher is the **author** of the following books:

> **High Marks: Regents Chemistry Made Easy**
>
> **High Marks: Regents Chemistry Made Easy - The Physical Setting**
>
> **High Marks: Regents Physics Made Easy - The Physical Setting**
>
> **High Marks: Regents Living Environment Made Easy**
>
> **High Marks: College Chemistry Made Easy**

Sharon Welcher is currently **writing**

> **Organic Chemistry** of Alkanes, Alkenes, and Alkynes

CHAPTER I: PHYSICAL BEHAVIOR OF MATTER

SECTION A

INTRODUCTION

CHEMISTRY is the study of matter. **Matter** is anything that takes up space and has mass–desk, book, salt, sugar, etc. Matter is made up of (pure) substances (elements and compounds) and mixtures.

SUBSTANCE: Any variety of matter that has the same (constant) properties and composition throughout. A substance is homogeneous, made up of only one thing. (A bag of sugar is homogeneous — made entirely of sugar.)

There are two types of substances: elements and compounds. Elements and compounds are homogeneous. An ELEMENT CANNOT BE DECOMPOSED (BROKEN DOWN) into anything simpler by chemical change. Examples are: hydrogen, oxygen, nitrogen. All elements are written down on the Periodic Table on page Reference Tables 20-21 in Appendix I. A COMPOUND CAN BE DECOMPOSED (BROKEN DOWN) by a chemical change. It is made of two or more different elements chemically united in a definite ratio. Examples of compounds are water (H_2O), sodium chloride (NaCl), magnesium oxide (MgO), sulfuric acid (H_2SO_4), and ammonia (NH_3). A binary compound consists of ONLY TWO ELEMENTS, e.g., NaCl. Different compounds (example carbon monoxide (CO) and carbon dioxide (CO_2)) have different physical and chemical properties, see Properties, bottom of page 1:2.

Question: Given the equation $2Na + 2H_2O \rightarrow 2NaOH + H_2$, which substance in this equation is a binary compound?

Solution: A BINARY COMPOUND has ONLY TWO ELEMENTS (two capital letters). H_2O has two elements, **hydrogen** and **oxygen**, capital H, capital O.

MIXTURE: In a MIXTURE, two or more substances are MIXED TOGETHER (not united). Example: salt and sugar mixed together. Mixtures can be **heterogeneous** (example: salt mixed with sugar, two different things) or **homogeneous** (example: *solutions*, which are considered like one thing; particles such as salt are spread out evenly in the solution (salt water)).

Compound	Mixture
Example: **Water (H₂O)**	Example: **Salt and sugar mixed together**
1. In a compound (example: H_2O) the elements hydrogen (H) and oxygen (O) are **chemically united**.	In a mixture, two or more substances (example: salt and sugar) are **mixed together** (not united).
2. In a compound, the elements unite in **definite proportions.** For example, H_2O is a compound, always made up of 2 H atoms and 1 O atom (definite ratio).	A mixture can have **different proportions** (different amounts): ½ cup salt mixed with ½ cup sugar or ¾ cup salt mixed with ¼ cup sugar.
3.In a compound (H_2O), the compound has **different properties** than the elements(H_2 and O_2) making up the compound. H_2O (water) has different properties than the H_2 (hydrogen), a colorless gas, and O_2 (oxygen) gas.	In a mixture, each substance still has its **own properties**.Salt still has the salty taste; sugar still has the sweet taste, etc.

MAKING MODELS: A **model shows** or **represents** what **something looks like** or **how** it **works.** You can use an organic chemistry model kit to show elements, compounds, and mixtures. The model kit has many different colored balls and sticks to attach any of the balls together.

    1a. Use one colored ball to show one atom of the element iron (Fe).
     b. Use two **identical** (same color, same size) attached balls to show
        an element that has two (of the same) atoms together (oxygen, O_2).
    2. Use two (or more) different balls that are attached to show a
        compound (FeS, made of one atom of iron (Fe) and
      one atom of sulfur(S)).
    3. Use two or more different balls that are not attached to show a
        mixture of iron (Fe) and sulfur (S), or any other mixture (such as
        Fe and S and FeS).

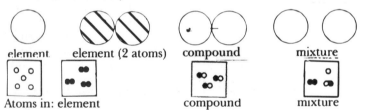

| element | element (2 atoms) | compound | mixture |

Atoms in: element   compound   mixture

PROPERTIES OF A SUBSTANCE
Properties are characteristics that can be used to identify a substance.
Physical Properties: A physical property can be recognized without
changing the substance to anything else. Examples are color, hardness,
phase(solid, liquid, or gas), solubility, odor, density, mass, volume,
boiling point, melting point, can it form a solution, and conductivity.
Chemical Properties: A chemical property describes how a substance
reacts to form new substances. Examples are does it burn, does it react
with water, with oxygen (oxidizes), does it react with acid. When a

candle burns, you have no more candle. The candle changes into new substances, carbon dioxide and water. Being able to burn is a chemical property of a candle.

Try Sample Questions #1-4, page 13, and do Homework Questions #1-7, page 16, and #39-40, page 19.

SEPARATING MIXTURES

The substances or parts of a **mixture** can be **separated** by physical methods such as **distillation, filtration, chromatography, being picked up by a magnet,** etc.

SEPARATION WITH A MAGNET: A mixture of iron and sulfur can be separated by its physical properties. Iron has the physical property of being picked up by a magnet; sulfur cannot be picked up. A magnet can be used to separate iron from sulfur.

DISTILLATION is a process in which a **mixture** of liquids or liquids and solids can be **separated by its boiling points.**

Examples:

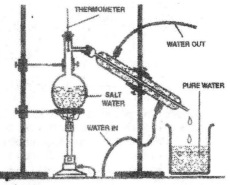

Distillation of salt water. Boil the salt water. Only the water (not the salt) forms vapor. The water vapor turns back to pure water.

Distillation of petroleum liquids.

Petroleum is a mixture that has gasoline, kerosine, fuel oil, etc. Each one has a different boiling point.
Gasoline has the lowest boiling point; therefore, it evaporates first. Gasoline vapor is collected and cooled to form gasoline. Now the gasoline is separated from the other liquids.

EVAPORATION is liquid changing into gas. You have a salt solution (salt and water). The water (liquid) forms water vapor (gas), which goes into the air. The salt remains in the beaker (or evaporating dish, etc.).

FILTRATION is a process that separates the solid and liquid parts of a mixture (see figure at right). Example: Take a piece of filter paper (in a funnel). Put a mixture of sand and water on top of the paper.

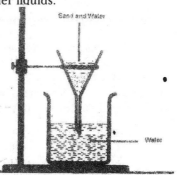

The water (liquid part) goes through the filter paper into the glass beaker while the sand (solid part) remains on top of the filter paper. Filtration can also separate larger particles (sand) from smaller dissolved ones (example, salt or sugar evenly spread out in water). Filter paper does not let the sand go through, but lets the water and dissolved (soluble) salt and sugar go through. Filtration cannot separate dissolved salt or sugar from the water, because all three went through the filter paper.

CHROMATOGRAPHY is a way to separate different molecules in a mixture. Example: Put a drop from a mixture of amino acids near one end of the chromatography paper or filter paper.

Put the end of the paper in a solvent like alcohol. The solvent alcohol moves up the paper. Different molecules in the amino acid mixture move different distances up the paper (some molecules are more soluble (dissolve more); some stick to the paper), separating the molecules in the mixture by molecular polarity (is the molecule polar or nonpolar).

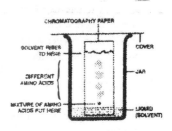

A **CENTRIFUGE** is an instrument that spins very fast. The (spinning) force, (like many times gravity) quickly pushes the most dense particles to the bottom of the tube. The less dense ones stay higher up; the least dense are at the top. A centrifuge separates the more dense and less dense particles in the mixture.

Do Homework Questions # 8-9, page 16, and #41, page 19.

There are three phases of matter, solid (s), liquid (ℓ), and gas (g). Let's compare:

Solids (s): 1. DEFINITE SHAPE AND DEFINITE VOLUME.
  2. **Particles** are **close together** in fixed positions, **vibrate** but **do not move** (strongest forces of attraction between particles.)
  3. Crystalline (crystal) structure.

Liquids (ℓ): 1. DEFINITE VOLUME (4 oz. or 8 oz., etc.) and TAKE THE SHAPE OF THE CONTAINER. (Water in a plate takes the shape of the plate; water in a cup takes the shape of the cup.)
  2. **Particles** are still **close together** and **move** while touching each other (less strong forces of attraction between particles).
  3. No regular pattern or arrangement of particles.

Gases (g): 1. NO DEFINITE VOLUME and NO DEFINITE SHAPE of their own.
  2. **Particles** are very **far apart,** spread all over, and fill evenly a closed container; particles **move all over** (least strong (weakest) forces of attraction).
  3. Total disorder - particles move in all directions.

Particle diagram: Each circle below represents one particle (1 atom or 1 molecule); 3 circles = 3 particles (example 3 atoms).

SOLID
regular pattern or arrangement

LIQUID
no regular pattern

GAS
particles spread all over

Disorder of the particles or **entropy increases** from solid to liquid to gas. Solids-fixed position, least entropy (disorder); liquids-move, more entropy (disorder); gases-move all over, most entropy (disorder).

MAKING MODELS: You can use an organic chemistry model kit to show solids, liquids, and gases. Use the colored balls to show water (H_2O) in the solid, liquid, and gas phases. One colored ball represents one water molecule. A molecule is the smallest part of a compound.

- To show a solid (ice), put colored balls close together and attach them with sticks so they cannot move.
- To show a liquid (water), put balls close together and have them move while touching each other.
- To show a gas (water vapor), put the colored balls far apart and move them in all directions.

ENERGY

Energy is the ability to do work. There are different forms of energy and energy can be changed (converted) from one form to another:

Chemical Energy: Energy released or absorbed in a chemical reaction. Burning coal, etc.: chemical energy changes to heat energy

Electrical Energy: Energy of flow of electrons (moving electrons).

Electromagnetic Radiation: Energy made up of waves, which include gamma rays, ultraviolet, light, radio waves, etc.

Heat (thermal) Energy: Random motion (movement) of atoms and molecules.

Mechanical Energy: Energy in moving objects. Example: rock falling, car moving.

Nuclear Energy: Energy given off when a nucleus breaks up into smaller nuclei or when smaller nuclei unite to form a larger nucleus.

Energy is absorbed or given off in chemical reactions:

EXOTHERMIC: Energy is GIVEN OFF in a chemical reaction.

ENDOTHERMIC: Energy is ABSORBED in a chemical reaction.

Do Homework Questions #10-13, pages 16-17, and #42-43, page 19.

How Do We Measure Energy? Energy is measured in **joules**. One joule = 1 newton-meter. One joule is a unit of energy, given on Table D, page Reference Tables-4. See Table D. One kilojoule is equal to 1000 joules.

Formula used to solve heat energy problems (how much heat, in joules, is absorbed or released) and to find change in temperature or temperature:
$$q = mC\Delta T$$
q = heat (in joules); m= mass (in grams); C= **specific heat capacity**, the amount of heat needed to raise 1 gram $1\,^\circ C$; ΔT = change in temperature.

Using Table T

Look at Table T, on page Reference Tables 26-27, in Appendix I.

Table T

Heat	$q = mC\Delta T$	q = heat
		m = mass
		C = specific heat capacity
		ΔT = change in temperature

TABLE T

C= Specific heat capacity of water (H_2O) = **4.18 joules/gram·°C**, which means 4.18 joules is needed to raise 1 gram of water 1°C. Specific heat of $H_2O(l)$ = 4.18 joules/g-°C is given on Table B, on page Reference Tables 2, in Appendix I.

Table B

Specific Heat Capacity of H_2O (ℓ)	4.18J/g•K

Question 1: How much heat energy in joules is absorbed by 30 grams of water (H_2O) when it is heated from 20°C to 30°C?

Solution:

This is a **heat energy** problem. Use the heat formula on Table T:
$$q = m \, C \, \Delta T$$
There is 30 grams of water, therefore write 30g under m (mass in grams) in the equation. Go to Table B. On Table B, it is written, specific heat capacity of $H_2O(l)$= 4.18 joules/g •°C. Write 4.18 under C in the formula. ΔT = change in temperature. Subtract the temperatures (final temperature minus starting temperature). 30°C-20°C = 10°C. Write 10°C under ΔT. Multiply 30 x 4.18 x 10 = 1254.

$$
\begin{aligned}
q = \quad & m \qquad\quad C \qquad\qquad \Delta T \\
= \quad & 30g \quad 4.18 \text{ Joules/g•°C} \quad 10°C \\
= \quad & 1254 \text{ joules}
\end{aligned}
$$

If a question asks for change in temperature *or final temperature* (instead of heat absorbed), solve for ΔT (change in temperature, example 5°C).

If heat is absorbed in this reaction, temperature goes up by ΔT (5°C); the *final temperature* is 5°C more (higher). If heat is given off, temperature goes down by ΔT (5°C); the *final temperature* is 5°C lower (less).

Question 2: How much heat (thermal) energy in joules is absorbed by 100 grams of water when it is heated from 20° to 30°C? (This is the same question as question 1, but you have more mass.)

Solution: q = mC ΔT
$$= 100 \text{ g} \times 4.18 \text{ Joules/g•°C} \times 10°C = 4180 \text{ joules}$$
As you can see, when you have more mass, you have more energy (joules).

A lot of heat energy or thermal energy is needed to raise the temperature of a swimming pool 5°C; it has a very large mass. Only a little heat energy is needed to raise the temperature of a cup of water 5°C; it has a small mass. As you know, heat energy depends on how much mass there is. q = mCΔT.

TEMPERATURE: Suppose you want to measure the temperature of warm milk. The thermometer might say 50°C, 60°C or 70°C. **Temperature** is a measure of the **average kinetic energy** of the molecules. The **higher** the temperature, the MORE KINETIC ENERGY the molecules have (the molecules move faster, at a higher speed, more molecular motion). The **lower** the temperature, the LESS KINETIC ENERGY (less molecular motion). Temperature does not depend on how much mass (matter) there is. You have a cup of milk at 80°C and a cup of milk at 20°C. Mix the two cups of milk together. **Rule**: Heat flows from a body at higher temperature to a body at lower temperature. This means heat flows from the cup of milk at 80°C to the cup of milk at 20°C until both cups are at the same temperature, about 50°C.

If you have a piece of aluminum at 80°C in contact with a piece of aluminum at 30°C, heat flows from the aluminum at 80°C to the aluminum at 30°C. Heat flows from higher temperature to lower temperature until both are at the same temperature (about 55°C).

SUMMARY: Heat travels from higher temperature to lower temperature until both temperatures are the same.

THERMOMETERS: Use a thermometer to measure temperature. Let's take a look at a Celsius thermometer. **BOILING POINT** and **FREEZING POINT** are the **two fixed points of the thermometer**. On the Celsius scale, there is 100°C between the freezing point and boiling point.

0°C = freezing point of water, also called the **ice-water equilibrium temperature**.

100°C = boiling point of water, also called the **water-steam equilibrium temperature**.

At equilibrium, liquid mercury (Hg) changes to solid Hg and solid Hg changes to liquid Hg at the same rate; this happens at one temperature, the freezing point or melting point. Look on Table S at the melting point of Hg.

Now let's understand the Kelvin (or absolute) temperature.

Kelvin = °Celsius + 273.

This formula is given in Table T, on page Reference Tables 25-26 in Appendix I. If you know the Celsius temperature, add 273 to that number and you get Kelvin. Use the formula K = °C + 273.

Table T

Temperature	K = °C + 273	K = Kelvin °C = degrees Celsius

Question: What is the Kelvin temperature for 0°C (freezing point of water)?

Solution: K = °C + 273   K = 0°C + 273 = 273K.

Question: What is the Kelvin temperature for 100°C (boiling point of water)?
Solution: K = °C + 273   K = 100°C + 273 = 373K.

Question: If the Celsius temperature is −40°C, what is the Kelvin temperature?
Solution: K = °C + 273   K = −40°C + 273 = 233K.

TABLE T

HEATING ICE

Let's see what happens to an ice cube as it changes from ice to water to gas:

You have some ice cubes. You leave them on a plate. The ice, which is solid water, takes in some heat from the air (air temperature might be $20°C$), and the ice (a solid) begins to melt and changes into water (a liquid).

The heat that the ice (solid) takes in (334 joules per gram) to become water (liquid) is called the heat of fusion. **HEAT OF FUSION** is the amount of heat needed to change a solid to a liquid at a constant temperature. Look at **TABLE B**, on page Reference Tables 2 in Appendix I.

Table B gives the heat of fusion (ice→water) = 334 joules/g. While the ice is melting, the temperature (average kinetic energy of the molecules of water) stays the same.

You want to heat the water to make hot tea. You put a pot of water on the fire. You put a thermometer in the pot of water. The temperature of the water is rising from $0°C$ (cold water) to $100°C$ (hot water).

Water (liquid) takes in heat and begins to boil and change into water vapor (gas). Heat added to a liquid at the boiling point overcomes the attractive forces of the liquid. The heat that the water (liquid) takes in (2260 joules per gram) to become water vapor (gas) is called heat of vaporization. **HEAT OF VAPORIZATION** is the amount of heat needed to change liquid to gas at constant temperature (same temperature). Look again at Reference **TABLE B**. Table B gives heat of vaporization (water→water vapor [gas])=2260 joules/g.

While the water is boiling (water changes to water vapor), the **temperature** (average kinetic energy of the molecules) **stays the same**. After **all** the water changes to water vapor (at $100°C$), the temperature of the gas rises as more heat is added.

Table B
Physical Constants for Water

Heat of Fusion	334 J/g
Heat of Vaporization	2260 J/g
Specific Heat Capacity of H_2O (ℓ)	4.18 J/g··K

HEATING CURVE DIAGRAM

Here is a **Heating Curve Diagram**, with explanations, to show what you just learned. **STUDY THIS AND KNOW IT.**

TABLE B
TABLE B
TABLE B

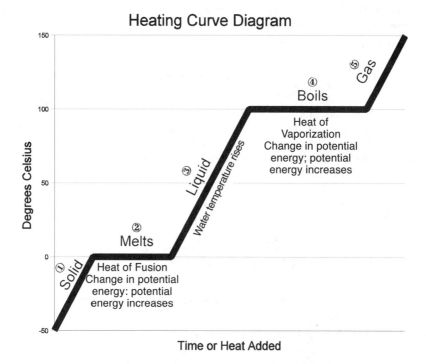

Heating Curve Diagram

Degrees Celsius (y-axis): 150, 100, 50, 0, -50

Time or Heat Added (x-axis)

⑤ Gas

④ Boils
Heat of
Vaporization
Change in potential
energy; potential
energy increases

③ Liquid — Water temperature rises

② Melts

① Solid
Heat of Fusion
Change in potential
energy: potential
energy increases

Look at ① in the diagram. The ice (solid) from a very cold freezer is at -50°C. The ice rises in temperature to 0°C. **Increase in temperature** means increase in kinetic energy.

Look at ② in the diagram. As you learned, at 0°C, the ice begins to melt. **Heat of fusion** is the amount of heat needed to change solid (example: ice) to liquid (water) at constant temperature. Heat of fusion of ice at 0°C and 1 atmosphere = 334 joules/g. (Ice takes in 334 joules per gram to become water.) In the diagram, melting is drawn as a horizontal line. During **melting, the temperature stays the same**. No change in temperature means no change in kinetic energy. During melting there is a **change in potential energy** (stored energy). Potential energy increases.

Look at ③ in the diagram (after all the ice melts). The **temperature** of the water now **rises** from 0°C to 100°C.

Look at ④ in the diagram. At 100°C, the water **boils** (changes to gas). **Heat of vaporization** is the amount of heat needed to change liquid (example: water) into gas (water vapor) at constant temperature. Heat of vaporization of water at 100°C and 1 atmosphere is 2260 joules/g. (Water takes in 2260 joules per gram to change into water vapor.) In the diagram, the horizontal line is drawn to show that water boils. When the **water boils, the temperature remains the same**. No change in temperature means no change in kinetic energy. When the water boils, there is a **change in potential energy** (**stored energy**). Potential energy increases. As you can see, line ④, heat of vaporization, is longer than line ②, heat of fusion. The longer line shows that heat is added for more time, for example more minutes, (therefore more heat is added) which means the heat of vaporization is greater than the heat of fusion.

Look at ⑤ in the diagram. After all the water has boiled, the **temperature** of the gas **rises**.

As you can see, solids are at the lowest temperature and have the least energy. Liquids are at a higher temperature and have more energy. Gases are at the highest temperature and have the most energy.

Do Sample Question #8, page 1:14

As you learned, in a solid the particles are very close together, therefore, the distance between the particles or molecules (intermolecular distance) is very small. Similarly, in a liquid, particles are very close together and intermolecular distance (distance between particles) is also very small. In a gas, particles are far apart, therefore, intermolecular distance is very large.

The shape of the heating curve is similar for most substances (example: alcohols), but the boiling points and freezing points are different.

Try Sample Questions #8-10, page 14, then do Homework Questions #23-28, pages 17-18, and #45, page 19.

COOLING CURVE DIAGRAM

Now let's discuss the **reverse way** of the Heating Curve, or the **Cooling Curve**. (Gas cools to liquid, and liquid cools to solid).

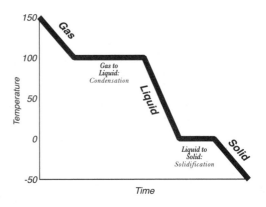

Comparison of Heating and Cooling Curves

The **Cooling Curve** looks like this. It is the **reverse of the heating curve**: You learned that, when water boils, it takes in (endothermic reaction) **2260 joules** per gram to change water to gas.

In the **Cooling** Curve, which is the reverse, or opposite, way, **gas changes to water (liquid)** and gives off (**exothermic** reaction) 2260 joules of heat per gram. This is the same amount as the heat of vaporization.

You also learned, in the Heating Curve, that, when ice melts, it takes in (**endothermic** reaction) 334 joules per gram to change ice (solid) into water (liquid).

In the Cooling Curve, which is the reverse, or opposite, way, **water (liquid) changes into ice (solid)** (freezes) and gives off (**exothermic** reaction) 334 joules of heat. This is the same amount as the heat of fusion.

You learned in the Heating Curve, when liquid changes into gas (boiling)), the **temperature stays the same**. Similarly, in the Cooling Curve, when **gas changes to liquid (condensation)**, the **temperature stays the same**. In the diagram, condensation is drawn as a horizontal line. No change in temperature means no change in kinetic energy. When gas changes to liquid (condensation), there is a **change in potential energy** (stored energy); potential energy decreases.

You learned in the Heating Curve, when solid changes to liquid, the **temperature stays the same**. Similarly, in the Cooling Curve, when **liquid changes to solid (solidification)**, the **temperature remains the same**. In the diagram, solidification (freezing) is drawn as a horizontal line. No change in temperature means there is no change in kinetic energy. When liquid changes to solid (solidification), there is only a **change in potential energy**; potential energy decreases.

You saw that during any **phase change** (solid to liquid, gas to liquid, etc.) there is no change in kinetic energy (temperature), only a **change in potential energy**. Evaporation (liquid to gas below the boiling point), sublimation (solid to gas) and deposition (gas to solid) are also phase changes; there is no change in kinetic energy, only a change in potential energy.

The shape of the cooling curve is similar for most substances (example: alcohols), but the boiling points and freezing points are different.

Note:If an example has 10 kilojoules of heat is given off (or absorbed) in one minute, obviously 20 kilojoules is given off (or absorbed) in two minutes and 30 kilojoules in three minutes.

Try Sample Question # 11, page 14, then do Homework Questions #29-33, page 18, and #46, page 19.

HEAT OF FUSION

HEAT OF FUSION: you already learned that heat of fusion is the amount of heat that the solid (example, ice) takes in (absorbs) to become a liquid (example, water). This happens **without** a change of temperature.

Look at **Reference Table B** on page Reference Tables 2 : heat of fusion for water (ice to water) is 334 joules per gram.

Question: How much heat energy is absorbed when 10 grams of ice melts to form liquid water at the same temperature?
  (1) 334   (2) 3340   (3) 33400   (4) 33

Solution: Answer *2*.
Heat Absorbed =
    Mass (Of Substance) X Heat of Fusion (Of Substance)
Mass of Ice in this Problem   = 10 g
Heat of fusion for water (Table B)   =334 joules per g
Heat absorbed = mass x heat of fusion
          = 10g X 334 joules/g =3340 joules
Answer *2*.

The formula heat absorbed = mass x heat of fusion is given on Table T, on page Reference Tables 25-26. Table T uses the letter **q** to mean heat absorbed, **m** to mean mass, and H_f to mean heat of fusion.

Table T

Heat	$q = m\ H_f$	q = heat H_f = heat of fusion
		m= Mass

HEAT OF VAPORIZATION

HEAT OF VAPORIZATION: You already learned that the heat of vaporization is the amount of heat the liquid (example, water) takes in (absorbs) to become a gas (for example, water vapor). This happens **without** a change of temperature. Look at **Reference Table B** on page Reference Tables 2: Heat of Vaporization for water (water to water vapor) is 2260 joules/g.

Look at **Reference Table B** on page Reference Tables 2

Question: How much heat is absorbed when 70.0 grams of water is completely vaporized at its boiling point?
    (1) 1582   (2) 15.82   (3) 15820 (4) 158200

Solution: Answer *4*.
Heat absorbed =
    mass (of substance) x Heat of Vaporization (of substance).
Mass of water in the problem = 70 grams
Heat of Vaporization for water (Table B) = 2260 joules/g

Heat absorbed = mass   x   Heat of Vaporization
             = 70 g   x   2260 joules/g
             = 158200 joules: Answer 4.

The formula heat absorbed = mass x heat of vaporization is given on Table T, on page Reference Tables 25-26. Table T uses the letter q to mean heat, m to mean mass, and H_v to mean heat of vaporization.

Table T

Heat	$q = m\,H_v$	q = heat H_v = heat of vaporization
		m= mass

(side tab) **TABLE B AND TABLE T**

PHYSICAL CHANGE

Physical change is a change in appearance (examples: smaller pieces or a different phase), but the substance itself is not changed (see examples below). **No new substance** is **produced**.

Examples of physical change:
1. Tear paper into pieces. Each piece of paper is still paper.
2. Chop carrots. Each little piece of carrot is still carrot.
3. Heat ice and ice becomes water and water vapor.

    The **three phases**, **ice, water,** and **vapor**, have the **same** water (H_2O) **molecules**, but they are **arranged differently**. In ice (solid), the water molecules are close together and rigid, in water they move a little, and in vapor they move far apart.
4. Dissolve sugar in water. The dissolved sugar is still the same sweet sugar it was originally.

CHEMICAL CHANGE

A **chemical change** (chemical reaction) produces a **new substance** with **different properties.**

Examples of chemical changes:

1.   Na   +   Cl   ⟶   NaCl
     Sodium   Chlorine   Sodium Chloride
     Metal   Poisonous gas   Salt that you eat

   Two poisonous substances (Na and Cl) produce a new and very different substance, the salt that we eat (NaCl).

2.   Burn magnesium. Magnesium (a silvery metal) unites with oxygen (a colorless gas) to produce magnesium oxide (a white powder). Magnesium oxide is a new and different substance.

3.   Combustion means burning (reacting with oxygen). Hydrocarbons (example gasoline) burn (unite with oxygen) to form new and different substances, carbon dioxide and water.

4.   Metals such as iron, aluminum or copper corrode (unite or combine with oxygen), forming new and different substances, iron oxide (rust), copper oxide, or aluminum oxide.

5.   If a cool metal is put into water or acid **and bubbles** are **formed**, you know it is a chemical change because a new substance (a gas, example hydrogen gas) was produced.

Try Sample Question #12, page 14, then do Homework Questions, #34-38, pages 18-19.

SAMPLE REGENTS & REGENT−TYPE QUESTIONS AND SOLUTIONS

1.   Which substance can not be decomposed into simpler substances?
     (1) ammonia   (2) aluminum   (3) methane   (4) methanol

2.   Which statement describes a characteristic of all compounds?
     (1) Compounds contain one element, only.
     (2) Compounds contain two elements, only.
     (3) Compounds can be decomposed by chemical means.
     (4) Compounds can be decomposed by physical means.

3.   Which formula represents a binary compound?
     (1) NH_4NO_3 (2) CH_4 (3) CH_3COCH_3 (4) $CaCO_3$

4.   An example of a heterogeneous mixture is:
     (1) soil   (2) sugar   (3) carbon monoxide   (4) carbon dioxide

5.   What occurs when the temperature of 10.0 grams of water is changed from 15.5 °C to 14.5 °C?
     (1) The water absorbs 103.5 joules.   (3) The water absorbs 26.9 joules.
     (2) The water releases 41.8 joules.   (4) The water releases 62.7 joules.

6. How many joules are equivalent to 35 kilojoules?
   (1) 0.035 joule   (3) 3,500 joules
   (2) 0.35 joule   (4) 35,000 joules

7. The diagrams below represent two solids and the temperature of each.

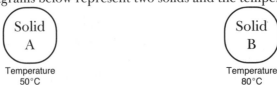

   Temperature   Temperature
   50°C   80°C

   What occurs when the two solids are placed in contact with each other?
   (1) Heat energy flows from solid *A* to solid *B*. Solid *A* decreases in temperature.
   (2) Heat energy flows from solid *A* to solid *B*. Solid *A* increases in temperature.
   (3) Heat energy flows from solid *B* to solid *A*. Solid *B* decreases in temperature.
   (4) Heat energy flows from solid *B* to solid *A*. Solid *B* increases in temperature.

8. The graph at right represents the uniform heating of a solid, starting below its melting point.
   Which portion of the graph shows the solid and liquid phases of the substance existing in equilibrium?
   (1) *AB*   (2) *BC*   (3) *CD*   (4) *DE*

   How much heat is added during interval DE? during interval CD?

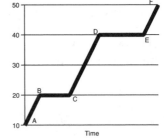

9. The graph at right represents the relationship between temperature and time as heat is added uniformly to a substance, starting when the substance is a solid below its melting point.
   Which portion of the graph represent times when heat is absorbed and potential energy increases while kinetic energy remains constant?
   (1) *A* and *B*   (3) *A* and *C*
   (2) *B* and *D*   (4) *C* and *D*

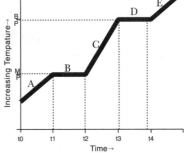

10. The amount of energy needed to change
    a given mass of ice to water at constant temperature is called the heat of
    (1) condensation   (3) fusion
    (2) crystallization   (4) formation

11. Which phase change results in a release of energy?
    (1) $Br_2(\ell) \rightarrow Br_2(s)$ (3) $H_2O(s) \rightarrow H_2O(\ell)$
    (2) $I_2(s) \rightarrow I_2(g)$ (4) $NH_3(\ell) \rightarrow NH_3(g)$

12. What is the total number of joules of heat energy absorbed when 10.0 grams of water is vaporized at its normal boiling point?
    (1) 33.4   (2) 226.0   (3) 22600   (4) 3340

SOLUTIONS

1. Answer *2*. You learned that **elements** cannot be decomposed into anything simpler. Aluminum is an element. Aluminum (Al) is on the Periodic Table of Elements.

2. Answer *3*. Remember, this is the definition of a compound.

3. Answer *2*. A binary compound consists of *only* two elements. Answer 2, CH_4, is made up of *only* two elements, C and H. The other choices are wrong because they have more than two elements. Each element starts with a capital letter.

4. Answer *1*. You learned that a mixture has two or more substances mixed together. Soil has minerals, rocks, dirt, bacteria, etc., all mixed together. Soil is heterogeneous because different parts of the soil have different amounts of materials. Some parts of the soil have more rocks; some parts of the soil have less rocks.

5. Answer *2*. Remember the formula: # joules = grams x C× ΔT. To find ΔT, subtract the two temperatures. 15.5°C - 14.5°C = 1°C ; # joules = grams x C × ΔT, # joules = 10 x 4.18 × 1 = 41.8. When the temperature of the water gets lower, 15.5°C to 14.5°C, it means water gives off (releases) joules (heat).

6. Answer *4*. 1 kilojoule = 1,000 joules. Therefore, 35 kilojoules = 35,000 joules.

7. Answer *3*. Remember, you learned that heat flows from a higher temperature to a lower temperature until both are at the same temperature. Heat flows from solid B (80°C) to solid A (50°C) until both are at the same temperature (about 65°C). Solid B's temperature decreases.

8. Answer *2*. BC represents melting. Solid and liquid exist in equilibrium. During melting, both solid and liquid can be at the same temperature. *Additional information: DE represents boiling. Liquid and gas exist in equilibrium.*

9. Answer *2*. Look at the heating curve. B represents melting and D represents boiling. During melting (B) and boiling (D), heat is absorbed, potential energy increases and there is no change in kinetic energy.

10. Answer *3*. Heat of Fusion is the amount of heat needed to change a solid to a liquid at constant temperature.

11. Answer *1*. Remember, when a solid (ice) melts and changes into liquid (water), ice takes in heat (334 joules/g). Solid → liquid takes in heat. The reverse way, liquid → solid, releases or gives off heat. Therefore, $Br_2(\ell)$ (ℓ means liquid) becoming $Br_2(s)$ (s means solid) releases heat energy.

12. Answer *3*. Remember the formula: energy absorbed = mass (of water) x heat of vaporization (of water)

    Mass of water = 10.0 grams.
    Heat of vaporization (for water) (look at Table B) = 2260 joules/g
    Heat absorbed = mass x Heat of Vaporization
    = 10 g x 2260 joules/g
    = 22600 joules   Answer *3*

1. Which pair are classified as chemical substances?
   (1) mixtures and solutions   (3) elements and mixtures
   (2) compounds and solutions   (4) compounds and elements

2. Which of the following substances can *not* be decomposed by chemical change?
   (1) sulfuric acid (2) ammonia   (3) water   (4) argon

3. Which substance can be decomposed by a chemical change?
   (1) beryllium   (2) boron   (3) methanol   (4) magnesium

4. Which substances can be decomposed chemically (by chemical means)?
   (1) CaO and Ca   (3) CO and Co
   (2) MgO and Mg   (4) CaO and MgO

5. Which formula represents a binary compound?
   (1) Ne   (2) Br_2 (3) C_3H_8 (4) H_2SO_4

6. Given the equation:
   $$2Na + 2H_2O \rightarrow 2NaOH + H_2$$
   which substance in this equation is a binary compound?
   (1) Na   (2) H_2 (3) H_2O (4) NaOH

7. When sugar is dissolved in water, the resulting solution is classified as a
   (1) homogeneous mixture   (3) homogeneous compound
   (2) heterogeneous mixture   (4) heterogeneous compound

8. The separation of petroleum into components based on their boiling points is accomplished by
   (1) cracking   (3) fractional distillation
   (2) melting   (4) addition polymerization

9. Solid and liquid parts of a mixture can be separated by
   (1) voltaic cell (2) filtration   (3) x-rays   (4) inoculation

10. In which sample are the particles arranged in a regular geometric pattern?
    (1) $HCl(\ell)$ (2) $NaCl(aq)$ (3) $N_2(g)$ (4) $I_2(s)$
    Hint: ℓ means liquid, aq means aqueous, g means gas, s means solid

11. Under the same conditions of temperature and pressure, a liquid differs from a gas because the particles of the liquid
    (1) are in constant straight-line motion
    (2) take the shape of the container they occupy
    (3) have no regular arrangement
    (4) have stronger forces of attraction between them

12. Which set of properties does a substance such as $CO_2(g)$ have?
    (1) definite shape and definite volume
    (2) definite shape but no definite volume
    (3) no definite shape but definite volume
    (4) no definite shape and no definite volume

13. Thermal energy is associated with
    (1) energy given off when a nucleus breaks up into smaller nuclei
    (2) energy of flow of electrons
    (3) random movement of atoms and molecules
    (4) energy made up of waves which include gamma rays

14. How many kilojoules are equivalent to 10 joules?
    (1) 0.001 kjoule (2) 0.01 kjoule (3) 1000 kjoules (4) 10,000 kjoules

15. What is the total number of joules of heat energy absorbed by 15 grams of water
    when it is heated from 30°C to 40°C?
    (1) 100   (2) 150   (3) 270   (4) 627

16. What occurs when the temperature of 10.0 grams of water is changed from 15.5°C
    to 14.5°C?
    (1) The water absorbs 41.8 joules
    (2) The water releases 41.8 joules
    (3) The water absorbs 155 joules
    (4) The water releases 145 joules

17. Compared to the average kinetic energy of 1 mole of water at 0°C, the average
    kinetic energy of 1 mole of water at 298 K is
    (1) the same, and the number of molecules is the same
    (2) the same, but the number of molecules is greater
    (3) greater, and the number of molecules is greater
    (4) greater, but the number of molecules is the same

18. Solid X is placed in contact with solid Y. Heat will flow spontaneously from X to Y
    when
    (1) X is 20°C and Y is 20°C   (3) X is -25°C and Y is -10°C
    (2) X is 10°C and Y is 5°C   (4) X is 25°C and Y is 30°C

19. The boiling point of water at standard pressure is
    (1) 0.000 K   (2) 100 K   (3) 273 K   (4) 373 K

20. The *minimum* number of fixed points required to establish the Celsius temperature
    scale for a thermometer is
    (1) 1   (2) 2   (3) 3   (4) 4

21. The temperature of a sample of a substance changes from 10°C to 20°C. How many
    Kelvin degrees does the temperature change?
    (1) 10   (2) 20   (3) 283   (4) 293

22. Which Kelvin temperature is equal to -73°C?
    (1) 100 K   (2) 173 K   (3) 200 K   (4) 346 K

23. At which point do a liquid and a solid exist at equilibrium?
    (1) sublimation point   (3) boiling point
    (2) vaporization point   (4) melting point

24. At 1 atmosphere of pressure, the steam-water equilibrium occurs at a temperature
    of
    (1) 0 K   (2) 100 K   (3) 273 K   (4) 373 K

25. The heat of fusion of a substance is the energy measured during a
    (1) phase change   (3) chemical change
    (2) temperature change   (4) pressure change

26. The heat of fusion is defined as the energy required at constant temperature to change 1 unit mass of a
(1) gas to a liquid
(2) gas to a solid
(3) solid to a gas
(4) solid to a liquid

27. Which phase change is endothermic?
(1) gas → solid (2) gas → liquid (3) liquid →solid (4) liquid → gas

28. Given the equilibrium at 101.3 kPa:
$H_2O(s) \rightleftharpoons H_2O(\ell)$ Hint: (s) means solid,   (ℓ) means liquid
At what temperature does this equilibrium occur?
(2) 100 K
(2) 273 K
(3) 298 K
(4) 373 K

29. Which change results in a release of energy?
(1) the melting of $H_2O(s)$
(2) the boiling of $H_2O(\ell)$
(3) the evaporation of $H_2O(\ell)$
(4) the condensation of $H_2O(g)$

30. Which change of phase is exothermic?
(1) $H_2O(s) \rightarrow H_2O(g)$
(2) $CO_2(s) \rightarrow CO_2(\ell)$
(3) $H_2S(g) \rightarrow H_2S(\ell)$
(4) $NH_3(\ell) \rightarrow NH_3(g)$

31. Which phase change results in a release of energy?
(1) $Br_2(\ell) \rightarrow Br_2(s)$
(2) $I_2(s) \rightarrow I_2(g)$
(3) $H_2O(s) \rightarrow H_2O(\ell)$
(4) $NH_3(\ell) \rightarrow NH_3(g)$

32. The graph at right represents the uniform cooling of a substance, starting with the substance as a gas above its boiling point.
   During which interval is the substance completely in the liquid state?
(1) AB
(2) BC
(3) CD
(4) DE

33. Which graph best represents a change of phase from a gas to a solid?

(1)   (2)   (3)   (4)

34. How many joules of heat are absorbed when 70.00 grams of water are completely vaporized at its boiling point?
(1) 1582   (2) 15.82   (3) 15820   (4) 158200

35. What is the total number of joules of heat energy absorbed when 10 grams of water is vaporized at its normal boiling point?
(1) 33.4   (2) 226.0   (3) 22600   (4) 3340

36. The heat of fusion of a compound is 30.0 joules per gram. What is the number of joules of heat that must be absorbed by a 15.0-gram sample to change the compound from solid to liquid at its melting point?
(1) 15.0 joules   (2) 45.0 joules   (3) 150.0 joules   (4) 450.0 joules.

37. An example of a physical change is
(1) burning magnesium
(2) boiling water
(3) combining sodium and chlorine to form sodium chloride
(4) exploding fireworks

38. An example of a chemical change is
(1) boiling water (2) cutting wood
(3) melting ice   (4) combining hydrogen and oxygen to form water

CONSTRUCTED RESPONSE QUESTIONS: Parts B-2 and C of NYS Regents Exam

Questions 15, 16, 22, 28, 32, 34, 35, 36, without giving the 4 choices, can also be used here.

39. Compare the properties of elements and compounds; give 2 examples of each.

40. a. Compare the properties of compounds and mixtures; give 2 examples of each.
b. Why does a magnet pick up iron (Fe) in a mixture of iron (Fe) and sulfur (S), but not in the compound iron sulfide (FeS)?

41. Explain 3 ways of separating mixtures.

42. Compare and contrast solids, liquids, and gases in terms of the spacing and movement of particles.

43. Using organic models, how would you demonstrate solids, liquids and gases.

44. Compare the heat needed to raise the temperature of 1 cup (250 mL) of water and a coffee urn with 10 gallons (38 liters) of water from 20°C to 90°C.

45. Draw a heating curve, label the axes, and mark the curve to show: solid, liquid, gas, melts, boils, heat of fusion, heat of vaporization.

46. Compare and contrast heating and cooling curves. Give 2 points of similarity and 2 differences.

CHAPTER QUESTIONS: Parts B-2 and C of NYS Regents Exam

47. By which process is a precipitate (insoluble substance) most easily separated from the liquid in which it is suspended?

48. The graph shows the heating curve of 1.0 gram of a solid as it is heated at a constant rate, starting at a temperature below its melting point. What is the heat of vaporization and along what line on the graph is it measured?

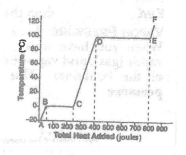

49. Draw (particle) models, showing at least 5 molecules of: a. hydrogen gas b. liquid hydrogen c. solid hydrogen. Let oo represent a molecule of hydrogen.

50. The graph shows a cooling curve, starting with the liquid above the freezing point.
A. What is the melting point?
B. Name the phase change that takes place during the 10 minute cooling time.
C. At which point do the particles have the highest kinetic energy?

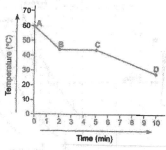

51. When a mixture of water, sand and salt is filtered, what passes through the filter paper?

(1) water, only   (2) water and sand, only

(3) water and salt, only   (4) water, sand, and salt

SECTION B

SOLIDS

SOLIDS: Definite shape, definite volume, and crystalline structure. Crystals have particles arranged in a regular geometric pattern. Particles vibrate but don't change position.

MELTING POINT: Temperature at which a solid changes to a liquid at one atmosphere pressure. (Solid and liquid are in equilibrium.) Example: temperature at which ice changes to water (melts).

SUBLIMATION: Change from solid to gas without passing through the liquid phase (no liquid phase). Examples of solids passing directly to gas:

Dry Ice: $CO_2(s)$ directly to ▸ $CO_2(g)$
Carbon dioxide solid   Carbon dioxide gas

Iodine: $I_2(s)$ directly to ▸ $I_2(g)$
Iodine solid   Iodine gas

When solids change directly to gas (just like when solids change to liquid) potential energy increases.

LIQUIDS

LIQUIDS: Definite volume and take the shape of the container it is in (example, water).

EVAPORATION is when the water changes into a gas (i.e., water vapor). Heat is absorbed (taken in) when liquid water changes to a gas.

VAPOR PRESSURE: Let's discuss a closed container. When you have water, some water is changed to vapor (gas), and vapor exerts a pressure on the sides of the container. This pressure is called **vapor pressure**.

Look at Table H: As the **temperature** of water or the other liquids (propanone, ethanol, ethanoic acid) **increases**, the **v a p o r p r e s s u r e increases**.

→ represents vapor pressure on sides of container

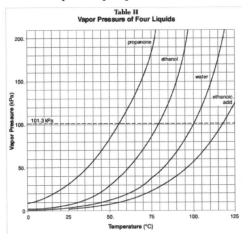

Table H
Vapor Pressure of Four Liquids

Look at the vapor pressure of water.

At 25°C, vapor pressure is 3 kPa.
At 50°C, vapor pressure is 12 kPa.
At 70°C, vapor pressure is 31 kPa.

Look at the vapor pressure of ethanol. The vapor pressure at 25°C is 7 kPa and at 75°C is 85 kPa.

TABLE H

For all liquids, as temperature increases, vapor pressure increases. Molecules have more kinetic energy and some can overcome the force holding the water molecules together and escape to form a gas. More gas causes a higher vapor pressure. (As gas increases, vapor pressure increases.)

You can also see on Table H that propanone, at 50°C, has more vapor pressure than ethanoic acid at 50°C, because propanone has weaker forces holding the propanone molecules together and the propanone molecules can escape more easily and form a gas.

Note: Table H also shows that, if ethanoic acid has a vapor pressure of 30 kPa, the temperature must be 82°C.

BOILING POINT: The atmosphere exerts a pressure of 101.3 kPa or 1 atmosphere of pressure on everything. Look at Table A, below. You see **1 atmosphere pressure = 101.3 kPa**, which is also called standard pressure.

Try Sample Questions #1-5, on page 32, then do Homework Questions, #1-10, page 35, and #28-29, page 37.

TABLE A

Name	Value	Unit
Standard Pressure	101.3 kPa	kilopascal
	1 atm	atmosphere

Water boils when the vapor pressure equals atmospheric pressure (the pressure of the atmosphere). Look at **Table H**, on the previous page. At 100°C, vapor pressure of water = 101.3 kPa. Atmospheric pressure = 101.3kPa. Therefore, AT THE **BOILING POINT**, **100°C**, **VAPOR PRESSURE (101.3 KPA) = ATMOSPHERIC PRESSURE (101.3 KPA)**. In short, water boils (boiling point) when vapor pressure equals atmospheric pressure.

GASES AND GAS LAWS

GASES have no definite volume and no definite shape. They take the shape and volume of the container.

You learned sublimation is when solids change to gas without passing through the liquid state; **deposition** is the reverse, when **gases change to solids without passing through the liquid phase.**

Iodine: $I_2(g)$ ____directly to► $I_2(s)$

Iodine gas   Iodine solid

BOYLE'S LAW: At constant temperature, the **volume of a gas is inversely proportional to pressure**. This means that, **the more pressure** you have on a gas, **the smaller the volume** of the gas. If you double the pressure (two times as much), the volume of the gas (how much space it takes up) is half.

2 times as much pressure on a gas, volume of gas is ½.

3 times as much pressure on a gas, volume of gas is 1/3.

The **formula** is:

$$P_1V_1 = P_2V_2$$

V_1 is the old volume; P_1 is the old pressure. V_2 is the new volume; P_2 is the new pressure.

Question: A gas has a volume of 50 mL, at a pressure of one atmosphere. Increase (raise) the volume to 100 mL, and the temperature remains constant. What is the new pressure?

There are two different ways of solving this question, by using either Boyle's Law (above) or Combined Gas Law. Use whichever method is easier.

Solution:

Old $\begin{cases} V_1 = \text{old volume} = 50 \text{ mL} \\ P_1 = \text{old pressure, 1 atm} \end{cases}$ New $\begin{cases} V_2 = \text{new volume} = 100 \text{ mL} \\ P_2 = \text{new pressure} = ? \end{cases}$

$P_1V_1 = P_2V_2$

Substitute numbers in the equation:

1 atm x 50 mL = P_2 x 100 mL
or P_2 x 100 mL = 1 atm x 50 mL

Divide both sides by 100,

$$\frac{100P_2}{100} = \frac{50}{100}$$

$$P_2 = 0.5 \text{ atmosphere.}$$

The new pressure is 0.5 atmosphere.

Or use this method:

Use the **Combined Gas Law**,

$$\frac{P_1V_1}{T_1} = \frac{P_2V_2}{T_2}$$

to solve gas problems that have to do with *1* or *2* or *3*:
  (1) Volume (mL or liters) and pressure (kPa or atmospheres)
  (2) Volume (mL or liters) and temperature (K)
  (3) Volume (mL or liters), temperature (K), and pressure (kPa or atmospheres)

This method has an **advantage** over the previous method because this method can be used to solve all 3 types of gas problems, while the previous method can only solve type #1.

In this example we had:

V_1 = old volume = 50 mL V_2= new volume = 100 mL
P_1 = old pressure = 1 atm P_2= new pressure = ?

Look at the **Combined Gas Law** in **Table T**:

Table T

Combined Gas Law	$\dfrac{P_1V_1}{T_1} = \dfrac{P_2V_2}{T_2}$	P = pressure V = volume T = temperature (K)

$$\frac{P_1V_1}{T_1} = \frac{P_2V_2}{T_2}$$

Substitute numbers in the equation:

$$1 \text{ atm} \times 50 \text{ mL} = P_2 \times 100 \text{ mL}$$
$$\text{or } P_2 \times 100 \text{ mL} = 1 \text{ atm} \times 50 \text{ mL}$$

(Leave out T_1 and T_2, because temperature is not given.)
In any problem, if temperature is not given, leave out T_1 and T_2.

Divide both sides by 100.

$$\frac{100P_2}{100} = \frac{50}{100}$$

$$P_2 = .5 \text{ atm}$$

---

*Try Sample Question #6, on page 32, and then do
Homework Questions, #11-15, on pages 35-36.*

CHARLES'S LAW: At constant pressure, **volume is directly proportional to Kelvin (absolute) temperature.** This means the **higher the temperature,** the **bigger the volume.** If you double the Kelvin temperature (two times as much), the volume of the gas doubles. The lower the temperature, the smaller the volume. The formula is:

$$\frac{V_1}{T_1 \text{ (Kelvin)}} = \frac{V_2}{T_2 \text{ (Kelvin)}}$$

0 K is called absolute zero = -273°C. The graph shows that the higher the Kelvin temperature, the bigger the volume.

At constant pressure, if the temperature increases from 0°C to 1°C (273K to 274K), the volume increases by $\frac{1}{273}$.

If the temperature increases from 0°C to 10°C, the volume increases by $\frac{10}{273}$.

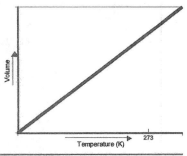

Question: At 200K, the volume of a gas is 100 mL. The temperature is raised to 300K. What is the new volume?

There are two different ways of solving this question, by using Charles's Law (above) or Combined Gas Law, which you used in the last example. Use whichever is easier.

Solution:

$V_1 = 100$ mL.
(Old volume = 100 ml.)

$V_2 = ?$
(New volume = ? ml.)

$T_1 = 200$K
(Old temperature = 200K)

$T_2 = 300$K
(New temperature = 300K)

$$\frac{V_1}{T_1 \ (Kelvin)} = \frac{V_2}{T_2 \ (Kelvin)}$$

Substitute numbers in the equation:

$$\frac{100}{200} = \frac{V_2}{300}$$

Cross multiply:

$$200 \ V_2 = 30,000$$

Divide both sides by 200:

$$\frac{200V_2}{200} = \frac{30,000}{200} \qquad V_2 = \frac{30,000}{200} = 150$$

The new volume is 150 mL.

Or use this method:

Use the **Combined Gas Law,**

$$\frac{P_1V_1}{T_1} = \frac{P_2V_2}{T_2}$$

This law has the **advantage** that it solves all types of gas problems having to do with pressure, volume, and temperature. This law combines Boyle's and Charles's Laws; it can be used instead of using either one.

In this example, we had:

$V_1 = $ old volume $= 100$ mL $V_2 = $ new volume $= ?$
$T_1 = $ old temperature $= 200$K $T_2 = $ new temperature $= 300$K

Look at the **Combined Gas Law** in **Table T**:

Table T

Combined Gas Law	$\dfrac{P_1V_1}{T_1} = \dfrac{P_2V_2}{T_2}$	P = pressure V = volume T = temperature (K)

$$\frac{P_1V_1}{T_1} = \frac{P_2V_2}{T_2}$$

Substitute numbers in the equation:

$$\frac{100}{200} = \frac{V_2}{300}$$

(Leave out P_1 and P_2, because pressure is not given.)
In any problem, when pressure is not given, leave out P_1 and P_2.

Cross-multiply:

$$200\,V_2 = 30000$$

Divide both sides by 200:

$$\frac{200V_2}{200} = \frac{30000}{200} \qquad V_2 = \frac{30000}{200} = 150\ mL$$

Remember: In the **Combined Gas Law** or Charles's Law, only use **Kelvin temperature.** If the problem gives you the temperature in degrees Celsius, like 40°C, you must change the temperature to Kelvin and then do the problem.

$$K = °C + 273\ K = 40° + 273 = 313K.$$

You learned that **Boyle's Law** deals with **volume** and **pressure**. (Hint: **B**oyle begins with **B**; take the bottom off the B and you get **P** for **p**ressure.)

Charles's Law deals with **volume** and **Kelvin temperature**.

COMBINED GAS LAW deals with any combination of **volume, pressure** and **temperature**. To solve problems, **you can always use the combined gas law** instead of Boyle's or Charles's Law.

Question: At a temperature of 273K, a 400 mL gas sample has a pressure of 101.3 kPa. If the pressure is changed to 50.65 kPa, at which temperature will this gas sample have a volume of 551 mL:
   (1) 100 K   (2) 188 K   (3) 273 K   (4) 546 K

Solution: Use the **combined gas law.**

Old $\begin{cases} P_1 = \text{pressure} = 101.3\ kPa \\ V_1 = \text{volume} = 400\ mL \\ T_1 = \text{temperature} = 273\ K \end{cases}$ New $\begin{cases} P_2 = \text{pressure} = 50.65\ kPa \\ V_2 = \text{volume} = 551\ mL \\ T_2 = \text{temperature} = ? \end{cases}$

Look at the combined gas law in Table T:

Table T

Combined Gas Law	$\dfrac{P_1V_1}{T_1} = \dfrac{P_2V_2}{T_2}$	P = pressure V = volume T = temperature (K)

$$\frac{P_1V_1}{T_1} = \frac{P_2V_2}{T_2}$$

Substitute numbers in the equation:

$$\frac{101.3 \times 400}{273} = \frac{50.65 \times 551}{T_2}$$

Cross-multiply:

$$101.3 \times 400 \times T_2 = 273 \times 50.65 \times 551$$

Divide both sides by 101.3 x 400:

$$\frac{101.3 \times 400 \times T_2}{101.3 \times 400} = \frac{273 \times 50.65 \times 551}{101.3 \times 400} \qquad T_2 = \frac{273 \times 50.65 \times 551}{101.3 \times 400} = 188 \ K$$

Answer 2

When you **increase** the **temperature** of a gas in a closed container, the particles have more kinetic energy, collide more with the walls of the container, and exert (cause) more pressure.

In a closed container, the number of particles remains the same, even if there is an increase or decrease in temperature or pressure.

STP, **STANDARD TEMPERATURE AND PRESSURE** (of a gas), is $0°C$ or 273K (temperature) and 1 atmosphere or 101.3 kPa (pressure).

TABLE A

Name	Value	Unit
Standard Pressure	101.3 kPa	kilopascal
	1 atm	atmosphere
Standard Temperature	273 K	kelvin
	0°C	degree Celsius

If you have a combined gas law problem, and a **gas** is at **STP**, use **temperature 273 K** and **pressure** (P_1 and P_2) **101.3 kPa** or **1 atm** from Table A in the combined gas law.

Try Sample Questions #7-8, page 32, then do Homework Questions, #16-20, pages 36-37, and #30, page 37.

NOTE:
DASHED LINE IN MARGIN MEANS THIS IS
NOT PART OF THE NYS REGENTS CURRICULUM

PARTIAL PRESSURE

The pressure exerted by each of the gases in a gas mixture is called the partial pressure of that gas. If you have a mixture of nitrogen and hydrogen, the pressure produced by nitrogen (partial pressure of nitrogen) and the pressure produced by hydrogen (partial pressure of hydrogen) are equal to the total pressure (pressure of nitrogen and pressure of hydrogen). Total pressure is equal to the sum of the partial pressures.

Question: A mixture has 1 mole of N_2 and 3 moles of H_2 at 25°C. If the total pressure is 8 atmospheres,
    A. What is the partial pressure of N_2?
    B. What is the partial pressure of H_2?

Solution:
A. Partial pressure of

$$N_2 = \frac{\text{Moles of } N_2}{\text{Total number of moles}} = \frac{1 \text{ Mole}}{4 \text{ Moles}}$$

N_2 has ¼ of the total number of moles, therefore N_2 has ¼ of all the pressure. ¼ x 8 atmospheres (the total pressure) = 2 atmospheres. The partial pressure of the **nitrogen** is **2** atmospheres.

B. You want to find the partial pressure of H_2: Since the total pressure is 8 atmospheres, and the partial pressure of nitrogen is 2 atmospheres, therefore the rest of the pressure is caused by hydrogen, and the partial pressure of **hydrogen** is **6** atmospheres.

or

If you want to figure out the partial pressure of hydrogen directly, again write:

$$\text{Partial Pressure of } H_2 = \frac{\text{Moles of } H_2}{\text{Total moles}} \times \text{Total Pressure}$$

$$= \frac{3 \text{ Moles } H_2}{4 \text{ Moles Total}} \times 8 \text{ Atmospheres} = 6 \text{ Atmospheres}$$

As you can see in this example, total pressure is equal to the sum of the partial pressures:

$$\underset{\text{total pressure}}{8 \text{ atmospheres}} = \underset{\text{partial pressure of } N_2}{2 \text{ atmospheres } N_2} + \underset{\text{partial pressure of } H_2}{6 \text{ atmospheres } H_2}$$

Try Sample Question #9, page 33, and then do Homework Questions, #21-22, page 37.

KINETIC MOLECULAR THEORY

KINETIC MOLECULAR THEORY is a **MODEL** that tells how gases should behave, also called "ideal gas laws."

1. A gas is composed of particles that are in continuous **random motion**.

---

2. There is a **transfer of energy** between colliding particles; the total energy remains constant.

3. The volume of gas particles is **negligible** in comparison with the volume of space they are in.There is a lot of space between the particles.

4. Gas particles are considered as having **no force of attraction** for each other.

SUMMARY:

The Four Points of the Kinetic Molecular Theory = ideal gas laws.

Hydrogen and helium follow the ideal gas laws or kinetic molecular theory *(4 points)*.

Real gases deviate from ideal gas laws. **Deviations from the gas laws** means how the gases are **different** from kinetic molecular theory (ideal gas laws).

Make the **OPPOSITE of points 3 and 4** of the kinetic molecular theory:

Point 3 of the Kinetic Molecular Theory: "The volume of gas particles is negligible."

Deviation: The volume of gas particles is significant. Gas particles do have some volume.

Point 4 of the Kinetic Molecular Theory: "Gas particles have no force of attraction."

Deviation: Gas particles *do* have a force of attraction.

These two deviations become important under **high pressure** and **low temperature**, where molecules are closer together.

[Obviously, the reverse way, low pressure and high temperature, gases don't deviate and follow the kinetic molecular theory (ideal gas laws.) This means, if you lower the pressure and raise the temperature of a gas, the gas will follow more the ideal gas laws (be more like an ideal gas).]

Explanation of Boyle's Law Using Kinetic Molecular Theory

Boyle's Law states that pressure and volume are inversely proportional when temperature is constant (stays the same).

1. You have 1 million molecules in a 1 liter container.

2. Decrease the size of the container to ½ liter (**½ the volume).**

3. Now the molecules will hit the wall twice as often (because they are in a smaller container); therefore, there will be **twice the pressure.**

You see that **pressure** and **volume** are **inversely proportional.**

EXPLANATION OF CHARLES'S LAW USING KINETIC MOLECULAR THEORY

Charles's Law states that as temperature increases, volume increases, while pressure is constant.

Put air in a balloon. Heat the air.

Higher temperature means molecules have **more kinetic energy** (**move faster** and hit the walls of the balloon more often and with more force) and therefore there is **more pressure** on the balloon.

The balloon increases in size (**gets bigger**) to offset the increase in pressure. In short, you see, as **temperature increases, volume increases**. This verifies Charles' Law.

If you **double** the (Kelvin) **temperature** of the air in a balloon, the **volume** will **double**.

Real-world application (not necessary to memorize): When you double the temperature of the air in a cylinder of an automobile engine (from 300K to 600K), the volume doubles (example: from 250 mL to 500 mL). This pushes the piston, giving the car the power to run.

You see that, as **temperature increases**, **volume increases**.

AVOGADRO'S HYPOTHESIS OR LAW: Equal volumes of gases under the same conditions of temperature and pressure have equal numbers of molecules. (But noble gases (He, Ne, etc.) are monatomic (made only of single atoms), therefore equal volumes have equal numbers of atoms).

One liter of hydrogen has the same number of molecules as one liter of oxygen, which has the same number of molecules as one liter of nitrogen, under the same conditions of temperature and pressure.

One mole (abbreviated as mol) has 6.02×10^{23} particles/mol (also called Avogadro's number). One mole of a gas occupies a volume of 22.4 liters at STP (standard temperature and pressure). *Note for New York State Regents Students: These two values are **not** part of the required curriculum but **may** be asked as an application on the exam.*

Question: How many molecules are in 11.2 liters of a gas at STP?

Solution: 1 mole occupies 22.4 liters. Therefore, there is ½ mole in 11.2 liters. 1 mole has 6.02×10^{23} molecules, therefore ½ mole has ½ that number = $½ \times 6.02 \times 10^{23} = 3.01 \times 10^{23}$, or about 3×10^{23} molecules.

KINETIC ENERGY VS ATTRACTIVE FORCE

You learned that temperature is a measure of the average kinetic energy. The higher the temperature, the more kinetic energy the particles have.

There are attractive forces that hold particles or molecules together.

Below 0°C, water is the form of ice. At a low temperature, particles have little kinetic energy to overcome the attractive forces holding the particles together and you have solid ice. (The particles have little energy to fight against the attractive force holding the particles together. The attractive force wins the fight and holds the particles very tightly together in solid form.)

At a higher temperature, 40°C, particles have more kinetic energy and can overcome some of the attractive forces and now be a liquid, example water.

At a very high temperature, 100°C, particles have more kinetic energy and can overcome the attractive forces of the water and be a gas. (At high temperature, particles have a lot of energy to fight against attractive forces holding the particles together. The kinetic energy of the particles is stronger and wins the fight against the attractive force holding the particles together. The particles escape as a gas.)

SUMMARY: The higher the temperature, the more kinetic energy the particles have and the more they can overcome the attractive forces holding the particles together. The particles can change from solid to liquid to gas.

Try Sample Questions #10-12, page 33, then do Homework Questions, #23-27, page 37, and #31-33, page 38.

GRAPHING EXPERIMENTAL **DATA**

Let's see how we can draw graphs based on experimental data for topics in matter and energy.

Problem 1:

A student obtained the following data showing how temperature affects the volume of a gas. Draw a graph to show the data.

Temperature, K	Volume, mL
50	100
100	200
200	400
400	800

How to draw the graph: (See graph on top of the next page.)

1. On the x axis, put "Temperature, K". The **thing you change** (in this case temperature) is always put on the **x axis**. What you change is called the independent variable. Space the lines along the axis equally; there must be an equal number of degrees between each two ines. (See graph on next page.)

2. On the Y axis, put "Volume, mL". The **result** you get (volume) because you changed the temperature is always put on the **y axis**. It is called the dependent variable. Space the lines along the axis equally; there must be an equal number of mL between each two lines. (See graph on next page.)

3. Plot the experimental data on the graph. Draw a circle around each point. Connect the points with a line (or draw a straight line that is the best fit between the points). Do not continue the line past the last point.

High Marks: Regents Chemistry Made Easy

4. Put a title on the graph which shows what the graph is about. Example: "Effect of temperature on gas volume".

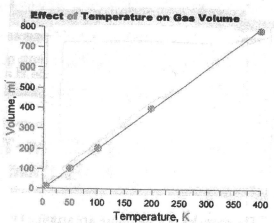

Problem 2:

A student obtained the following data for the heating curve of water, starting with ice. Draw a graph to plot the data.

Time, Minutes	Temp., °C
0	-20
1	0
3	0
5	100
16	100
17	120

How to draw the graph:

1. On the x axis, put "Time, minutes". **The thing you change** (in this case time) is always put on the **x axis**. This is the independent variable. Space the lines along the axis equally. There must be an equal number of minutes between lines. (See graph below.)

2. On the y axis, put "Temperature, °C". The **result** you get (temperature) is always put on the **y axis**. This is the dependent variable. Space the lines along the axis equally. There must be an equal number of degrees between lines. (See graph below.)

3. Plot the experimental data on the graph. Draw a circle around each point. Connect the points with a line. Do not continue the line past the last point.

4. Put a title on the graph which shows what the graph is about. Example: "Effect of heating time on temperature of water."

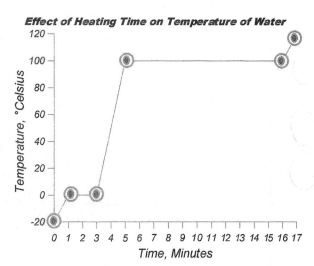

REMEMBER: When you answer Regents questions on MATTER AND ENERGY, use Table A , Table B, Table H and Table T.

SAMPLE REGENTS & REGENT-TYPE QUESTIONS AND SOLUTIONS

1.   The particles of a substance are arranged in a definite geometric pattern and are constantly vibrating. This substance can be in
   (1) the solid phase, only
   (2) the liquid phase only
   (3) either the liquid or the solid phase
   (4) neither the liquid nor the solid phase

2.   Which two compounds readily sublime at room temperature (25°C)?
   (1) $CO_2(s)$ & $I_2(s)$ (3) $NaCl(s)$ & $I_2(s)$
   (2) $CO_2(s)$ & $C_6H_{12}O_6(s)$ (4) $NaCl(s)$ & $C_6H_{12}O_6(s)$

3.   Under the same conditions of temperature and pressure, a liquid differs from a gas because the particles of the liquid
   (1) are in constant straight-line motion
   (2) take the shape of the container they occupy
   (3) have no regular arrangement
   (4) have stronger forces of attraction between them

4.   If the pressure on the surface of propanone in the liquid state is 90 kPa,, the liquid will boil at about
   (1) 0°C   (2) 55°C   (3) 80°C   (4) 100°C

5.   Water will boil at a temperature of 70°C when the pressure on its surface is
   (1) 121 kPa   (2) 91 kPa   (3) 31 kPa   (4) 61 kPa

6.   A gas sample has a volume of 25.0 milliliters at a pressure of 1.0 atmosphere. If the volume increases to 50.0 milliliters and the temperature remains constant, the new pressure will be
   (1) 1.00 atm   (2) 2.00 atm   (3) 0.250 atm   (4) 0.500 atm

7.   Which temperature represents absolute zero?
   (1) 0 K   (2) 0°C   (3) 273 K   (4) 273°C

8.   The volume of a given mass of an ideal gas at constant pressure is:
   (1) directly proportional to the Kelvin temperature
   (2) directly proportional to the Celsius temperature
   (3) inversely proportional to the Kelvin temperature
   (4) inversely proportional to the Celsius temperature

9. What is the pressure of a mixture of CO_2, SO_2 and H_2O gases, if each gas has a partial pressure of 30 kPa?
   (1) 30 kPa   (2) 60 kPa   (3) 90 kPa   (4) 120 kPa

10. Under which conditions does a real gas behave most nearly like an ideal gas?
    (1) high pressure and low temperature
    (2) high pressure and high temperature
    (3) low pressure and low temperature
    (4) low pressure and high temperature

11. According to the kinetic theory of gases, which assumption is correct?
    (1) Gas particles strongly attract each other.
    (2) Gas particles travel in curved paths.
    (3) The volume of gas particles prevents random motion.
    (4) Energy may be transferred between colliding particles.

12. The total quantity of molecules contained in 5.6 liters of a gas at STP is
    (1) 1.0 mole   (2) 0.75 mole   (3) 0.50 mole   (4) 0.25 mole

---

SOLUTIONS

1. Answer *1*. Only solids are particles arranged in a regular geometric pattern and vibrating.

2. Answer *1*. Sublimation is a change from solid to gas without passing through liquid phase (no liquid phase). You had two examples of sublimation: $CO_2(s)$ and $I_2(s)$.

3. Answer *4*. A liquid has stronger forces of attraction holding the particles together. (The other three choices are correct for both liquids and gases.)

4. Answer *2*. Step 1: Look at the Table of Vapor Pressures of 4 Liquids. Find 90 kPa on the Y axis (labeled vapor pressure). Step 2: Go across until it touches the vapor pressure curve of propanone. Step 3: Draw a line straight down to the x axis. It touches the x axis at 53°C.
   Explanation: Step 1: Pressure on surface = 90 kPa–Y axis. Step 2: Go across and touch vapor pressure curve; this is the point where **vapor pressure equals pressure on surface**. Step 3: Draw a line down and you see it is 53°C. **53°C is the boiling point because vapor pressure equals pressure on surface.**

5. Answer *3*. Look at the vapor pressure curve for water. In this example you are given temperature; in the last question, you were given vapor pressure. Step 1: Look at 70° (temperature is on the x axis). Step 2: Go up to the vapor pressure curve for water. Step 3: Now go across to the y axis (vapor pressure). You see that the vapor pressure of water

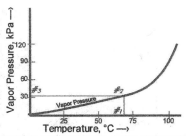

is 31 kPa. When the outside pressure equals the vapor pressure (31 kPa), water boils. Therefore, the outside pressure must be 31 kPa for water to boil (at 70°C).

6. Answer *4*. Use the combined gas law fromTable T.

$$\frac{P_1V_1}{T_1} = \frac{P_2V_2}{T_2}$$

V_1 = old volume = 25 mL
V_2 = new volume = 50 mL

P_1 = old pressure = 1 atmosphere
P_2 = new pressure = question

Substitute numbers in the equation:

1 atm x 25mL = P_2 x 50mL
Or P_2 x 50mL = 1 atm x 25mL
(Leave out T_1 and T_2 because temperature is not given).

Divide both sides by 50:

$$\frac{50\ P_2}{50} = \frac{25}{50}$$

$$P_2 = 0.5\ atm$$

7. Answer *1*. You learned that absolute zero = 0K.

8. Answer *1*. Remember Charles's Law. Volume is directly proportional to Kelvin temperature.

9. Answer *3*. You learned that the total pressure is equal to the sum of the partial pressures.

		CO_2		SO_2		H_2O
90	=	30	+	30	+	30
Total Pressure	+	Partial Pressure of CO_2		Partial Pressure of SO_2		Partial Pressure of H_2O

Therefore, the answer is 90.

10. Answer *4*. Use this trick. Remember, you memorized that deviations from the kinetic theory are important under high pressure and low temperature. The question asks for an ideal gas, which is the opposite of deviations, therefore low pressure and high temperature.

11. Answer *4*. You should remember the four points of the kinetic theory. One point of the kinetic theory is the transfer of energy between particles.

12. Answer *4*. One (1) mole of gas at STP occupies 22.4 liters. 22.4 liters has 1 mole of molecules. 5.6 liters is (¼ x 22.4 liters) = ¼ (1 mole), therefore 5.6 liters = ¼ mole.

or

Set up a proportion:

$$\frac{1\ Mole}{22.4\ L} = \frac{x\ Moles}{5.6\ L}$$

Cross multiply:

$$22.4X = 5.6$$

Divide by 22.4:

$$\frac{22.4}{22.4} = \frac{5.6}{22.4}$$

$$X = 0.25$$

1. The particles in a crystalline solid are arranged
    (1) randomly and far apart   (3) regularly and far apart
    (2) randomly and close together   (4) regularly and close together

2. The characteristic which distinguishes a true solid from other phases of matter at STP is that, in a true solid, the particles are
    (1) vibrating and changing their relative positions
    (2) vibrating without changing their relative positions
    (3) motionless but changing their relative positions
    (4) motionless without changing their relative positions

3. In which sample are the particles arranged in a regular geometric pattern?
    (1) HCl(ℓ)   (2) NaCl(aq)   (3) N₂(g)   (4) I₂(s)

4. Which process occurs when dry ice, CO_2(s) is changed into CO_2(g)?
    (1) crystallization   (3) sublimation
    (2) condensation   (4) solidification

5. Which two compounds readily sublime at room temperature (25°C)?
    (1) CO_2(s) and I_2(s)   (3) NaCl(s) and I_2(s)
    (2) CO_2(s) and $C_6H_{12}O_6$(s)   (4) NaCl(s) and $C_6H_{12}O_6$(s)

6. Which property of a sample of mercury is different at 320 K than at 300 K?
    (1) atomic mass   (3) vapor pressure
    (2) atomic radius   (4) melting point

7. Which sample of water has the greatest vapor pressure?
    (1) 100 mL at 20°C   (3) 20 mL at 30°C
    (2) 200 mL at 25°C   (4) 40 mL at 35°C

8. At which temperature will water boil when the external pressure is 31 kPa?
    (1) 25°C   (2) 50°C   (3) 70°C   (4) 100°C

9. What is the vapor pressure of a liquid at its normal boiling temperature?
    (1) 1 atm   (2) 2 atm   (3) 273 atm   (4) 760 atm

10. As the temperature of H_2O(ℓ) in a closed system decreases, the vapor pressure of the H_2O(ℓ)
    (1) decreases   (2) increases   (3) remains the same

11. Which set of properties does a substance such as CO_2(g) have?
    (1) definite shape and definite volume
    (2) definite shape but no definite volume
    (3) no definite shape but definite volume
    (4) no definite shape and no definite volume

12. When the pressure exerted on a confined gas at a constant temperature is doubled, the volume of the gas is
    (1) halved   (2) doubled   (3) tripled   (4) quartered

13. The pressure on 30.0 milliliters of an ideal gas increases from 101.3 kPa to 202.6 kPa at constant temperature. The new volume is

(1) $30 \text{ mL} \times \dfrac{101.3\ kPa}{202.6\ kPa}$

(2) $30 \text{ mL} \times \dfrac{202.6\ kPa}{101.3\ kPa}$

(3) $\dfrac{101.3\ kPa}{30\ mL} \times 202.6 \text{ kPa}$

(4) $\dfrac{202.6\ kPa}{30\ mL} \times 101.3 \text{ kPa}$

14. A sample of gas has a volume of 2.0 liters at a pressure of 1.0 atmosphere. When the volume increases to 4.0 liters, at a constant temperature, the pressure will be
    (1) 1.0 atm   (2) 2.0 atm   (3) 0.50 atm   (4) 0.25 atm

15. A gas sample has a volume of 25.0 milliliters at a pressure of 1.0 atmospheres. If the volume increases to 50.0 milliliters and the temperature remains constant, the new pressure will be
    (1) 1.0 atm   (2) 2.00 atm   (3) 0.250 atm   (4) 0.500 atm

16. Which graph represents the relationship between volume and Kelvin temperature for an ideal gas at a constant pressure?

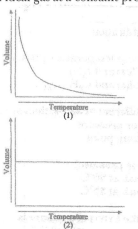

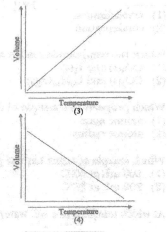

17. A sample of gas occupies 4 liters at STP. The volume is changed to 2 liters and the temperature is changed to 25°C. The new pressure of the gas is approximately
    (1) 220 kPa   (2) 75 kPa   (3) 400 kPa   (4) 27 kPa

18. At a temperature of 273 K, a 400-milliliter gas sample has a pressure of 101.3 kilopascals. If the pressure is changed to 50.65 kPa, at which temperature will this gas sample have a volume of 551 milliliters?
    (1) 100 K   (2) 188 K   (3) 273 K   (4) 546 K

19. A gas has a volume of 2 liters at 323 K and 3 atmospheres. When the temperature is changed to 273 K and the pressure is changed to 1 atmosphere, the new volume of the gas would be equal to

(1) $2L \times \dfrac{273K}{323K} \times \dfrac{1\ atm}{3\ atm}$

(2) $2L \times \dfrac{323K}{273K} \times \dfrac{1\ atm}{3\ atm}$

(3) $2L \times \dfrac{273K}{323K} \times \dfrac{3\ atm}{1\ atm}$

(4) $2L \times \dfrac{323K}{273K} \times \dfrac{3\ atm}{1\ atm}$

20. The volume of a 1.00-mole sample of an ideal gas will decrease when the
    (1) pressure decreases and the temperature decreases.
    (2) pressure decreases and the temperature increases
    (3) pressure increases and the temperature decreases

    (4) pressure increases and the temperature increases

21. When 7.00 moles of gas A and 3.00 moles of gas B are combined, the total pressure exerted by the gas mixture is 100 kPa. What is the partial pressure exerted by gas A in this mixture?
    (1) 10 kPa   (3) 70 kPa
    (2) 30 kPa   (4) 100 kPa

22. A sealed container has 1 mole of helium and 2 moles of nitrogen at 30°C. When the total pressure of the mixture is 90 kPa, what is the partial pressure of the nitrogen?
    (1) 15 kPa   (2) 30 kPa   (3) 60 kPa   (4) 90 kPa

23. One reason that a real gas deviates from an ideal gas is that the molecules of the real gas have
    (1) a straight-line motion
    (2) no net loss of energy on collision
    (3) a negligible volume
    (4) forces of attraction for each other

24. At STP, which gas has properties most similar to those of an ideal gas?
    (1) NH_3 (2) CO_2 (3) O_2 (4) H_2

25. A real gas would behave most like an ideal gas under conditions of
    (1) low pressure and low temperature
    (2) low pressure and high temperature
    (3) high pressure and low temperature
    (4) high pressure and high temperature

26. At STP, 1 liter of $O_2(g)$ and 1 liter of $Ne(g)$ have the same
    (1) mass   (3) number of atoms
    (2) density   (4) number of molecules

27. Flask A contains 2 liters of $CH_4(g)$ and Flask B contains 2 liters of $O_2(g)$. Each gas sample has the same
    (1) density   (3) number of molecules
    (2) mass   (4) number of atoms

CONSTRUCTED RESPONSE QUESTIONS: Parts B-2 and C of NYS Regents Exam

Questions 12, 13, 14, 15, 16, 18, 19, 21, 22, 27, without giving the 4 choices, can also be used here.

28. Explain sublimation and give 2 examples.

29. A.   How does temperature affect vapor pressure?
    B.   Why does temperature affect vapor pressure?

30. A.   At 300K, the volume of a gas is 100 mL. The temperature is raised to 500 K at constant pressure. What is the new volume?
    B.   The temperature of a 2.0 liter sample of helium gas at STP is increased to 27°C and the pressure is decreased to 80. kPa What is the new volume of the helium sample?

31. Explain Boyle's and Charles's Laws according to the kinetic molecular theory.

32. How many molecules are there in 44.8 liters of a gas at STP?

33. Explain the 4 parts of the kinetic molecular theory.

<table>
<tr><td colspan="2">34. Construct a graph of the cooling curve for water, using the following data:</td><td colspan="2">35. Construct a graph of the heating curve for water, using the following data:</td></tr>
<tr><td>Time, min.</td><td>Temp., °C</td><td>Time, Min.</td><td>Temp., °C</td></tr>
<tr><td>0</td><td>120</td><td>0</td><td>-10</td></tr>
<tr><td>3</td><td>100</td><td>2</td><td>0</td></tr>
<tr><td>5</td><td>100</td><td>4</td><td>0</td></tr>
<tr><td>15</td><td>0</td><td>6</td><td>40</td></tr>
<tr><td>20</td><td>0</td><td>8</td><td>80</td></tr>
<tr><td>25</td><td>-20</td><td>10</td><td>100</td></tr>
<tr><td></td><td></td><td>12</td><td>100</td></tr>
<tr><td></td><td></td><td>14</td><td>100</td></tr>
<tr><td></td><td></td><td>16</td><td>100</td></tr>
<tr><td></td><td></td><td>18</td><td>120</td></tr>
</table>

CHAPTER QUESTION: Parts B-2 and C of NYS Regents Exam

36. The table below shows the data collected by a student as heat was applied at a constant rate to a solid below its freezing point.

Time, min.	Temperature, °C	Time, Min.	Temperature, °C
0	20	18	44
2	24	20	47
4	28	22	51
6	32	24	54
8	32	26	54
10	32	28	54
12	35	30	54
14	38	32	58
16	41	34	62

What is the boiling point of this substance?

37. A. Which graph shows the pressure-temperature relationship expected for an ideal gas?(The x-axis shows temperature). Explain your answer.
   B. Which graph best shows the relationship between the pressure of a gas and its average kinetic energy at constant volume? (The x-axis shows average kinetic energy). Explain your answer.

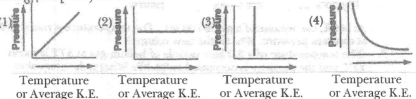

| Temperature or Average K.E. | Temperature or Average K.E. | Temperature or Average K.E. | Temperature or Average K.E. |

CHAPTER 2: ATOMIC CONCEPTS

SECTION A

TO THE STUDENT: Study and know the book. Review the **bold words.**

ATOMIC MODELS

(Note: Names of the scientists do not have to be memorized.)

John Dalton stated that elements are made of atoms. Atoms of one element are alike; atoms of different elements are different.

J.J. Thomson discovered electrons, particles with a negative charge. To him, the atom was a hard sphere of positive charge, with electrons in it.

Ernest Rutherford **bombarded gold foil** with **alpha particles** (nuclei of helium atoms). Most of the alpha **particles went** straight **through** the foil, **showing** that most of the **atom is empty space.** Alpha particles are positively charged. Some **alpha particles** that hit the gold foil **bounced back.** This **showed** that most of the mass of the atom is in the center, the **nucleus,** which is **positive.** Rutherford's model had electrons going around the nucleus.

Niels Bohr found that electrons are in different orbits around the nucleus.

THE ATOM

The **NUCLEUS** is in the center of the atom and has **protons** and **neutrons.** The **electrons** are around the nucleus. (Since protons, neutrons and electrons are inside the atom, they are called subatomic particles.) Most of the atom is **empty space.**

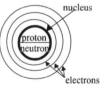

	PROTON	**NEUTRON**	**ELECTRON**
CHARGE	positive	none	negative
MASS	1 atomic mass unit	1 atomic mass unit	hardly any: 1/1836 atomic mass unit

Atomic mass unit is also written as u. Mass of one proton is one atomic mass unit (1 u), therefore mass of ten protons equals ten atomic mass units (10 u).

TABLE O lists the neutron, $_0^1n$; proton, $_1^1H$; and electron, $_{-1}^0e$. The top number is the mass. Mass of neutron = 1, mass of proton = 1, mass of electron = 0 (which means very, very little).

Table O
Symbols Used in Nuclear Chemistry

Name	Notation	Symbol
alpha particle	$_2^4He$ or $_2^4\alpha$	α
beta particle (electron)	$_{-1}^0e$ or $_{-1}^0\beta^-$	β^-
neutron	$_0^1n$	n
proton	$_1^1H$ or $_1^1p$	p

The **ATOMIC NUMBER** of any atom is equal to the number of **protons**, which is equal to the number of **electrons**.

SUMMARY:

Atomic Number = number of protons = number of electrons.

The atom lithium, Li, has an atomic number of 3. **Atomic number 3 = 3 protons** (3 positive charges) = **3 electrons** (3 negative charges). The **first** circle or **shell** can only hold **2 electrons**. Therefore, the **third electron** goes into the **next** circle or **shell**.

Lithium: $_3$Li

atomic number

Mass # – 7
Atomic # – 3 Li

The **MASS NUMBER** is equal to the total number of **protons and neutrons** in the nucleus. Lithium has a **mass number** of 7, which is **equal to 3 protons** and **4 neutrons** in the nucleus. To find the **number of neutrons**, take the **mass number minus the atomic number**.

 # neutrons in Lithium = 7 (mass number) - 3 (atomic number) = 4 neutrons

As you see, lithium is written as $_3^7$Li. Li is the symbol for lithium; 3 is the atomic number, which is written at the bottom of the symbol of the element; and 7 is the mass number, which is written at the top of the symbol of the element

---

Question: Lithium has three protons.
   1. What is the atomic number?
   2. How many electrons does a lithium atom have?

Solution:
   1. The atomic number equals the number of protons. Lithium has 3 protons, therefore the atomic number = 3.
   2. An **atom** (example lithium) **is** electrically **neutral,** which means, in an atom, the **number** of **protons equals** the **number** of **electrons.** A lithium atom has 3 protons, therefore a lithium atom has 3 electrons.

Question: Lithium has 3 protons and 4 neutrons.
   A. What is the mass number?
   B. What is the charge of the lithium nucleus?

Solution:
   A. Mass number equals the sum of protons and neutrons. A lithium atom has 3 protons and 4 neutrons. Therefore, lithium has a mass number of 7.

   B. The lithium nucleus has 3 protons (3 positive charges) and 4 neutrons (neutrons have zero charge); therefore, charge of the lithium nucleus is +3.

Question: An atom has a mass number of 9 and has 5 neutrons. How many protons does it have?

Solution:

Mass number = number of protons (p) + number of neutrons (n).

$$
\begin{array}{rl}
\text{Mass Number} & = p + n \\
9 & = p + 5 \\
\underline{-5} & \quad \underline{-\ 5} \\
4 & = p
\end{array}
$$

Answer: 4 protons.

Check: $4p + 5n = 9$ mass number

Question: An atom has a mass number 11 and has 5 protons. How many neutrons does it have?

Solution:

mass number = number of protons (p) + number of neutrons (n).

$$
\begin{array}{rl}
\text{Mass Number} & = p + n \\
11 & = 5 + n \\
\underline{-5} & \quad \underline{-\ 5} \\
6 & = \quad n
\end{array}
$$

Answer: 6 neutrons.

Check: $6n + 5p = 11$ mass number

Now let's look at hydrogen. The element hydrogen, **H**, has an atomic number of 1 and a mass number of 1.

You do not have to memorize these numbers. If you know the element and want to know the atomic number, look at the Periodic Table, page Reference Tables 20-21. Hydrogen is written $_1^1$H. Below the symbol of the element is the atomic number; therefore, you can see that hydrogen has an atomic number of 1. Each element has its own atomic number. Each element with its atomic number is also given in Table S.

Mass # → $_1^1$H
Atomic # →

The atomic number is equal to the number of protons, which identifies the element (tells you what element it is). Atomic number 1 has 1 proton, and that tells you it is hydrogen (see drawing above).

Try Sample Questions #1-4, page 12, then do Homework Questions 1-10, pages 13-14, and #28, page 15.

ISOTOPES of the same element have the **same atomic number** but **different mass numbers.** There are three **isotopes** of hydrogen: $_1^1$H, $_1^2$H, and $_1^3$H. They all have the same atomic number, 1, which is the atomic number of hydrogen, but they have different top (mass) numbers. (These isotopes have different numbers of neutrons.)

Use this equation:
mass # - atomic # = number of neutrons
Therefore:

^1_1H: mass # - atomic # = number of neutrons; 1 - 1 = 0 neutrons

^2_1H: mass # - atomic # = number of neutrons; 2 - 1 = 1 neutrons

^3_1H: mass # - atomic # = number of neutrons; 3 - 1 = 2 neutrons

As you can see, isotopes differ in the number of neutrons.

You can **identify** (know) and describe which hydrogen **isotope** it is by the **mass number** (sum of protons and neutrons).

Isotope ^3_1H can be written as ^3H, ^3_1H, **Hydrogen-3** or **H-3**.

There are three isotopes of carbon, $^{12}_6C$, $^{13}_6C$, and $^{14}_6C$. Isotopes have the same atomic number, different mass number, and different number of neutrons. Again you can **identify** (know) and describe which carbon atom it is by the **mass number** (12, 13, or 14).

Isotope $^{14}_6C$ can be written as ^{14}C (common notation), $^{14}_6C$, **carbon-14**, or **C-14**. Mass # is written after the dash: C-14.

One (1) atomic mass unit equals $1/12^{th}$ the mass of ^{12}C.

$$1\ u\ (atomic\ mass\ unit) = 1.66\ x\ 10^{-24}g.$$

The **ATOMIC MASS** of an element is the weighted average mass of the naturally occurring isotopes of that element. The average is weighted according to the proportions in which the isotopes occur.

There are two isotopes of chlorine, $^{35}_{17}Cl$ and $^{37}_{17}Cl$. They have the same atomic number but different mass numbers. Look at the box for the element $_{17}Cl$ from the Periodic Table. The atomic mass of $_{17}Cl$ is 35.5, which is the average weight of all the isotopes. Atomic mass is given in atomic

| 35.453 |
| **Cl** |
| 17 |

mass units (u); atomic mass of $^{35.5}_{17}Cl$ is 35.5 u (atomic mass units).

Since the atomic mass, average weight 35.5 u, is closer to atomic mass 35 (of $^{35}_{17}Cl$) than to atomic mass 37 (of $^{37}_{17}Cl$), there is more of the isotope $^{35}_{17}Cl$ and it is the most abundant isotope.

Question: There is 25% of the naturally occurring isotope $^{37}_{17}Cl$ and 75% of the naturally occurring $^{35}_{17}Cl$. What is the atomic mass of the element?

Solution: Method 1 (Can be used when percent is easily changed to a fraction): 25% or one quarter of chlorine has an atomic mass of 37. 75% or 3/4 of chlorine has a mass of 35. Take the average of the masses in the proportion which they are.

$$37 \quad \text{1/4 or 1 out of 4}$$

$$\left.\begin{matrix} 35 \\ 35 \\ 35 \end{matrix}\right\} \text{3/4 or 3 out of 4}$$

$$4\ )\ \overline{142}$$

$$35.5 = \text{Atomic mass}$$
Average

Atomic mass = 35.5 u

Method 2: Take the percentage of each isotope times its mass; add the numbers.

Isotope		Percentage	Times	Mass		
$^{37}_{17}Cl$	Take 25%:	$\frac{25}{100}$ or .25	X	37 u	=	9.25 u
$^{35}_{17}Cl$	Take 75%:	$\frac{75}{100}$ or .75	X	35 u	=	26.25 u
				Add:		35.50 u

Atomic
mass

Atomic mass = 35.50 u

Do Sample Question #6B, page 2:12: Find atomic mass

Question: Carbon-14 differs from hydrogen-3 in that carbon-14 has

(1) 6 more neutrons   (2) 6 more protons

(3) 6 more electrons   (4) 2 more neutrons

Solution: Carbon-14 means the mass number is 14. Carbon has an atomic number of 6. See Periodic Table. $_6C$

Number of neutrons = mass # - atomic # = 14 - 6 = **8 neutrons**.

Hydrogen-3 means mass # is 3. Hydrogen has an atomic number of 1.

See Periodic Table. $_1H$

# of neutrons = mass # - atomic # = 3-1 = **2 neutrons**.

Answer 1. Carbon-14 has 6 more neutrons than hydrogen-3.

Try Sample Questions #5-6, page 12, then do Homework Questions, #11-14, page 14, and #29, page 15.

CHANGES IN ATOMIC MODELS

THOMSON MODEL OF THE ATOM: HARD SPHERE MODEL. The atom is a hard sphere of positive charge with electrons (negative charges) in it.

RUTHERFORD MODEL OF THE ATOM. Most of the mass of the atom is in the center, the nucleus, which is positive. Protons are in the nucleus. Most of the atom is empty space. Electrons go around the nucleus.

BOHR MODEL OF THE ATOM. Protons are in the nucleus, which is positive. Electrons revolve (go around) the nucleus in concentric circular orbits.

MODERN MODEL: WAVE MECHANICAL MODEL (ELECTRON CLOUD). Protons are in the nucleus. The electron cloud model (based on the work done by many scientists over a very long time) shows that an electron is in **an orbital**, (which is not the exact location of the electron, but) which is the most probable place (location) where the electron is. It shows the electron as a diffuse (spread out) cloud of negative charge. The thickest (most dense) part of the cloud is the most probable place to find the electron. The thinnest part of the cloud is the least likely place to find the electron. When an electron goes from an orbital which has more energy to an orbital which has less energy, a spectrum (colors, energy) is given off.

PRINCIPAL ENERGY LEVELS

PRINCIPAL ENERGY LEVELS(SHELLS or PRINCIPAL QUANTUM NUMBERS) can be shown as 1, 2, 3, 4:

$$\ominus\ )_1\ )_2\ )_3\ )_4$$

Principal energy level shows how far the electron is from the nucleus. The first energy level (Shell #1) is closest to the nucleus, while other energy levels are further away from the nucleus. Electrons in the first energy level have the lowest energy. Those in the 2nd energy level have more energy; those in the 3rd have still more energy, etc.

First principal energy level can only hold **2** electrons.
Second principal energy level can only hold **8** electrons.
Third principal energy level can only hold **18** electrons.
Fourth principal energy level can only hold **32** electrons.

Maximum Number of Electrons in Each Principal Energy Level

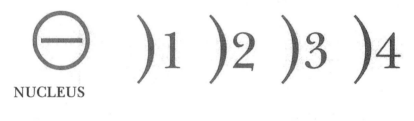

NUCLEUS

Maximum # of electrons	2	8	18	32

Try Sample Question #7, page 12, then do Homework Questions #15-18, page 14, and #30, page 15.

ELECTRON CONFIGURATION

Electron configuration shows how many electrons are in each principal energy level.

$_9F\ \ominus\ \begin{matrix})) \\ )_2\ )_7 \\ )) \end{matrix}$

Fluorine

Fluorine ($_9^{19}F$) has an atomic number of 9, therefore fluorine has 9 electrons.

1. You put 2 electrons in principal energy level 1. Principal energy level 1 can only hold 2 electrons.

2. The next 7 electrons are in principal energy level 2.

The **electron configuration** to describe the electrons in **fluorine** is **2-7**: 2 electrons in the first principal energy level, 7 electrons in the second principal energy level.

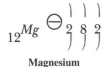

Magnesium

Magnesium ($_{12}^{24}$Mg) has an atomic number of 12; therefore, it has 12 electrons:

1. You put 2 electrons in principal energy level 1. Principal energy level 1 can only hold 2 electrons.

2. The next 8 electrons are in principal energy level 2. The second principal energy level only holds 8 electrons.

3. The next two electrons are in principal energy level 3.

The **electron configuration** to describe the electrons in **magnesium** is **2-8-2** : 2 electrons in the first energy level, 8 in the second energy level, and 2 electrons in the third energy level.

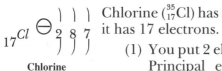

Chlorine

Chlorine ($_{17}^{35}$Cl) has an atomic number of 17; therefore, it has 17 electrons.

(1) You put 2 electrons in principal energy level 1. Principal energy level 1 can only hold 2 electrons.

(2) The next 8 electrons are in principal energy level 2. Principal energy level 2 can only hold 8 electrons.

(3) The next 7 electrons then are in principal energy level 3.

The **electron configuration** of **chlorine** is **2-8-7**; 2 electrons in the first energy level, 8 in the second energy level, and 7 electrons in the third energy level.

USE THE *PERIODIC TABLE* FOR ELECTRON CONFIGURATIONS

On the Regents, you will be given the Periodic Table, on Page Reference Tables 20-21. Look at the element C at the top of the Periodic Table. In the lower left hand corner of that box, it says "electron configuration." This is the electron configuration of C. Similarly, for each of the elements, the electron configuration is in the lower left hand corner of the box in the table. Look at $_{17}$Cl on the Periodic Table. The electron configuration is given as 2-8-7, just like you figured out in the example above. By looking at the electron configuration in the Periodic Table, you see chlorine has 3 principal energy levels (2 electrons in the first, 8 electrons in the second, and 7 electrons in the third energy level).

Do Homework Questions #19-21, pages 14-15.

PERIODIC TABLE

GROUND AND EXCITED STATES

An atom is in the **GROUND STATE** when the electrons are filling the atom in order: 2-8-18-32, like it is written in the Periodic Table. The Periodic Table shows the ground state electron configurations.

An atom is **excited** when the electrons have absorbed energy or gotten more energy. The **electrons jump ahead** to a higher energy level, leaving one of the inner principal energy levels partly empty.

Question: Which is the electron configuration of an atom in the excited state?
   (1) 2-8-2   (2) 2-8-1   (3) 2-7-1   (4) 2-8-3

Solution: Answer 3. The first energy level can hold 2 electrons. The second energy level can hold 8 electrons. In choice 3, there are only 7 electrons in the second principal energy level, because one electron jumped ahead to the third principal energy level. The second principal energy level is now partially empty (only 7 electrons in the second principal energy level), therefore the atom is excited.

You can also use the Periodic Table, page Reference Tables 20-21, to realize that the atom 2-7-1 is excited. Look at the **electron configuration** of sodium, $_{11}Na$, **2-8-1**, or any element after it on the Periodic Table. You realize that the first principal energy level must have 2 electrons before electrons can go into the second energy level. The second principal energy level must have 8 electrons before electrons go into the third energy level. In the example, **2-7-1**, you notice that the **eighth electron is missing** in the second principal energy level, therefore the atom is **excited**.

Question: Which electron configuration represents an atom in the excited state?
   (1) 1-2   (2) 2-1   (3) 2-3   (4) 2-7

Solution: Answer 1. The first energy level is missing an electron, therefore the atom is excited. (The first energy level can hold 2 electrons.)

Or, look in the Periodic Table at the electron configuration of sodium, $_{11}Na$, 2-8-1. The first energy level has 2 electrons. In this example, 1-2, an electron is missing from the first principal energy level; therefore, the atom is excited.

When the **excited electrons** (the electrons that jumped ahead) **go back to lower energy levels**, they give off energy (in specific amounts called **QUANTA**), which produces a **spectrum** of colors, or **BRIGHT LINE SPECTRUM**.

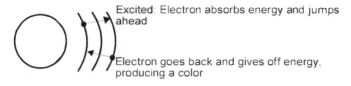

Excited: Electron absorbs energy and jumps ahead

Electron goes back and gives off energy, producing a color

For example, in a sodium atom, $^{23}_{11}Na$, when an **excited electron** goes back to a **lower energy level**, it gives off energy, which **produces a color** (colored light).

Note: When the colored light from an element (example sodium) goes through a prism, it produces a spectrum.

Excited electrons from **different atoms** of the **same element** (example, different atoms of Na) return to different energy levels. In some sodium atoms, the excited electrons go back from energy level 3 to energy level 2, while in other sodium atoms the excited electrons go back from energy level 3 to energy level 1 (called energy transitions), producing a **spectrum of color** or **bright-line spectrum.**

A **flame test helps** to **identify** an **element.** In a flame test, different metal ions are heated in a flame, and each element (metal ion) produces its own color. (Colors are produced because electrons jump back and give off energy). Lithium always produces a crimson (red) color (red light) when heated in a flame. If you heat an element (metal ion) in a flame and you get a red color (red light), you know it is lithium.

A **bright-line spectrum** also **helps** to **identify** an **element** (figure out which element it is). The (bright-line) spectrum of lithium looks like this:

blue green   orange red
| | | |
SPECTRUM OF LITHIUM

Each element has its own bright-line spectrum, different from any other element. If you have an element and you want to see if it is lithium, **compare the bright-line spectrum** of the sample that you have with the bright line spectrum of lithium. If you have lithium, your sample's bright-line spectrum and lithium's spectrum will be the same. If you do not have lithium, your sample's spectrum and lithium's spectrum will be different.

BRIGHT-LINE SPECTRA

Lithium	| |	| |
Sample 1	| |	| |
Sample 2	|	| | |

Sample 1 is lithium; Sample 2 is not lithium.

Mixture (samples 1 and 2) | | | | | | |

Mixture (samples 1 and 2) has all the lines of sample 1 and sample 2 (see above).

The spectrum of any element (examples lithium, hydrogen) has lines showing colors (see lithium spectrum above) or lines showing wavelength. The spectrum of hydrogen can be drawn either way:

violet blue blue-green   red   OR

| | | | 400   500   600   700

wavelength (nm)

Lines drawn at (draw lines):410.2 434.1 486.1 656.3

Question: The excited atom has an electron configuration of 1-6. What atom is this?

Solution: To find the number of electrons an atom has, add the number of electrons in each principal energy level. You know that the number of electrons equals the atomic number. In this case, 1 + 6 = 7 **electrons = atomic number** 7, which is nitrogen. In Table S, on page Reference Tables 23-24, you see that **atomic number 7 (first column)** is **nitrogen.** Or, look at the **Periodic Table** to the right or on page Reference Tables 20-21. In the box in the Periodic Table with **atomic** number 7 is N; **N** is **nitrogen.**

$_6C$	$_7N$	$_8O$
↑ atomic number	↑ atomic number	↑ atomic number

ELECTRON DOT METHOD

If you want to show how many electrons are in the last principal energy level (valence electrons), you can use the Lewis electron dot structures. Electrons in the last principal energy level or valence electrons affect the chemical properties of the element. If an atom has 1 or 2 electrons in the last principal energy level that atom is very active, which means the atom easily unites(reacts) with other atoms to form compounds.If an atom has 8 electrons in the last principal energy level, the atom does not react or reacts very little with other atoms.

Look at the electron configuration of magnesium, $_{12}Mg$, on the Periodic Table.

↓ last principal energy level

The electron configuration is 2-8-2.

There are 2 electrons in the last principal energy level, 2 valence electrons. All the other electrons in the atom are non-valence electrons (not in the last principal energy level). Mg has (2 + 8 =) 10 non-valence electrons.

LEWIS ELECTRON DOT METHOD: Look at the electron configuration for the element in the Periodic Table on page Reference Tables 20-21. Find the number of **electrons in the last principal energy level**. If there are **three electrons** in the **last** principal **energy level**, put **three dots** around the symbol of the element, X (X means any element).

How to Write Lewis Electron Dot Structures

There can only be 2 electrons on each side of the symbol of the element.

1. Put the 1st and 2nd valence electrons on one side of the symbol of the element. You can put the 1 or 2 electrons on any side of X.

$$X\cdot \qquad X\colon$$

2. For the 3rd, 4th, and 5th electrons, put each electron on a different side of the symbol. $\overset{\cdot}{X}\colon \quad \cdot\overset{\cdot}{X}\colon \quad \cdot\overset{\cdot}{X}\colon$

3. For the 6th, 7th and 8th electrons, add the electron to any side with 1 electron. $\cdot\overset{\cdot\cdot}{X}\colon \quad \colon\overset{\cdot\cdot}{X}\colon \quad \colon\overset{\cdot\cdot}{\underset{\cdot}{X}}\colon$

Element	Valence Electrons	Placement of Dots	Electron Dot
$_1$H Hydrogen	1	1 dot on one side	H·
$_2$He Helium	2	2 dots on one side	He:
$_3$Li Lithium	1	1 dot on one side	Li·
$_4$Be Beryllium	2	2 dots on one side	Be:
$_5$B Boron	3	2 dots on one side, 1 dot on another side	Ḃ:
$_6$C Carbon	4	2 dots on one side, 1 dot on a second side, 1 dot on a third side	·Ċ:
$_7$N Nitrogen	5	2 dots on one side, 1 dot on each of the other three sides	·N̈:
$_8$O Oxygen	6	2 dots on each of two sides, 1 dot on each of the other two sides	·Ö:
$_9$F Fluorine	7	2 dots on each of three sides, 1 dot on the fourth side	:F̈:
$_{10}$Ne Neon	8	2 dots on each side	:N̈e:
$_{16}$S Sulfur	6	2 dots on each of two sides, 1 dot on each of the other two sides	·S̈:

PERIODIC TABLE

Try Sample Question #8, page 12, then do Homework Questions #22-27, page 15, and #31-32, page 15.

REMEMBER: When you answer Regents questions on ATOMIC CONCEPTS, use the Periodic Table, Table O and Table S.

SAMPLE REGENTS & REGENT-TYPE QUESTIONS AND SOLUTIONS

1. A neutron has approximately the same mass as
   - (1) an alpha particle
   - (2) a beta particle
   - (3) an electron
   - (4) a proton

2. What is the total number electrons present in an atom of $_{27}^{56}Co$?
   - (1) 27
   - (2) 32
   - (3) 59
   - (4) 86

3. A particle of matter contains 6 protons, 7 neutrons and 6 electrons. This particle must be a
   - (1) neutral carbon atom
   - (2) neutral nitrogen atom
   - (3) positively charged carbon ion
   - (4) positively charged nitrogen ion

4. What is the mass number of an atom which contains 28 protons, 28 electrons and 34 neutrons?
   - (1) 26
   - (2) 56
   - (3) 62
   - (4) 90

5. Neutral atoms of ^{35}Cl and ^{37}Cl differ with respect to their number of
   - (1) electrons
   - (2) protons
   - (3) neutrons
   - (4) positrons

6A. One atomic mass unit equals
   - (1) 1/12 the mass of $_{6}^{12}C$
   - (2) 1/12 the mass of $_{1}^{1}H$
   - (3) 1/7 the mass of $_{7}^{14}N$
   - (4) 1/8 the mass of $_{8}^{16}O$

6B. Naturally occurring boron is composed of two isotopes. The percent abundance and the mass of each isotope are listed below:
   19.9% of the boron atoms have a mass of 10.013 atomic mass units.
   80.1% of the boron atoms have a mass of 11.009 atomic mass units.
   Calculate the atomic mass of boron. Your response must include *both* a correct numerical setup and the calculated result.

7. An element has an atomic number of 18. What is the principal quantum number (n) of its outermost electrons?
   - (1) 1
   - (2) 2
   - (3) 3
   - (4) 4

8. Which is the electron dot symbol of an atom of boron in the ground state?
   - (1) $\cdot \underset{.}{B} \vdots$
   - (2) $B \cdot$
   - (3) $\cdot B \vdots$
   - (4) $B \vdots$

SOLUTIONS

1. Answer 4. You can get a higher mark on the Regents by just knowing how to use the Tables. Look at Table O, on page Reference Tables 17. On Table O, it is written neutron $_{0}^{1}n$ and proton $_{1}^{1}H$. You know the top number is the mass number and you see that the neutron and proton have the same mass.

2. Answer 1. You learned that the atomic number is equal to the number of protons, which is equal to the number of electrons. The question is what is the number of electrons in $_{27}^{56}Co$. The bottom number is the atomic number. $_{27}Co$ means the atomic number is 27 = 27 protons = 27 electrons.

3. Answer 1. 6 protons means atomic number 6. 6 protons (positive) = 6 electrons (negative), which means a neutral atom. Look at the Periodic Table. Atomic number 6 is a carbon atom. The answer is Answer 1, a neutral carbon atom.

4. Answer *3*. Mass number is equal to the total number of protons and neutrons: 28 + 34 = 62.

5. Answer *3*. Atoms of the same element have the same atomic number. On the Periodic Table, you see chlorine (Cl) has an atomic number of 17. $^{35}_{17}Cl$ and $^{37}_{17}Cl$ are isotopes. They have the same atomic number, but different mass numbers and differ in the number of neutrons. Remember, you learned that the number of neutrons = mass number - atomic number.

$$\text{In } ^{35}_{17}Cl, \text{ number of neutrons} = 35 - 17 = 18.$$

$$\text{In } ^{37}_{17}Cl, \text{ number of neutrons} = 37 - 17 = 20.$$

6A. Answer *1*. One atomic mass unit is defined as 1/12 the mass of $^{12}_{6}C$.

6B. Take the percentage of each isotope times its mass; add the numbers.

Two Isotopes		Percentage	Times	Mass		
B	Take 19.9%:	0.199	X	10.013 u	=	1.99 u
B	Take 80.1%:	0.801	X	11.009 u	=	8.82 u
				Add:		10.81 u

Atomic mass

Atomic mass = 10.81 u

7. Answer *3*. Look at the Periodic Table. Look at atomic number 18. Look at the electron configuration of atomic number 18, $_{18}Ar$: 2-8-8. You see that the outermost electrons are in the third shell, or third principal energy level.

8. Answer *4*. Electron dot notation shows electrons in the last shell. Look at the electron configuration of boron, $^{11}_{5}B$, on the Periodic Table:

Electron configuration: 2 - 3

In the second (last) shell, there are 3 electrons. Next to the letter B, show three dots, which represent the 3 electrons in the last shell, $\overset{\bullet}{B}{:}$.

NOW LET'S TRY SOME HOMEWORK QUESTIONS:

---

1. The total number of electrons in a neutral atom of every element is always equal to the atom's
   (1) mass number
   (2) number of neutrons
   (3) number of protons
   (4) number of nucleons

2. The mass of an electron is approximately equal to $\dfrac{1}{1836}$ of the mass of
   (1) a positron   (2) a proton   (3) a beta particle   (4) an alpha particle

3. Which particle has a mass of approximately one atomic mass unit and a unit positive charge?
   (1) a neutron   (2) a proton   (3) a beta particle   (4) an alpha particle

4. The atomic number of an atom is equal to the number of
   (1) neutrons in the atom
   (2) protons in the atom
   (3) neutrons plus protons in the atom
   (4) protons plus electrons in the atom

---

5. Compared to the entire atom, the nucleus of the atom is
   (1) smaller and contains most of the atom's mass
   (2) smaller and contains little of the atom's mass
   (3) larger and contains most of the atom's mass
   (4) larger and contains little of the atom's mass

6. The mass number of an atom is always equal to the total number of its
   (1) electrons only   (3) electrons plus protons
   (2) protons only   (4) protons plus neutrons

7. The nucleus of an atom of $^{127}_{53}$I contains
   (1) 53 neutrons and 127 protons   (2) 53 protons and 127 neutrons
   (3) 53 protons and 74 neutrons   (4) 53 protons and 74 electrons

8. What is the mass number of an atom which contains 21 electrons, 21 protons and 24 neutrons?
   (1) 21   (2) 42   (3) 45   (4) 66

9. Which of the following particles has the *least* mass?
   (1) an electron   (2) a proton   (3) a deuteron   (4) a neutron

10. An atom of carbon-14 contains
    (1) 8 protons, 6 neutrons and 6 electrons
    (2) 6 protons, 6 neutrons and 8 electrons
    (3) 6 protons, 8 neutrons and 8 electrons
    (4) 6 protons, 8 neutrons and 6 electrons

11. The atomic mass of an element is defined as the weighted average mass of that element's
    (1) most abundant isotope   (3) naturally occurring isotopes
    (2) least abundant isotope   (4) radioactive isotopes

12. All isotopes of a given element must have the same
    (1) atomic mass   (3) mass number
    (2) atomic number   (4) number of neutrons

13. Which symbols represent atoms that are isotopes of each other?
    (1) ^{14}C and ^{14}N (2) ^{16}O and ^{18}O (3) ^{131}I and ^{131}I (4) ^{222}Rn and ^{222}Ra

14. Neutral atoms of the same element can differ in their number of
    (1) neutrons   (2) positrons   (3) protons   (4) electrons

15. What is the maximum number of electrons in an energy level with a principal quantum number of 3?
    (1) 6   (2) 9   (3) 3   (4) 18

16. What is the maximum number of electrons that may be present in the fourth principal energy level of an atom?
    (1) 8   (2) 2   (3) 18   (4) 32

17. An atom contains a total of 29 electrons. When the atom is in the ground state, how many different principal energy levels will contain electrons?
    (1) 1   (2) 2   (3) 3   (4) 4

18. What is the total number of principal energy levels or shells in an atom with an atomic number of 30?
    (1) 3   (2) 4   (3) 5   (4) 6

19. The electron configuration of an atom in the ground state is 2-4. The total number of occupied principal energy levels in this atom is
    (1) 1   (2) 2   (3) 3   (4) 4

20. In an atom that has an electron configuration of 2-7, what is the total number of electrons in its principal energy level of highest energy?
    (1) 2   (2) 5   (3) 6   (4) 7

---

21. What is the total number of occupied principal energy levels in an atom of $_{11}$Na? (Hint: See Periodic Table)
    (1) 1   (2) 2   (3) 3   (4) 4

22. An atom with the electron configuration 2-8-11-2 has an incomplete
    (1) first principal energy level   (3) third principal energy level
    (2) second principal energy level (4) fourth principal energy level

23. Which is the electron configuration of an atom in the excited state?
    (1) 2-4   (2) 2-3   (3) 2-7-2   (4) 2-8-1

24. Which electron configuration represents an atom in an excited state?
    (1) 2-8-2   (2) 2-8-1   (3) 2-8   (4) 2-7-2

25. Which is an electron configuration of a fluorine atom in the excited state? (Hint: In Table S, find the atomic number (column 1) for fluorine (column 3))
    (1) 2-6   (2) 2-7   (3) 2-6-1   (4) 2-7-1

26. Energy is released when an electron changes from a principal energy level of
    (1) 1 to 2   (2) 2 to 3   (3) 3 to 2   (4) 3 to 5

27. The characteristic bright-line spectrum of sodium is produced when its electrons
    (1) return to lower energy levels
    (2) jump to higher energy levels
    (3) are lost by the neutral atoms
    (4) are gained by the neutral atoms

CONSTRUCTED RESPONSE QUESTIONS: Parts B-2 and C of NYS Regents Exam

28. Compare (how is it similar) and contrast (how is it different) mass and charge of a proton, neutron, and electron.

29. A. Explain how isotopes of hydrogen are similar and are different.
    B. State, in terms of subatomic particles, how an atom of C-13 is different from an atom of C-12.

30. Compare and contrast the amount of energy and number of electrons in the 4 principal energy levels.

31A. How can you tell whether an electron is in an excited state (What information do you need to know)?
31B. Draw four vertical lines showing the wavelengths of the spectral lines for the Balmer Series of hydrogen: 410.2, 434.4, 486.1, and 656.3 nm.

400.   500.   600.   700.

Wavelength (nm)

32. Write the electron configuration and Lewis electron dot structure of $_2$He, $_4$Be, $_6$C, $_7$N, $_9$F, $_{13}$Al, $_{14}$Si, $_{15}$P, $_{17}$Cl, $_{18}$Ar.

CHAPTER QUESTION: Parts B-2 and C of NYS Regents Exam

33. A. In the gold-foil experiment, alpha particles were directed toward the foil. Most of the alpha particles passed directly through the foil with no effect. This result did not agree with the "hard spheres model" for the atom. What conclusion about the internal structure of the atom did this evidence show?
    B. In the same experiment, some of the alpha particles returned toward the source. What does this evidence indicate about the charge of the atom's nucleus?

SECTION B

Principal energy levels may be divided into sublevels.

Principal energy level 1 has ONE ENERGY SUBLEVEL: s.

Principal energy level 2 has TWO SUBLEVELS: s and p.

Principal energy level 3 has THREE SUBLEVELS: s, p and d.

Principal energy level 4 has FOUR SUBLEVELS: s, p, d and f.

Four Principal Energy Levels, with Their Sublevels

ELECTRON CONFIGURATION

Electron configuration shows how many electrons are in each sublevel.

$_1H$ ⊝ 1

Hydrogen

Hydrogen has an atomic number of 1; hydrogen has 1 electron. The electron is in the **principal energy level 1**, therefore, you **write "1"**. Then you **write the sublevel**; you write "s" because the first energy level only has an "s" sublevel. (*Now you have "1s."*) You write **superscript 1 ("1s^1")** because there is **one electron** in the "s" sublevel. **The electron configuration** is:

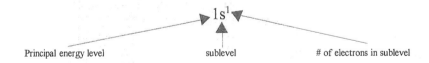

Hydrogen diagram: $1s^1$ with labels — Principal energy level, sublevel, # of electrons in sublevel

$_2He$ ⊝ 2

Helium

Helium has an atomic number of 2; helium has 2 electrons. The 2 electrons are in **principal energy level 1. Write "1".** Then you write the **sublevel**; you **write "s"**, because the first energy level has only an "s" sublevel. (*Now you have "1s."*) You **write superscript 2 ($1s^2$)** because there are **two electrons** in the s sublevel. The electron configuration is $1s^2$.

　　　　　　High Marks: Regents Chemistry Made Easy

Lithium

In lithium, $_3$Li, the **first shell** (principal energy level), with two electrons, is **just like He**, and the **first shell** has the same electron configuration as helium, $1s^2$. The **first shell** can **only hold two electrons. After that, put the electrons in** the next shell, **principal ENERGY LEVEL 2**, because there is no more room in the first shell. Let's now describe the **second shell**:

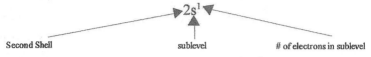

Second Shell   sublevel   # of electrons in sublevel

Electron configuration of Li is $1s^2 2s^1$.

first shell   second shell

Beryllium

In beryllium, $_4$Be, the **first shell**, with 2 electrons, is just like helium, with an electron configuration of $1s^2$. The **second** shell is $2s^2$. The electron configuration of Be is $1s^2 2s^2$.

second shell   # of electrons
     sublevel

Boron

Boron, $_5$B, has 2 electrons in the first shell; the first shell is $1s^2$. Boron has 3 electrons in the second shell. The **SECOND shell** is **divided into "s" and "p" sublevels:** $1s\ 2sp$. "s" CAN ONLY HOLD TWO ELECTRONS. AFTER THAT'S ALL FILLED UP, THEN PUT 1 ELECTRON IN THE "p" SUBLEVEL OF THE SECOND SHELL.

First Shell   Second Shell
$1s^2$ $2s^2$ $2p^1$
All filled up   All filled up   Put the next
electron in the
p sublevel (of
the second
shell)

Each "s" sublevel can only hold 2 electrons. Electron configuration of B is $1s^2 2s^2 2p^1$.

Carbon

In carbon, $_6$C, the **first shell is $1s^2$**. The **second shell** has **"s" and "p" sublevels**. "s" CAN ONLY HOLD 2 ELECTRONS. AFTER THE $1s^2$ AND $2s^2$ ARE ALL FILLED UP, THEN PUT 2 ELECTRONS IN THE "p" SUBLEVEL OF THE SECOND SHELL. $1s^2\ 2s^2\ 2p^2$.

Nitrogen

In $_7$N, the **$1s^2$** and **$2s^2$ are all filled up**, then PUT 3 ELECTRONS IN THE "p" SUBLEVEL OF THE SECOND SHELL, $1s^2 2s^2 2p^3$. ("p" CAN HOLD 6 ELECTRONS).

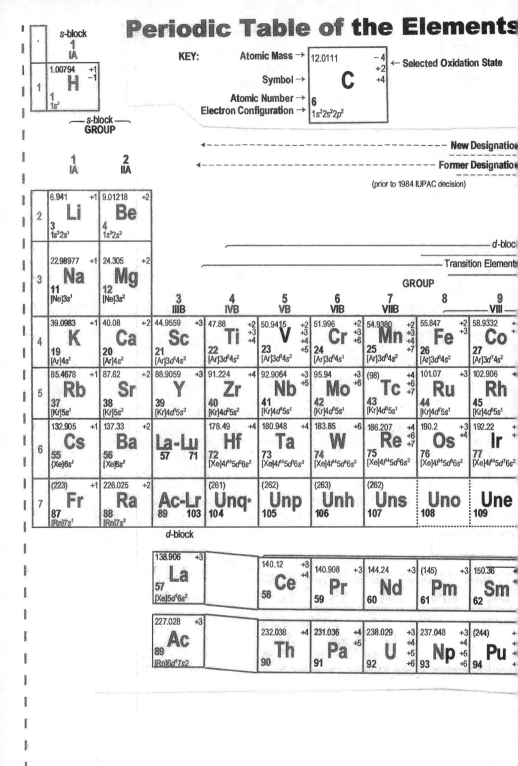

Periodic Table of the Elements

Relative atomic masses are based on $_{12}C = 12.00000$

4.00260	0
He	
2	
$1s^2$	

18
O

----- *p*-block -----
GROUP

13 IIIA	14 IVA	15 VA	16 VIA	17 VIIA	18 O
10.81 +3 **B** 5 $1s^22s^22p^1$	12.0111 −4 +2 +4 **C** 6 $1s^22s^22p^2$	14.0067 −3 +1 −2 +2 −1 +3 +4 +5 **N** 7 $1s^22s^22p^3$	15.9994 −2 **O** 8 $1s^22s^22p^4$	18.998403 −1 **F** 9 $1s^22s^22p^5$	20.179 0 **Ne** 10 $1s^22s^22p^6$
26.98154 **Al** 13 [Ne]$3s^23p^1$	28.0855 −4 +2 +4 **Si** 14 [Ne]$3s^23p^2$	30.97376 −3 +3 +5 **P** 15 [Ne]$3s^23p^3$	32.06 −2 +4 +6 **S** 16 [Ne]$3s^23p^4$	35.453 −1 +1 +3 +5 +7 **Cl** 17 [Ne]$3s^23p^5$	39.948 0 **Ar** 18 [Ne]$3s^23p^6$

10	11 IB	12 IIB					
8.69 +2 +3 **Ni** r]$3d^84s^2$	63.546 +1 +2 **Cu** 29 [Ar]$3d^{10}4s^1$	65.39 +2 **Zn** 30 [Ar]$3d^{10}4s^2$	69.72 +3 **Ga** 31 [Ar]$3d^{10}4s^24p^1$	72.59 − + **Ge** 32 [Ar]$3d^{10}4s^24p^2$	74.9216 −3 +3 +5 **As** 33 [Ar]$3d^{10}4s^24p^3$	78.96 −2 +4 +6 **Se** 34 [Ar]$3d^{10}4s^24p^4$	79.904 −1 +5 **Br** 35 [Ar]$3d^{10}4s^24p^5$	83.80 0 +2 **Kr** 36 [Ar]$3d^{10}4s^24p^6$
06.42 +2 +4 **Pd** [Ar]$4d^{10}5s^0$	107.868 +1 **Ag** 47 [Kr]$4d^{10}5s^1$	112.41 +2 **Cd** 48 [Kr]$4d^{10}5s^2$	114.82 +3 **In** 49 [Kr]$4d^{10}5s^25p^1$	118.71 +2 +4 **Sn** 50 [Kr]$4d^{10}5s^25p^2$	121.75 − + **Sb** 51 [Kr]$4d^{10}5s^25p^3$	127.60 −2 +4 +6 **Te** 52 [Kr]$4d^{10}5s^25p^4$	126.905 −1 +1 +5 +7 **I** 53 [Kr]$4d^{10}5s^25p^5$	131.29 0 +2 +4 +6 **Xe** 54 [Kr]$4d^{10}5s^25p^6$
5.08 +2 +4 **Pt** e]$4f^{14}5d^96s^1$	196.967 +1 +3 **Au** 79 [Xe]$4f^{14}5d^{10}6s^1$	200.59 +1 +2 **Hg** 80 [Xe]$4f^{14}5d^{10}6s^2$	204.383 +1 +3 **Tl** 81 [Xe]$4f^{14}5d^{10}6s^2$ $6p^1$	207.2 +2 +4 **Pb** 82 [Xe]$4f^{14}5d^{10}6s^2$ $6p^2$	208.980 +2 +5 **Bi** 83 [Xe]$4f^{14}5d^{10}6s^2$ $6p^3$	(209) + **Po** 84 [Xe]$4f^{14}5d^{10}6s^2$ $6p^4$	(210) **At** 85 [Xe]$4f^{14}5cf^{10}6s^2$ $6p^5$	(222) 0 **Rn** 86 [Xe]$4f^{14}5d^{10}6s^2$ $6p^6$

*The systematic names and symbols for elements of atomic numbers greater than 103 will be used until the approval of trivial names by IUPAC.

ASS NUMBERS IN PARENTHESES ARE THE MASS NUMBERS OF THE MOST STABLE OR COMMON ISOTOPES

1.96 +2 +3 **Eu**	157.25 +3 **Gd** 64	158.925 +3 **Tb** 65	162.50 +3 **Dy** 66	164.930 +3 **Ho** 67	167.26 +3 **Er** 68	168.934 +3 **Tm** 69	173.04 +2 +3 **Yb** 70	174.967 +3 **Lu** 71	Lanthanoid

43) +3 +4 +5 +6 **Am** 5	(247) +3 **Cm** 96	(247) +3 +4 **Bk** 97	(251) +3 **Cf** 98	(252) **Es** 99	(257) **Fm** 100	(258) **Md** 101	(259) **No** 102	(260) **Lr** 103	Actinoid

In $_8$O, the $1s^2$ and $2s^2$ **are all filled up**, then PUT 4 ELECTRONS IN THE "p" SUBLEVEL OF THE SECOND SHELL, $1s^2 2s^2 2p^4$.

Oxygen

In $_9$F, the $1s^2$ and $2s^2$ **are all filled up**, then PUT 5 ELECTRONS IN THE "p" SUBLEVEL OF THE SECOND SHELL, $1s^2 2s^2 2p^5$.

Fluorine

In $_{10}$Ne, the $1s^2$ and $2s^2$ **are all filled up**, then PUT 6 ELECTRONS IN THE "p" SUBLEVEL OF THE SECOND SHELL, $1s^2 2s^2 2p^6$.

Neon

SUMMARY:

Electron Configurations

$_1$H	$1s^1$
$_2$He	$1s^2$
$_3$Li	$1s^2 2s^1$
$_4$Be	$1s^2 2s^2$
$_5$B	$1s^2 2s^2 2p^1$
$_6$C	$1s^2 2s^2 2p^2$
$_7$N	$1s^2 2s^2 2p^3$
$_8$O	$1s^2 2s^2 2p^4$
$_9$F	$1s^2 2s^2 2p^5$
$_{10}$Ne	$1s^2 2s^2 2p^6$

USE THE *PERIODIC TABLE* FOR ELECTRON CONFIGURATIONS

Look at the element C at the top of the Periodic Table on pages Chap. 2:18-19 and on page Additional Tables 2-3. In the lower left hand corner of that box, it says "electron configuration." This is the electron configuration of C, just like you figured it out. Similarly, for each of the elements, the electron configuration is in the lower left hand corner of the box on the Periodic Table. If you want to know the **electron configuration** of any element, just **look at the Periodic Table**, on pages Chap. 2:18-19 and on page Additional Tables 2-3. Remember, if it is $1s^2$, 1 is the shell, "s" is the sublevel, and 2 is the number of electrons in the "s" sublevel of shell #1 (first shell).

In the Periodic Table, the electron configuration of $_{11}$Na is written **(Ne)$3s^1$**. The electron configuration of Ne is $1s^2\ 2s^2\ 2p^6$. **Add $3s^1$** to the electron configuration of Ne, and you have $1s^2 2s^2 2p^6 3s^1$, which is the **electron configuration of Na**.

In the Periodic Table, the electron configuration of $_{17}Cl$ is written $(Ne)3s^2 3p^5$. Look at $_{10}Ne$. The electron configuration of Ne is $1s^2 2s^2 2p^6$. Add $3s^2 3p^5$ to the electron configuration of neon, and you have $1s^2 2s^2 2p^6 3s^2 3p^5$, which is the electron configuration of Cl.

Question: How many sublevels do the electrons of chlorine occupy?

Solution: Look at the Periodic Table. $_{17}Cl$ has $(Ne)3s^2 3p^5$. Ne is $1s^2 2s^2 2p^6$; add to this $3s^2 3p^5$. Therefore, chlorine has five sublevels: 1s, 2s, 2p, 3s, and 3p.

Try Sample Questions #1-3, on page 26, then do Homework Questions, #1-7, pages 27-28.

An atom is in the **GROUND STATE** when the electrons are filling the atom in order: $1s^2 2s^2 2p^6 3s^2 3p^6$, etc., like it is written in the Periodic Table.

An atom is **excited** when the electrons have absorbed energy or gotten more energy. The **electrons jump ahead** to a higher energy level, leaving one of the previous orbitals half empty.

Question: Which is the electron configuration of an atom in the excited state?
    (1) $1s^2 2s^2 2p^2$ (2) $1s^2 2s^2 2p^1$ (3) $1s^1 2s^2 2p^4$ (4) $1s^2 2s^2 2p^3$

Solution: Answer *3*. The $1s^2$ electron jumped ahead to the 2p sublevel. (The first electron in the s sublevel is called $1s^1$. The second electron in the s sublevel is called $1s^2$.) The $1s^2$ electron is missing, therefore the atom is excited.

You can also use the Periodic Table to realize that the atom, $1s^1 2s^2 2p^4$, is excited. Look at the electron configurations of $_7N$, $1s^2 2s^2 2p^3$, and $_8O$, $1s^2 2s^2 2p^4$. In this example, $1s^1 2s^2 2p^4$, you notice that **$1s^2$ is missing**, therefore the atom is **excited**.

Question: Which electron configuration represents an atom in the excited state?
    (1) $1s^2 2s^1 2p^5$ (2) $1s^2 2s^2 2p^5$ (3) $1s^2 2s^2 2p^6$ (4) $1s^2 2s^1$

Solution: Answer *1*. The $2s^2$ is missing. The $2s^2$ electron jumped ahead, therefore the atom is excited. Or, look at the Periodic Table, at the electron configurations: $_8O$ is $1s^2 2s^2 2p^4$, and $_9F$ is $1s^2 2s^2 2p^5$. You can see that, in this example, $1s^2 2s^1 2p^5$, the $2s^2$ is missing.

Try Sample Questions #4-5, on page 26, then do Homework Questions, #8-11, page 28.

Periodic Table

ORBITAL NOTATION

By using the electron configuration, you saw how many electrons are in each sublevel: $1s^2 2s^2 2p^6$. Now let's see how the electrons fill up the **ORBITALS** of each sublevel.

The ORBITAL is the place where the electron may be found. An **orbital can only hold two (2) electrons**. The **s-sublevel** has ONE s-ORBITAL (SPHERICAL SHAPED). The **p-sublevel** has 3 p-ORBITALS (DUMBBELL-SHAPED). The **d-sublevel** has 5 ORBITALS, and the **f-sublevel** has 7 ORBITALS.

Draw a box to show each orbital:

The "s" sublevel has one orbital; therefore, make one box, ☐. The "p" sublevel has 3 orbitals; make 3 boxes, ☐☐☐.

You can draw the orbitals in the first and second shells as follows:

1s	2s	2p

1 in 1s means first shell   *2 in 2s and 2p means second shell*

Element	Explanation of Orbital Notation	Orbital Notation
₁H *Hydrogen*	Figure out the electron configuration of hydrogen, or look at the Periodic Table, on page Additional Tables 2-3. The electron configuration of hydrogen is $1s^1$. Hydrogen has one electron (superscript 1) in the 1s sublevel; therefore put one arrow in the 1s orbital to represent one electron.	1s [↑]
₂He *Helium*	Figure out the electron configuration, or look at the Periodic Table on page Additional Tables 2-3. The electron configuration of ₂He is $1s^2$. Helium has **two electrons** (superscript 2) in the 1s sublevel or 1s orbital. Therefore, **put 2 arrows (electrons) in the 1s orbital**. But the two electrons in the same orbital must have **opposite** spin; one electron goes clockwise, and one electron goes counterclockwise. Therefore, **make 1 arrow going up, ↑, and one arrow going down, ↓**.	1s [↑↓]
₃Li *Lithium*	Figure out the electron configuration. The electron configuration of ₃Li is $1s^2 2s^1$. Two electrons are in the 1s; therefore, put **two arrows in the 1s orbital, one arrow up, ↑, and one arrow down, ↓**. There is one more electron in the 2s subshell (or orbital); therefore, **put one arrow in the 2s orbital**.	1s 2s [↑↓] [↑]
₄Be *Beryllium*	Figure out the electron configuration. The electron configuration is $1s^2 2s^2$. Two electrons are in the 1s-orbital (one arrow up, ↑, and one arrow down, ↓) and two electrons are in the 2s orbital (one arrow up, ↑, and one arrow down, ↓).	1s 2s [↑↓] [↑↓]

Periodic Table

Element	Explanation of Orbital Notation	Orbital Notation
$_5$B *Boron*	Figure out the electron configuration. The electron configuration of $_5$B is $1s^2 2s^2 2p^1$, two electrons are in the 1s orbital, two electrons in the 2s orbital, and 1 electron in the 2p orbital. The p-sublevel has 3 orbitals at the same energy.	1s 2s 2p 2p 2p [↑↓] [↑↓] [↑][ ][ ]
$_6$C *Carbon*	Figure out the electron configuration. $1s^2 2s^2 2p^2$ is the electron configuration of carbon. Two electrons are in the 1s orbital, two electrons in the 2s orbital, and 2 electrons in the 2p orbital. The **p-sublevel has 3 orbitals. You must put 1 electron in each p-orbital before putting a** *second* **arrow (electron) in any p-orbital.**	1s 2s 2p 2p 2p [↑↓] [■] [↑][↑][ ]
$_7$N *Nitrogen*	Figure out the electron configuration, $1s^2 2s^2 2p^3$ is the electron configuration of nitrogen. Two electrons are in the 1s orbital, two electrons in the 2s orbital, and 3 electrons in separate p-orbitals. **The p-sublevel has 3 orbitals. You must put 1 electron in each p-orbital before putting a second arrow (electron) in any p-orbital.**	1s 2s 2p 2p 2p [↑↓] [↑↓] [↑][↑][↑]
$_8$O *Oxygen*	$1s^2 2s^2 2p^4$ is the electron configuration of oxygen. Two electrons are in the 1s orbital, two electrons in the 2s orbital. Put 1 electron in each p-orbital, and then add a second electron to a p-orbital (arrow in opposite direction).	1s 2s 2p 2p 2p [↑↓] [↑↓] [↑↓][↑][↑]
$_9$F *Fluorine*	$1s^2 2s^2 2p^5$ is the electron configuration of fluorine. Two electrons are in the 1s orbital, two electrons in the 2s orbital, and 5 electrons in the 2p orbitals. Put 1 electron in each 2p orbital and a second electron in two of the 2p orbitals, as shown in the orbital notation to the right.	1s 2s 2p 2p 2p [↑↓] [↑↓] [↑↓][↑↓][↑]
$_{10}$Ne *Neon*	$1s^2 2s^2 2p^6$ is the electron configuration of neon. Two electrons are in the 1s orbital, two electrons in the 2s orbital, and 6 electrons in the 2p orbitals.	1s 2s 2p 2p 2p [↑↓] [↑↓] [↑↓][↑↓][↑↓]

For elements after neon, the same pattern holds. Sublevels are filled in order: 3s, then 3p.

1s   2s   2p   3s   3p

You must put one electron in each p-orbital before putting a second electron in any p-orbital.

Element	Explanation of Orbital Notation	Orbital Notation
$_{15}$P *Phosphorus*	The electron configuration of phosphorus is $1s^2 2s^2 2p^6 3s^2 3p^3$. Two electrons are in the 1s orbital, two electrons in the 2s orbital, 6 electrons in the 2p orbitals, two electrons in the 3s orbital, and three electrons in separate 3p orbitals.	1s 2s 2p 2p 2p [↑↓] [↑↓] [↑↓][↑↓][↑↓] 3s 3p 3p 3p [↑↓] [↑][↑][↑]
$_{16}$S *Sulfur*	$1s^2 2s^2 2p^6 3s^2 3p^4$ is the electron configuration of sulfur. Two electrons are in the 1s orbital, two electrons are in the 2s orbital, 6 electrons in the 2p orbitals, and 2 electrons in the 3s orbital. Put one electron in each 3p-orbital, and then add a second electron to a 3p-orbital (arrow in opposite direction).	1s 2s 2p 2p 2p [↑↓] [↑↓] [↑↓][↑↓][↑↓] 3s 3p 3p 3p [↑↓] [↑↓][↑][↑]

You learned that the p-sublevel can hold 6 electrons. A completely filled p-sublevel has 6 electrons. A half-filled p-sublevel has 3 electrons.

ELECTRON DOT METHOD

If you want to show how many electrons are in the last shell (valence electrons), you can use the electron dot method.

ELECTRON DOT METHOD (Using Periodic Table that shows sublevels): Look at the **electron configuration** for the element in the Periodic Table on page Additional Tables 2-3. Find the number of **electrons in the last shell**. If there are three electrons in the last shell, put three **dots** around the symbol of the element.

Or

Calculate the number of electrons in the last shell from the electron configuration or atomic number. If there are three electrons in the last shell, put three **dots** around the symbol of the element.

Element	Explanation of Electron Dot Method	Electron Dot
$_1$H *Hydrogen*	Electron configuration is $1s^1$ ←# of electrons. (Remember the superscript ^{1} means one electron. There is **1 electron in the last shell**; therefore, put **1 dot** around the H symbol.	$\dot{H}$
$_2$He *Helium*	Electron configuration is $1s^2$ ←# of electrons; there are **2 electrons in the last shell** (valence electrons). Put **2 dots** around the He symbol.	$\ddot{He}$
$_3$Li *Lithium*	Electron configuration is $1s^2 2s^1$; there is **1 electron** in the 2nd shell (**last shell**). Put **1 dot** around the Li symbol.	$\dot{Li}$

Periodic Table

Element	Explanation of Electron Dot Method	Electron Dot
$_6$C *Carbon*	electrons in *last* shell ↓ Electron configuration is $1s^2 2s^2 2p^2$ (2s and 2p are both in the *last* shell). There are **4 electrons in the last shell.** Put **4 dots** around the C symbol.	·Ċ·
$_9$F *Fluorine*	electrons in *last* shell ↓ Electron configuration is $1s^2 2s^2 2p^5$. There are **7 electrons in the last shell.** Put **7 dots** around the F symbol.	:F̈:

SUMMARY

Principal energy levels (shells): 1, 2, 3 and 4.
The first principal energy level (shell) has an s-sublevel.
The second principal energy level (shell) has s- and p-sublevels.
The third principal energy level has s-, p- and d-sublevels.
The fourth principal energy level has s-, p-, d- and f-sublevels.

$$ \circ \, \Big) \, 1\,s \, \Big) \, 2\,s\,p \, \Big) \, 3\,s\,p\,d \, \Big) \, 4\,s\,p\,d\,f $$

Four Principal Energy Levels
with their sublevels

PRINCIPAL ENERGY LEVELS, SUBLEVELS AND ORBITALS

Principal Energy Levels *(Shells)*	Sub-level	Orbitals *(each orbital can hold 2 electrons)*	Number of Electrons
1	s	s	2 (2 in s-sublevel)
2	sp	s p	8 (2 in s-sublevel, 6 in p-sublevel)
3	spd	s p d	18 (2 in s-sublevel, 6 in p-sublevel, 10 in d-sublevel)
4	spdf	s p d f	32 (2 in s-sublevel, 6 in p-sublevel, 10 in d-sublevel, 14 in f-sublevel)

Try Sample Questions #6-7, page 26, then do Homework Questions #12-16, page 28.

SAMPLE REGENTS & REGENT-TYPE QUESTIONS AND SOLUTIONS

1. An atom of an element has an electron configuration of $1s^2 2s^2 2p^2$. What is the total number of valence electrons in this atom?
   (1) 6   (2) 2   (3) 5   (4) 4

2. Which principal energy level has no "f" sublevel?
   (1) 5   (2) 6   (3) 3   (4) 4

3. In an atom that has an electron configuration of $1s^2 2s^2 2p^3$, what is the total number of electrons in its sublevel of highest energy?
   (1) 1   (2) 2   (3) 3   (4) 4

4. Which electron configuration represents an atom in an excited state?
   (1) $1s^2 2s^2$ (3) $1s^2 2s^2 2p^6$
   (2) $1s^2 2s^2 3s^1$ (4) $1s^2 2s^2 2p^6 3s^1$

5. Which electron transition represents the release of energy?
   (1) 1s to 3p   (2) 2s to 2p   (3) 3p to 1s   (4) 2p to 3s

6. Which orbital notation represents an atom in the ground state with 6 valence electrons?

   (1)
   (2)
   (3)
   (4)

7. The atom of which element in the ground state has 2 unpaired electrons in the 2p sublevel?
   (1) fluorine   (2) nitrogen   (3) beryllium   (4) carbon

SOLUTIONS

1. Answer 4: $1s^2 2s^2 2p^2$. Valence electrons are electrons in the last shell. $1s^2 \underline{2s^2 2p^2}$. (The raised number is # of electrons.) In the last shell (shell 2), there are 2 electrons + 2 electrons = 4 electrons.

2. Answer 3. Shell 1, or principal energy level 1, has one sublevel:"s."Shell 2 has two sublevels: "s" and "p." Shell 3 has three sublevels: "s," "p" and "d." Shell 3 doesn't have an "f" sublevel. Or, you can also get the answer by looking at the electron configurations on the Periodic Table, such as $1s^2 2s^2 2p^6 3s^2 3p^6 \dots 4s^2$. In the third shell, after the number 3, there is no "f."

3. Answer 3. $\ominus$ 1s 2s 2p The further away the electrons are from the nucleus, the more energy (higher energy) the electrons have. In the example, $1s^2 2s^2 2p^3$, the sublevel furthest away from the nucleus (highest energy) is the 2p sublevel. $2p^3$ means there are 3 electrons in the p sublevel.

4. Answer *2*. As you learned, excited means that an electron jumps ahead to higher energy levels: $1s^2\ 2s^2\ 3s^1$. An electron from the second shell jumps ahead to the third shell. The 2p electrons are missing. You can look at electron configurations on the Periodic Table and see that 2p is missing.

5. Answer *3*. When electrons go from higher energy levels to lower energy levels, energy is released. The electron in Answer 3 goes from the 3p (third shell) to 1s (first shell), giving off energy.

6. Answer *2*. As you learned, if there are only 6 electrons in the last shell, it has to be $s^2\ p^4$. Or, look at the electron configurations on the Periodic Table. Six valence electrons means six electrons in the last shell, like $2s^2\ 2p^4$.

    Look at Answer 2. Remember, you learned that, when filling he "s" orbital, one arrow goes up and one arrow goes down.

    Now, there are p^4, four electrons in the "p" sublevel. You must put 1 electron in each p orbital before putting a second electron (arrow) in any p orbital:

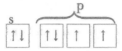

    The fourth electron goes back to the first "p" orbital and fills it, with an opposite spin (opposite direction).

7. Answer *4*. Look at the Periodic Table. Look at the electron configuration. You want two (unpaired) electrons in the 2p sublevel: $2p^2$ or $2p^4$ (The superscript represents the number of electrons). What element on the Periodic Table says $2p^2$? Carbon does. Its electron configuration in the Periodic Table is $1s^2\ 2s^2\ 2p^2$.

Now Let's Try A Few Homework Questions:

1. What is the total number of valence electrons in an atom with the electron configuration $1s^2\ 2s^2\ 2p^6\ 3s^2\ 3p^3$?
    (1) 6   (2) 2   (3) 3   (4) 5

2. An atom with the electron configuration $1s^2\ 2s^2\ 2p^6\ 3s^2\ 3p^6\ 3d^3\ 4s^2$ has an incomplete
    (1) 2p sublevel   (3) third principal energy level
    (2) second principal energy le el   (4) 4s sublevel

3. The electron configuration of an atom in the ground state is $1s^2\ 2s^2\ 2p^2$. The total number of occupied principal energy levels in this atom is
    (1) 1   (2) 2   (3) 3   (4) 4

4. Given an atom with the electron configuration $1s^2\ 2s^2\ 2p^3$, how many orbitals are completely filled?
    (1) 1   (2) 2   (3) 3   (4) 4

---

5. Which principal energy level has a maximum of three sublevels?
   (1) 1   (2) 2   (3) 3   (4) 4

6. In an atom that has an electron configuration of $1s^2 2s^2 2p^5$, what is the total number of electrons in its sublevel of highest energy?
   (1) 3   (2) 4   (3) 5   (4) 6

7. What is the total number of occupied sublevels in an atom of sodium in the ground state?
   (1) 3   (2) 4   (3) 5   (4) 6

8. Which is the electron configuration of an atom in the excited state?
   (1) $1s^2\ 2s^2\ 2p^2$ (3) $1s^2\ 2s^2\ 2p^5\ 3s^2$
   (2) $1s^2\ 2s^2\ 2p^1$ (4) $1s^2\ 2s^2\ 2p^6\ 3s^1$

9. Which electron configuration represents an atom in an excited state?
   (1) $1s^2 2s^2 2p^6 3s^2$ (2) $1s^2 2s^2 2p^6 3s^1$ (3) $1s^2 2s^2 2p^6$ (4) $1s^2 2s^2 2p^5 3s^2$

10. Which is an electron configuration of a fluorine atom in the excited state?
    (1) $1s^2\ 2s^2\ 2p^4$ (3) $1s^2\ 2s^2\ 2p^4\ 3s^1$
    (2) $1s^2\ 2s^2\ 2p^5$ (4) $1s^2\ 2s^2\ 2p^5\ 3s^1$

11. Energy is released when an electron changes from a sublevel of
    (1) 1s to 2p   (2) 2s to 3s   (3) 3s to 2s   (4) 3p to 5s

12. The m ximum number of electrons that a single orbital of the 3d sublevel may contain is
    (1) 5   (2) 2   (3) 3   (4) 4

13. What is the total number of occupied "s" orbitals in an atom of nickel in the ground state?
    (1) 1   (2) 2   (3) 3   (4) 4

14. Which atom in the ground state contains one completely filled "p"-orbital?
    (1) N   (2) O   (3) He   (4) Be

15. The total number of orbitals in a "d" sublevel is
    1. 1   (2) 5   (3) 3   (4) 7

16. Which is the orbital notation for the electrons in the third principal energy level of an argon atom in the ground state?

CHAPTER 3: PERIODIC TABLE

SECTION A

THE PERIODIC TABLE

The **PERIODIC TABLE**, on the next page or page Reference Tables 20-21, gives us a lot of information. You must know how to use the Periodic Table to ANSWER THE REGENTS QUESTIONS. Turn to the Periodic Table on the next page. The elements in the Periodic Table are arranged in order of increasing atomic number.

I. **LOOK AT THE PERIODIC TABLE** on the next page. Look at the box at the top of the Periodic Table.

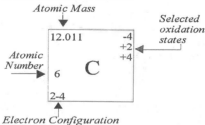

The letter "C" is the element carbon.

The bottom number, 6 in the lower left hand corner, represents the atomic number. The atomic number of carbon is 6.

The top left number, 12.011, is the atomic mass.

In the upper right hand corner are the oxidation numbers for carbon, -4, +2 and +4. In the bottom left hand corner, 2-4 is the electron configuration of carbon.

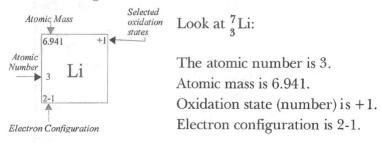

Look at $^{7}_{3}$Li:

The atomic number is 3.

Atomic mass is 6.941.

Oxidation state (number) is +1.

Electron configuration is 2-1.

Periodic Table

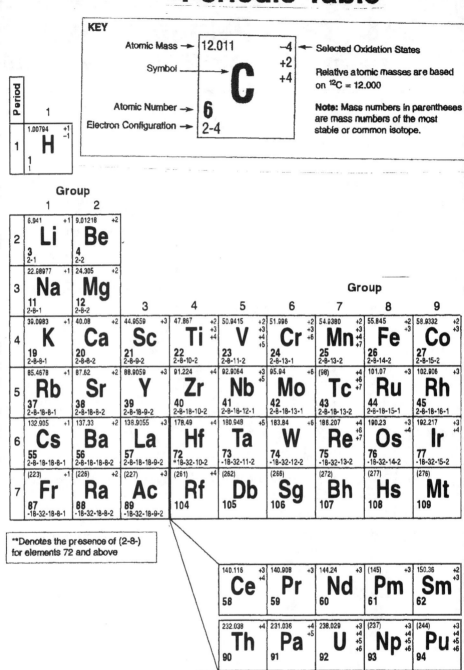

of the Elements

Group

18
4.00260   0
He
2
2

13	14	15	16	17	18
10.81 +3 **B** 5 2-3	12.011 -4 +2 +4 **C** 6 2-4	14.0067 -3 -2 -1 +1 +2 +3 +4 +5 **N** 7 2-5	15.9994 -2 **O** 8 2-6	18.9984 -1 **F** 9 2-7	20.180 0 **Ne** 10 2-8
26.98154 +3 **Al** 13 2-8-3	28.0855 -4 +2 +4 **Si** 14 2-8-4	30.97376 -3 +3 +5 **P** 15 2-8-5	32.065 -2 +4 +6 **S** 16 2-8-6	35.453 -1 +1 +5 +7 **Cl** 17 2-8-7	39.948 0 **Ar** 18 2-8-8

10	11	12						
.693 +2 +3 **Ni** -16-2	63.546 +1 +2 **Cu** 29 2-8-18-1	65.409 +2 **Zn** 30 2-8-18-2	69.723 +3 **Ga** 31 2-8-18-3	72.64 +2 +4 **Ge** 32 2-8-18-4	74.9216 -3 +3 +5 **As** 33 2-8-18-5	78.96 -2 +4 +6 **Se** 34 2-8-18-6	79.904 -1 +1 +5 **Br** 35 2-8-18-7	83.798 0 +2 **Kr** 36 2-8-18-8
6.42 +2 +4 **Pd** -18-18	107.868 +1 **Ag** 47 2-8-18-1	112.41 +2 **Cd** 48 2-8-18-18-2	114.818 +3 **In** 49 2-8-18-18-3	118.71 +2 +4 **Sn** 50 2-8-18-18-4	121.760 -3 +3 +5 **Sb** 51 2-8-18-18-5	127.60 -2 +4 +6 **Te** 52 2-8-18-18-6	126.904 -1 +1 +5 +7 **I** 53 2-8-18-18-7	131.29 0 +2 +4 +6 **Xe** 54 2-8-18-18-8
5.08 +2 +4 **Pt** -32-17-1	196.967 +1 +3 **Au** 79 -18-32-18-1	200.59 +1 +2 **Hg** 80 -18-32-18-2	204.383 +1 +3 **Tl** 81 -18-32-18-3	207.2 +2 +4 **Pb** 82 -18-32-18-4	208.980 +3 +5 **Bi** 83 -18-32-18-5	(209) +2 +4 **Po** 84 -18-32-18-6	(210) -1 **At** 85 -18-32-18-7	(222) 0 **Rn** 86 -18-32-18-8
1) **Ds** 10	(280) **Rg** 111	(285) **Cn** 112	(284) **Uut** 113**	(289) **Uuq** 114	(288) **Uup** 115	(292) **Uuh** 116	(?) **Uus** 117	(294) **Uuo** 118

.964 +2 +3 **Eu**	157.25 +3 **Gd** 64	158.925 +3 **Tb** 65	162.500 +3 **Dy** 66	164.930 +3 **Ho** 67	167.259 +3 **Er** 68	168.934 +3 **Tm** 69	173.04 +2 +3 **Yb** 70	174.9668 +3 **Lu** 71
3) +3 +4 +5 +6 **Am**	(247) +3 **Cm** 96	(247) +3 +4 **Bk** 97	(251) +3 **Cf** 98	(252) +3 **Es** 99	(257) +3 **Fm** 100	(258) +2 +3 **Md** 101	(259) +2 +3 **No** 102	(262) +3 **Lr** 103

*The systematic names and symbols for elements of atomic numbers 113 and above will be used until the approval of trivial names by IUPAC.

PERIODIC TABLE

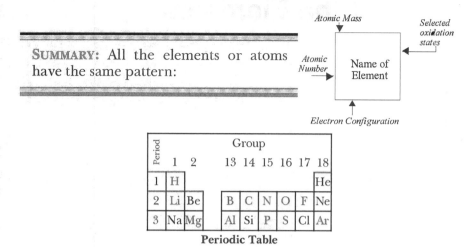

SUMMARY: All the elements or atoms have the same pattern:

Atomic Mass

Atomic Number

Name of Element

Selected oxidation states

Electron Configuration

Periodic Table

II. PERIODS

Periods are **horizontal rows** on the Periodic Table. Look at the word "period" at the top left of the Periodic Table.

Period 1 has hydrogen (H) and helium (He). Electrons in Period 1 are only in one principal energy level. Valence electrons are in the first principal energy level.

Period 2: Electrons are in 2 principal energy levels; valence electrons are in the second principal energy level (example, $_3^7 \text{Li}$.)

Period 3: Electrons are in 3 principal energy levels; valence electrons are in the third principal energy level. (For example, $_{11}^{23} \text{Na}$).

III. GROUPS

Vertical columns are Groups. Look at the Periodic Table, above and on the two previous pages. Groups 1-18 are labeled. **Elements in a group** have similar properties (**react similarly**), because the outermost (last) principal energy level of each element in the group has the same number of valence electrons. Look at the elements in Groups 1, 2 and 17:

Group 1		Group 2		Group 17	
Li Na K Rb Cs Fr	have 1 valence electron, 1 electron in last principal energy level	Be Mg Ca Sr Ba Ra	have 2 valence electrons; 2 electrons in last principal energy level	F Cl Br I At	have 7 valence electrons; 7 electrons in last principal energy level

The **PERIODIC LAW** states that the properties of the elements are a periodic function of their atomic number. This means that, when the elements are arranged by atomic number, those with similar properties will be at regular intervals; these elements will be in the same group.

PERIODIC TABLE

IV. **BLOCKS (spdf)**

Columns 1 and 2 are called "s" block. They contain elements in which "s" orbitals are being filled up.

Columns 3-12 are called "d" block, with "d" orbitals being filled. "d" block includes transition elements.

Columns 13-18 are called "p" block, with "p" orbitals being filled.

Bottom: Lanthanide series and Actinide series are called "f" block.

V. **METALS, NONMETALS, METALLOIDS & NOBLE GASES**

Look at the outline of the Periodic Table, below, and the full Periodic Table on pages Chap. 3:2-3. All elements to the *left of the zigzag line are metals*. More than two-thirds of all elements are metals. All **elements** touching the **zigzag line** are **metalloids**. Elements to the *right of the zigzag line are nonmetals*, except Group 18. *Noble gases* are *Group 18*.

VI. **MOST ACTIVE METAL**: Bottom, left hand corner of the Periodic Table, **Fr (francium)**. (Fr is more active than Cs; Cs more active than Rb, as will be explained later in the chapter.)

MOST ACTIVE NONMETAL: upper right-hand corner, **F** (fluorine).

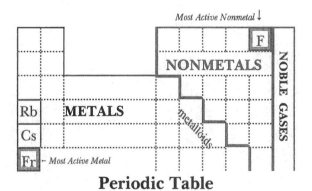

Periodic Table

VII. In the same period, like Period 2, elements in Group 1 are more active than elements in Group 2. It is easier (less energy needed) to lose one valence electron than to lose two valence electrons.

Question: Which element in Period 2 of the Periodic Table is the most reactive nonmetal?

(1) carbon　　(2) nitrogen　　(3) oxygen　　(4) fluorine

Solution: Answer *4*. You learned that fluorine, in the upper right hand corner of the Periodic Table, is the most reactive nonmetal.

METALS, METALLOIDS, NONMETALS AND NOBLE GASES

METALS: Elements to the **left of the zigzag line** on the Periodic Table are metals. **Metals** have LOW IONIZATION ENERGY (little energy is needed to remove an electron from an atom) and LOW ELECTRONEGATIVITY (little attraction for electrons). (Shown in **Table S.**) Metals tend to LOSE ELECTRONS to form positive ions. (See next page.)

Metals have metallic LUSTER (shine), and are MALLEABLE (can be made into sheets), and DUCTILE (can be made into wires.)

Metals are GOOD CONDUCTORS OF HEAT AND ELECTRICITY.

You learned metals are left of the zigzag line on the Periodic Table.

Aluminum is considered a metal because it has characteristics of a metal (example aluminum pots).

Most metals (examples: $_{11}Na$, $_{47}Ag$, and $_{72}Hf$) are solids at room temperature. **Exception**: The metal mercury (Hg) is a liquid at room temperature.

NONMETALS: Elements to the **right of the zigzag line** on the Periodic Table are nonmetals, except Group 18 which are noble gases. Nonmetals have HIGH IONIZATION ENERGY (a lot of energy is needed to remove an electron from an atom) and HIGH ELECTRONEGATIVITY (big attraction for electrons). (Shown in **Table S.**)

Nonmetals tend to GAIN ELECTRONS when they react with metals, becoming negative ions, or to share electrons when they combine with nonmetals.

Nonmetals LACK METALLIC LUSTER AND ARE BRITTLE in the solid phase.

Nonmetals are POOR CONDUCTORS OF HEAT AND ELECTRICITY.

Nonmetals are usually gases, molecular solids or network solids. The exception is bromine, which is a volatile liquid at room temperature.

METALLOIDS: Elements touching the zigzag line on the Periodic Table are metalloids. These elements have some **properties** of both **metals and nonmetals**. Examples of metalloids are B (boron), Si (silicon), As (arsenic), Sb (antimony) and Te (tellurium).

B
Si
Ge As
Sb Te

Question: Which set of elements contains a metalloid?
(1) K, Mn, As, Ar	(3) Ba, Ag, Sn, Xe
(2) Li, Mg, Ca, Kr	(4) Fr, F, O, Rn

Solution: Answer *1*. This answer has the element As, which **touches the zigzag line** on the Periodic Table and therefore is a metalloid.

NOBLE GASES: Group 18 is called **noble gases**. Elements of Group 18 have 8 valence electrons (helium is an exception).Look at Group 18 in the Periodic Table. In the upper right hand corner of the box for each element is the oxidation number. $_2$He, $_{10}$Ne and $_{18}$Ar have oxidation numbers of zero, which means they do not form compounds. $_{36}$Kr, $_{54}$Xe and $_{86}$Rn have some positive oxidation numbers, therefore, they can form compounds. ($_{86}$Rn is very radioactive and difficult to work with.) The old name, inert gases, for Group 18 is no longer applicable, because it is possible to form compounds of krypton, xenon and radon with fluorine and oxygen.

Noble gases are monatomic gases: Atoms in this group have a **complete outer principal energy level** and are **stable**.

In Group 18, there are forces of attraction called **dispersion forces** between the atoms. You will learn more about these forces in the chapter on bonding. For example, in helium, there are dispersion forces (forces of attraction) holding the helium atoms close together. This force allows the noble gas to become a liquid or solid.

When you **go down Group 18** (from He to Ne to Rn), **dispersion forces** (forces that hold the atoms together) **increase**; therefore, **the boiling point increases**. The boiling point of Kr (krypton) is higher than the boiling point of He (helium).

Question: Classify each element as a metal, nonmetal, metalloid, or noble gas:
  1. Element has luster and is a good conductor of heat and electricity.
  2. Element does not react.
  3. Element is brittle in the solid phase and a poor conductor of heat and electricity.
  4. Element has some characteristics of a metal and some of a nonmetal.

Solution: Look above at the properties of metals, metalloids, nonmetals and noble gases.
  1. Metal-has luster, is a good conductor of heat and electricity.
  2. Noble gas-some noble gases do not react.
  3. Nonmetal-brittle in solid phase, is a poor conductor.
  4. Metalloid-has characteristics of metal and nonmetal.

IONS OF METALS AND NONMETALS

When a metal such as **lithium loses** an **electron**, a nonmetal like **fluorine gains** the **electron**.

The lithium atom loses an electron and becomes an Li^+ ion. See next page. This causes the lithium ion to have a complete outer shell (1st shell is complete with 2 electrons). Ions always have a complete outer shell.

$$_3Li \ominus \overset{)\,)}{\underset{)}{2}}$$

Lithium atom has atomic #3, 3 protons(+) and **3 electrons(−)**.

Electron configuration 2-1

$$_3Li \ominus \overset{)}{2} \quad +$$

Lithium ion has atomic #3, 3 protons(+) and **2 electrons(−)**, therefore Li ion is +1.

Electron configuration 2

Li atom and Li ion always have atomic #3 = 3 protons. **Ion and atom differ** only in the **number of electrons**.

A fluorine atom gains an electron and becomes an F⁻ ion. See diagram below. This causes the F⁻ ion to have a complete outer shell. (The 2nd shell is complete with 8 electrons).Ions always have a complete outer shell.

$$_9F \ominus \overset{)\,)}{\underset{)}{2}}$$

Fluorine atom has atomic #9, 9 protons(+) and **9 electrons(−)**.

Electron configuration 2-7

$$_9F \ominus \overset{)\,)}{\underset{)}{2\,8}} \quad -$$

Fluorine ion has atomic #9, 9 protons(+)and 10 electrons(−), therefore F ion is −1.

Electron configuration 2-8

F atom and F ion always have atomic #9 = 9 protons. **Ion and atom differ** only in the **number of electrons**.

Any ion has a different number of protons than electrons. Metals lose electrons and their ions are positive. A +1 ion (or an ion with a charge of 1+) has 1 less electron than proton; a +2 ion has 2 less electrons than protons, etc. Nonmetals gain electrons and form negative ions with more electrons than protons. A 1- ion (example F⁻) has 1 more electron than proton, a 2- ion has 2 more electrons than protons, and a 3- ion has 3 more electrons than protons.

	Metals				Nonmetals		
Atom	# protons = electrons	ion	#electrons	Atom	#protons = electrons	ion	#electrons
Li	3	Li⁺	2	F	9	F⁻	10
Ca	20	Ca²⁺	18	S	16	S²⁻	18
				P	15	P³⁻	18

Question: What is the total number of electrons in a Mg^{2+} ion?

Solution: Look at Mg (magnesium) in the Periodic Table. The atomic number is 12. Mg has 12 electrons.

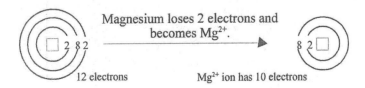

Magnesium loses 2 electrons and becomes Mg²⁺.

2 8 2   12 electrons   8 2   Mg²⁺ ion has 10 electrons

Mg^{2+} means Mg loses 2 electrons and becomes Mg^{2+}.

Question: An O^{2-} ion has 10 electrons. How many protons does it have?

Solution: A negative ion has more electrons than protons. An ion with a 2- charge has 2 more electrons than protons. Therefore, a 2-ion with 10 electrons has 8 protons.

Or, Method 2: oxygen atom or oxygen ion has an atomic number of 8 (see Periodic Table), therefore oxygen ion has 8 protons.

Question: How many neutrons are in an O^{2-} ion from oxygen-18?

Solution: Oxygen-18 means mass number = 18. All atoms of oxygen have atomic number 8 (see Periodic Table).

neutrons = mass # –atomic # = 18 –8 = 10 neutrons.

ALLOTROPES

The element oxygen exists in two different forms, O_2 and O_3. Different forms of the same element are called allotropes.

Allotropes are forms of the **same element** that have **different molecular formulas** (O_2 and O_3) or **different** crystalline **structures** (forms of carbon). Allotropes have different properties.

Types of allotropes:

1. **Oxygen** has 2 **allotropes.** O_2 is the oxygen we breathe; O_3 makes up the ozone layer.

   Models: Let 1 black ball (•) represent 1 oxygen (O) atom, therefore O_2 (2 oxygen atoms) is (••) and O_3 (3 oxygen atoms) is (•••).

2. **Carbon** has **many allotropes**, which differ in their crystalline structure (arrangement of atoms):
   a. Diamond: Every carbon atom is bonded to four other carbon atoms and forms a tetrahedron. Diamond is very hard.
   b. Graphite: Carbon atoms are arranged in sheets or layers. Graphite is used in lead pencils.
   c. Coal: No definite pattern.
   d. Buckminsterfullerene: Made up of rings of 5 and 6 carbon atoms connected into a ball. (It looks like the outside of a soccer ball.) Each buckminsterfullerene has 60 to 70 carbon atoms.

3. **Phosphorus** has **allotropes,** (white, black, or red phosphorus) which differ in their molecular structure (arrangement of atoms).

4. **Arsenic** has two allotropes (two forms) (gray or yellow arsenic). One form of arsenic (gray) has atoms (particles) closer together, therefore this allotrope (gray) has more mass (grams) and more density (density = $\frac{mass}{volume}$) than the other allotrope (yellow).

Try Sample Questions #5-6, on pages 17-18, and then do Homework Questions, #16-27, pages 21-22, and #53-56, page23.

PHYSICAL PROPERTIES OF ELEMENTS

Physical properties are color, odor, solubility, density, boiling point, melting point, hardness, malleability, conductivity, phase (solid, liquid, or gas), etc. A physical property is a characteristic of a substance (example: diamond is very hard) that you can notice without changing the substance (diamond) into anything else. You can notice the boiling point of water without changing the water into anything else. Some physical properties are listed in Table S.

Different elements have different physical properties. As you know, the elements left of the zigzag line are **metals** and therefore are **malleable** and **conduct heat** and **electricity**. Metals have the physical properties of malleability and conductivity.
Elements on the right of the zigzag line in the Periodic Table are nonmetals. Nonmetals are not malleable and do not conduct heat and electricity. (These elements do not have malleability and conductivity.) Elements in **Group 1** form compounds that are **soluble**. They have the property of solubility. Examples are sodium chloride, potassium sulfide, and lithium sulfate.
Elements in Group 2 form some compounds that are soluble and some compounds that are not soluble. Barium sulfate is insoluble (not soluble). Barium bromide is soluble. Calcium chloride is soluble. Calcium phosphate is insoluble. You will learn in Chapter 6 how to use Table F to determine which compounds are soluble.
Elements differ in density, $\frac{mass}{volume}$. Elements in **Group 2** have **higher density** (more dense) than elements in Group 1 in the same period. You can look at **Table S** and you can find the density of the elements on the Periodic Table.
Elements differ in hardness. Elements of **Group 1** are **soft** and they can be cut with a knife. Elements of **Group 2** are **harder** but can be cut with a hard steel knife. Elements in Groups 3-12 (transition elements) are much harder.
Most of the **elements** are **solid** at STP. Only **mercury** and **bromine** are **liquids** at STP.
Elements of **Group 18** and **nitrogen, oxygen, fluorine**, and **chlorine** are **gases** at STP.

CHEMICAL PROPERTIES OF ELEMENTS

Being able to burn is a chemical property. The chemical property of an element describes how an element reacts in a chemical reaction. Does it burn? Does it react with water, acid, etc.?
You learned that elements in Group 1 have 1 valence electron; elements in Group 2 have 2 valence electrons. Elements in Group 1 react faster than elements in Group 2 because it is easier to lose one valence electron than to lose two valence electrons. $_{11}Na$, an **element** in **Group 1**, **reacts vigorously** in cold water. $_{12}Mg$, an element in **Group 2**, does not react in cold water, only in hot water and **slowly**.

As you go down Groups 1 and 2, reactivity (activity) increases. $_{87}$Fr (**Francium**), **bottom left hand corner** of the **Periodic Table, is the most** active. It is more active than Cs; Cs (cesium) is more active than Rb (rubidium).

Group I
Li
Na
K
Rb
Cs
Fr

← Activity Increases

most active ↑

Look at the Periodic Table. In Group 16, sulfur is less active than oxygen. For **nonmetals** in a group, the rule is that the **top one is more active** than the elements beneath it. Oxygen is more active than sulfur.

Look at the Periodic Table. Group 17 is very active. Elements in Group 17 react very fast and occur in compounds only. Fluorine is the most active nonmetal in Group 17 and in the entire Periodic Table. Chlorine is less active than fluorine. Remember, we said for **nonmetals** in a group, the **top one is more active** than the elements beneath it. F is more active than Cl; Cl is more active than Br; Br is more active than I.

In Group 18, **small noble gases,** $_2$He, $_{10}$Ne and $_{18}$Ar **do not react. Heavier** noble gases, $_{36}$Kr, $_{54}$Xe and $_{86}$Rn can **react** with fluorine and oxygen.

Look at **TABLE S**, page Reference Tables 24-25. **Table S** lists **physical** and **chemical properties** of the **elements,** such as **melting point** (example: Fe, 1811 K), **boiling point** (Fe, 3134 K), density, electronegativity, etc. The heading of Table S has melting point (K) and boiling point (K); K means Kelvin temperature.

Boiling point of Fe (temperature when liquid changes to a gas) is 3134 K, therefore, at any temperature above 3134 K (boiling point), Fe (iron) is a gas.

Melting point (temperature when solid changes to liquid) = **Freezing point** (temperature when liquid changes to solid). Table S gives melting point of Fe = 1811 K, therefore you know freezing point of Fe = 1811 K. The **melting point** of **Fe** (see Table S) is **1811 K**, therefore, at any **temperature** below 1811 K, Fe (iron) is a solid. At any temperature above the melting point (1811 K) and below the boiling point (3134 K), Fe (iron) is a liquid. You can use the equation K = °C + 273 (given on Table T) to convert 1811 K into Celsius, which is 1538 °C.

If a question gives you density, melting point, or boiling point of an element on the Periodic Table, you can look at Table S and see which element it is.

PROPERTIES FROM LOCATION IN THE PERIODIC TABLE

When you look at the Periodic Table and you see an element in a specific location or placement, it tells you about the physical and chemical properties of that element. Look at $_{11}$Na in the Periodic Table on page Reference Tables 20-21. By **looking** at the **Periodic Table**, what can **you find out about Na** :

   1. It is left of the zigzag line, therefore Na is a metal.

2. You learned that metals are malleable, ductile, good conductors of heat and electricity, and most metals are solids (except mercury). Therefore, you know that sodium is malleable, ductile, a good conductor of heat and electricity, and is a solid.

3. You learned that elements in Group 1 are very reactive. Na is in Group 1 and is very reactive. (Sodium reacts with cold water.)

4. You learned in Group 1, $_{87}Fr$, in the bottom left hand corner of the Periodic Table, is the most active. Fr is more active than Cs; Cs is more active than Rb, etc. Therefore, you know by looking at the Periodic Table that K is more active than Na and Na is more active than Li.

5. You learned, in the same period, Group 1 is more active than Group 2. Remember, $_{11}Na$ (Group 1) reacts in cold water and $_{12}Mg$ does NOT react in cold water, but only reacts in warm water, and very slowly.

From the Periodic Table you know:

1. Sodium is a metal.
2. It is malleable, ductile, a good conductor of heat and electricity, and a solid.
3. Na is very active.
4. K is more active than Na and Na is more active than Li.
5. Na is more active than Mg.

As a scientist, in the 3 questions below, where would you place the elements in the Periodic Table?

Question: A poor conductor of heat and electricity and the most active element.

Solution: Nonmetals are poor conductors of heat and electricity. Very active nonmetals are in Group 17. The most active nonmetal is at the top of Group 17. Therefore, you would put the element at the top of Group 17. ($_9F$ is at the top of Group 17).

Question: Most reactive element with 1 valence electron.

Solution: 1 valence electron means the element is in Group 1. Most reactive means the element is at the bottom of Group 1. Therefore, put the element at the bottom of Group 1.

Question: Element has a complete outer shell and does not react.

Solution: Group 18 has a complete outer shell. Small noble gases do not react, therefore put the element near the top of Group 18.

Question: Compare and contrast elements in Group 14 with metallic and nonmetallic properties, based on their location in the Periodic Table.

Solution: Look at the Periodic Table. Elements left of the zigzag line are metals. Those touching the zigzag line are metalloids. Elements to the right of the zigzag line are nonmetals.

Carbon (C), right of the zigzag line, is a nonmetal with, of course, nonmetallic properties.

Silicon (Si) and germanium (Ge), touching the zigzag line, are metalloids. They have some metallic and some nonmetallic properties.

Tin (Sn) and lead (Pb), left of the zigzag line, are metals with metallic properties.

The names and symbols of the elements are given in Table S, page Reference Tables 24-25.

GROUPS OF THE PERIODIC TABLE

Now that you learned about physical and chemical properties and you see where elements are placed based on these properties, let's discuss these groups in a little more detail.

CHEMISTRY OF A GROUP: Elements in the **same group react similarly**. They have the **same number of valence electrons**. As you go down a group of the Periodic Table, radius increases, electronegativity (attraction for electrons) decreases, and ionization energy generally decreases (because valence electrons are further away from positive protons and more electrons are repelling the valence electrons).Ionization energy, electronegativity, and atomic radius are given in Table S. You will learn about them later in the chapter.

LEWIS ELECTRON DOT STRUCTURES FOR GROUPS

As you learned in Chapter 2,

There can only be 2 electrons on each side of the symbol of the element

(1) Put the 1st and 2nd valence electrons on one side of the symbol of the element. You can put the 1 or 2 electrons on any side of X.

$$X\cdot \qquad X:$$

(2) For the 3rd, 4th, and 5th electrons, put each electron on a different side of the symbol. $\quad \dot X: \quad \cdot \dot X: \quad \cdot \dot X:$

(3) For the 6th, 7th and 8th electrons, add the electron to any side with 1 electron. $\quad \cdot \ddot X: \quad :\ddot X: \quad :\ddot X:$

Look at the table below:

Lewis Electron Dot Structures

X means elements (nucleus and non-valence electrons); • means 1 valence electron.							
Group	1	2	13	14	15	16	17	18
Element	$X\cdot$	$X:$	$\dot X:$	$\cdot\dot X:$	$\cdot\dot X:$	$\cdot\ddot X:$	$:\ddot X:$	$:\ddot X:$

GROUPS 1 AND 2: Elements in **Group 1** have one electron in the last principal energy level: one valence electron. One valence electron is shown by 1 dot next to X. Remember, in Chapter 2, the electron dot

symbol for lithium (an element in Group 1) was Li•. Elements in Group 1 can be shown by X· or drawn as Li•, Na•, K•, etc.

Elements in **Group 2** have 2 electrons in the last principal energy level: 2 valence electrons. Look again at the table on the previous page. Elements in Group 2 each have 2 valence electrons, which can be shown by X:, or drawn as Be:, Mg:, Ca:, etc.

1. Because they are very reactive, elements in Groups 1 and 2 occur in nature only in compounds. Elements in both groups are reduced to their free state by **electrolysis** of their fused (melted) compounds. (To get sodium from its compounds, electrolysis is used).

2. Elements in Group 1 unite with Cl_2 (chlorine) to form chlorides having the formula NaCl (sodium chloride), KCl (potassium chloride), LiCl, etc. and having the general formula MCl. M stands for any element in Group 1. Elements in Group 2 unite with Cl_2 (chlorine) to form compounds having the formula $MgCl_2$ (magnesium chloride), $BeCl_2$, etc.; the general formula is MCl_2. M stands for any element in Group 2. (Formulas will be explained in Chapter 4, Section B.)

3. Low ionization energy, low electronegativity: Look at Table S, on page Reference Tables 24-25.

4. Elements in Group 1 and Group 2 lose electrons to form ionic compounds that are relatively stable. An ionic compound is formed when an atom (example: Li) loses an electron and another atom (example: F) gains the electron.

Group I
Li
Na
K
Rb
Cs
Fr

↑ Activity Increases

most active ↑

5. Look at the Periodic Table. As you go down Groups 1 and 2, reactivity (activity) increases. $_{87}$**Fr** (**Francium**), **bottom left hand corner of the Periodic Table, is the most active**. It is **more active than Cs**; Cs (cesium) is more active than Rb (rubidium).

Question: Determine the group of an element that forms a compound XCl_2.

Solution: Look at Cl in the Periodic Table. It has 7 valence electrons. It will gain 1 electron to have 8 electrons (a complete outer shell). Two Cl atoms will gain 2 electrons. Therefore, X must lose 2 electrons; it must have 2 electrons in the last shell, and must be in Group 2.

Later you will learn how to use the Oxidation Number to determine which group X is in.

GROUP 13: Elements of Group 13 have 3 valence electrons, which can be shown by X:.

GROUP 14: Elements of Group 14 have 4 valence electrons, which can be shown by ·X: , or drawn as ·C:, ·Si: , Ge: ,etc.

GROUP 15: Elements of Group 15 have 5 valence electrons, which can be shown by $\cdot \ddot{X}$:, or drawn as $\cdot \ddot{N}$:, $\cdot \ddot{P}$:, etc.

Look at group 15 in the Periodic Table. N (nitrogen) **and P** (phosphorus) **are nonmetals**, found to the **right of the zigzag line** on the Periodic Table. **As** (arsenic) and **Sb** (antimony) are **metalloids**, touching the zigzag line. **Bi** (bismuth) is a **metal, left of the zigzag line.** In Group 15, there is a **progression** (a **gradual change**) from nonmetal to metal with increasing atomic number.

N (nitrogen) is a diatomic molecule with a triple bond, $N \equiv N$ (3 pairs of electrons are shared). A lot of energy is required to break this bond. Nitrogen compounds are important in all living things. Some nitrogen compounds are unstable; they are used as explosives.

P (phosphorus) is found in DNA, RNA, bones and teeth. Phosphorus exists as a tetra-atomic molecule, P_4.

Question: In the ground state, atoms of the elements in Group 15 of the Periodic Table all have the same number of
  (1) filled principal energy levels
  (2) occupied principal energy levels
  (3) neutrons in the nucleus
  (4) electrons in the valence shell
Solution: Answer **4**. All elements in the same group have the same number of valence electrons (electrons in the last shell). All elements in Group 15 have 5 valence electrons.

GROUP 16: Elements of Group 16 have 6 valence electrons, which can be shown by $\ddot{X}$:, or drawn as $\cdot \ddot{O}$:, $\cdot \ddot{S}$:, etc.

1. There is a progression from nonmetal to metal with increasing atomic number: Oxygen (O) and sulfur (S) are nonmetals; selenium (Se) and tellurium (Te) are metalloids; polonium (Po) is a metal.

2. O_2 is a very active nonmetal. Because of its high electronegativity, oxygen has a negative oxidation state. Exception: In compounds with fluorine (F), oxygen (O) has a positive oxidation number (because F is even more electronegative).

3. Sulfur is less active than oxygen. For **nonmetals** in a group, the rule is that the **top one is more active** than the elements beneath it. Oxygen is more active than sulfur.

4. Polonium is a decay product of uranium. Polonium is radioactive and is an alpha emitter.

GROUP 17: Group 17 is called the halogen group. Elements of Group 17 have 7 valence electrons, which can be shown by $\ddot{X}$:, or drawn as $\ddot{F}$:, $\ddot{Cl}$:, etc.

1. This group has relatively high electronegativity. **F (fluorine)** is the **most active nonmetal** and is the **most electronegative** (Table S: 4.0 electronegativity). F has only a negative oxidation state. Other elements of Group 17 show positive oxidation states when they form compounds with elements that are more electronegative than themselves. (Oxidation numbers are in the upper right hand corner of each box (element) on the Periodic Table).

2. F (fluorine) and Cl (chlorine) are gases; Br (bromine) is a liquid; I (iodine) is a solid. **As you go down Group 17, dispersion forces increase**; therefore, elements go from **gases (F, Cl) to liquid (Br) to solid (I).**

   Group 17

F	↘ gases
Cl	↗
Br	←liquid
I	← solid

3. They are very reactive and occur in nature as compounds only.

4. Diatomic F_2 can be prepared only by electrolysis of fused compounds. Cl_2, Br_2 and I_2 can be prepared by chemical means (chemical reactions).

5. F (fluorine) is a nonmetal (right of the zigzag line on the Periodic Table), At (astatine) is a metalloid (touches the zigzag line).As you go down Group 17, the metallic character increases.

Question: Which group contains a liquid that is a nonmetal at STP?
   (1) 14   (2) 15   (3) 16   (4)   17

Solution: Answer **4**. Group 17 has Br (bromine), a liquid and a nonmetal. Nonmetals are to the right of the zigzag line on the Periodic Table.

GROUP 18: Group 18 is called Noble gases. Elements of Group 18 have 8 valence electrons (helium is an exception). 8 valence electrons can be shown by ⠰Ẍ⠆, or drawn as ⠰N̈e⠆, ⠰Ä̈r⠆ , etc.

Group 18 elements are monatomic gases: Atoms in this group have a **complete outer principal energy level** and are **stable**.

As you learned, for Groups 1, 2 and 13-18 on the Periodic Table, elements within the same group have the same number of valence electrons (helium is the exception) and therefore similar chemical properties

Try Sample Questions #7-13, page 18, then do Homework Questions #28-43, pages 22-23, and #57-58, page 23.

TRANSITION ELEMENTS: Elements in Groups 3-11 of the Periodic Table are called **TRANSITION ELEMENTS**.

Colored ions: Ions of transition elements are usually colored, both in solid compounds and in solution. For **example**, if you are told that "a **solution has blue ions,**" you know it must be a **transition element**.

These are elements in which the **"d" orbitals** of the next to outermost principal energy level are being filled. See Periodic Table (with sublevels), on page Additional Tables 2-3.

Look again at the transition elements on the Periodic Table. These elements have **multiple positive oxidation states** (can use electrons from the "s" and "d" sublevels).

PERIODIC TABLE

Question: In which classification is an element placed if the outermost three sublevels of its atoms have a ground state electron configuration of $3p^6 3d^5 4s^2$?
  (1) alkaline earth metals   (3) metalloids
  (2) transition metals   (4) nonmetals

Solution: Answer 2. $3p^6 3d^5 4s^2$ is a transition element, because the d sublevel is being filled up (now partly filled). (The d sublevel can hold up to ten electrons.) Or, look for an electron configuration of $3d^5 4s^2$ on the Periodic Table. You find it is Mn, which is a transition element. All transition elements are labeled on the Periodic Table.

Table S, on page Reference Tables 24-25, lists the names of the elements, atomic numbers and symbols of the elements so you can easily find the elements on the Periodic Table.

Try Sample Questions #14-15, on page 18, then do Homework Questions, 44-50, page 23.

TABLE S

SAMPLE REGENTS & REGENT-TYPE QUESTIONS AND SOLUTIONS

1. The chemical properties of the elements are periodic functions of their atomic
   (1) masses   (2) weights   (3) numbers   (4) radii

2. According to the Periodic Table, which element has more than one positive oxidation state?
   (1) cadmium   (2) iron   (3) silver   (4) zinc

3. Boron and arsenic are similar in that they both
   (1) have the same ionization energy
   (2) have the same covalent radius
   (3) are in the same family of elements
   (4) are metalloids (semimetals)

4. An atom of an element has 28 innermost electrons and 7 outermost electrons. In which period of the Periodic Table is this element located?
   (1) 5   (2) 2   (3) 3   (4) 4

5. Which of the following substances is the best conductor of electricity?
   (1) NaCl(s)   (2) Cu(s)   (3) $H_2O(\ell)$ (4) $Br_2(\ell)$

---

6. Group 18 elements Kr and Xe have selected oxidation states of other than zero. These oxidation states are an indication that these elements have
   (1) no chemical reactivity   (3) stable nuclei
   (2) some chemical reactivity   (4) unstable nuclei

7. Which two elements have the most similar chemical properties?
   (1) aluminum and barium   (3) chlorine and sulfur
   (2) nickel and phosphorus   (4) sodium and potassium

8. As the atoms of the metals of Group 1 in the ground state are considered in order, from top to bottom, the number of occupied principal energy levels
   (1) decreases   (2) increases   (3) remains the same

9. Which metal is most likely obtained by the electrolysis of its fused salt?
   (1) Au   (2) Ag   (3) Li   (4) Zn

10. Which statement best describes Group 2 elements?
    (1) They have one valence electron, and they form ions with a 1+ charge.
    (2) They have one valence electron, and they form ions with a 1- charge.
    (3) They have two valence electrons, and they form ions with a 2+ charge.
    (4) They have two valence electrons, and they form ions with a 2- charge.

11. Which element is a liquid at room temperature?
    (1) K   (2) I_2 (3) Hg   (4) Mg

12. Which group contains elements in three phases of matter at STP?
    (1) noble gases   (3)   Group 2
    (2) transition elements   (4) halogens (Group 17)

13. Which electron configurations represent the first two elements in Group 17 of the Periodic Table?
    (1) 2-1 and 2-2   (3)   2-7 and 2-8-7
    (2) 2-2 and 2-3   (4)   2-8 and 2-8-7

14. A transition element in the ground state could have an electron configuration of
    (1) $1s^2 2s^2 2p^6 3s^2 3p^6 4s^2$ (3) $1s^2 2s^2 2p^6 3s^2 3p^6 3d^{10} 4s^2 4p^3$
    (2) $1s^2 2s^2 2p^6 3s^2 3p^6 3d^5 4s^2$ (4) $1s^2 2s^2 2p^6 3s^2 3p^6 3d^{10} 4s^2 4p^6$

15. Which of the following aqueous solutions is blue?
    (1) Na_2SO_4(aq)   (3) $MgSO_4$(aq)
    (2) K_2SO_4(aq)   (4) $CuSO_4$(aq)

SOLUTIONS

1. Answer 3. You learned that properties of the elements are periodic functions of the atomic number.

2. Answer 2. Oxidation numbers are in the upper right hand corner of each box in the Periodic Table. Iron (Fe) has two positive oxidation numbers: Fe^{2+} and Fe^{3+}.

3. Answer *4*. Look at the Periodic Table. Boron (B) and arsenic (As) are touching the zigzag line. The elements touching the zigzag lines are metalloids.

4. Answer *4*. This atom has 28 and 7, = 35, electrons. 35 electrons = 35 protons = 35 atomic number. Look at the Periodic Table: Atomic number 35 is $_{35}Br$. Go across to the left of the Periodic Table: you see "4," which means period four.

$$\boxed{4 \quad | \quad _{19}K \; \ldots \ldots \ldots \ldots \ldots \; _{35}Br}$$

5. Answer *2*. Look at the Periodic Table. Metals are on the left of the zigzag line and are good conductors of electricity. Since $_{29}Cu$ is to the left of the zigzag line, Cu is a good conductor of electricity.

6. Answer *2*. If Kr and Xe have positive oxidation numbers, it means they lose electrons and take part in some chemical reactions.

7. Answer *4*. Elements in a group react similarly because they have the same number of valence electrons. Look at the Periodic Table. Sodium (Na) and potassium (K) are both in Group 1. They have 1 valence electron. The other choices do not have two elements from the same group.

8. Answer *2*. As you go down a group on the Periodic Table, each element further down has one more shell or principal energy level. Go down Group 1 of the Periodic Table: $_3Li$, 2-1, has 2 shells (principal energy levels); $_{11}Na$ 2-8-1, has 3 shells (principal energy levels); $_{19}K$, 2-8-8-1, has 4 shells (principal energy levels); $_{37}Rb$, 2-8-18-8-1, has 5 shells (principal energy levels); $_{55}Cs$, 2-8-18-18-8-1, has 6 shells (principal energy levels).

9. Answer *3*. You learned that elements in Group 1 and Group 2 are obtained by electrolysis of their fused (melted) salts. Li is in Group 1.

10. Answer *3*. Group 2 has 2 electrons in the last shell, or 2 valence electrons. Group 2 elements lose 2 electrons and have a 2+ charge. An example of a Group 2 element is calcium (Ca):

    Ca 2 8 8 2 *loses 2 electrons and becomes* Ca 2 8 8, *a* Ca^{2+} *ion*

    Or, look at the Periodic Table. The oxidation number in the upper right hand corner of each box in Group 2 is +2, which means a 2+ charge. Answer 3 is the correct answer.

11. Answer *3*. Hg (mercury) is liquid at room temperature.

12. Answer *4*. Look in the Periodic Table at Group 17 (halogens). As you go down in Group 17, dispersion forces increase, and therefore elements go from gases (F (fluorine) and Cl (chlorine)), to liquid (Br (bromine)), to solid (I (iodine)). Answer 4: Halogens (Group 17) have three phases: gas, liquid and solid.

F
Cl
Br
I
At

13. Answer *3*. Look at the Periodic Table. Look at the electron configuration of the first two elements in Group 17.

$$_9F: 2\text{-}7; \; _{17}Cl \; 2\text{-}8\text{-}7: \text{Answer } 3$$

14. Answer **2**. Look at Answer 2: The electron configuration is $1s^2\ 2s^2\ 2p^6\ 3s^2\ 3p^6\ \mathbf{3d^5}\ 4s^2$. This is a **transition element**, because the d sublevel is **being filled up** (now partly filled). The d sublevel can hold up to ten electrons.

Or, add the exponents of the electron configuration in Answer 2:

$1s^2\ 2s^2\ 2p^6\ 3s^2\ 3p^6\ 3d^5\ 4s^2$, $2+2+6+2+6+5+2 = 25$ electrons

25 electrons = 25 protons = atomic number 25. Atomic number 25 is Mn, which is a transition element. It is labeled "transition element" in the Periodic Table.

Or, look again at the electron configuration in the Periodic Table of Answer 2: $3d^5 4s^2$. On top of that element on the Periodic Table is the label "transition element."

15. Answer **4**. A blue solution or any colored solution has compounds of **transition elements**. Groups 3-11 in the Periodic Table are "transition elements." Look at the Periodic Table. Copper is an element in Group 11. Answer 4, $CuSO_4$, has $_{29}Cu$, a transition element.

NOW LET'S TRY A FEW HOMEWORK QUESTIONS:

---

1. Which is the electron configuration of an atom of a Period 3 element?
   (1) 2-1   (2) 2-3   (3) 2-5   (4) 2-8-1

2. In which category of elements in the Periodic Table do a l of the atoms have valence electrons in the second principal energy level (second shell)?
   (1) Group 2   (3) the alkaline earth family
   (2) Period 2   (4) Group 1

3. Which elements have the most similar chemical properties?
   (1) K and Na   (2) K and Cl   (3) K and Ca   (4) K and S

4. Because of its high reactivity, which element is *never* found free in nature?
   (1) O   (2) F   (3) N   (4) Ne

5. According to the Periodic Table, which element has more than one positive oxidation state?
   (1) cadmium   (2) iron   (3) silver   (4) zinc

6. An atom of an element has 28 innermost electrons and 7 outermost electrons. In which period of the Periodic Table is this element located?
   (1) 5   (2) 2   (3) 3   (4) 4

7. In which set do the elements exhibit the most similar chemical properties?
   (1) H, O and F   (3) Li, Na and K
   (2) Hg, Br and Rn   (4) Al, S and P

8. Atoms of elements in a group of the Periodic Table have similar chemical properties. This similarity is most closely related to the atoms'
   (1) number of principal energy levels
   (2) number of valence electrons
   (3) atomic structure
   (4) atomic masses

9. Which group below contains elements with the greatest variation in chemical properties?
   (1) Li, Be, B   (2) Li, Na, K   (3) B, Al, Ga   (4) Be, Mg, Ca

---

10. The chemical properties of the elements are periodic functions of their atomic
    (1) masses   (2) weight   (3) numbers   (4) radii

11. Which element is in Group 2 and Period 7 of the Periodic Table?
    (1) magnesium (2) manganese (3) radium   (4) radon

12. Bromine has chemical properties most similar to
    (1) fluorine   (2) potassium (3) krypton   (4) mercury

13. In which category of elements in the Periodic Table do all the atoms have valence
    electrons in the third principal energy level?
    (1) Group 3   (2) Period 3   (3) Group 2   (4) Group 1

14. All of the atoms of the elements in Period 2 have the same number of
    (1) protons   (3) valence electrons
    (2) neutrons   (4) occupied principal energy levels

15. Which is the electron configuration of an atom of a Period 3 element?
    (1) 2-1   (2) 2-3   (3) 2-5   (4) 2-8-5

16. When a metal reacts with a nonmetal, the metal will
    (1) lose electrons and form a positive ion
    (2) lose protons and form a positive ion
    (3) gain electrons and form a negative ion
    (4) gain protons and form a negative ion

17. At STP, which substance is the best conductor of electricity?
    (1) nitrogen   (2) neon   (3) sulfur   (4) silver

18. Which element is classified as a metalloid (semimetal)?
    (1) sulfur   (2) silicon   (3) barium   (4) bromine

19. Which element is classified as a semimetal (metalloid)?
    (1) Sn   (2) Sb   (3) Pb   (4) P

20. A property of most nonmetals in the solid state is that they are
    (1) good conductors of heat   (3) brittle
    (2) good conductors of electricity (4) malleable

21. Which set of elements contains a metalloid?
    (1) K, Mn, As, Ar   (3) Ba, Ag, Sn, Xe
    (2) Li, Mg, Ca, Kr   (4) Fr, F, O, Rn

22. Boron and arsenic are similar in that they both
    (1) have the same ionization energy
    (2) have the same covalent radius
    (3) are in the same family of elements
    (4) are metalloids (semi-metals)

23. A characteristic of most nonmetallic solids is that they are
    (1) brittle   (3) malleable
    (2) ductile   (4) conductors of electricity

24. Which Group 18 element is most likely to form a compound with the element
    fluorine?
    (1) He   (2) Ne   (3) Ar   (4) Kr

25. Which noble gas would most likely form a compound with fluorine?
    (1) He   (2) Ne   (3) Ar   (4) Kr

26. Which is not an allotrope of carbon?
    (1) graphite   (3) buckminsterfullerene
    (2) diamond   (4) ozone

27. Allotropes
    (1) are different forms of the same element
    (2) are different mixtures
    (3) have the same atomic number but different mass number
    (4) only radioactive elements

28. A. At standard pressure, which element has a freezing point *below* standard
    temperature?
    (1) Hg   (2) Hf   (3) Ir   (4) In
        Hint: See Table A and Table S.
    B. At which Celsius temperature does lead change from a solid to a liquid?
    (1) 328 °C   (2) 0°C   (3) 874 °C   (4) 601 °C

29. A. Which properties are characteristic of the Group 1 (IA) metals?
    (1) low reactivity and the formation of unstable compounds
    (2) high reactivity and the formation of unstable compounds
    (3) low reactivity and the formation of stable compounds
    (4) high reactivity and the formation of stable compounds

    B. Which of these Group 14 elements has the most metallic properties?
    (1) C   (2) Ge   (3) Si   (4) Sn

30. In the ground state, atoms of the elements in Group 15 of the Periodic Table all
    have the same number of
    (1) filled principal energy levels   (3) neutrons in the nucleus
    (2) occupied principal energy levels   (4) electrons in the valence shell

31. The metalloids that are included in Group 15 are antimony (Sb) and
    (1) N   (2) P   (3) As   (4) Bi

32. Which of the Group 15 elements can lose an electron most readily?
    (1) N   (2) P   (3) Sb   (4) Bi

33. A. Which element in Group 15 would most likely have luster and good electrical
    conductivity?
    (1) N   (2) P   (3) Bi   (4) As
    B. Which element in Group 15 has the most metallic character?
    (1) As   (2) N   (3) Bi   (4) P

34. In the formula X_2O_5, the symbol X could represent an element in Group
    (1) 1   (2) 2   (3) 18   (4) 15

35. Which element is a member of the halogen family (Group 17)?
    (1) K   (2) B   (3) I   (4) S

36. Which group contains elements in the solid, liquid and gaseous phase at room
    temperature?
    (1) 17   (2) 2   (3) 18   (4) 4

37. Which element is a solid at room temperature and standard pressure?
    1. bromine   (2) iodine   (3) mercury   (4) neon

38. Which group contains a liquid that is a nonmetal at STP?
    1. 14   (2) 15   (3) 16   (4) 17

39. In which group does each element have a total of four electrons in the outermost
    principal energy level?
    1. 1   (2) 18   (3) 16   (4) 14

40. Which element is so active chemically that it occurs naturally only in compounds?
    1. potassium   (2) silver   (3) copper   (4) sulfur

41. Because of its high reactivity, which element is *never* found free in nature?
    1. O   (2) F   (3) N   (4) Ne

42. The electron dot structure for elements in Group 14 is
    1. $X\cdot$ (2) $X\!:$ (3) $\cdot X\!:$ (4) $\cdot X\!:$

43. Elements in Group 18
  1. have 3 valence electrons in the last principal energy level (shell)
  (2) have 18 valence electrons in the first principal energy level
  3. have a complete outer principal energy level (shell)
  4. have an incomplete outer energy level

44. Which aqueous salt solution has a color?
  1. $BaSO_4(aq)$ (2) $CuSO_4(aq)$ (3) $SrSO_4(aq)$ (4) $MgSO_4(aq)$

45. The color of Na_2CrO_4 is due to the presence of
  1. a noble gas   (3) a transition element
  2. a halogen   (4) an alkali metal

46. A solution of $Cu(NO_3)_2$ is colored because of the presence of the ion
  1. Cu^{2+} (2) N^{5+} (3) O^{2-} (4) NO_3^-

47. Which aqueous solution is colored?
  1. $CuSO_4(aq)$ (2) $BaCl_2(aq)$ (3) $KCl(aq)$ (4) $MgSO_4(aq)$

48. In which classification is an element placed if it has a ground state electron configuration of 2-8-13-2 and is colored?
  1. alkaline earth metals   (3) metalloids (semimetals)
  2. transition metals   (4) nonmetals

49. An atom that has multiple positive oxidation states is:
  1. an alkali metal   (3) a transition element
  2. an alkaline earth metal   (4) a noble gas element

50. Which compound forms a colored aqueous solution?
  1. $CaCl_2$ (2) $CrCl_3$ (3) $NaOH$ (4) KBr

CONSTRUCTED RESPONSE QUESTIONS: Parts B-2 and C of NYS Regents Exam

51. Why do elements in the same group react similarly?

52. In a period, why are elements in Group 1 more active than elements in Group 2?

53. Explain why Si and As are metalloids (properties).

54. Compare and contrast the properties of metals, nonmetals, and metalloids.

55. As you go down Group 18, why does the boiling point increase?

56. Compare and contrast the allotropes of carbon.

57. By looking at the location of F in the Periodic Table, what can you know about fluorine?

58. How is the Lewis electron dot structure different for elements in Groups 1, 2, 13, 14, 15, 16, 17, 18.

CHAPTER QUESTION: Parts B-2 and C of NYS Regents Exam

59. Given the particle diagram: At 101.3 kPa and 298 K, which element could this diagram represent?

  (1) Rn (2) Xe (3) Ag (4) Kr

  Hint: Look for a similar diagram in Chapter 1, Section A.

IONIZATION ENERGY

IONIZATION ENERGY is the amount of energy needed to remove an electron. First ionization energy is the amount of energy needed to remove the first electron. The smaller the amount of ionization energy, the easier it is to lose an electron.

IONIZATION ENERGY DOWN A GROUP

Let's see what happens to **ionization energy** as you go **down a group**. Look at Group 1.

Step 1: List the **Group 1 elements** with their **atomic numbers** in a vertical column as shown below (just **like** the **Periodic Table** has the elements in Group 1):

Atomic #	Symbol
3	Li
11	Na
19	K
37	Rb

Step 2: Look at **part** of Reference **Table S**, below. The elements in Table S are listed in order of atomic number. **Copy** the ionization energy for elements listed in Step 1 ($_3$Li, $_{11}$Na, $_{19}$K and $_{37}$Rb) from Table S, below, **and full Table S** on pages Reference Tables 24-25, and **write** the **ionization energy** next to the element.

Part of Table S

Atomic #	Symbol	Name	First Ionization Energy
1	H	Hydrogen	1312
2	He	Helium	2372
3	Li	Lithium	520
.	.	.	.
.	.	.	.
11	Na	Sodium	496

Atomic #	Symbol	Copy Ionization Energy
3	Li	520
11	Na	496
19	K	419
37	Rb	403

As you can see, in Group 1, ionization energy **decreases** as you go down the group. Similarly, as you **go down** any **group**, **ionization energy decreases.**

Electrons are negative; protons (in the nucleus) are positive. As you go down a group, ionization energy decreases because the valence electrons (electrons in the last shell) are further away from the positive protons (less attraction) and therefore, less energy is needed to remove valence electrons.

PERIODIC TABLE

TABLE S

Question: What happens to ionization energy as you go down Group 17?

Solution: Look at the Periodic Table. List the elements in Group 17, just as the Periodic Table does. Go to Table S on pages Reference tables 24-25 and copy the ionization energy for each element.

Atomic #	Symbol	Copy Ionization Energy
9	F	1681
17	Cl	1251
35	Br	1140

Ionization energy decreases as you go down Group 17.

NOTE: Table S is right after the Periodic Table in the Reference Table booklet given out with the Regents.

IONIZATION ENERGY ACROSS A PERIOD

Let's see what happens to **ionization energy** as you go **across a period**. **Look** at the **Periodic Table** below or on page Reference Tables 20-21. Remember: periods are horizontal rows on the Periodic Table. Period 1 has H and He; Period 2 has Li, Be, B, C, N, O, F and Ne.

Period								
1	H							He
2	Li	Be	B	C	N	O	F	Ne

Let's see what happens to ionization energy as you go across Period 2.

Step 1: Look at **Table S. Start** with $_3$Li and **end** with $_{10}$Ne. These **elements** in **Table S** are listed in the **same order** as on the Periodic Table. (All elements in Table S are listed in the same order as periods on the Periodic Table.)

TABLE S

Atomic #	Symbol	Name	First Ionization Energy
3	Li	Lithium	520
4	Be	Beryllium	900
5	B	Boron	801
6	C	Carbon	1086
7	N	Nitrogen	1402
8	O	Oxygen	1314
9	F	Fluorine	1681
10	Ne	Neon	2081
11	Na	Sodium	496

Step 2: **Look** at the **ionization energy** column.

Step 3: If it makes it easier, put a piece of **paper** across Table S **above atomic #3** (lithium) and another piece of paper **below atomic #10** (neon), to separate Period 2 from the other periods.

We see that the **ionization energy** generally **increases** as you go **across a period**, from 520 for Li to 2081 for Ne. (There are some exceptions (due to stability), but this is the general trend.)

ELECTRONEGATIVITY

ELECTRONEGATIVITY is the **attraction for electrons**. The larger the electronegativity, the more the atom attracts electrons.

ELECTRONEGATIVITY DOWN A GROUP

Let's see what happens to the **electronegativity** as you go **down** a **group.** Look at Group 1.

Step 1: List the **Group 1 elements** with their **atomic numbers** in a vertical column as shown below (just **like** the **Periodic Table** has the Group 1 elements):

Atomic #	Symbol
3	Li
11	Na
19	K
37	Rb

Step 2: Look at **part** of Reference **Table S**, below. The elements in Table S are listed in order of atomic number. **Copy** the electronegativity for the elements listed in Step 1 ($_3$Li, $_{11}$Na, $_{19}$K and $_{37}$Rb) from Table S, below, and **full Table S** on pages Reference Tables 24-25, and **write** the **electronegativity** next to the element.

Part of Table S

Atomic #	Symbol	Name	First Ionization Energy	Electronegativity
1	H	Hydrogen	1312	2.1
2	He	Helium	2372	—
3	Li	Lithium	520	1.0
.	.	.	.	.
.	.	.	.	.
11	Na	Sodium	496	0.9

Atomic #	Symbol	Copy Electronegativity
3	Li	1.0
11	Na	0.9
19	K	0.8
37	Rb	0.8

As you can see, in Group 1, electronegativity decreases as you go down the group. As you go **down** any **group, electronegativity decreases**.

ELECTRONEGATIVITY ACROSS A PERIOD

Let's see what happens to **electronegativity** as you go **across a period**. Look at the Periodic Table below or on page Reference Tables 20-21. Remember: periods are horizontal rows on the Periodic Table. Period 1 has H and He; Period 2 has Li, Be, B, C, N, O, F and Ne.

Period								
1	H							He
2	Li	Be	B	C	N	O	F	Ne

Let's see what happens to electronegativity as you go across Period 2.

Step 1: **Look** again at **Table S**, below. Start with $_3$Li and end with $_{10}$Ne (Period 2). These elements in Table S are listed in the same order as Period 2 on the Periodic Table. (All elements in Table S are listed in the same order as periods on the Periodic Table.)

TABLE S

Atomic #	Symbol	Name	First Ionization Energy	Electronegativity
3	Li	Lithium	520	1.0
4	Be	Beryllium	900	1.6
5	B	Boron	801	2.0
6	C	Carbon	1086	2.6
7	N	Nitrogen	1402	3.0
8	O	Oxygen	1314	3.5
9	F	Fluorine	1681	4.0
10	Ne	Neon	2081	----
11	Na	Sodium	496	0.9

Step 2: **Look** at the **electronegativities**.

Step 3: Again, if it makes it easier, put a piece of **paper** across Table S **above atomic #3** (lithium) and another piece of paper **below atomic #10** (neon).

Step 4: Again, you can see that **electronegativity increases** as you go **across a period** from $_3$Li to $_9$F.

Try Sample Questions # 1-2, page 32, then do Homework Questions # 1-10, pages 33-34.

ATOMIC RADIUS

An **ATOMIC RADIUS** is half the distance between adjacent nuclei, (or, you can say, the distance from the nucleus to the outer valence electrons). In the Periodic Table, as you go across a period from left to right, the radius decreases. The increase in positive charge in the nucleus pulls electrons closer, therefore the atomic radius gets smaller. As you go down a group, each element has an extra shell, and therefore a larger atomic radius. (Shown in **Table S**.)

There are three types of atomic radius:

1. **Covalent atomic radius**: The distance from the nucleus to the outer valence electrons when atoms are covalently bonded; or, half the distance between covalently bonded nuclei of the same atom.

2. **Van der Waals radius**: Half the distance between nuclei of the same atom that have not formed a chemical bond.

3. **Atomic radius in metals**: Half the distance from nucleus to nucleus in a crystalline metal.

ATOMIC RADIUS DOWN A GROUP

Let's see what happens to **atomic radius** as you go **down** a **Group**. Look at Group 1.

Step 1: Again list the **Group 1 elements** with their **atomic numbers** in a vertical column as shown below (just as they are in the Periodic Table, and as we did for ionization energy and electronegativity).

Atomic #	Symbol
3	Li
11	Na
19	K
37	Rb

Step 2: Look at part of Reference Table S, below. The elements in Table S are listed in order of atomic number. Copy the atomic radius for the elements listed in Step 1 ($_3$Li, $_{11}$Na, $_{19}$K and $_{37}$Rb) from **Table S**, below, and on **pages Reference Tables 24-25**, and write the radius next to the element.

Table S

Atomic #	Symbol	Name	First Ionization Energy	Electro-negativity	...	...	Atomic Radius
1	H	Hydrogen	1312	2.1			37
2	He	Helium	2372	–			32
3	Li	Lithium	520	1.0			155
.	.	.	.	.			.
.	.	.	.	.			.
11	Na	Sodium	496	0.9			190

Atomic #	Symbol	Copy Atomic Radius
3	Li	155
11	Na	190
19	K	235
37	Rb	248

As you can see, in Group 1, atomic radius increases as you go down the group. Similarly, as you go **down** any **group**, the **atomic radius increases.**

Question: What happens to atomic radius as you go down Group 14?

Solution: Look at the Periodic Table. List the elements in Group 14, just like the Periodic Table does. Go to Table S on page reference Tables 24-25 and copy the atomic radius for each element.

Atomic #	Symbol	Atomic Radius
6	C	91
14	Si	132
32	Ge	137

As you go down Group 14, atomic radius increases.

Note: As you go down a group, you have more shells (a larger atomic radius) and the valence electrons are further away from the positive protons, therefore electronegativity and ionization energy decrease.

ATOMIC RADIUS ACROSS A PERIOD

Let's see what happens to the atomic radius as you go across a period.

Look at the Periodic Table below or on page Reference Tables 20-21. Remember: periods are horizontal rows on the Periodic Table. Period 1 has H and He; Period 2 has Li, Be, B, C, N, O, F and Ne.

Period								
1	H							He
2	Li	Be	B	C	N	O	F	Ne

Let's see what happens to atomic radius as you go across Period 2.

Step 1: Look again at **Table S**. Start with $_3$Li and end with $_{10}$Ne (Period 2). These elements in Table S are listed in the same order as Period 2 on the Periodic Table. (All elements in Table S are listed in the same order as periods on the Periodic Table.)

Table S

Atomic #	Symbol	Name	First Ionization Energy	Electro-negativity	...	...	Atomic Radius
3	Li	Lithium	520	1.0			155
4	Be	Beryllium	900	1.6			112
5	B	Boron	801	2.0			98
6	C	Carbon	1086	2.6			91
7	N	Nitrogen	1402	3.0			92
8	O	Oxygen	1314	3.5			65
9	F	Fluorine	1681	4.0			57
10	Ne	Neon	2081	—			51
11	Na	Sodium	496	0.9			190

Step 2: Look at the **atomic radius.**

Step 3: Again, if it makes it easier, put a piece of **paper** across Table S **above atomic #3** (lithium) and another piece of paper **below atomic #10** (neon).

Step 4: Again, you can see that **atomic radius decreases** as you go **across a period** from $_3$Li to $_{10}$Ne.

Try Sample Question #3, page 32, then do Homework Questions #11-17, page 34, and #28-29, page 35.

IONIC RADIUS OF METALS AND NONMETALS

IONIC RADIUS: When a **metal**, such as Li (lithium), loses an electron, the lithium atom becomes a Li$^+$ ion and becomes smaller. The ionic radius (radius of the ion) is **smaller**. (When anything loses something, it becomes smaller.)

Atomic Radius   *Ionic Radius*

$_3^7 Li$

Li Atom   *Li Ion*

Metals lose electrons, therefore, the **ionic radius is smaller** than the atomic radius.

When a **nonmetal**, such as F (fluorine), $_9$F 2) 7), gains an electron, it becomes an F$^-$ ion, which is larger. The ionic radius (radius of the ion) is **larger**. (When anything gains something, it becomes larger.)

Nonmetals gain electrons, therefore the **ionic radius is larger** than the atomic radius.

Ionic radius increases as you go **down** a **group**, because more principal energy levels are added.

The chart below shows how ionic radius changes as you go across a period.

Elements in Period 3 and their ionic radii
(in picometers)

Group	1	2	13	14		15	16	17
	$_{11}$Na	$_{12}$Mg	$_{13}$Al	$_{14}$Si		$_{15}$P	$_{16}$S	$_{17}$Cl
Ionic radius	95	65	50	41		212	184	181

Ionic radius decreases ➤   Ionic radius decreases

When you go **across** the **metals** (Groups 1-14) of a period, each element has **higher atomic number**, which means **more protons** (positive charges) that pull the electrons (negative charges) closer to the nucleus, therefore **ionic radius** of **metals decreases.** When you go across the nonmetals (Groups 15-17) of a period, each element has **higher atomic number**, which means **more protons** that pull the electrons closer to the nucleus, therefore **ionic radius** of **nonmetals decreases.**

Nonmetals in a **period** have a **larger ionic radius** than metals. Nonmetals form ions by gaining electrons and completing the last energy level. (A sulfur ion has 3 complete energy levels.) Metals form ions by losing electrons and losing the last energy level. (An aluminum ion has 2 complete energy levels.) Therefore, nonmetals have a larger ionic radius than metals, because they have one additional energy level.

Try Sample Question #4, page 33, then do Homework Questions #18-23, pages 34-35, and #30, page 35.

HINT: An easy way to compare the ionization energy of metals and nonmetals is to use Table S and look at lithium (Group 1, metal), ionization energy 520, and fluorine (Group 17, nonmetal), ionization energy 1681. **Nonmetals have high ionization energy.**

An easy way to compare the electronegativity of metals and nonmetals is to use Table S and look again at lithium (Group 1, metal), electronegativity 1.0, and fluorine (Group 17, nonmetal), electronegativity 4.0 . **Nonmetals have high electronegativity.**

CHEMISTRY OF A PERIOD

As you go across a period from left to right (for example, Na, Mg, Al, Si, P, S, Cl), you notice that atomic number increases (see Periodic Table), radius generally decreases, ionization energy and electronegativity generally increase (see Table S), and there is a transition from positive to negative oxidation states (numbers) (see Periodic Table). You learned that protons are positive and electrons are negative. You can figure out that as you go across a period atomic number increases, there are **more protons** (positive), **more attraction** for electrons, (greater electronegativity), it is harder to lose electrons (greater ionization energy), and smaller atomic radius. The elements change from active metals to less active metals, to metalloids, to moderately active nonmetals, to very active nonmetals, and to a noble gas. As you can see, when you go across a period, the properties of the elements vary greatly.

Do Homework Questions, #24-27, page 35, and #31, page 35.

SUMMARY: PERIODIC TABLE: The Periodic Table shows **atomic number** (number of electrons), **atomic mass, oxidation states** (oxidation numbers), and **electron configuration** (electrons in each principal energy level). The Periodic Table also shows **periods** and **groups**, as well as **metals, metalloids, nonmetals and noble gases**.

REMEMBER: When you answer Regents questions on the **PERIODIC TABLE,** use the Periodic Table and Table S.

SAMPLE REGENTS & REGENT–TYPE QUESTIONS
AND SOLUTIONS

1. A diatomic element with a high first ionization energy would most likely be a
   (1) nonmetal with a high electronegativity
   (2) nonmetal with a low electronegativity
   (3) metal with a high electronegativity
   (4) metal with a low electronegativity

2. The table below shows some properties of elements *A*, *B*, *C*, and *D*.

Element	Ionization Energy	Electro-negativity	Conductivity of Heat and Electricity
A	low	low	low
B	low	low	high
C	high	high	low
D	high	high	high

   Which element is most likely a nonmetal?
   (1) *A* (2) *B* (3) *C* (4) *D*

3. An atom of which of the following elements has the *smallest* atomic radius?
   (1) Li (2) Be (3) C (4) F

4. Which atom has a radius larger than the radius of its ion?
    (1) Cl   (2) Ca   (3) S   (4) Se

SOLUTIONS

1. Answer *1*. Compare the ionization energies of Group 1 (metal) and Group 17 (nonmetal) in Table S. Nonmetals have high ionization energy. Look at Table S again. Compare the electronegativities of Group 1 (metal) and Group 17 (nonmetal). Nonmetals also have high electronegativity. (By the way, diatomic elements, such as N_2, O_2 and Cl_2, are in groups 15, 16 and 17, to the right of the zigzag line in the Periodic Table, and are nonmetals.)

2. Answer *3*. Look at Table S. As you learned, an easy way to compare the ionization energy of metals and nonmetals is to look at Li (Group 1, metal), ionization energy 520, and F (Group 17, nonmetal), ionization energy 1681 **Nonmetals have high ionization energies.**

    You also learned that an easy way to compare the electronegativities of metals and nonmetals is to look again at Li (Group 1, metal), electronegativity 1.0, and F (Group 17, nonmetal), electronegativity 4.0. **Nonmetals have high electronegativities.**

    You also know that **nonmetals** are **poor (low) conductors of heat and electricity**. Answer 3, element C, is a nonmetal because it has high ionization energy, high electronegativity, and low conductivity of heat and electricity.

3. Answer *4*. Look at Table S. Look at the right hand column, the atomic radius, in pm:

Li	Be	C	F
155	112	91	57

    F has the smallest covalent radius.

4. Answer *2*. Ca is a metal. A metal loses electrons and becomes an ion. Ca loses two electrons and becomes Ca^{2+}, calcium ion.

$$Ca\ 2\ 8\ 8\ 2 \xrightarrow[\text{2 electrons}]{\text{loses}} Ca\ 2\ 8\ 8 \quad Ca^{2+}\ ion$$

The radius of the calcium atom is bigger than the radius of the calcium ion.

NOW LET'S TRY A FEW HOMEWORK QUESTIONS:

1. Most nonmetals have the properties of
    (1) high ionization energy and poor electrical conductivity
    (2) high ionization energy and good electrical conductivity
    (3) low ionization energy and poor electrical conductivity
    (4) low ionization energy and good electrical conductivity

2. Which element has the highest first ionization energy?
    (1) sodium   (2) aluminum   (3) calcium   (4) phosphorus

3. Which element in Group 1 has the greatest tendency to lose an electron?
    (1) cesium   (2) rubidium   (3) potassium   (4) sodium

4. Which of the Group 15 elements can lose an electron most readily?
    (1) N   (2) P   (3) Sb   (4) Bi

5. Which first ionization energy is the most probable for a very reactive metal?
    (1) 380 kjoules/mol   (3) 1681 kjoules/mol
    (2) 1086 kjoules/mol   (4) 2372 kjoules/mol

6. Which element in Group 1 has the highest tendency to lose an electron?
    (1) cesium   (2) rubidium   (3) potassium   (4) sodium

7. A nonmetal could have an electronegativity of
    (1) 1.0   (2) 2.0   (3) 1.6   (4) 2.6

8. Of all the elements, the one with the highest electronegativity is found in Period
    (1) 1   (2) 2   (3) 3   (4) 4

9. A metal can have an electronegativity of
    (1) 4.0   (2) 1.0   (3) 3.5   (4) 3.2

10. Which electronegativity is possible for a Group 1 metal?
    (1) 1.0   (2) 2.0   (3) 3.0   (4) 4.0

11. According to Reference Table S, which of the following elements has the smallest atomic radius?
    (1) nickel   (2) cobalt   (3) calcium   (4) potassium

12. Which electron configuration represents the atom with the largest covalent radius?
    (1) 2-1   (2) 1   (3) 2-2   (4) 2-3

13. Which of the following elements has the largest atomic radius?
    (1) beryllium   (2) magnesium   (3) calcium   (4) strontium

14. Which element in Period 3 has the largest atomic radius?
    (1) Cl   (2) Al   (3) Na   (4) P

15. Which of the following atoms has the largest atomic radius?
    (1) Na   (2) K   (3) Mg   (4) Ca

16. An atom of which of the following elements has the *smallest* atomic radius?
    (1) Li   (2) Be   (3) C   (4) F

17. As the elements Li to F in Period 2 of the Periodic Table are considered in succession, how do the relative electronegativity and the covalent radius of each successive element compare?
    (1) the relative electronegativity decreases, and the covalent radius decreases
    (2) the relative electronegativity decreases, and the covalent radius increases
    (3) the relative electronegativity increases, and the covalent radius decreases
    (4) the relative electronegativity increases, and the covalent radius increases

18. Which element's ionic radius is smaller than its atomic radius?
    (1) neon   (2) nitrogen   (3) sodium   (4) sulfur

19. An ion of which element is smaller than its atom?
    (1) F   (2) O   (3) Cl   (4) Na

20. An ion of which element is larger than its atom?
    (1) Al   (2) Br   (3) Ca   (4) Sr

21. Which atom has a radius larger than the radius of its ion?
    (1) Cl   (2) Ca   (3) S   (4) Se

22. When a sodium atom becomes an ion, the size of the atom
    (1) decreases by gaining an electron
    (2) decreases by losing an electron
    (3) increases by gaining an electron
    (4) increases by losing an electron

23. When a metal atom combines with a nonmetal atom, the nonmetal atom will
    (1) lose electrons and decrease in size
    (2) lose electrons and increase in size
    (3) gain electrons and decrease in size
    (4) gain electrons and increase in size

24. Which group below contains elements with the greatest variation in chemical properties?
    (1) Li, Be, B   (2) Li, Na, K   (3) B, Al, Ga   (4) Be, Mg, Ca

25. As the elements in Period 3 are considered from left to right, they tend to
    (1) lose electrons more readily and increase in metallic character
    (2) lose electrons more readily and increase in nonmetallic character
    (3) gain electrons more readily and increase in metallic character
    (4) gain electrons more readily and increase in nonmetallic character.

26. As the elements Li to F in Period 2 of the Periodic Table are considered in succession, how do the relative electronegativity and the covalent radius of each successive element compare?
    (1) The relative electronegativity decreases and the covalent radius decreases.
    (2) The relative electronegativity decreases and the covalent radius increases.
    (3) The relative electronegativity increases and the covalent radius decreases.
    (4) The relative electronegativity increases and the covalent radius increases.

27. As the elements in Period 2 are considered in sequence from left to right on the Periodic Table, the nuclear charge
    (1) increases   (2) decreases   (3) remains the same

CONSTRUCTED RESPONSE QUESTIONS: Parts B-2 and C of NYS Regents Exam

28. Compare and contrast what happens to electronegativity, ionization energy, and atomic radius as you go down Group 2. Use the Periodic Table and data from Table S to show how you reached your conclusion.

29. Compare and contrast what happens to electronegativity, ionization energy, and atomic radius as you go across period 3. Use the Periodic Table and data from Table S to show how you reached your conclusion.

30. Explain why, in the same period, the ionic radius of nonmetals is larger than the ionic radius of metals.

31. Explain how the properties of elements vary (change) as you go across a period.

CHAPTER QUESTION: Parts B-2 and C of NYS Regents Exam

32. Base your answers to question 32 on the reading passage below and on your knowledge of chemistry.

OBTAINING METALS FROM COMPOUNDS (ORES)

To obtain metals from compounds (ores):
In the earth, the metal is in a compound. The manufacturers want to get the metal alone out of the ore (compound).
1. You learned that metals in Group 1 and Group 2 are obtained by electrolysis of their fused compounds (salts). Examples are sodium (Na) and potassium (K) (both in Group 1), obtained by electrolysis of their fused salts. Aluminum metal (from aluminum oxide, Al_2O_3, in the mineral bauxite) is also obtained by electrolysis.
2. Moderately active metals like zinc (Zn) and iron (Fe) are obtained by reduction of their oxides by carbon (coke) or carbon monoxide (CO).

$$ZnO + C + \text{heat} \rightarrow Zn + \quad CO$$

(zinc oxide + carbon   →   zinc + carbon monoxide)

---

3. **Chromium** can be obtained from its oxide, a relatively stable compound, by **reduction with aluminum.**

$$2\ Al\ +\ Cr_2O_3\ \rightarrow\ Al_2O_3\ +\ 2\ Cr$$

(aluminum + chromium oxide → aluminum oxide + chromium)

Now you have chromium alone, not in a compound.

32.A. Why are metals in Group 1 and Group 2 obtained by electrolysis?

B. How does the ionic radius of aluminum compare with the atomic radius of aluminum?

C. How does the ionization energy of sodium compare with the ionization energy of zinc?

D. How is zinc obtained from zinc oxide?

CHAPTER 4: CHEMICAL BONDING

SECTION A

ELECTRONEGATIVITY AND BONDING BETWEEN ATOMS

A **CHEMICAL BOND** is an **attraction between** the **protons** (positive) of **one atom** and the **electrons** (negative) of the **next atom** that attaches the atoms together. A chemical bond is formed by a sharing or transferring of electrons. A chemical bond has stored or potential energy. After a chemical bond forms, all the atoms have a complete outer shell and are stable. Like noble gases, they do not react easily. When a **bond** is **broken, energy** is **absorbed** (taken in). In $Br_2 \longrightarrow Br + Br$, the bond between the two Br (bromine) atoms in Br_2 is broken and energy is absorbed, producing two separate Br atoms (Br and Br). When a **bond** is **formed, energy** is **released**. The compounds have lower energy.

ELECTRONEGATIVITY is the atom's **attraction for electrons** in a bond. The higher the electronegativity, the more the atom attracts electrons. The lower the electronegativity, the weaker is the attraction for electrons.

Look at Table S, below, or on page Reference Tables 24-25. Electronegativity is given in the 5th column. Electronegativity is measured on an arbitrary scale of 0-4.

Table S

Atomic Number	Symbol	Name	Ionization Energy	Electro-negativity
1	H	hydrogen		2.1
2	He	helium		----
3	Li	lithium		1.0
--------	-------	-------		-----
8	O	oxygen		3.5
9	F	fluorine		4.0
10	Ne	neon		----
11	Na	sodium		0.9
12	Mg	magnesium		1.3
13	Al	aluminum		1.6
14	Si	silicon		1.9
15	P	phosphorus		2.2
16	S	sulfur		2.6
17	Cl	chlorine		3.2

TABLE S

Now look at atomic number 9 (column 1) for fluorine (column 3). The electronegativity (column 5) of fluorine is 4.0 (the biggest electronegativity). This means that F has the greatest attraction for electrons.

Question: An atom of which of the following has the greatest ability to attract electrons:
    (1) silicon   (2) sulfur   (3) oxygen   (4) sodium
Solution: Answer *3*. Look at Table S, on page Chap. 4:1. The fifth column has the electronegativity of all the elements. You can see that:
    Si = 1.9   S = 2.6   O = 3.5   Na = 0.9
O has the biggest electronegativity of the four choices and therefore has the greatest ability to attract electrons.

*Try Sample Question #1, on page 14, then do
Homework Questions, #1-6, page 17.*

BONDS BETWEEN ATOMS

The **octet rule** states that atoms tend to lose, gain, or share electrons in order to have a complete outer shell, just like the noble gases.

I. IONIC BONDS

Ionic bonds are formed when a metal (example: Na or Ca) **transfers one (1) or more electrons** to a nonmetal (example: Cl) to form ions (charged particles).

Na 2 8 1
11

Cl 2 8 7
17

Na **transfers** (gives away) an electron

Na (sodium) gives away (**transfers**, or loses) an electron to Cl (chlorine) and becomes Na$^+$ (positive ion). Cl gains an electron and becomes Cl$^-$ (negative ion) (see drawing below):

Both Na$^+$ and Cl$^-$ now have a complete outer shell just like the noble gases. Na$^+$ has an electron configuration of 2-8 (2 electrons in the first shell, 8

Na 2 8
Na$^+$ ion

Cl 2 8 8
Cl$^-$ ion

electrons in the second shell), the same as Ne, the noble gas nearest to it (see Periodic Table on page Reference Tables 20-21). Cl$^-$ has an electron configuration of 2-8-8, the same as Ar, the noble gas nearest to it (see Periodic Table).

You learned **opposite** charges (**positive charge** of Na and **negative charge** of Cl) **attract** one another, **forming** an **ionic bond. In short**, an ionic bond is **formed** by the **transfer of electrons**.

A **MODEL** of NaCl (ionic bonding)
can be shown by:

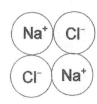

Lewis electron dot structures show what happens to the **valence electrons (electrons in** the **last shell).** There are several ways to show the Lewis electron dot structure. Electron dot diagrams use a •, an x or an ° to show valence electrons. Na has one valence electron (see drawing on the previous page), so you write Na•, Na^x, or Na°. Cl has 7 valence electrons (see drawing on the previous page), so you write Cl as:

$$:\overset{\bullet\bullet}{\underset{\bullet}{Cl}}: \quad \text{or} \quad \overset{x\,x}{\underset{x}{x\,Cl\,x}} \quad \text{or} \quad \overset{\circ\circ}{\underset{\circ\circ}{\circ\,Cl\,\circ}}$$

Let's use Lewis electron dot structures to show Na giving away an electron to Cl and forming Na$^+$ and Cl$^-$.

$$\underset{\substack{\text{Sodium}\\\text{Atom}}}{Na_x} \quad + \quad \underset{\substack{\text{Chlorine}\\\text{Atom}}}{:\overset{\bullet\bullet}{\underset{\bullet\bullet}{Cl}}:} \quad \Rightarrow \quad Na_x\overset{\bullet\bullet}{\underset{\bullet\bullet}{Cl}}: \quad \Rightarrow \quad \underset{\substack{\text{Sodium}\\\text{Ion}}}{Na^+}\underset{\substack{\text{Chloride}\\\text{Ion}}}{[:\overset{\bullet\bullet}{\underset{x\;\bullet}{Cl}}:]^-}$$

Let's look at one way of showing the electron dot structure of Group 1 elements (Na, Li) and Group 17 elements (Cl, F) in the table below. You learned in Chapter 3, Periodic Table, that elements in Group 1 (Na, Li, K) have 1 valence electron.

Lewis Electron Dot Structures

X means the element (nucleus and non-valence electrons). • means one valence electron.	
Group	1	17
Element	X•	:X:

One valence electron is shown by 1 dot next to X. Any element in Group 1 has 1 valence electron, which can be shown by X• (Na•, Li•, K•).

Elements of Group 17 (Cl, F, Br) have 7 valence electrons. Any element in Group 17 has 7 valence electrons, which can be shown as :X: (:Cl:, :F:, :Br:).

As sodium (or any element) gives away an electron to chlorine (or any element), Na$_x$ + :Cl: → Na$^+$[:Cl:]$^-$, both Na$^+$ and Cl$^-$ (like all ions) have a complete outer shell, just like the noble gases (in this case, Na$^+$ is like Ne, Cl$^-$ is like Ar). The noble gases have a stable valence electron configuration (the electron configuration tends not to change) because the noble gases tend not to bond with other atoms.

By looking at the electronegativities in **Table S** at the beginning of this chapter, you can tell when an ionic bond is formed.

TABLE S

ELECTRONEGATIVITY is the attraction for the electrons that form a chemical bond. Look at the electronegativity of Na and Cl in Table S. The electronegativity of $_{11}$Na is 0.9 and the electronegativity of $_{17}$Cl is 3.2 (column 5). Subtract one electronegativity from the other:

3.2 *(for Cl)* - 0.9 *(for Na)* = 2.3. **Ionic bond.**

If, when you **subtract** the **electronegativities** of two atoms (for example, Na and Cl), you get **2.0 or more**, then an **ionic bond** is formed. (Some chemists count a bond as ionic if the difference in electronegativity is 1.7 or more). This is a generalization (a general trend) for figuring out (predicting) if a bond is ionic.

Question: Does magnesium oxide (MgO) have an ionic bond? Explain.

Solution: Answer *yes*. Look at Table S, on page Chap. 4:1. Mg (magnesium) has an electronegativity of 1.3. O (oxygen) has an electronegativity of 3.5. **Subtract the electronegativities:**

3.5 - 1.3 = 2.2   Ionic bond.

You learned that an **ionic bond** is formed when the **difference in electronegativity** is **2.0** or **more**. (Some chemists count a bond as ionic if the difference in electronegativity is 1.7 or more).

The **EXCEPTION** to this rule is METAL HYDRIDES (compounds of an active metal and hydrogen [example: Na and H]). **An electronegativity difference of 1.2 or more** is mostly **ionic**. Look at **Table S**. Na (sodium) has an electronegativity of 0.9; H (hydrogen) has an electronegativity of 2.1. There is an electronegativity difference of 1.2, therefore it is ionic.

SUMMARY: An **ionic bond**
1. is formed by the **transfer of electrons**
2. is formed when a **positive ion** (positively charged ion) such as Na$^+$ and a **negative ion** (negatively charged ion) such as Cl$^-$ **attract each other, and**
3. has an **electronegativity difference of 2.0 or more.** (Some chemists count a bond as ionic if the electronegativity difference is 1.7 or more).

Question: In which compound does the bond between the atoms have the *least* ionic (*least* polar) character?

1. HF   (2) HCl   (3) HBr   (4) HI

Solution: Answer *4*. When the question asks for the LEAST ionic character (LEAST polar bond), it means the smallest difference in electronegativity.

---

1) HF   Look at Table S. The electronegativity of H is 2.2; the electronegativity of F is 4.0. Subtract the electronegativities: $4 - 2.2 = 1.8$. The difference in electronegativity is 1.8.

2) HCl   The electronegativity of H is 2.2; the electronegativity of Cl is 3.2. Subtract the electronegativities: $3.2 - 2.2 = 1.0$. The difference in electronegativity is 1.0.

3) HBr   The electronegativity of H is 2.2; the electronegativity of Br is 3.0. Subtract the electronegativities: $3.0 - 2.2 = 0.8$ The difference in electronegativity is 0.8.

4) HI   The electronegativity of H is 2.2; the electronegativity of I is 2.7. Subtract the electronegativities: $2.7 - 2.2 = 0.5$. The difference in electronegativity is 0.5.

Answer 4 is the **smallest difference in electronegativity**, which means the **least ionic character** (least polar, smallest degree of polarity).

Properties of **Ionic Substances** (have **Ionic Bonds**):
1. **Hard**
2. **Good conductors** of **electricity**, because ions can move, **in solution** and in **liquid** form [example: KCl(ℓ) – ℓ means liquid – KCl is melted and becomes a liquid], but **not** in solid form.
3. **High melting and boiling points** because of the strong attraction of the ions.
4. Dissolve in polar substances, such as water.

LEWIS ELECTRON DOT STRUCTURES FOR ELEMENTS

Let's review what you learned about the Lewis electron dot structures for elements and how to place the valence electrons (electron arrangement) for different groups. There can only be 2 electrons on each side of the symbol of the element. Let X represent the element.

1. Put the 1st and 2nd valence electrons on one side of the symbol of the element. You can put the first 2 electrons on any side of the symbol of the element.   X·   X:
2. For the 3rd, 4th, and 5th electrons, put each electron on a different side of the symbol.   X: ·X: ·X:
3. For the 6th, 7th and 8th electrons, add the electron to any side with 1 electron.   ·X: :X: :X:

One way of showing Lewis electron dot structures is as follows:

Lewis Electron Dot Structures for Elements

X means elements (nucleus and non-valence electrons); • means 1 valence electron.							
Group	1	2	13	14	15	16	17	18
Element	X·	X:	X:	·X:	·X:	·X:	:X:	:X:

Examples: lithium Li•, carbon ·C:, neon :Ne:, sulfur ·S:

LEWIS ELECTRON DOT STRUCTURES FOR IONIC BONDING
(Transferred Electrons)

You can figure out Lewis electron dot structures for ions.

Group 1 elements have 1 valence electron. They lose 1 electron (negative) to become an ion. The ion has a 1+ (positive) charge because it lost an electron (negative) and became X^+. Examples: Li^+, Na^+.

Group 2 elements have 2 valence electrons (electrons in the outer shell). They lose 2 electrons (negative) to become an ion. The ion has a 2+ (positive) charge because it lost 2 electrons (negative), becoming X^{2+}. Examples: Be^{2+}, Mg^{2+}.

Group 13 elements have 3 valence electrons. They lose 3 electrons (negative) to become an ion, forming X^{3+}. Example: Al^{3+}.

Group 15 elements have 5 valence electrons. They want to have a complete shell (with 8 electrons). Therefore, Group 15 elements gain 3 electrons (negative), forming $[:\overset{..}{\underset{..}{X}}:]^{3-}$. Example: $[:\overset{..}{\underset{..}{N}}:]^{3-}$.

Group 16 elements have 6 valence electrons. They gain 2 electrons (negative) to have a complete outer shell, forming $[:\overset{..}{\underset{..}{X}}:]^{2-}$. Example: $[:\overset{..}{\underset{..}{O}}:]^{2-}$.

Group 17 elements have 7 valence electrons. They gain 1 electron (negative) to have a complete outer shell, forming $[:\overset{..}{\underset{..}{X}}:]^-$. Example: $[:\overset{..}{\underset{..}{Cl}}:]^-$.

Look at the table below for Lewis Electron Dot Structures for ionic bonding. It is the same as we figured out.

Lewis Electron Dot Structures

Group	1	2	13	14	15	16	17
Ionic Bonds	X^+	X^{2+}	X^{3+}		$[:\overset{..}{\underset{..}{X}}:]^{3-}$	$[:\overset{..}{\underset{..}{X}}:]^{2-}$	$[:\overset{..}{\underset{..}{X}}:]^-$

Any ion in Group 1 has the electron dot structure X^+. Examples of ions from Group 1 are Li^+ and K^+.

Any ion in Group 2 has the electron dot structure X^{2+}. Examples are Mg^{2+} and Ca^{2+}.

Any ion in Group 16 has electron dot structure $[:\overset{..}{\underset{..}{X}}:]^{2-}$. Examples are: $[:\overset{..}{\underset{..}{O}}:]^{2-}$, $[:\overset{..}{\underset{..}{S}}:]^{2-}$.

DRAWING LEWIS ELECTRON DOT STRUCTURES FOR IONIC BONDING (EASY METHOD)

METALS (left of the zigzag line on the Periodic Table) lose valence electrons to form ions:

1. Write all **metal ions** as **X.** (X represents the symbol of the element, example: Li).

2. The **oxidation number** is in the upper right hand corner of the box on the Periodic Table. (If there is more than one oxidation number, write the first number). Rewrite the oxidation number,

if it is +2, write 2+ next to the ion; if it is +3, write 3+, etc. Examples: Mg has an oxidation number of +2 (see Periodic Table), therefore write 2+ next to the **ion, Mg^{2+}**. Al has an oxidation number of +3 (see Periodic Table), therefore write 3+ next to the **ion, Al^{3+}**. If the oxidation number is +1, just write + (the 1 is understood). Example: Li has an oxidation number of +1; just write **Li$^+$** (the 1 is understood).

Question: What is the electron dot structure for Na ion?

Solution: Na is a metal. Metals lose valence electrons, therefore just **write Na** (symbol of the element). The **oxidation number** is **+1**; therefore, write **Na$^+$**.

Question: What is the electron dot structure for Ca ion?

Solution: Ca is a metal. Ca loses valence electrons, therefore just write the symbol **Ca**. The **oxidation number** is **+2** (see Periodic Table); write 2+ next to the **ion Ca^{2+}**.

NONMETALS (right of the zigzag line on the Periodic Table) gain electrons to form ions; they get a complete outer shell (octet):
1. Write all **nonmetal ions** as $\ddot{\underset{..}{X}}$ (Example, $\ddot{\underset{..}{Cl}}$).
2. **Oxidation numbers** are in the upper right hand corner of the box on the Periodic Table. Rewrite the oxidation number; if it is -3, write 3- next to the ion; if it is -2, write 2-, etc. If it is -1, just write - (the 1 is understood). Examples: O has oxidation number -2 (see Periodic Table), therefore write $[\ddot{\underset{..}{O}}]^{2-}$. Cl has oxidation number -1, therefore write $[\ddot{\underset{..}{Cl}}]^-$.

Examples of Lewis electron dot structures of ionic compounds:

Lithium chloride Li$^+$ $[\ddot{\underset{..}{Cl}}]^-$ Calcium chloride Ca$^{2+}[\ddot{\underset{..}{Cl}}]^- [\ddot{\underset{..}{Cl}}]^-$

Magnesium oxide Mg^{2+} $[\ddot{\underset{..}{O}}]^{2-}$.

Try Sample Questions #2-5, on pages 14-15, then do Homework Questions, #7-16, pages 17-18, and #31-32, page 19.

You learned about **ionic bonds** (transfer of electrons) between atoms. Now you will learn about **covalent bonds** (sharing of electrons).

II. COVALENT BONDS

A COVALENT BOND is formed when two atoms (both nonmetals) **share electrons.** (There is **no** transfer of electrons from one atom to another.)

In **covalent bonds, the electronegativity difference is less than 2.0.** (Some chemists count a bond as covalent if the difference in electronegativity is less than 1.7.)

LEWIS ELECTRON DOT STRUCTURES FOR COVALENT BONDING
(Shared Valence Electrons)

Let's draw electron dot structures for atoms in covalent bonds.

Group 2 has two valence electrons. Put **one dot on each side** of the symbol (shares 2 valence electrons to form 2 bonds). •X•.

Group 13 has 3 valence electrons. Put 1 **dot on each of 3 sides** (shares 3 valence electrons to form 3 bonds) •X• .

Group 14 has 4 valence electrons. Put one dot on each side of the symbol (shares 4 valence electrons to form 4 bonds) •X• .

Group 15 has 5 valence electrons. Put 1 dot on each of the 4 sides, and 1 more electron on any side. •X:

Group 16 has 6 valence electrons. Put 1 dot on each side (4 sides). There are 2 more electrons. **A side can only have 2 electrons**; therefore, put 1 more dot on each of 2 sides •X: .

Group 17 has 7 valence electrons. Put 1 dot on each of the 4 sides. There are 3 more electrons. A side can only have 2 electrons; therefore, put 1 more dot on each of 3 sides :X: .

Lewis Electron Dot Structures for Covalent Bonds

Group	1	2	13	14	15	16	17
Covalent		•X•	•X•	•X•	•X:	•X:	:X:

In short, when you have **covalent** bonds, Step 1: put one electron on each side of the atom. Step 2: Then (if the atom has more electrons), put a second electron on each side of the atom. See Lewis electron dot structure for ammonia (NH_3) on page Chap 4:10.

TYPES OF COVALENT BONDS

Here are three types of covalent bonds, and substances that have these bonds:

A. **NONPOLAR COVALENT**: When electrons are shared between atoms of the **same** element (therefore, same electronegativity), the **electrons** are **shared equally** and the bond is nonpolar. Electrons are in the middle between atoms.:F:F: , :Cl:Cl: , :Br:Br: , :I:I:

Let's now explain **how a nonpolar covalent bond** is formed between the two atoms. Each **fluorine** atom has **7 valence electrons**. Look at the **electron configuration** of fluorine on the Periodic Table. It is **2-7** (2 electrons in 1st shell, 7 valence electrons in the last shell).

F has **7 electrons** in the **outer shell**: :F• . The **octet rule** is that **atoms tend to gain, lose, or share electrons** in order to have a **complete outer shell like** the **noble gases**. Noble gases are **stable** and generally do not form bonds (gaining, losing, or sharing electrons). Two F atoms bond together to form F_2 (see figure).

Since the 2 electrons in the middle (1 from each F atom) are shared by both F atoms, **each F atom now has 8 electrons**, 6 electrons of its own on the outside and 2 (in the middle, shared). **F now** has a **complete outer shell** (more stable) with 8 electrons, just **like** the **noble gas neon (Ne).** As you know, the electron configuration of F is 2-7. Now, in F_2, each F atom has 8 electrons in the outer shell.

X X   O O
X**F**X **F**O
X O O
X X   O O

6 electrons 2 elec- 6 electrons
on outside trons on outside
that
bond

Lewis electron dot diagram for F_2

Therefore, the **electron configuration** of **F** in F_2 is **2-8**, just **like** the electron configuration of neon, $_{10}Ne$, (2-8), the **noble gas nearest to fluorine.** You see on the Periodic Table that the electron configuration of Ne, (noble gas) is 2-8. Cl_2, Br_2, and I_2 (Group 17) have similar electron dot diagrams to F_2 (above), two electrons in the middle shared between two Cl, two Br, or two I.

B. **POLAR COVALENT:** When electrons are shared between atoms of **different elements** (different electronegativities); they are **shared unequally.** H:C̈l: The element with higher electronegativity will attract electrons more strongly, causing a part of the molecule to have a negative charge. The other part has a positive charge.

Look at Table S, at the beginning of the chapter. The electronegativity of Cl is 3.2; the electronegativity of H is 2.2. Subtract one electronegativity from the other:

$$3.2 - 2.2 = 1.0.$$

The difference in electronegativity is 1.0; therefore the bond is **polar covalent**: *covalent* because the difference in electronegativity is less than 2.0 (some say less than 1.7); and *polar* because the two elements have different electronegativities (H = 2.2; Cl = 3.2).

Let's explain **how** a **polar covalent bond is formed.** H has 1 electron in the outer shell and Cl has 7 electrons in the outer shell. (See the electron configurations on the Periodic Table.) In a covalent compound, **atoms share electrons** to have **8 electrons** in the **last shell (octet** rule). Exceptions: **Hydrogen and helium**, which have valence electrons only in the first principal energy level, have a **complete shell** with **2 electrons.**

The 2 electrons in the middle (1 from H, 1 from Cl) are shared by H and Cl. **H in HCl** now has **2 electrons**, electron configuration 2, just **like** the **electron configuration of helium** (see Periodic Table), the noble gas nearest to H. The electron configuration of Cl is 2-8-7. **Cl in HCl** has 7 electrons of its own and shares 1

O O
H X **Cl** O
O O
O O

Electron dot diagram
(Lewis structures)

electron from hydrogen to have **8 electrons (octet rule).** The electron configuration of Cl in HCl is 2-8-8, just **like** the **electron configuration of argon**, the **noble gas closest to it.**

TABLE S

More polar covalent Lewis electron dot diagrams:

$$H \overset{..}{\underset{\overset{\bullet x}{H}}{\overset{x}{N}}} x H \qquad \overset{\overset{x x}{..}}{\underset{\overset{x x}{\underset{\overset{x x}{Cl}}{Cl}}}{Cl} \overset{..}{P} Cl} \qquad H\overset{..}{\underset{\overset{..}{H}}{O}} \qquad \left[\overset{..}{\underset{..}{O}} H\right]^{-} \qquad H\overset{\overset{H}{..}}{\underset{\overset{..}{H}}{C}}\overset{..}{\underset{..}{Br}}$$

NH_3(ammonia) PCl_3 H_2O (water) OH^-(hydroxide) CH_3Br

Question: Of the 4 compounds below:
1. Which compound has bonds with the greatest degree of polarity (most polar)?
2. Which compound has bonds with the least degree of polarity (least polar)?
(1) CO_2 (2) SiO_2 (3) NO_2 (4) NCl_3

INFORMATION FOR QUESTIONS 1 AND 2:

(1) CO_2 Electronegativity of C = 2.6
Electronegativity of O = 3.4
Subtract the electronegativities: 3.4 - 2.6 = 0.8

(2) SiO_2 Electronegativity of Si = 1.9
Electronegativity of O = 3.4
Subtract the 2 electronegativities: 3.4 - 1.9 = 1.5

(3) NO_2 Electronegativity of N = 3.0
Electronegativity of O = 3.4
Subtract the electronegativities: 3.4 - 3.0 = 0.4

(4) NCl_3 Electronegativity of N = 3.0
Electronegativity of Cl = 3.2
Subtract the electronegativities 3.2 - 3.0 = 0.2

Solution to Question 1: The greatest degree of polarity (most polar) means the biggest difference in electronegativity between the 2 atoms. Look at Table S.

Choice 2, SiO_2, has the largest difference in electronegativity, 1.5, therefore the greatest degree of polarity in the bond.

Solution to Question 2: The least (smallest) degree of polarity (least polar) means the smallest difference in electronegativity between the 2 atoms.

Choice 4, NCl_3, has the least difference in electronegativity, 0.2, therefore the smallest degree of polarity in the bond.

C. **COORDINATE COVALENT**: A coordinate covalent bond is formed when **one atom donates both electrons that are shared.**

$$\left[H\overset{\overset{H}{\overset{oo}{..}}}{\underset{\overset{x x}{H}}{\overset{o}{\underset{o}{N}}}}\overset{o}{x}H\right]^{+} \text{coordinate covalent bond}$$

NH_4^+ (ammonium ion)

N (nitrogen) donates a pair of electrons to share with H^+, forming a coordinate covalent bond between nitrogen and hydrogen. Therefore, **NH_4^+ (ammonium ion)** has a **coordinate covalent bond**.

H_2O + H^+ produces H_3O^+. H_3O^+ (hydronium ion) also has a coordinate covalent bond.

SUBSTANCES THAT HAVE COVALENT BONDS

D. **MOLECULAR SUBSTANCES**: Molecular substances (elements or compounds) are made up of molecules; a molecule is the smallest discrete particle of an element or compound that has **covalent bonds** between atoms. An **element that has covalent bonds** between its atoms (examples O_2, H_2, N_2) is called a **molecular element**. A **compound that has covalent bonds** (example H_2O) is called a **molecular compound**. Molecular substances may be gases, liquids or solids, depending on the force of attraction between the molecules.

Examples of molecules are H_2O, H_2, HCl, CCl_4, $C_6H_{12}O_6$ and CO_2.

Electron-dot diagram for carbon dioxide (CO_2):   Ö::c::Ö

Properties of **molecular substances** (have **covalent bonds**):
1. **Soft**
2. **Poor conductors** of **heat, poor conductors** of **electricity** because there are **no** charged particles (no ions or mobile electrons).
3. **Low melting points and boiling points** because of weak attraction between molecules.

Question: What type of substance is soft, a poor conductor of heat and electricity, and low-melting? What type of bond does it have?

Solution: Molecular substance.

Hint: To help you remember the characteristics of a molecular substance, think of the word *SPLASH*. (You can splash in H_2O, which is a molecular substance.) *S* stands for *soft*; *P* stands for *poor* conductor of heat and electricity; *L* stands for *low* melting point. Molecular substances have covalent bonds.

E. **NETWORK SOLIDS**: These are solids that have **covalent bonds** between atoms linked in **one big network** or one big **macromolecule** with no discrete particles. This gives them some properties different from most covalent compounds. Network solids are hard, poor conductors of heat and electricity, and have high melting points. Examples are: Diamond (C), silicon carbide (SiC), and silicon dioxide (SiO_2). Diamond (C) does not have discrete (separate) particles. The carbon atoms in diamond are connected by covalent bonds to form one big macromolecule.

SUMMARY: A covalent bond (1) is formed by **sharing of electrons** and (2) has an electronegativity difference of less than 2.0. (Some chemists count a bond as covalent if the electronegativity difference is less than 1.7.)

NUMBER OF COVALENT BONDS: A bond is made up of 1 pair of electrons:

Single covalent bond: Atoms share 1 pair of electrons. Example: 2 F atoms share 2 electrons, which make up 1 pair.
(2 shoes are 1 pair of shoes.)

:F:F:

One pair of
shared electrons

Double covalent bonds: 2 atoms share 2 pairs of electrons (which have 4 electrons). Example: 2 carbon atoms share 2 pairs of electrons:

H H
H:C::C:H

Triple covalent bonds: Atoms share 3 pairs of electrons. Example: Two carbon atoms share 3 pairs of electrons:

H : C ::: C : H

Double covalent bonds and triple covalent bonds are multiple covalent bonds. In a **multiple covalent bond**, atoms share more than 1 pair of electrons, such as 2 or 3 pairs of electrons.

Saturated organic compounds are carbon compounds that have only single bonds. A single bond is 1 pair of electrons shared between 1 carbon and 1 hydrogen atom or between two carbon atoms, etc.

H H
H:C:C:H
H H

Unsaturated organic compounds have at least 1 double or 1 triple bond. C_2H_4 and C_2H_2 are unsaturated organic compounds.

Other electron dot diagrams:

CF_4: Carbon has 4 valence electrons(x); fluorine has 7 valence electrons(•). Now C and each F has 8 electrons.

:F:
:F:C:F:
:F:

O_2: There are two lines (two bonds, two pairs of shared electrons or four electrons) between the two atoms. Now each oxygen atom has eight electrons.

:O=O:

SOLUBILITY:

Polar substances (like HCl) dissolve in polar liquids, such as water.
Nonpolar substances (like wax) dissolve in nonpolar liquids, such as CCl_4, carbon tetrachloride.

SUMMARY OF IONIC AND COVALENT BONDING:

1. Metals react with nonmetals to form **ionic** compounds (have ionic bonds). NaCl has **ionic** bonds.
2. Nonmetals react with other nonmetals to form **molecular** (covalent) compounds (have **covalent** bonds). HCl has covalent bonds. (H is a nonmetal; H is on top of Group 1 because it has 1 valence electron).
3. Compounds like $CaCO_3$ and NH_4Cl (or NH_4NO_3) have **both** covalent and ionic bonds. CO_3^{2-} has covalent bonds; Ca^{2+} ion bonds CO_3^{2-} ion by an ionic bond.

   NH_4^+, a polyatomic ion (made of many atoms) has covalent bonds. NH_4^+ ion is bonded to Cl^- ion (or NO_3^- ion) by an ionic bond.

Look at Table E (Selected Polyatomic Ions) on page Reference Tables 5. Polyatomic ions (listed on Table E, examples NH_4^+, NO_3^-, and PO_4^{3-}) have covalent bonds. These polyatomic ions are bonded to other ions by an ionic bond.

Reminder: How Bonds Are Formed

As you learned, when atoms bond together, some **atoms gain** or **lose electrons (ionic bond)** or **share (covalent)** electrons. Each atom gets a **complete outer shell**, generally **8 electrons** like the noble gases. This is called the **octet rule**. Exception: hydrogen and helium have valence electrons in the 1st shell, and the 1st shell can only hold 2 electrons, therefore a complete outer 1st shell has 2 electrons.

Try Sample Questions #6–11, page 15, then do Homework Questions, #17-28, pages 18-19, and #33-35, page 19.

III. METALLIC BONDING

Metallic bonding occurs in metals (example: copper). A metal consists of positive ions (atoms without valence electrons) surrounded by a "sea" of **mobile electrons** (moving valence electrons). In **metallic bonding**, the **attraction** between the **negative electrons (from many atoms)** and the **positive ions** holds the metal together.

Metals conduct heat and electricity because **electrons** can **move far away** from an atom to other atoms, carrying the heat and electricity with them.

Metals are malleable (can be hammered into sheets) and ductile (made into wires) because the **moving valence electrons** are attracted to the positive ions, forming metallic bonds (hold the metal together in any shape).

Metals have high melting points and high boiling points. In Table S, compare the melting points and boiling points of a metal (example: $_{11}Na$) and nonmetal (example: $_{17}Cl$) in the same period. You see both the melting point and boiling point of the metal (Na) are higher than for a nonmetal (Cl).

Properties of Metals (have Metallic Bonds):
1. **Hard.**
2. **Good conductors** of **heat and electricity** because of mobile electrons. Examples: aluminum, copper (solids), mercury (liquid).
3. **High melting and boiling points** because of strong attraction.

Try Sample Question #12, on page 15, then do Homework Questions #29-30, page 19.

SUMMARY: Bonds Between Atoms (between one atom and another atom):

1. **Ionic bonds**: Transfer of electrons, difference in electronegativity of 2.0 or more. (Some say 1.7 or more.)
2. **Covalent bonds**: Sharing of electrons, difference in electronegativity (generally) less than 2.0. (Some say less than 1.7.)
   a. *nonpolar covalent*
   b. *polar covalent*
   c. *coordinate covalent bonds*
   d. *molecular substances*
   e. *network solids*
3. **Metallic bonds**: mobile electrons attracted to positive ions.

SUMMARY: Properties:

1. Ionic	Hard	Good conductor of electricity (solution and liquid, not solid) because of moving ions	High melting and boiling points
2. Covalent	Soft	Poor conductor of electricity	Low melting and boiling points
3. Metallic	Hard	Good conductor of electricity (solid and liquid) because of mobile (moving) electrons	High melting and boiling points

HINT: **Molecular Substances** (have covalent bonds) Write the word spl**ash**-**soft**, **poor** conductor of electricity, **low** melting point.
Ionic and Metallic Substances-make it the opposite of soft, poor conductor, and low melting point, which is **hard, good conductor and high melting point.**
But, **ionic** substances are good conductors of electricity as a **liquid** and in **solution** (not in solid form); **metallic** substances are good conductors in **liquid** and also in **solid** forms.

M O D E L S : You can make a model of

$:\ddot{F}:\ddot{F}:$, $\left[H\overset{..}{\underset{H}{\overset{H}{\times}}}N\overset{o}{\times}H \right]^+$, (bent molecule with O^-, H^+, H^+) , $H:\ddot{C}l:$, etc. by using an organic

chemistry model kit, modeling clay, or polystyrene balls. Use balls or pieces of clay to show atoms. Use toothpicks or sticks to show the bonds connecting the atoms.

SAMPLE REGENTS & REGENT-TYPE QUESTIONS AND SOLUTIONS

1. Atoms of which element have the weakest attraction for electrons?
   (1) Na (2) P (3) Si (4) S
2. In which compound have electrons been transferred to the oxygen atom?
   (1) CO_2 (2) NO_2 (3) N_2O (4) Na_2O

3. When ionic bonds are formed, metallic atoms tend to

(1) lose electrons and become negative ions
(2) lose electrons and become positive ions
(3) gain electrons and become negative ions
(4) gain electrons and become positive ions

4. In which compound does the bond between the atoms have the *least* polar character?
(1) H_2O (2) H_2S (3) HCl   (4) HF

5. Atoms of nonmetals generally react with atoms of metals by:
(1) gaining electrons to form ionic compounds
(2) gaining electrons to form covalent compounds
(3) sharing electrons to form ionic compounds
(4) sharing electrons to form covalent compounds

6. Which structural formula represents a nonpolar molecule?
(1) H–Cl   (2)   H–O   (3) H–H   (4) H–N–H
                     | |
                     H   H

7. Which molecule contains a polar covalent bond?
(1) $\overset{xx}{\underset{xx}{x}}I \overset{..}{\underset{..}{:}} I :$ (2)$H \overset{x}{\cdot} H$ (3) $H \overset{..}{\underset{\overset{x}{\underset{H}{}}}{x N x}} H$ (4) $: N \overset{x}{\underset{x}{::}} N :$

8. In the diagram of an ammonium ion below, why is bond *A* considered to be coordinate covalent?

$$\left[\begin{array}{c} H \\ H \ \overset{xx}{\underset{o}{x N x}} \ H \\ \underset{H}{\overset{xo}{o}} \end{array} \right]^{+} \quad \xleftarrow{\quad} bond\ A$$

(1) Hydrogen provides a pair of electrons to be shared with nitrogen.
(2) Nitrogen provides a pair of electrons to be shared with hydrogen.
(3) Hydrogen transfers a pair of electrons to nitrogen.
(4) Nitrogen transfers a pair of electrons to hydrogen.

9. Which formula represents a molecular substance?
(1) CaO   (2)   CO   (3) Li_2O (4) Al_2O_3

10. What type of bonding is present within a network solid?
(1) hydrogen   (2) covalent   (3) ionic   (4) metallic

11. Which substance is an example of a network solid?
(1) nitrogen dioxide   (3) carbon dioxide
(2) sulfur dioxide   (4) silicon dioxide

12. The ability to conduct electricity in the solid state is a characteristic of metallic bonding. This characteristic is best explained by the presence of
(1) high ionization energies   (3) mobile electrons
(2) high electronegativities   (4) mobile protons

SOLUTIONS

1. Answer *1*. You learned that electronegativity is the attraction for electrons (in a bond). You learned that, the smaller the electronegativity, the weaker is the attraction for electrons. Look on Table S. Look at H (hydrogen). The electronegativity for hydrogen, 2.2, is given in the 5th column. The electronegativities are:

$$Na_{0.9} \quad P_{2.2} \quad Si_{1.9} \quad S_{2.6}$$

Na has the smallest electronegativity, 0.9, which is the weakest attraction for electrons.

2. Answer **4**. Transfer of electrons means ionic bonds, which have an electronegativity difference of 2.0 or more. Look at the electronegativities in Table S, on page Chap. 4:1. Subtract the electronegativities of the two elements and find the difference.

Choice	1. CO_2	2. NO_2	3. N_2O	4. Na_2O
Electronegativities	C O 2.6 3.5	N O 3.0 3.5	N O 3.0 3.5	Na O 0.9 3.5
Difference in Electronegativity	3.5 - 2.6 = 0.9	3.5 - 3.0 = 0.5	3.5 - 3.0 = 0.5	3.5 - 0.9 = 2.6
Bonding	covalent	covalent	covalent	ionic

A difference in electronegativity of 2.0 or more is ionic (transfer of electrons). Na_2O has a difference in electronegativity of 2.6, therefore Na_2O is formed by a transfer of electrons.

3. Answer **2**. You learned that an ionic bond is formed when Na (which is a metal) loses an electron and becomes a Na^+ (positive ion.) *Metallic* atoms lose electrons and become positive ions.

4. Answer **2**. When the question asks for the **LEAST polar** character, it means the **smallest difference** in electronegativity.
   1) H_2O Look at Table S. The electronegativity of H is 2.1; the electronegativity of O is 3.5. Subtract the electronegativities: 3.5 - 2.1 = 1.4. The difference in electronegativity is 1.4.
   2) H_2S The electronegativity of H is 2.1; the electronegativity of S is 2.6. Subtract the electronegativities: 2.6 - 2.1 = 0.5. The difference in electronegativity is 0.5.
   3) HCl   The electronegativity of H is 2.1; the electronegativity of Cl is 3.2. Subtract the electronegativities: 3.2 - 2.1=1.1. The difference in electronegativity is 1.1.
   4) HF   The electronegativity of H is 2.1; the electronegativity of F is 4.0. Subtract the electronegativities: 4.0 -2.1 = 1.9. The difference in electronegativity is 1.9.
   Answer 2 is the **smallest difference in electronegativity**, which means the **least polar character**.

5. Answer **1**. Nonmetals have 5, 6 or 7 electrons in the last shell, and they tend to gain electrons in order to have eight electrons in the last shell. The electronegativity difference between most metals and nonmetals is 2.0 or more. Therefore, ionic compounds are formed.

6. Answer **3**. You learned that nonpolar means **equal sharing of electrons**. A nonpolar molecule is answer 3. It has 2 atoms of the same element.
   Additional Information: A nonpolar molecule can also have two different elements with the **same electronegativity**.

7. Answer **3**. Polar covalent is where electrons are shared between atoms of different elements (different electronegativities). The difference in electronegativity is less than 2.0 (Some say less than 1.7.)
   Look at Table S. Answer 3 is NH_3. The electronegativity of N is 3.0. The electronegativity of H is 2.1. The difference in electronegativity is 3.0 - 2.1 = 0.9. It is polar covalent.

8. Answer **2**. You learned that a coordinate covalent bond is formed when one atom donates a pair of electrons to share.
   Electrons of any atom can be represented by either x's or o's. The five x's in the diagram represent the five valence electrons of nitrogen. Nitrogen is providing the pair of electrons to share with H^+. You learned that NH_4^+ is an example of coordinate covalent bonding.

9. Answer *2*. Remember, five types of covalent bonds were described: nonpolar covalent, polar covalent, coordinate covalent, *molecular substance* and network solid. You know that a molecular substance must have covalent bonds (electronegativity difference of less than 2.0). Look at Table S.

Choice	1. CaO		2. CO		3. LiO		4. Al_2O_3	
Electronegativities	Ca	O	C	O	Li	O	Al	O
	1.0	3.5	2.6	3.5	1.0	3.5	1.6	3.5
Difference in Electronegativity	2.5		0.9		2.5		1.9	
Bonding	ionic		covalent		ionic		ionic	

Remember: Something that is ionic cannot be a molecular substance.
Answer 2, CO, is a molecular substance because it has covalent bonds.

10. Answer *2*. In the book, five types of covalent bonds were described: nonpolar covalent, polar covalent, coordinate covalent, molecular substance and *network solid*.

11. Answer *4*. An example of a network solid is silicon dioxide.

12. Answer *3*. You learned that metallic bonding has mobile electrons.

NOW LET'S TRY A FEW HOMEWORK QUESTIONS:

1. Which kind of energy is stored in a chemical bond?
   (1) potential energy   (3) activation energy
   (2) kinetic energy   (4) ionization energy

2. Which statement is true concerning the reaction $N(g) + N(g) \rightarrow N_2(g)$ + energy?
   (1) A bond is broken and energy is absorbed.
   (2) A bond is broken and energy is released.
   (3) A bond is formed and energy is absorbed.
   (4) A bond is formed and energy is released.

3. Atoms of which element have the *weakest* attraction for electrons?
   (1) Na   (2) P   (3) Si   (4) S

4. Which atom has the greatest attraction for a pair of electrons in a bond between the atom and oxygen?
   (1) Al   (2) Si   (3) S   (4) P

5. Given the electron dot formula: $H \colon \overset{..}{\underset{..}{X}} \colon$
                                        H
   The attraction for the bonding electrons would be greatest when X represents an atom of
   (1) S   (2) O   (3) Se   (4) Te

6. Which atom has the *least* attraction for the electrons in a bond between that atom and an atom of hydrogen?
   (1) carbon   (2) nitrogen   (3) oxygen   (4) fluorine

7. Which compound contains ionic bonds?
   (1) $MgCl_2$ (2) CO_2 (3) NO_2 (4) CH_4

8. Which compound contains ionic bonds?
   (1) N_2O (2) Na_2O (3) CO   (4) CO_2

9. Which bond has the greatest ionic character?
   (1) H–Cl   (2) H–F   (3) H–O   (4) H–N

10. Which compound has the greatest degree of ionic character?
    (1) NaF   (2) MgF_2 (3) AlF_3 (4) SiF_4

11. In which compound does the bond between the atoms have the *least* ionic character?
    (1) HF (2) HCl (3) HBr (4) HI

12. A strontium atom differs from a strontium ion in that the atom has a greater
    (1) number of electrons   (3) atomic number
    (2) number of protons   (4) mass number

13. A white crystalline salt conducts electricity when it is melted and when it is dissolved in water. Which type of bond does this salt contain?
    (1) ionic (2) metallic (3) covalent (4) network

14. Which substance is a conductor of electricity?
    (1) NaCl(s) (2) NaCl(ℓ) (3) $C_6H_{12}O_6(s)$ (4) $C_6H_{12}O_6(ℓ)$

15. Which compound contains a bond with the *least* ionic character?
    (1) CO (2) CaO (3) K_2O (4) Li_2O

16. In which compound have electrons been transferred to the oxygen atom?
    (1) CO_2 (2) NO_2 (3) N_2O (4) Na_2O

17. In a nonpolar covalent bond, electrons are
    (1) located in a mobile "sea" shared by many ions
    (2) transferred from one atom to another
    (3) shared equally by two atoms
    (4) shared unequally by two atoms

18. Which molecule is nonpolar and contains a nonpolar covalent bond?
    (1) CCl_4 (2) F_2 (3) HF (4) HCl

19. Two atoms with an electronegativity difference of 0.4 form a bond that is
    (1) ionic, because electrons are shared
    (2) ionic, because electrons are transferred
    (3) covalent, because electrons are shared
    (4) covalent, because electrons are transferred

20. The correct electron dot formula for hydrogen chloride is

    (1) $H \colon Cl$ (2) $\colon \ddot{H} \colon Cl$ (3) $H \colon \ddot{\ddot{Cl}} \colon$ (4) $\colon H \colon \ddot{Cl} \colon$

21. Which combination of atoms can form a polar covalent bond?
    (1) H and H (2) H and Br (3) N and N (4) Na and Br

22. The bond between hydrogen and oxygen in a water molecule is classified as
    (1) ionic and nonpolar   (3) covalent and nonpolar
    (2) ionic and polar   (4) covalent and polar

23. What kind of a bond is formed in the reaction shown below?

    $$H \colon \ddot{N} \colon H + H^+ \rightarrow \left[H \colon \underset{H}{\overset{H}{N}} \colon H \right]^+$$

    (1) metallic bond   (3) hydrogen bond
    (2) network bond   (4) coordinate covalent bond

24. Which species contains a coordinate covalent bond?

    (1) $\overset{xx}{\underset{xx}{Cl}} \colon \ddot{Cl} \colon$ (2) $H \colon \ddot{Cl} \colon$ (3) $Na^+ \left[\colon \ddot{Cl} \colon \right]^-$ (4) $\left[H \colon \underset{H}{\overset{H}{N}} \colon H \right]^+$

25. Which is a molecular substance?
    (1) CO_2 (2) CaO (3) KCl (4) $KClO_3$

26. A. Which formula represents a molecular solid?
   (1)NaCl(s)　　　(2) $C_6H_{12}O_6$(s)　(3) Cu(s)　　　(4) KF(s)

   B. A substance was found to be a soft, nonconducting solid at room temperature. The substance is most likely
   (1)　a network solid　　　　　(3)　a metallic solid
   (2)　a molecular solid　　　　(4)　an ionic solid

27. The bonds in all network solids are
   (1)　covalent　　(2)　ionic　　　(3)　metallic　　(4)　nonpolar

28. The chemical bonding in sodium phosphate, Na_3PO_4, is classified as
   (1)　both covalent and ionic　　(3)　ionic, only
   (2)　metallic, only　　　　　　(4)　both covalent and metallic

29. A solid substance conducts electricity. The bonding in the substance is
   (1)　metallic　　(2)　ionic　　　(3)　nonpolar covalent　(4) polar covalent

30. Which element consists of positive ions immersed in a "sea" of mobile electrons?
   1.　sulfur　　　(2)　nitrogen　　(3)　calcium　　(4)　chlorine

CONSTRUCTED RESPONSE QUESTIONS:　Parts B-2 and C of NYS Regents Exam

31. Show, using Lewis electron dot structures, how:
   $_{56}Ba + {}_{16}S \longrightarrow BaS$.
   $_{20}Ca + {}_{8}O \longrightarrow CaO$
   $_{19}K + {}_{9}F \longrightarrow KF$
   Hint: Look at Periodic Table and Table S.

32. Show, using Lewis electron dot structures, how:
   $2 {}_{11}Na + {}_{16}S \longrightarrow Na_2S$
   $_{20}Ca + 2 {}_{17}Cl \longrightarrow CaCl_2$

33. Compare and contrast polar, nonpolar, and coordinate covalent bonds.

34. Draw a Lewis electron dot structure to show:
   a.　one compound with polar covalent bonds.
   b.　one compound with nonpolar covalent bonds.
   c.　one compound with coordinate covalent bonds.

35. Compare and contrast properties of ionic and molecular(covalent) substances.

CHAPTER QUESTIONS:　Parts B-2 and C of NYS Regents Exam

36. Use Lewis electron dot structures to show ionic compounds:
   MgO, $MgCl_2$, LiCl, AlF_3.

37. Use Lewis electron dot structures to show
   a. covalent compounds: Cl_2, H_2S, NH_3, H_2O; b.NH_4^+ (ammonium) ion
   c. more covalent compounds: CF_4, CCl_4

38. For each set of properties, give the type of bonding (ionic, covalent, or metallic) in the substance.
   a.　Hard, high melting point, conducts electricity in solid.
   b.　Hard, high melting point, conducts in solution.
   c.　Soft, low melting point, poor conductor.

SECTION B

In Section A, you learned about attraction between atoms. Now let's learn about attraction between molecules.

INTERMOLECULAR FORCES
(Forces of Attraction Between Molecules)

1. **DIPOLES**: A molecule composed of only two atoms will be a **dipole** if the bond between the atoms is polar. For example, "$^+$H$-$Cl$^-$." Cl has a slightly negative charge because the Cl is more electronegative (has a greater attraction for electrons) and pulls the shared electrons closer to the chlorine atom. H has a slightly positive charge because the chlorine pulls the shared electrons closer to Cl and therefore further away from H, giving the H a slightly positive charge. The positive end of the molecule is attracted to the negative part of the next molecule.

MOLECULAR POLARITY: Whether or not a molecule is polar depends on its shape, whether its bonds are polar, and how the charge is spread out (distributed) over the molecule.

POLAR BONDS

A. Symmetrical Molecules (SYMMETRY). Because of **symmetry**, molecules that have **polar bonds** are **overall NONPOLAR** (+ and - charges cancel or balance out) .

1. Carbon dioxide (CO$_2$) can be shown by $\underset{-}{O} = \underset{+}{C} = \underset{-}{O}$. The valence electrons of C are pulled equally by both oxygen atoms; therefore, no end is more negative. (Another similar example is CS$_2$, which is $\underset{-}{S} = \underset{+}{C} = \underset{-}{S}$.)

2. Methane (CH$_4$) can be shown by H$\overset{\overset{\text{H}}{\bullet\bullet}}{\underset{\underset{\text{H}}{\bullet\bullet}}{\bullet\bullet\text{C}\bullet\bullet}}$H . Carbon pulls the valence electrons equally from all hydrogen atoms. Therefore, no end of the molecule is more positive. (Another similar example is CCl$_4$, which is Cl$\overset{\overset{\text{Cl}}{\bullet\bullet}}{\underset{\underset{\text{Cl}}{\bullet\bullet}}{\bullet\bullet\text{C}\bullet\bullet}}$Cl.)

In short, **CH$_4$, CCl$_4$, CO$_2$** and **CS$_2$** have **polar bonds** but, because of **symmetry (symmetrical charge distribution)**, they are **overall nonpolar** (nonpolar molecules).

B. Asymmetrical Molecules (NO SYMMETRY). If a molecule has polar bonds (and there is **no symmetry** to cancel out + and - charges), the molecule is polar.

1. Look at H_2O: Water is asymmetrical.

    Water is an asymmetrical molecule (no symmetry) with polar bonds, therefore water is a **polar molecule.**

2. Ammonia: H:N:H is a **polar molecule.**

3. Hydrochloric Acid: H-Cl is a **polar molecule.**

Nonpolar Bonds. Molecules with **nonpolar** bonds are **always nonpolar.** Both atoms pull the electrons equally. Example: diatomic molecules, O_2, Cl_2, N_2.

Shape: CS_2 linear, CH_4 tetrahedral, H_2O bent, NH_3 trigonal pyramid.

2. **HYDROGEN BONDING**

    a. **HYDROGEN BONDING** is a strong intermolecular force that connects **one water molecule with another water molecule.** (There are also other examples of hydrogen bonding.)

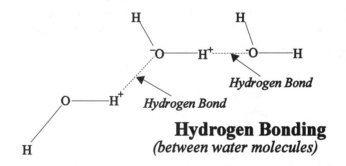

Hydrogen Bonding
(between water molecules)

The positive part, H^+, of one water molecule is attracted to the negative part of the next water molecule. The dotted lines show the hydrogen bonds.

The hydrogen bond in H_2O (it holds the molecules tighter together) and not in H_2S causes the **boiling point of H_2O** to be **higher** than the boiling point of H_2S.

b. Hydrogen bonds are formed between molecules when hydrogen is covalently bonded to a small, highly electronegative atom, such as fluorine $(_9F)$, oxygen $(_8O)$ or nitrogen $(_7N)$. Therefore, **hydrogen bonding** occurs in **compounds with N, O or F**, such as **H_2O, NH_3** and **HF.**

---

3. **DISPERSION FORCES**

    a. **DISPERSION forces** are **weak intermolecular forces** (attractive forces) **between nonpolar molecules.** Example: dispersion forces **between hydrogen molecules.** There are also dispersion forces **between CCl_4 molecules** or **CO_2 molecules,** which are overall **nonpolar** (see page 4:20)

    b. Dispersion forces make it possible for small, nonpolar molecules (such as hydrogen, helium or oxygen) to exist in the **liquid or solid phase** under conditions of low temperature and high pressure.

    c. Dispersion forces increase with increasing molecular size and with decreasing distance between molecules. For example, in the halogen group (Group 17), as the **elements get heavier** ($^{19}_{9}F$, $^{35}_{17}Cl$, $^{80}_{35}Br$, $^{127}_{53}I$), the **dispersion forces increase** and the **boiling point increases**. Similarly, as you go down Group 1 (Li, Na, K, etc.), the elements get heavier, dispersion forces increase, and the boiling points and melting points increase.

    Again, in compounds of one type (example: alkanes), the heavier they are, the more dispersion forces and the higher the boiling point.

4. **MOLECULE-ION ATTRACTION**

    a. **MOLECULE-ION ATTRACTION** is the **attraction** between the **ions** of an ionic compound, such as salt, and **molecules** of water (or other polar liquids. When you put salt, an ionic compound, into water, the **Na^+ positive ion** (from the salt) IS ATTRACTED TO THE NEGATIVE PART (O^{2-}) OF THE WATER; the **Cl^- negative ion** (from the salt) is ATTRACTED TO THE POSITIVE PART (H^+) OF THE WATER. Opposite charges attract: positive to negative and negative to positive. **NaCl in water** is written as **NaCl(aq).**

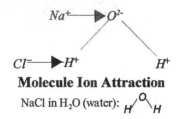

Molecule Ion Attraction

NaCl in H_2O (water): H–O–H

    b. Similarly, when KCl dissolves in H_2O, or any ionic substance dissolves in H_2O, there is molecule-ion attraction. The positive ion of salt goes to the negative part of water, and the negative ion of salt goes to the positive part of water.

    c. **HYDRATED IONS** are positive and negative ions surrounded by water.

SUMMARY: The **intermolecular forces** (forces of attraction between molecules) are: (1) **dipoles**; (2) **hydrogen bonds**; (3) **dispersion forces**; and (4) **molecule-ion attraction**.

INTERMOLECULAR FORCES AND VAPOR PRESSURE

You learned that forces of attraction between molecules are called intermolecular forces. Look at Table H, below or on page

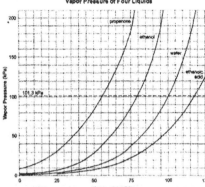

Table H
Vapor Pressure of Four Liquids

Reference Tables 9. Ethanoic acid has a vapor pressure of 8 kPa at 50°C. Of the four liquids shown in Table H, ethanoic acid has the **lowest vapor pressure** because it has the **strongest intermolecular** (between molecules) **forces** holding the molecules together. The molecules cannot easily become a gas (vapor).

Propanone has a vapor pressure of 83 kPa at 50°C. Of the four liquids shown in Table H, propanone has the **highest vapor pressure** because it has the **weakest intermolecular** (between molecules) **forces** holding the molecules together. The molecules can easily become a gas (vapor).

Try Sample Questions #1-6, on page 26, then do Homework Questions, #1-16, pages 27-29, and #27-29, page 29.

CHEMICAL FORMULAS

CHEMICAL FORMULAS describe the composition of elements or compounds. Formulas can be molecular formulas or empirical formulas.

a. A **MOLECULAR FORMULA** indicates the **total number of atoms of each element** needed to form the molecule:

C_2H_6 2 atoms of carbon, 6 atoms of hydrogen

H_2O 2 atoms of hydrogen, 1 atom of oxygen

$C_6H_{12}O_6$ 6 atoms of carbon, 12 atoms of hydrogen,
(glucose)   6 atoms of oxygen.

What is the molecular formula for acetic acid, CH_3COOH. Acetic acid (CH_3COOH) has 2 C atoms, 4 H atoms, and 2 O atoms, therefore write total number of C, H, and O atoms. Write $C_2H_4O_2$ (**molecular formula** of acetic acid).

b. An **EMPIRICAL FORMULA** is the **simplest ratio** in which atoms combine to form a compound. If the formula is C_2H_6, divide the C and H by the largest number that all of the elements can be divided by, in this case, 2:

$$C_2/2 = C_1; \qquad H_6/2 = H_3.$$

Therefore, the **empirical formula** is CH_3.

Now, let's take another example: $C_6H_{12}O_6$ is the molecular formula. Divide the C, the H and the O by the largest number that all the elements can be divided by, 6:

$$C_6/6 = C; \quad H_{12}/6 = H_2; \quad O_6/6 = O.$$

Therefore, the **empirical formula** is CH_2O.

Let's see where you put the elements in a chemical formula. In a **chemical formula** made up of two elements, you put the **metal first** and the **nonmetal last**. Look at the formula NaCl (sodium chloride). **Na** (sodium), the **metal**, is written **first**; and **Cl** (chlorine), the **nonmetal**, is written **last**. The name of the compound is sodium chlor**ide**. Compounds made of **two elements** end in **-ide**.

HOW TO WRITE FORMULAS:

In a compound, **the sum of all the oxidation numbers must equal zero**. Look at the **Periodic Table** on page Reference Tables 20-21. Look at the C atom on the top of the Periodic Table. In every box of the Periodic Table, in the upper right hand corner, is the oxidation number (oxidation state). Carbon has oxidation numbers (oxidation states) of -4, $+2$ and $+4$.

$$\begin{array}{r} -4 \\ +2 \\ +4 \end{array}$$

C

What is the formula for sodium chloride? Look at sodium, Na, in the Periodic Table. In the upper right hand corner of the box for $_{11}$Na is $+1$. Look at chlorine, Cl, in the Periodic Table. In the upper right hand corner of the box for $_{17}$Cl is -1. Write Na^+Cl^-. **+1 and -1 oxidation numbers = 0** (sum of the oxidation numbers must equal zero).

Magnesium chloride: **Rule**: In a compound, the **sum of the oxidation numbers must equal zero**.

Look at the oxidation number of Mg on the Periodic Table. The oxidation number (oxidation state) of **Mg** is **+2**.

Look at the oxidation number of Cl. The oxidation number of Cl is -1. $2Cl^- = 2$ times $Cl^- = (2$ times $-1) = -2$.

$+2$ and $-2 = 0$. The formula is $MgCl_2$.

STOCK SYSTEM: A Roman numeral after the element shows the oxidation number of the element.

Nitrogen (I) oxide: **Oxygen** has an oxidation number of **-2**. The (I) shows you that nitrogen has an oxidation number of $+1$; therefore, you need $2N^+$ (2 times $+1$) = $+2$. $+2$ and -2 = 0. The formula is N_2O.

Nitrogen (II) oxide: N $+2$ oxidation #, O -2 oxidation #: Formula NO.

Nitrogen (IV) oxide: N $+4$ oxidation #, O -2 oxidation #: NO_2.

Vanadium (V) oxide: V $+5$ oxidation #, O -2 oxidation #: V_2O_5.

Iron (II) oxide: Fe $+2$ oxidation #, O -2 oxidation #: FeO.

Iron (III) oxide: Fe $+3$ oxidation #, O -2 oxidation #.

(2 x ($+3$)) and (3 x (-2)) = 0. Fe_2O_3.

See Table E, on page Reference Tables 5, for the oxidation number of polyatomic ions.

Write the formula for magnesium sulfate. **Mg** has an oxidation number of $+2$ (see Periodic Table, page Reference Tables 20-21). Look at sulfate in Table E: SO_4^{2-}. The oxidation number of SO_4 is -2.

Part of Table E	
Selected Polyatomic Ions	
CrO_4^{2-}	Chromate
NO_3^-	Nitrate
OH^-	Hydroxide
SO_4^{2-}	Sulfate

TABLE E

$+2$ and -2 = 0. In a compound, the sum of the oxidation numbers must equal zero. The formula is $MgSO_4$.

Write the formula for calcium hydroxide: **Ca** has an oxidation number of $+2$ (see Periodic Table). Hydroxide, OH^-, has an oxidation number of -1 (see Table E); therefore, you need **2 OH^-**. (2 times -1) = -2. $+2$ and -2 = 0. The formula is $Ca(OH)_2$.

EASY CRISS-CROSS METHOD TO WRITE FORMULAS

The oxidation number of Mg on the Periodic Table is $+2$. The oxidation number of Cl on the Periodic Table is -1. **Criss-cross only** $Mg_1^{2+} \quad Cl_2^{1-}$ the **numbers** of the two charges, the 2 of Mg and the 1 of Cl. **Write** these **numbers below** and to the **right of** the **elements**; do not use plus and minus signs. The 2 below and to the right of Cl means there are 2 atoms of Cl; the 1 below and to the right of Mg means there is 1 Mg atom. The formula is $MgCl_2$. (A 1 in a formula is not written.) Just write $MgCl_2$.

In the example MgO, the oxidation number of Mg is $+2$ and the oxidation number of O is -2 (see Periodic Table). Criss-cross only $Mg_2^{2+} \quad O_2^{2-}$ the numbers but not the plus and minus signs (like in the example above). The formula Mg_2O_2 should be reduced to lowest terms. Use the smallest whole number ratio; write MgO.

In the example $Mg(OH)_2$, the oxidation number of Mg is $+2$ and the oxidation number of OH^- is -1 (see Table E). Criss-cross only the numbers but not the plus and minus signs. $Mg_1^{2+} \quad (OH)_2^{1-}$ The formula is $Mg(OH)_2$.

The name of the compound $Mg(OH)_2$ is magnesium hydroxide. Mg is given on Table S as magnesium. OH on Table E (see above) is named hydroxide.

Try Sample Questions #7-8, on page 26, then do Homework Questions, #17-26, page 29, and #30-31, page 29.

REMEMBER: When you answer Regents questions on BONDING, use Table S, the Periodic Table, and Table E.

SAMPLE REGENTS & REGENT-TYPE QUESTIONS AND SOLUTIONS

1. Which molecule is nonpolar and has a symmetrical shape?
   (1) HCl   (2) CH_4 (3) H_2O (4) NH_3

2. Which type of bonding accounts for the unusually high boiling point of water?
   (1) ionic bonding   (3) hydrogen bonding
   (2) covalent bonding   (4) network bonding

3. Oxygen, nitrogen and fluorine bond with hydrogen to form molecules. These molecules are attracted to each other by
   (1) ionic bonds   (3) electrovalent bonds
   (2) hydrogen bonds   (4) coordinate covalent bonds

4. The attractions that allow molecules of krypton to exist in the solid phase are due to
   (1) ionic bonds   (3) molecule-ion forces
   (2) covalent bonds   (4) van der Waals forces

5. The dispersion forces of attraction between molecules always become stronger as molecular size
   (1) increases, and the distance between the molecules increases
   (2) increases, and the distance between the molecules decreases
   (3) decreases, and the distance between the molecules increases
   (4) decreases, and the distance between the molecules decreases

6. In which chemical system are molecule-ion attractions present?
   (1) KCl(g)   (2) KCl(ℓ)   (3) KCl(s)   (4) KCl(aq)

7. Which is an empirical formula?
   (1) C_2H_2 (2) H_2O (3) H_2O_2 (4) $C_6H_{12}O_6$

8. What is the formula for sodium oxalate?
   (1) NaClO   (2) Na_2O (3) $Na_2C_2O_4$ (4) $NaC_2H_3O_2$

SOLUTIONS

1. Answer **2.** You learned that CH_4 has polar bonds but, because of symmetry, the overall molecule is nonpolar.

2. Answer *3*. The hydrogen bond in water causes the boiling point of water to be higher than you would expect.

3. Answer *2*. You learned that hydrogen bonds are formed between molecules of hydrogen with O, N, or F.

4. Answer *4*. You learned that van der Waals forces are between nonpolar molecules. Van der Waals forces make it possible for molecules to exist in the solid phase. Answer 4, krypton with another krypton, is nonpolar.

5. Answer *2*. You learned that dispersion forces increase with increasing molecular size and decreasing distance between molecules.

6. Answer *4*. As you learned, an ionic salt (example: NaCl) in water has molecule-ion attraction. KCl is an ionic salt (the difference in electronegativity between K and Cl is more than 2.0). Answer 4 is KCl(aq). "aq" means water. Molecule-ion attraction is present in KCl in H_2O.

7. Answer *2*. An empirical formula is the smallest ratio of atoms. Answer 2, H_2O, means 2 atoms of hydrogen and 1 atom of oxygen, the smallest ratio of atoms.
   The other choices are wrong. Choice 1, C_2H_2 is a molecular formula. Divide by 2: $C_2/2$, $H_2/2$ becomes CH, the empirical formula. H_2O_2, divided by 2 becomes HO, its empirical formula, Choice 3. If you divide Choice 4, $C_6H_{12}O_6$, by 6, you get $C_6/6$, $H_{12}/6$ and $O_6/6$, which equals CH_2O, again an empirical formula.

8. Answer *3*. $Na_2C_2O_4$. Look at Table E. In Column 1 of Table E is written $C_2O_4^{2-}$ oxalate. This tells you that the formula for oxalate is $C_2O_4^{2-}$. The only choice with sodium and oxalate is Answer 3.

NOW LET'S TRY A FEW HOMEWORK QUESTIONS:

1. Which molecule is nonpolar and has a symmetrical shape?
   (1) HCl   (2) CH_4 (3) H_2O (4) NH_3

2. Which electron dot formula represents a polar molecule?

   (1) $\overset{\cdot\cdot}{\underset{\cdot\cdot}{O}}::C::\overset{\cdot\cdot}{\underset{\cdot\cdot}{O}}$ (2) $H:\overset{H}{\underset{H}{C}}:H$ (3) $H:\overset{\cdot\cdot}{\underset{\cdot\cdot}{O}}:$ (4) $:\overset{\;\;:\overset{\cdot\cdot}{Cl}:}{\underset{:\underset{\cdot\cdot}{Cl}:}{Cl}}:\overset{\cdot\cdot}{\underset{\cdot\cdot}{C}}:\overset{\cdot\cdot}{\underset{\cdot\cdot}{Cl}}:$

3. Hydrogen bonds are formed between molecules when hydrogen is covalently bonded to an element that has a
   (1) small atomic radius and low electronegativity
   (2) large atomic radius and low electronegativity
   (3) small atomic radius and high electronegativity
   (4) large atomic radius and high electronegativity

4. Hydrogen bonding is strongest between molecules of
   (1) H_2S (2) H_2O (3) H_2Se (4) H_2Te

5. Compared to the boiling point of H_2S, the boiling point of H_2O is relatively high. Which type of bonding causes this difference?
   (1) covalent   (2) hydrogen   (3) ionic   (4) network

6. Which of the following liquids has the weakest dispersion forces of attraction between its molecules?
   (1) Xe(ℓ)   (2) Kr(ℓ)   (3) Ne(ℓ)   (4) He(ℓ)

7. Given the phase change: $H_2(g) \rightarrow H_2(\ell)$, which kind of force acts between the molecules of H_2 during this phase change?
   (1) hydrogen bond   (3) molecule-ion
   (2) ionic bond   (4) dispersion

8. Which sequence of Group 18 elements demonstrates a gradual *decrease* in the strength of the dispersion forces?
   (1) Ar(ℓ), Kr(ℓ), Ne(ℓ), Xe(ℓ)   (3) Ne(ℓ), Ar(ℓ), Kr(ℓ), Xe(ℓ)
   (2) Kr(ℓ), Xe(ℓ), Ar(ℓ), Ne(ℓ)   (4) Xe(ℓ), Kr(ℓ), Ar(ℓ), Ne(ℓ)

9. The kind of attractions that result in the dissolving of sodium chloride in water are:
   (1) ion-ion   (3) atom-atom
   (2) molecule-ion   (4) molecule-atom

10. When calcium chloride is dissolved in water, to which end of the adjacent water molecules will a calcium ion be attracted?
    (1) the oxygen end, which is the negative pole
    (2) the oxygen end, which is the positive pole
    (3) the hydrogen end, which is the negative pole
    (4) the hydrogen end, which is the positive pole

11. Which type of attraction is directly involved when KCl dissolves in water?
    (1) molecule-molecule   (3) molecule-ion
    (2) molecule-atom   (4) ion-ion

12. The diagrams below represent an ionic crystal being dissolved in water.

    According to the diagrams, the dissolving process takes place by
    (1) hydrogen bond formation   (3) dispersion forces
    (2) network bond formation   (4) molecule-ion attractions

13. Which diagram best represents the structure of a water molecule?
    (1) O⟨^H_H (2) H⟨^O_O (3) O–H–O   (4) H–H–O

14. Molecule-ion attractions are found in
    (1) K(s)   (2) Kr(g)   (3) KCl(ℓ)   (4) KCl(aq)

15. In an aqueous solution of an ionic salt, the oxygen atom of the water molecule is attracted to the
    (1) negative ion of the salt, due to the oxygen's partial positive charge
    (2) negative ion of the salt, due to oxygen's partial negative charge
    (3) positive ion of the salt, due to oxygen's partial positive charge
    (4) positive ion of the salt, due to oxygen's partial negative charge

16. In which system do molecule-ion attractions exist?
    (1) NaCl(aq)   (2) NaCl(s)   (3) $C_6H_{12}O_6$(aq) (4) $C_6H_{12}O_6$(s)

17. A chemical formula is an expression used to represent
    (1) mixtures only   (3) compounds only
    (2) elements only   (4) compounds and elements

18. Which formula is an empirical formula?
    (1) K_2O (2) H_2O_2 (3) C_2H_6 (4) C_6H_6

19. Which hydrocarbon formula is also an empirical formula?
    (1) CH_4 (2) C_2H_4 (3) C_3H_6 (4) C_4H_8

20. An example of an empirical formula is
    (1) C_4H_{10} (2) $C_6H_{12}O_6$ (3) $HC_2H_3O_2$ (4) CH_2O

21. Which is an empirical formula?
    (1) H_2O_2 (2) H_2O (3) C_2H_2 (4) C_3H_6

22. What is the empirical formula of the compound whose molecular formula is P_4O_{10}?
    (1) PO (2) PO_2 (3) P_2O_5 (4) P_8O_{20}

23. Which formula represents lead (II) phosphate?
    (1) $PbPO_4$ (2) Pb_4PO_4 (3) $Pb_3(PO_4)_2$ (4) $Pb_2(PO_4)_3$

24. The correct formula for calcium phosphate is
    (1) $CaPO_4$ (2) $Ca_2(PO_4)_3$ (3) Ca_3P_2 (4) $Ca_3(PO_4)_2$

25. Which element is found in both potassium chlorate and zinc nitrate
    (1) hydrogen   (2) oxygen   (3) potassium   (4) zinc
                           Hint: See Table E

26. Which is the correct formula for nitrogen (I) oxide?
    (1) NO (2) N_2O (3) NO_2 (4) N_2O_3

CONSTRUCTED RESPONSE QUESTIONS: Parts B-2 and C of NYS Regents Exam

27. a. Compare and contrast asymmetrical and symmetrical molecules.
    b. Explain why CH_4, CCl_4, CF_4, CO_2 and CS_2 molecules are nonpolar.

28. Why does the hydrogen bond in water cause the boiling point of water to be higher than the boiling point of H_2S?

29. a. When you go down Group 17, $_9F$, $_{17}Cl$, $_{35}Br$, $_{53}I$, or down Group 18, why does the boiling point increase?
    b. Why in Group 17 is F a gas, Br a liquid, and I a solid?
    c. Which liquid in Table H has the weakest intermolecular forces? Explain how you reached your answer.

30. Write the formula for:
    a.   barium sulfide   f. sodium sulfate
    b.   sodium oxide   g. calcium hydroxide
    c.   calcium bromide   h. aluminum nitrate
    d.   aluminum oxide   i. tin(II) fluoride
    e.   potassium iodide   j. copper (II) hydroxide

31. Given the following molecular formulas, write the empirical formula :
    a. C_4H_8
    b. C_6H_{14}
    c. C_8H_{14}
    d. $C_6H_{12}O_6$
    e. $C_6H_5(OH)_3$

CHAPTER 5: MOLES / STOICHIOMETRY

SECTION A

COMPOUNDS

1. A compound is a substance made of 2 or more different elements that are **chemically united** (bonded together) in a **definite proportion**.

   Water, H_2O, is a compound made up of 2 elements, hydrogen and oxygen, chemically united together. A molecule of H_2O is always made of two atoms of hydrogen to one atom of oxygen (definite proportion).

2. Chemical **compounds** (example: H_2O) can be **broken down** by **chemical means** (chemical changes); new substances are produced. H_2O was broken down into new substances, H_2 and O_2, (by using an electric current).

   Table salt (sodium chloride) NaCl, is made up of 2 elements, Na and Cl chemically combined or united. The elements in NaCl are in definite proportion. There is always 1 Na to 1 Cl.   NaCl can be broken down by chemical means (chemical changes) into Na and Cl, 2 new substances. Electric current can break NaCl into Na and Cl.

3. Formulas and naming. A chemical compound can be represented by a **specific chemical formula** and given a **name** based on the **IUPAC** system of naming.

   Write the **formula** for the compound **sodium chloride**.

   Look at Table S. Sodium has the symbol Na. Chlorine has the symbol Cl. Look at sodium, Na, in the Periodic Table. In the upper right hand corner of the box for $_{11}Na$ is $+1$. Look at chlorine, Cl, in the Periodic Table. In the upper right hand corner of the box for $_{17}Cl$ is -1.

   Write Na^+Cl^-. **$+1$ and -1 oxidation numbers $= 0$**. (Sum of the oxidation numbers must equal zero.) The formula is NaCl.

NAMES OF COMPOUNDS BASED ON IUPAC SYSTEM

In ionic compounds, **metals** are written **first** (example: Na) and **nonmetals afterwards** (example: Cl). NaCl is called sodium chloride. Rule: The name of a compound made of only 2 elements, or NH_4^+ (ammonium) and one other element (example NH_4Cl), ends in -ide. Examples: NaCl is named sodium chloride, NH_4Cl ammonium chloride, KCl potassium chloride, KBr potassium bromide.

Use Table S and Table E to name compounds.

To name $NaClO_2$: Na is given on Table S as **sodium**. ClO_2 (see Table E) is named **chlorite**. Therefore, $NaClO_2$ is sodium chlorite.

Part of Table E

ClO_2^-	chlorite
NH_4^+	ammonium
SO_4^{2-}	sulfate

To name NH_4Cl: NH_4^+ (see Table E) is ammonium. Cl is given on Table S as chlorine. NH_4Cl is called ammonium chloride (see rule above: NH_4^+ and one other element (Cl) ends in **-ide**).

Compounds that contain a metal, a nonmetal, and oxygen (as the third element) are named to show the amount of oxygen.

Number of Oxygens	Compound Formula	Compound Begins with	Compound Ends with	Name of Compound
most ▲	$NaClO_4$	per-	-ate	sodium **perchlorate**
more	$NaClO_3$	–	-ate	sodium chlor**ate**
more	$NaClO_2$	–	-ite	sodium chlor**ite**
least	$NaClO$	hypo-	-ite	sodium **hypo**chlor**ite**

CHEMICAL FORMULAS

CHEMICAL FORMULAS describe the **composition** of **elements** or compounds. Formulas can be molecular formulas, empirical formulas or structural formulas. Let's review molecular and empirical formulas:

a. A **MOLECULAR FORMULA** indicates the total number of atoms of each element needed to form the molecule (**actual ratio** of atoms):

C_4H_{10} 4 atoms of carbon, 10 atoms of hydrogen

$C_4H_8Br_2$ 4 atoms of carbon, 8 atoms of hydrogen, 2 atoms bromine.

$(NH_4)_2S$ 2 x 1 atom of nitrogen = **2 atoms N,** 2 x 4 atoms of hydrogen = **8 atoms H,** 1 atom S (sulfur)

$Mg(ClO_3)_2$ 1 atom Mg, **2 atoms Cl,** 2 x 3 atoms of oxygen = **6 atoms O**

b. An **EMPIRICAL FORMULA** is the **simplest ratio** in which atoms combine to form a compound.

If the formula is H_2O_2, divide the H and the O by largest number that all the elements can be divided by, 2:

$$H_2/2 = H \qquad O_2/2 = O.$$

Therefore, the **empirical formula** is **HO**.

Now, let's take another example: $C_4H_8Br_2$ is the molecular formula. Divide the C , the H and the Br by the largest number that all the elements can be divided by, 2:

$$C_4/2 = C_2; \quad H_8/2 = H_4; \quad Br_2/2 = Br.$$

Therefore, the **empirical formula is C_2H_4Br**.

c. A **STRUCTURAL FORMULA** shows how the atoms are joined or connected to each other in a molecule. Example: propanone.

The lines or bonds connect the atoms together.

Propanone

You will learn more about structural formulas in the organic chemistry chapter.

Try Sample Questions #1-2, page 14, then do Homework Questions #1-7, page 17.

FINDING MOLECULAR FORMULA FROM EMPIRICAL FORMULA

Question: The empirical formula is CH and the molecular mass (the mass of the *molecular formula*) is 26. What is the molecular formula?

   (1) C_2H_2 (2) C_3H_3 (3) C_4H_4

There are two different ways of solving this question. Use whichever method is easier:

Solution: Answer *1*: Molecular formula is a multiple of the empirical formula (for example, 2, 3 or 4, etc., times the empirical formula).

Step 1: Find the mass of the empirical formula.
The empirical formula is CH. Look at the Periodic Table for the atomic masses of carbon (C) and hydrogen (H). C = 12, H = 1, therefore the mass of CH = 13 (empirical formula).

Step 2: Divide the mass of the molecular formula by the mass of the empirical formula:

$$\frac{Mass\ of\ Molecular\ Formula}{Mass\ of\ Empirical\ Formula} = \frac{26\ (given)}{13} = 2$$

Step 3: Multiply the empirical formula by the answer found above in Step 2:

$$\underset{\text{CH}}{\text{Empirical Formula}} \times \underset{2}{\text{Answer from Step 2}} = \underset{C_2H_2}{\text{Molecular Formula}}$$

Or use this method: As you know, the molecular formula has to be a multiple of CH (such as C_2H_2, C_3H_3). An **easy** way to solve the same problem is to test the choices:

Molecular Formula	Atomic Mass of Carbon (C) *times* # of Atoms		Atomic Mass of Hydrogen (H) *times* # of Atoms	Total
(1) $C_2H_2 =$	(12×2)	+	(1×2)	$= 26$, the correct answer
(2) $C_3H_3 =$	(12×3)	+	(1×3)	$= 39$, the wrong answer
(3) $C_4H_4 =$	(12×4)	+	(1×4)	$= 52$, the wrong answer

The question asked for the formula that has a molecular mass of 26; Answer 1, C_2H_2, has a molecular mass of 26.

To find the **molecular mass** of H_2 (hydrogen), you can say

atomic mass		# of atoms		
1	x	2	=	2

or you can say

# of atoms		atomic mass		
2	x	1	=	2

Use whichever order you find easier.

FINDING EMPIRICAL FORMULA FROM A MOLECULAR FORMULA

Question: The molecular formula is C_3H_6. What is the empirical formula?

Solution: C_3H_6 is the molecular formula. Divide the C and the H by the largest number that all the elements can be divided by, in this case, 3:

$$C_3/3 = C \quad H_6/3 = H_2.$$

Therefore, the empirical formula is CH_2.

CHEMICAL EQUATION

A chemical equation shows which bonds are broken and which bonds are built. In an equation, the number of atoms on the left side of the arrow must equal the number of atoms on the right side of the arrow.

After the elements or compounds are correctly written, you can only change the (coefficient) number in front of the element or compound.

If there is no coefficient in front of H_2, it means one H_2 = one H_2 molecule. The smallest part of an element is an atom; the smallest part of a compound is a molecule.

$$H_2 \quad + \quad O_2 \quad \rightarrow \quad H_2O$$

2 atoms	2 atoms	2 hydrogen atoms
H	O	1 oxygen atom

Left Side | Right Side

$$H_2 + O_2 \quad \rightarrow \quad H_2O$$

You have two **oxygen** atoms on the left side. You want two oxygen atoms on the right side. Therefore, put the coefficient 2 in front of H_2O (2 H_2O):

$$H_2 + O_2 \rightarrow 2\ H_2O$$

Now you have 2 H_2O: 2 times 1 oxygen atom = 2 oxygen atoms on the right side — good.

Now let's look at the **hydrogen** atoms.

On the right side, now you have 2 H_2O, which means 2 atoms of H x 2 coefficient = *4 H atoms on the right side*.

But on the *left side* of the equation $H_2 + O_2 \rightarrow 2\ H_2O$, you only have *2 H atoms*. Therefore put a coefficient of 2 *in front of H_2* on the left side.

$$2\ H_2 + O_2 \rightarrow 2\ H_2O.$$

Now it is good.

Let's make sure the equation is **balanced**:

Left side: You have 2 H_2. That equals 2 H atoms x 2 coefficient = *4 H atoms on the left side*. You have 4 H atoms on the right side, too — good.

Left Side | Right Side

$$2\ H_2 + O_2 \quad \rightarrow \quad 2\ H_2O$$

You have *2 atoms of oxygen on the left side*, and you have 2 atoms of oxygen on the right side.

Remember, the number **2** in front of H_2O on the right side, $2H_2O$, means 2x2 hydrogen atoms and **2x1 oxygen atoms = 2 oxygen atoms** on the **right side**. The equation is balanced.

Look at the equation: $2H_2 + O_2 \rightarrow 2H_2O$.
The smallest part of a compound is a molecule.
No number in front of O_2 = 1.

$2H_2$	+	O_2	$\rightarrow$	$2H_2O$
2 molecules H_2	+	1 molecule O_2	$\rightarrow$	2 molecules of water

or

(1 mole = 6.02 x 10^{23} molecules or particles)

2 moles H_2	+	1 mole O_2	$\rightarrow$	2 moles of water

The **mole ratio** (ratio of moles) of H_2 and O_2 = $\dfrac{2\ moles\ H_2}{1\ mole\ O_2}$ = 2 to 1 mole ratio.

The **mole ratio** (ratio of moles) **of H$_2$ and H$_2$O** (see mole equation above) $= \dfrac{2 \; moles \; H_2}{2 \; moles \; H_2O} = 2$ to 2 mole ratio, which is reduced to **1 to 1** mole ratio.

Do Homework Questions #13-17, pages 17-18.

CONSERVATION OF MASS, ENERGY, AND CHARGE

In all chemical reactions, there is **conservation of mass, energy**, and **charge**. In a chemical reaction, the amount of **matter stays** the **same** (matter cannot be created or destroyed).

In a chemical reaction, energy may be changed from one form to another, but the **total amount** of energy is the **same**. (Energy cannot be created or destroyed.) (**Energy** on **one side** of the equation **equals energy** on the **other side** of the equation.)

In a chemical reaction, the number of **positive charges** (or **negative charges**) on **one side** of the equation must **equal** the number of **positive charges** (or **negative charges**) on the **other side** of the equation.

Conservation of Matter: In the chemical equation:

$$2H_2 + O_2 \longrightarrow 2H_2O$$

there is **conservation of matter**; there is the same number of oxygen atoms (2) and hydrogen atoms (4) on both sides of the equation. (The total **matter** (example number of grams) on **one side** of the equation **equals** the total **matter** (number of grams) on the **other side** of the equation). Example: Given the equation $2H_2 + O_2 \longrightarrow 2H_2O$. 10 grams of H$_2$ and 80 grams of O$_2$ react completely (totally used up). Find grams of H$_2$O. Since you have 10 grams + 80 grams = 90 grams on the left side, you must have 90 grams on the right side, therefore 90 grams of H$_2$O.

Do Now Sample Question #3 on page 14.

Conservation of Energy: There is also **conservation of energy**. Energy cannot be created or destroyed. The total amount of energy on one side of the equation equals the total amount of energy on the other side of the equation.

In the equation $H^+ + OH^- \longrightarrow H_2O +$ energy,

energy in H^+ **+ energy** in OH^- **= energy** in H$_2$O **+ energy**.

Conservation of Charge: The total amount of charge is equal on both sides of an equation.

Example: $Na^\circ \longrightarrow Na^+ + e^-$

Before the arrow: Na or any element by itself has a **charge of 0.**

After the arrow: Na, charge of 1+ and 1 electron, (charge $= 1-) = 0.$

Example: $Al^\circ \longrightarrow Al^{3+} + 3e^-$

Before the arrow: Al or any free element has a **charge of 0.**

After the arrow: Al^{3+}, charge of 3+, and

$3e^-$, charge of 3 x (1-) = 3-   3+ + 3- = 0

MODEL OF CONSERVATION OF MATTER

Let's use a **model** to show the **conservation of matter**:

EXAMPLE 1:

$$Mg \quad + \quad Cl_2 \quad \longrightarrow \quad MgCl_2$$
magnesium + chlorine ⟶ magnesium chloride

Use 1 **black ball** to show **1 Mg** atom. Use 1 white ball to show 1 Cl atom; therefore, use **2 white balls** to show **2 chlorine** atoms (which is a molecule of Cl_2.)

Use **1 black ball** and **2 white balls** to show **$MgCl_2$**. Use toothpicks or sticks to put the 3 balls together.

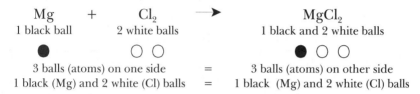

There are the same number of black balls (Mg atoms) and white balls (Cl atoms) on each side of the equation (balanced equation). This reaction **shows** the **Law of Conservation of Matter**: The **same number** and **type** of atoms are on **both sides** of the equation.

EXAMPLE 2:

$$Ca \quad + \quad 2HCl \quad \longrightarrow \quad CaCl_2 \quad + \quad H_2$$

| Calcium | Hydrochloric Acid | Calcium Chloride | Hydrogen |

Reminder: The number "2" in front of HCl means two molecules of HCl, each with one atom of hydrogen and one atom of chlorine. In 2HCl, there are two atoms of hydrogen and two atoms of chlorine.

Use **1 white ball** to show **1 Ca** atom. In **2HCl**, use **2 black balls** to show **2 hydrogen** atoms and **2 striped balls** to show **2 chlorine** atoms. Use **1 white ball** and **2 striped balls** to show **$CaCl_2$**, and use **2 black balls** to show **2 atoms** of hydrogen, **H_2**.

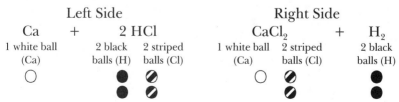

There is the same number and type of balls (atoms) on each side of the equation (balanced equation). This reaction **shows** the **Law of**

Conservation of Matter: The **same number** and **type** of atoms are on **both sides** of the equation.

Try Sample Questions #4-5, page 14, then do Homework Question #8, page 17, and #18-23, page 18.

MOLE-MOLE PROBLEMS

Mole-mole problems answer how many **moles** of one element or compound **react** with a given number (for example, 4 or 5 or 6) of **moles** of another element or compound.

Question: How many **moles of Ca** are needed to react completely with **6 moles of H₂O** in the following reaction:

$$Ca + 2H_2O \rightarrow Ca(OH)_2 + H_2$$

Solution: The equation shows that 1 mole of calcium reacts with 2 moles of H_2O to produce 1 mole of calcium hydroxide and one mole of hydrogen. (If there is no number in front of an element or compound, it means one mole.)

Now let's solve the problem. You are given:

$$Ca + 2H_2O \rightarrow Ca(OH)_2 + H_2$$

Step 1: **Cross out** anything (element or compound) that is **not** in the problem. **Cross out Ca(OH)₂ + H₂**, on the right hand side of the equation. Then, **write the coefficient** (the number in front of an element or compound) **under the element or compound.** If there is no number in front of the element or compound, it means 1.

$$\underset{\substack{1\ mole \quad\ 2\ moles}}{Ca\ +\ 2H_2O}\ \rightarrow\ \cancel{Ca(OH)_2}\ +\ \cancel{H_2}$$

Step 2: On **top** of the element or compound, **write the number of moles given in the PROBLEM.** In this problem, you have 6 moles of H_2O and "how many moles of Ca." Let "x" = number of moles of Ca:

Number of Moles given in Problem: X   6

$$Ca\ +\ 2H_2O\ \rightarrow\ \cancel{Ca(OH)_2}\ +\ \cancel{H_2}$$

Coefficient of Equation: 1 mole   2 moles

Step 3: **Set up a proportion:**

$$\frac{Top\ \#}{Bottom\ \#}\ \frac{x}{1} = \frac{6}{2}\ \frac{Top\ \#}{Bottom\ \#}$$

Cross multiply:

$$2x = 6;\ x = 3.$$

Answer: 3 moles of Ca.

Note: Sometimes a question can give you an extra fact (example 6 moles of H_2), which you do not need to use to solve the problem. Use only the facts you need.

Try Sample Questions #6-7, page 14, then do Homework Questions #24-25, page 18.

FORMULA MASS

FORMULA MASS is the sum of the atomic masses in the molecule.

REMINDER: On top of the Periodic Table, on page Reference Tables 20 is the element carbon. The upper left hand corner of the box is labeled "atomic mass." Similarly, the upper left hand corner of each box on the Periodic Table has the atomic mass of that element.

Question: What is the formula mass (or molecular mass) of K_2CO_3?

Solution: Look at the atomic masses of the elements (in the Periodic Table, on page Reference Tables 20-21):

The atomic mass of K is 39. You have K_2, which means two atoms of K. Therefore, you have to multiply two times the atomic mass of K:

$$\text{Atomic mass of K} = 39$$
$$2 \times 39 = 78 = K_2 \qquad\qquad K_2 = 78$$

Look at C in the Periodic Table.

$$\text{Atomic mass of C} = 12 \qquad\qquad C = 12$$

Look at O on the Periodic Table.

$$\text{Atomic mass of O} = 16.$$

In K_2CO_3, you have O_3, it means three oxygen atoms, therefore you have to multiply 3 times the atomic mass of oxygen:

$$3 \times 16 = 48 \qquad\qquad O_3 = 48$$

Add all the atomic masses together to get formula mass:____

$$\text{Formula mass} = 138$$

Or, you might like to make a chart to find the formula mass of K_2CO_3:

Element	Atomic Mass	Number of Atoms	Formula Mass
K	39	2	78
C	12	1	12
O	16	3	48
		Mass of K_2CO_3:	138

Use whichever method is easier for you to find formula mass.

PERIODIC TABLE

PER. TABLE

Question: What is the formula mass of $Ca(NO_3)_2$?

Solution: Look at the atomic masses of the elements (in the Periodic Table): Atomic mass of Ca = 40.   **Ca = 40**

Find mass of (NO_3): atomic mass of N = 14, atomic mass of O = 16, therefore mass of (NO_3) = 14 + 3(16) = 62

Mass of $(NO_3)_2$ = 62 x 2 = 124 $(NO_3)_2 = \underline{124}$

Formula mass of $Ca(NO_3)_2$ = 40 + 124 = 164 **Formula mass = 164**

GRAM FORMULA MASS

Gram formula mass is the mass of 6.02×10^{23} particles (1 mole of particles).

$$6.02 \times 10^{23} \text{ particles} = 1 \text{ mole of particles}$$

If you weigh 6.02×10^{23} particles (1 mole of particles) of K_2CO_3, the weight on the scale will be 138 grams.

Compound	Formula mass, amu	Mass of 6.02×10^{23} particles (on scale)
K_2CO_3	138 *(from last example)*	138 grams
H_2O	18	18 grams
O_2	32	32 grams

You saw K_2CO_3 has a formula mass of 138 and the mass of 6.02×10^{23} particles (1 mole) = 138 **grams**. This is called the **gram formula mass** or **molar mass** (because it is the mass of 1 mole of particles and you know mass is given in grams).

Rule for gram formula mass is the formula mass in grams.

To find the gram formula mass, take the **formula mass** (example: for K_2CO_3 **138**), **add the word "grams"** (K_2CO_3 138 **grams**), and you have **gram formula mass** (K_2CO_3 **138 grams**).

     gram formula mass = molar mass (K_2CO_3 138 grams)

(Similarly, if the formula mass of H_2O = 18, **gram** formula mass = 18 **grams** = molar mass. If the formula mass of carbon dioxide, CO_2, = 44, the **gram** formula mass of CO_2 = 44 **grams** = molar mass.)

You saw that the gram formula mass of K_2CO_3 = 138 grams. If they ask you the **mass of 1 mole** or **mass of 6.02×10^{23} molecules** (Avogradro's number), it is the **same answer as gram formula mass**, 138 grams for K_2CO_3. In gases like oxygen or helium, 1 **gram formula mass** occupies **22.4 liters**.

RULE:

1 mole (of molecules)

equals

1 gram formula mass

equals

6.02×10^{23} molecules

equals

22.4 liters (for gases at STP).

Note for New York State Regents students: These last two values are **not** *part of the required curriculum but* **may** *be asked as an application on the exam.*

Question: What is the mass of **1 mole** of K_2CO_3?

Solution: When they ask you the mass of 1 mole of a compound, you find **gram formula mass**, like you did in the previous example, or 138 grams.

Question: What is the mass of 6.02×10^{23} molecules of K_2CO_3?

Solution: It means, find the gram formula mass, as you did in the previous example, or 138 grams.

Question: What is the mass of H_2 gas that occupies 22.4 liters?

Solution: You learned that one gram formula mass (of a gas at STP) occupies 22.4 liters. Therefore, find the gram formula mass of H_2. Look at the Periodic Table, on page Reference Tables 20. Atomic mass of hydrogen = 1; H_2 = 2 hydrogen atoms; $2 \times 1 = 2$. $H_2 = 2$. The gram formula mass of H_2 is 2 grams, which occupies 22.4 liters.

REMEMBER THE RULE: 1 mole = 1 gram formula mass = 6.02×10^{23} molecules = (occupies) 22.4 liters (for gases at STP).

Question: What is the gram formula mass of NO_2?

Solution:

Look at the Periodic Table. The atomic mass of N is 14.   N = 14

Look again at the Periodic Table. The atomic mass of of O is 16. You have O_2, which means 2 oxygen atoms. $2 \times 16 = 32$.

$O_2 = 32$

Total Mass =   46 grams

All of these are the same:

Gram formula mass of NO_2 = 46 grams
Mass of 1 mole of NO_2 = 46 grams
Mass of 6.02×10^{23} molecules = 46 grams
Mass that occupies 22.4 liters = 46 grams

Question: What is the mass of 2 moles of NO_2?

Solution: 1 mole = 46 grams (gram formula mass). Therefore, **2 moles** = 2 times gram formula mass = 2 x 46 = 92 grams.

Question: What is the mass of 12×10^{23} molecules of NO_2?

Solution: The mass of 6×10^{23} molecules (gram formula mass) = 46 grams. You need 2 times that number of molecules, $2 \times 6 \times 10^{23}$, or 12×10^{23} or (2 times gram formula mass) 2 x 46 grams = 92 grams.

Question: What is the mass of 44.8 liters of NO_2?

Solution: The mass of 22.4 liters (gram formula mass) of NO_2 = 46 grams. The mass of 44.8 liters is twice as much, therefore 2 x 46 = 92 grams.

To be more exact, molar mass or gram formula mass is given in g/mol (grams in 1 mole). Molar mass or gram formula mass of K_2CO_3 = 138 g/mol.

You **must know** how to **round numbers** and round atomic masses to the **nearest tenth**. Read and know page **12**:8 in Chapter 12.

Try Sample Questions #8-10, page 15, then do Homework Question #9-10, page 17, and #26-29, page 18.

You **must know** how to use significant figures in solving math problems. Read and know pages **12**:8-**12**:12 in Chapter 12 and answer questions 8-15 on page **12**:25.

DENSITY

Table T

	Density		
DENSITY $= \dfrac{\textbf{\textit{mass}}}{\textbf{\textit{volume}}}$	**Density**	$d = \dfrac{m}{V}$	d = density m = mass V = volume

In the equation $d = \dfrac{m}{V}$, when mass (m) is constant, the bigger the density (d) of the element, the smaller is the volume (V).

Question: The mass of an object is 900 grams (g), and its volume is 30 cm^3. What is the density?

Solution: Use the density formula on Table T: $d = \dfrac{m}{V}$

$$d = \dfrac{m}{V}$$

$$d = \dfrac{900 \ g}{30 \ cm^3} = 30 \ \text{g/cm}^3.$$

Question: Mass of Fe = 15.9 g; Density = 7.87 g/cm³. Find volume.

Solution: $d = \dfrac{m}{V}$ d = density, m = mass, V = volume

$$7.87 \ \text{g/cm}^3 = \dfrac{15.9 \ g}{V}$$

Cross multiply: $7.87 \ \text{g/cm}^3 \bullet V = 15.9$ g

$$V = \dfrac{15.9 \ g}{7.87 \ \text{g/cm}^3} \qquad V = 2.0 \ \text{cm}^3$$

TABLE S

Density of the **elements** is given in **Table S**; see page Reference Tables 24-25. **Density** is at **STP** (means 273 K, 101.3 kPa); see bottom of Table S.

Question 1: At STP, a piece of metal has a mass of 12.54 g and a volume of 4.6 mL (cm³). What is its density (express in significant figures)?
Question 2: The metal is most likely
    (1) aluminum   (2) sodium   (3) sulfur   (4) iron

Solution 1: $d = \dfrac{m}{V}$ d = density, m = mass, V = volume

$$d = \dfrac{12.54 \ g \ \text{4 significant figures}}{4.6 \ cm^3 \ \text{2 significant figures}} = 2.726 \ \text{g/cm}^3$$

The answer must have the same number of significant figures as the number with the smallest number of significant figures = 2 significant figures. Density is 2.7 g/cm³.

Solution 2: From mass and volume, you found density = 2.7. The element with a density of about 2.7 g/cm³ is aluminum (see Table S). Answer *1*

$density = \dfrac{mass}{volume}$ **Red phosphorus** ($^{31}_{15}$P) has a density of **2.16 g/cm³** (2.16 grams in 1 cubic cm). **White phosphorus** ($^{31}_{15}$P) has a density of **1.823 g/cm³**. **Red P** has a **bigger density,** (2.16 grams/cm³) **more g (grams)** in the same volume, therefore **red phosphorus** has **more atoms.**

Gram formula mass (of a gas) = **density at STP** (g/L) **x 22.4 liters** (L)

Question: The density of a gas is 1.35 g/L at STP. Calculate the gram formula mass.

Solution: Gram formula mass = density x 22.4 liters (L)
(of a gas)   = 1.35 g/L x 22.4 L = 30.24 g (grams).

Question: Which gas has a density of 1.70 grams/liter (g/L) at STP?

   (1) F_2 (2) He   (3) N_2 (4) SO_2

Solution: Gram formula mass = density (grams/liter) x 22.4 liters
        (of a gas)   = 1.7 g/L   x 22.4 L (liters)

     Gram formula mass = 38.08 g (grams)
                         = 38.1 grams (rounded to the nearest tenth)

Which gas has a **gram formula mass** of 38 grams? Look at the Periodic Table. The atomic mass of F = 19. Answer 1, F_2, has two F atoms, therefore 2 atomic masses: 2 x 19 = 38 grams.

The gram formula mass of F_2 = 38 grams.

You saw before that a gas with a gram formula mass of 38 grams has a density of 1.70 g/l.   **Answer 1**.

Put oil and water in a beaker. The water (or any substance), with more density, goes to the bottom, while the oil (less density) goes to the top.

> **Try Sample Question #11, page 15, then do Homework Question #11-12, page 17, and #30, page 18.**

SAMPLE REGENTS & REGENT–TYPE QUESTIONS AND SOLUTIONS

1. Which statement describes a characteristic of all compounds?
    (1) Compounds contain one element, only.
    (2) Compounds contain two elements, only.
    (3) Compounds can be decomposed by chemical means.
    (4) Compounds can be decomposed by physical means.

2. Which is an empirical formula?
    (1) C_2H_2 (2) H_2O (3) H_2O_2 (4) $C_6H_{12}O_6$

3. In the reaction $CaCO_3 \longrightarrow CaO + CO_2$, 50 grams of $CaCO_3$ reacted completely (totally used up), producing 28 grams of CaO. Find grams of CO_2

4. Given the incomplete equation: $2N_2O_5(g) \rightarrow$
    which set of products completes and balances the incomplete equation?
    (1) $2N_2(g) + 3H_2(g)$ (3) $4NO_2(g) + O_2(g)$
    (2) $2N_2(g) + 2O_2(g)$ (4) $4NO(g) + 5O_2(g)$

5. What is the total number of moles of atoms in one mole of $(NH_4)_2SO_4$?
    (1) 10   (2) 11   (3) 14   (4) 15

6. Given the reaction: $CH_4(g) + 2O_2(g) \rightarrow CO_2(g) + 2H_2O(g)$.
    How many moles of oxygen are needed for the complete combustion of 3.0 moles of $CH_4(g)$?
    (1) 6.0 moles   (2) 2.0 moles   (3) 3.0 moles   (4) 4.0 moles

7. Given the reaction $CH_4 + 2O_2 \rightarrow CO_2 + 2H_2O$.
    What amount of oxygen is needed to completely react with 1 mole of CH_4?

---

(1) 2 moles   (2) 2 atoms   (3) 2 grams   (4) 2 molecules

8. Which gas sample contains a total of 3.0×10^{23} molecules?
   (1) 71 g of Cl_2 (2) 2.0 g of H_2 (3) 14 g of N_2 (4) 38 g of F_2

9. Which quantity represents 0.500 mole at STP?
   (1) 22.4 liters of nitrogen   (3) 32.0 grams of oxygen
   (2) 11.2 liters of oxygen   (4) 28.0 grams of nitrogen

10. At STP, 32.0 liters of O_2 contains the same number of molecules as
    (1) 22.4 liters of Ar   (3) 32.0 liters of H_2
    (2) 28.0 liters of N_2 (4)   44.8 liters of He

11. A sample of an unknown gas at STP has a density of 0.630 gram per liter. What is the gram molecular mass of this gas?
    (1) 2.81 g   (2) 14.1 g   (3) 22.4 g   (4) 63.0 g

SOLUTIONS

1. Answer *3*. Remember, this is the definition of a compound.

2. Answer *2*. An empirical formula is the smallest ratio of atoms. Answer 2, H_2O, means 2 atoms of hydrogen and 1 atom of oxygen, the smallest ratio of atoms.

   The other choices are wrong. Choice 1, C_2H_2 is a molecular formula. Divide by 2: $C_2/2$, $H_2/2$ becomes CH, the empirical formula. H_2O_2, divided by 2 becomes HO, its empirical formula, Choice 3. If you divide Choice 4, $C_6H_{12}O_6$, by 6, you get $C_6/6$, $H_{12}/6$ and $O_6/6$, which equals CH_2O, again an empirical formula.

3. $CaCO_3 \longrightarrow CaO + CO_2$ You can find the number of grams of CO_2 (using conservation of matter) because you know the grams of everything else ($CaCO_3$ and CaO). Since the left side of the equation has 50 grams, the right side (CaO and CO_2, see equation) must have 50 grams. Since CaO has 28 grams, therefore CO_2 must have (50 – 28) = 22 grams. The right side has 28 + 22 = 50 grams.

4. Answer *3*. Make sure the number of atoms on the left side is equal to the number of atoms on the right side. In Answer 3,
$$2N_2O_5(g) \rightarrow 4NO_2(g) + O_2(g).$$

On the left side:

N_2 means 2 atoms of nitrogen

$2N_2$: A coefficient of 2 in front of nitrogen means 2x2 atoms of nitrogen = 4 nitrogen atoms on the left side.

O_5 means 5 atoms of oxygen

$2N_2O_5$: coefficient 2 in front of N_2O_5 means 2 x 5 atoms of oxygen = 10 atoms of oxygen

Total: 10 oxygen atoms

Left side total:
   4 nitrogen atoms
   10 oxygen atoms

On the right side:

$4NO_2$ has 4 nitrogen atoms

Now count the oxygen atoms
NO_2 means 2 atoms of oxygen.
4 NO_2 means 4 x 2 atoms of oxygen = 8 oxygen atoms.

$+O_2$, **2** oxygen atoms

Total: 10 oxygen atoms

Right side total:
   4 nitrogen atoms
   10 oxygen atoms

Balanced. Good!

5. Answer *4* Let's understand what a formula means.

The formula NH_4 can mean either one atom of nitrogen (N) and four atoms of hydrogen (H), or it can mean one mole of nitrogen atoms and four moles of hydrogen atoms.

In **$(NH_4)_2SO_4$**:

NH_4 = one (mole of) nitrogen atoms and four (moles of) hydrogen atoms = 5 (moles of) atoms

$(NH_4)_2$ = 5 x 2 = 10 (moles of) atoms

SO_4 = one (mole of) sulfur (S) atoms and four (moles of) oxygen (O) atoms.

Add all the atoms together:

$(NH_4)_2SO_4$

(1+4)

5 x 2 + 1 + 4 = 15 (moles of) atoms.

6. Answer *1*. Follow the steps you learned to solve a mole-mole problem. The problem gives you the number of moles of an element or compound (CH_4), and you need to find the number of moles of the other element (oxygen) or compound.

Step 1: Cross out anything (element or compound) that is **not** in the problem, and write the coefficient (number in front of an element or compound) under the element or compound:

$$CH_4 + 2O_2 \rightarrow \cancel{CO_2} + 2\cancel{H_2O}$$
$$\textit{1 mole} \quad \textit{2 moles}$$

Step 2: On top of the element, write the number of moles given in the problem. Let "x" = moles of O_2:

# of moles given in problem: 3   X

coefficient of equation: $\underset{\textit{1 mole}}{CH_4} + \underset{\textit{2 moles}}{2O_2} \rightarrow \cancel{CO_2} + 2\cancel{H_2O}$

Step 3: Set up a proportion:

$$\frac{Top\ \#}{Bottom\ \#}\ \frac{3}{1} = \frac{x}{2}\ \frac{Top\ \#}{Bottom\ \#}$$

Cross multiply:

$$x = 6.$$

Answer: 6 moles of oxygen.

Or, if 1 mole of CH_4 reacts with 2 moles of O_2, then 3 moles of CH_4 (3 times as many moles) reacts with 3 x 2 = 6 moles of O_2.

7. Answer *1*. $CH_4 + 2O_2 \rightarrow CO_2 + 2H_2O$. No number in front of C or any other element means 1. Let's read the equation: 1 mole of CH_4 reacts with **2 moles of oxygen** to produce 1 mole of CO_2 and 2 moles of water.

8. Answer *3*. You learned that 1 mole = 1 gram molecular mass = 6.02×10^{23} molecule. As you can see, 3.0×10^{23} molecules = ½ mole = ½ gram molecular mass.

Look at the Periodic Table. The atomic mass of N = 14. N_2 = 2 nitrogen atoms = 2 x 14 = 28 grams = gram molecular mass of N_2. 14 grams = ½ gram molecular mass = 3.0×10^{23} molecules.

The other three choices listed each show a full mole.

9. Answer **2**. 1 mole of any gas at STP occupies 22.4 liters. Therefore, 0.5 mole = ½ mole of any gas at STP occupies 11.2 liters.

10. Answer **3**. (Avogadro's hypothesis states that) equal volumes of gases at the same temperature and pressure have (contain) the same number of molecules.

    Every liter of gas has the same number of molecules:

    32 liters of O_2 contains the same number of molecules as 32 liters of H_2.

11. Answer **2**. $Density = \dfrac{mass}{volume} = \dfrac{0.63\ g}{liter}$.

    Gram molecular mass = density x 22.4 liters:

    $$\dfrac{0.63\ g}{liter}\ X\ 22.4\ liters\ =\ 14.1\ g.$$

Now Let's Try a Few Homework Questions:

1. Which substances can be decomposed chemically (by chemical means)?
   (1) CaO and Ca (2) MgO and Mg   (3) CO and Co (4) CaO and MgO

2. A chemical formula is an expression used to represent
   (1) mixtures only   (3) compounds only
   (2) elements only   (4) compounds and elements

3. Which formula is an empirical formula?
   (1) K_2O (2) H_2O_2 (3) C_2H_6 (4) C_6H_6

4. Which hydrocarbon formula is also an empirical formula?
   (1) CH_4 (2) C_2H_4 (3) C_3H_6 (4) C_4H_8

5. An example of an empirical formula is
   (1) C_4H_{10} (2) $C_6H_{12}O_6$ (3) $HC_2H_3O_2$ (4) CH_2O

6. Which is an empirical formula?
   (1) H_2O_2 (2) H_2O (3) C_2H_2 (4) C_3H_6

7. What is the empirical formula of the compound whose molecular formula is P_4O_{10}?
   (1) PO   (2) PO_2 (3) P_2O_5 (4) P_8O_{20}

8. Which is a correctly balanced equation for a reaction between hydrogen gas and oxygen gas?
   (1) $H_2(g) + O_2(g) \rightarrow H_2O(\ell) + heat$ (3) $2H_2(g) + 2O_2(g) \rightarrow H_2O(\ell) + heat$
   (2) $H_2(g) + O_2(g) \rightarrow 2H_2O(\ell) + heat$ (4) $2H_2(g) + O_2(g) \rightarrow 2H_2O(\ell) + heat$

9. What is the total mass of 3.01×10^{23} atoms of helium gas?
   (1) 8.00 g   (2) 2.00 g   (3) 3.50 g   (4) 4.00 g

10. What is the total number of atoms contained in 2.00 moles of nickel?
    (1) 58.9   (2) 118   (3) 6.02×10^{23} (4) 1.20×10^{24}

11. Which gas has a density of 1.70 grams per liter at STP?
    (1) $F_2(g)$ (2) He(g)   (3) $N_2(g)$ (4) $SO_2(g)$

12. Which gas could have a density of 2.05 grams per liter at STP?
    (1) N_2O_3 (2) NO_2 (3) HF   (4) HBr

CONSTRUCTED RESPONSE QUESTIONS: Parts B-2 and C of NYS Regents Exam

13. What is the empirical formula of the compound whose molecular formula is P_4O_{10}?

14. The molecular formula is C_2H_4. What is the empirical formula?

15. The molecular formula is H_2O_2. What is the empirical formula?

16. A compound has an empirical formula of CH_2 and a molecular mass of 56. What is its molecular formula?

17. The empirical formula is CH_3 and the formula mass is 30. What is the molecular formula?

18. Given the unbalanced equation: $\_Ag(s) + \_H_2S(g) \rightarrow \_Ag_2S(s) + \_H_2(g)$
    What is the sum of the coefficients when the equation is completely balanced using the smallest whole-number coefficients?

19. When the equation $\_Al_2(SO_4)_3 + \_ZnCl_2 \rightarrow \_AlCl_3 + \_ZnSO_4$
    is correctly balanced using the smallest whole-number coefficients, the sum of coefficients is

20. What is the total number of hydrogen atoms required to form one molecule of $C_3H_5(OH)_3$?

21. What is the total number of moles of oxygen atoms present in 1 mole of $Mg(ClO_3)_2$?

22. Given the unbalanced equation: $\_CaSO_4 + \_AlCl_3 \rightarrow \_Al_2(SO_4)_3 + \_CaCl_2$
    What is the coefficient of $Al_2(SO_4)_3$ when the equation is completely balanced using the smallest whole-number coefficients?

23. Balance the following equations:
    (1) aluminum + bromine $\longrightarrow$ aluminum bromide
    (2) zinc + hydrochloric acid $\longrightarrow$ zinc chloride + hydrogen
    (3) $Zn + HCl \longrightarrow ZnCl_2 + H_2$
    (4) $Na + H_2O \longrightarrow Na(OH) + H_2$
    (5) $Mg + O_2 \longrightarrow MgO$
    (6) $H_2O \longrightarrow H_2 + O_2$

24. Given the reaction: $Ca + 2H_2O \rightarrow Ca(OH)_2 + H_2$.
    What is the total number of moles of Ca needed to react completely with 4.0 moles of H_2O?

25. Given the *unbalanced* equation: $N_2(g) + H_2(g) = NH_3(g)$
    When the equation is balanced using the smallest whole-number coefficients, the ratio of moles of hydrogen compared to moles of ammonia produced is

26. What is the total number of atoms in 1 mole of calcium?

27. A. What is the gram formula mass of $CaSO_4$?
    B. What is the gram formula mass of $Ca_3(PO_4)_2$?

28. Find the gram formula mass of NaCl, MgO, and $Ca(NO_3)_2$. Round atomic masses to the nearest tenth. (Hint: See pages **12:7-12:8**).

29. A. What is the total mass in grams of 0.75 mole of SO_2?
    B. What is the total mass in grams of 0.013 mole of I_2?

30. A 1.00-mole sample of neon gas has a volume of 24.4 liters at 298 K and 101.3 kPa. Calculate the density of this sample.

CHAPTER QUESTIONS: PARTS B-2 AND C OF NYS REGENTS EXAM

31. In a laboratory experiment, a student reacted 2.8 grams of Fe(s) (steel wool) in excess $CuSO_4(aq)$, according to the following balanced equation:
    $Fe(s) + CuSO_4(aq) \longrightarrow FeSO_4(aq) + Cu(s)$
    When the Fe(s) was completely consumed, the precipitated Cu had a mass of 3.2 grams. Did the student's result in this experiment verify the mole ratio of Fe(s) to Cu(s) as predicted by the equation? Calculate the mole ratio of Fe to Cu and explain.

SECTION B

PERCENT COMPOSITION

Question: What is the percent by mass of magnesium in magnesium oxide?

Solution: Look on Table T, below or on page Reference Tables 25-26. Use the Percent Composition formula:

Table T

Percent Composition	% composition by mass = $\dfrac{mass\ of\ part}{mass\ of\ whole}$ x 100

Look in the Periodic Table at the atomic masses of magnesium (24) and oxygen (16). The gram formula mass of MgO (Mg=24 + O=16) = 40 grams. When they ask you what percentage of Mg is in MgO, use the percentage composition formula in Table T:

$$\% \ composition \ by \ mass \ = \ \frac{mass\ of\ part}{mass\ of\ whole} \ x \ 100$$

Put the mass of Mg on top and the total mass of MgO on the bottom:

$$\text{Percent Composition of Magnesium} = \frac{Mass\ of\ Mg}{Mass\ of\ MgO} \ X \ 100 = \frac{24}{40} \ X \ 100 = \frac{6}{10} \ X \ 100 = 60 \ percent$$

Question: Here is the data from an experiment:

1. Mass of empty crucible (container) + cover   11.70 grams
2. Mass of crucible (container) + cover + hydrated salt (salt with water in the crystals) before heating   14.90 grams
3. Mass of crucible + cover + anhydrous salt (salt with no water in the crystals) after thorough heating   14.53 grams

Realize: After thorough heating, mass is constant (in this example 14.53 grams) because all the water went out.

What is the approximate percent by mass of the water in the hydrated salt (salt with water in the crystals?

Solution: Use the Percent Composition formula from Table T:

Table T

Percent Composition	% composition by mass = $\dfrac{mass\ of\ part}{mass\ of\ whole}$ x 100

To find **"mass of part"** (water) take:

#2 (mass of container + cover + salt with water) minus

#3 (mass of container + cover + salt without water) =

$$14.90 - 14.53 = \textbf{0.37 gram}$$

To find **"mass of whole"** (salt + water) take:

#2 (mass of container + cover + salt with water) minus

#1 (mass of container + cover) =

14.90 − 11.70 = **3.20 gram.**

Substitute in the formula:

% composition by mass = $\dfrac{mass\ of\ part}{mass\ of\ whole}$ x 100 = $\dfrac{.37}{3.20}$ x 100 = 11.56%

Do Sample Question #1, page 26:
Find percent of H₂O in BaCl₂•2H₂O

MOLE PROBLEMS
CHANGING GRAMS TO MOLES

Question: You have 54 grams of LiF. How many moles of LiF do you have?

Solution: Look at Table T:

Table T

Mole Calculations	number of moles = $\dfrac{given\ mass\ (g)}{gram\text{-}formula\ mass}$

To find the **number of moles** of LiF, take the number of grams in the problem and divide it by the gram formula mass.

Formula:

$$number\ of\ moles = \frac{given\ mass\ (grams)}{gram\text{-}formula\ mass}$$

Grams of LiF = 54, as given in the problem. For the gram formula mass of LiF, look at the Periodic Table. The mass of Li is 7, and the mass of F is 19. Therefore, the gram formula mass of LiF = 7 + 19 = 26:

$$number\ of\ moles = \frac{given\ mass\ (grams)}{gram\text{-}formula\ mass} = \frac{54\ grams}{26\ grams} = 2\ moles\ of\ LiF$$

Question: How many moles is 108 grams of H₂O?

Solution: Use the mole calculation formula on Table T.
To find the **number of moles** of H₂O, take the number of grams in the problem and divide it by the gram formula mass.
Grams of H₂O = 108, as given in the problem.

$$number\ of\ moles = \frac{given\ mass\ (g)}{gram\ formula\ mass}$$

For the gram formula mass of H₂O, look at the Periodic Table. The mass of hydrogen = 1; H₂ is 2 hydrogen atoms; the mass of H₂ = atomic mass (1) x # of atoms (2) = 2. The mass of oxygen = 16.

Therefore, the gram formula mass of water =

$$\underset{1}{\underset{\text{(atomic mass of H)}}{}} \quad \underset{x}{\underset{\text{(# of H atoms)}}{}} \quad \underset{2}{} \quad + \quad \underset{16}{\underset{\text{(atomic mass of O)}}{}} \quad = 18g$$

$$number\ of\ moles = \frac{given\ mass\ (grams)}{gram\ formula\ mass} = \frac{108}{18} = 6\ moles\ of\ H_2O.$$

CHANGING MOLES TO GRAMS

Question: What is the mass of 2 moles of NO_2?

Solution: You learned the mass of 1 mole of NO_2 = 1 Gram formula mass of NO_2 = 46 grams.

2 moles = 2 times gram formula mass = 2 x 46 grams = 92 grams

FORMULA FROM PERCENT COMPOSITION

If you know the percent of each element in the compound, you can find the formula of the compound.

Question: A compound is 86% C and 14% H by mass. What is the **empirical** formula for this compound?

Solution: In *this type of problem*, assume that you had 100 gram. If there are 100 gm and 86% is carbon, 86 grams is carbon. If there are 100 gm and 14% is hydrogen, 14 grams is hydrogen.

In order to get the empirical formula (smallest ratio of atoms), you have to find out the number of moles of carbon and hydrogen.

$$\#\ moles\ (of\ atoms) = \frac{grams}{gram\ atomic\ mass}$$

\# moles of C = $\frac{86}{12}$ = 7.17 C

\# moles of H = $\frac{14}{1}$ = 14 H

You have 7.17 C and 14 H: $C_{7.17}H_{14}$. In a formula, you want each element (C and H) to have whole numbers. Here there are 7.17 carbon and 14 hydrogen. Divide the number of each element by the smallest number (7.17):

For carbon: $\frac{7.17}{7.17}$ = 1 carbon.

For hydrogen: $\frac{14}{7.17}$ = about 2 hydrogen.

Therefore, the formula is CH_2 (1 carbon, 2 hydrogen). (No number after C in the formula means 1 C.)

Try Sample Question #2, page 26, then do Homework Questions #5-12, page 27.

TYPES OF CHEMICAL REACTIONS

1. **Synthesis Reaction:** Two or more elements or simpler compounds unite to form a compound.

$$Na \; + \; Cl \longrightarrow NaCl$$
Sodium + Chlorine   Sodium chloride (salt)

$$H_2O \; + \; CO_2 \longrightarrow H_2CO_3$$
water + Carbon dioxide   carbonic acid

2. **Decomposition Reaction:** A compound is broken down into 2 or more elements or simpler compounds.

$$NaCl \longrightarrow Na \; + \; Cl$$
sodium chloride   sodium + Chlorine

$$H_2CO_3 \longrightarrow H_2O \; + \; CO_2$$
carbonic acid   water + carbon dioxide

As you can see, a decomposition reaction is the opposite of synthesis.

3. **Single Replacement Reaction:** A free element (an element alone, example: iron, Fe) replaces an element that is part of a compound.

$$Fe \; + \; CuSO_4 \longrightarrow FeSO_4 \; + \; Cu$$
iron + copper sulfate   iron sulfate + copper

4. **Double Replacement Reaction:** Two elements replace each other or switch partners. In the example below, the Na and Ag replace each other or switch partners. Two new compounds, $NaNO_3$ and $AgCl$, are formed.

$$NaCl \; + \; AgNO_3 \longrightarrow NaNO_3 \; + \; AgCl$$
sodium chloride + silver nitrate   sodium nitrate + silver chloride

Do Homework Question #13, page 28.

MASS-MASS PROBLEMS

Mass-mass problems give you the **grams** of one element or compound and ask you to find the **grams** of another element or compound.

Question: How many **grams** of calcium are needed to react completely with 108 **grams** of H_2O in the equation

$$Ca + 2H_2O \rightarrow Ca(OH)_2 + H_2$$

Solution: These steps are very similar to those in the last problem. (Mole-mole problem: How many moles of calcium react with 6 moles of water.)

PERIODIC TABLE

Step 1: **Cross out** anything (element or compound) that is **not** in the **problem and write the coefficient under the element or compound:**

$$Ca + 2H_2O \rightarrow Ca(OH)_2 + \cancel{H_2}$$
$$\text{1 mole} \quad \text{2 moles}$$

Step 2: There are 108 grams of water given in this problem. **Change** the **grams** of H_2O **into moles of H_2O** and **put** that number of **moles** of H_2O **on top** of the compound H_2O in the equation:

$$\frac{\text{number of moles}}{\text{of water}} = \frac{\text{grams}}{\substack{\text{gram molecular mass} \\ \text{(gram formula mass)}}} = \frac{108}{\substack{18\ (H_2O) \\ ((1x2)+16=18)}} = 6 \text{ moles } H_2O$$

Therefore, put 6 moles on top of H_2O. The question asks about calcium. Let "X" = number of moles of calcium and write X on top of calcium:

# of moles given in problem: X   6

coefficient of equation:

$$\underset{\text{1 mole} \quad \text{2 moles}}{Ca + 2H_2O \rightarrow Ca(OH)_2 + \cancel{H_2}}$$

Step 3: **Set up a proportion:**

$$\frac{Top \#}{Bottom \#} \frac{x}{1} = \frac{6}{2} \frac{Top \#}{Bottom \#}$$

Cross multiply:

$$2x = 6;\ x = 3 \text{ moles of Ca.}$$

Step 4: The question is **how many grams** of Ca are needed. **1 mole has 1 gram formula mass.** Therefore, 1 mole of Ca (see Periodic Table, $^{40}_{20}Ca$; the top number, 40, is the atomic mass) = 40 grams. 3 moles of Ca = 3 x 40 = **120 grams**.

or

The answer using a different method (factor label method) is:

$$108\ \cancel{g\text{-}H_2O} \times \frac{1\ \cancel{mole\ H_2O}}{18\ \cancel{g\text{-}H_2O}} \times \frac{1\ \cancel{mole\ Ca}}{2\ \cancel{moles\ H_2O}} \times \frac{40\ g\ Ca}{1\ \cancel{mole\ Ca}} = 120\ g\ Ca$$

If you are given extra information, such as grams of three substances (Ca, H_2O, and H_2) instead of just grams of two substances (Ca and H_2O), you just can set up a proportion with only two substances, the unknown (Ca) and any one other substance (either H_2O or H_2).

MASS-VOLUME PROBLEMS

Mass-volume problems deal with one element or compound in **grams** (mass) and one element or compound in **milliliters or liters** (volume). Both grams and ml or liters are in the same problem. In **mass-volume** problems, you use the **same Steps 1, 2 and 3** as you just had in mass-mass problems:

Question: How many **liters** of N_2 will react with 36 grams of H_2 at STP, given

$$N_2 + 3 H_2 \rightarrow 2 NH_3$$

Solution:

Step 1: **Cross out** anything (element or compound) that is **not** in the problem, and **write the coefficient under the element or compound**:

$$N_2 \; + \; 3H_2 \; \rightarrow \; \cancel{2NH_3}$$
1 mole   3 moles

Step 2: You have 36 grams of H_2. **Change the grams into moles** of hydrogen and **put** that number of **moles on top** of H_2.

$$moles\ of\ H_2 = \frac{grams\ of\ H_2}{gram\ formula\ mass} = \frac{36}{2} = 18\ moles$$

Therefore, put 18 moles on top of hydrogen. The question is about nitrogen. Let "x" = number of moles of nitrogen and write x on top of nitrogen:

Number of Moles given in Problem: X   18
$$N_2 + 3H_2 \rightarrow 2N\cancel{H_2}$$
Coefficient of Equation: 1 mole   3 moles

Step 3: **Set up a proportion**:

$$\frac{Top\ \#}{Bottom\ \#}\ \frac{x}{1} = \frac{18}{3}\ \frac{Top\ \#}{Bottom\ \#}$$

Cross multiply:

$$3x = 18;\ x = 6\ moles\ of\ N_2.$$

Step 4: The question in this problem is how many **liters** of nitrogen. You know that **1 mole of gas at STP occupies 22.4 liters**, therefore, 6 moles of gas is 6 x 22.4 liters = **134.4 liters**.

or

The answer using a different method (factor label method) is:

$$36\ \cancel{g\ H_2} \times \frac{1\ \cancel{mole\ H_2}}{2\ \cancel{g\ H_2}} \times \frac{1\ \cancel{mole\ N_2}}{3\ \cancel{moles\ H_2}} \times \frac{22.4\ L\ N_2}{1\ \cancel{mole\ N_2}} = 134.4\ L\ N_2$$

VOLUME-VOLUME PROBLEMS

A **volume-volume problem** is very easy. The problem gives you **liters or ml** of one element or compound and asks you to **find liters or ml** of the other element or compound. Use the same unit, either **ml or liters** throughout the problem:

Question: What is the total number of liters of carbon dioxide formed by the complete combustion of 28 liters of $C_2H_6(g)$?

$$2\ C_2H_6(g) + 7\ O_2 \rightarrow 4\ CO_2 + 6\ H_2O$$

Solution:

Step 1: Always **cross out** anything (element or compound) that is **not** in the problem, **and write the coefficient under the element or compound**:

$$2\ C_2H_6(g) + 7\ \cancel{O_2} \rightarrow 4\ CO_2 + 6\ \cancel{H_2O}$$

   2 moles   4 moles

Step 2: Since the problem gives both C_2H_6 and CO_2 in liters, write the **number of liters** on the **top of the equation**. Write 28 on top of C_2H_6. Let "x" = the number of liters of CO_2. Write x on top of CO_2.

(Volume) Liters: 28   X

$$2\ C_2H_6 + 7\ \cancel{O_2} \rightarrow 4\ CO_2 + 6\ \cancel{H_2O}$$

Coefficient of Equation: 2 moles   4 moles

Step 3: Set up a proportion:

$$\frac{Top\ \#}{Bottom\ \#}\ \ \frac{28}{2} = \frac{x}{4}\ \ \frac{Top\ \#}{Bottom\ \#}$$

Cross multiply:

$$2x = 112;\ x = 56.$$

Answer: 56 liters of CO_2.

or

The answer using a different method (factor label method) is:

$$28\ \cancel{L\ C_2H_6} \times \frac{1\ \cancel{mole\ C_2H_6}}{22.4\ \cancel{L\ C_2H_6}} \times \frac{4\ \cancel{mole\ CO_2}}{2\ \cancel{moles\ C_2H_6}} \times \frac{22.4\ L\ CO_2}{1\ \cancel{mole\ CO_2}} = 56\ L\ CO_2$$

GRAHAM'S LAW

GRAHAM'S LAW states that under the same conditions of temperature and pressure, gases diffuse (spread) at a rate inversely (opposite) proportional to the square roots of the molecular masses.

RULE: Gases diffuse at a faster rate when the molecular mass is smaller (when it is lighter). Gases diffuse at a slower rate when the molecular mass is bigger (when it is heavier).

Question: At the same temperature and pressure, which gas will diffuse through air at the fastest rate?

   (1) H_2 (2) O_2 (3) CO   (4) CO_2

Solution: Answer *1*. Find the molecular masses of each of the gases. Look at the Periodic Table, on page Reference Tables 20:

1. H: Atomic mass is 1, therefore $H_2 = 1 \times 2 = 2$. The mass of the H_2 molecule, or molecular mass is 2.

2. O: Atomic mass is 16, therefore $O_2 = 16 \times 2 = 32$. The mass of the O_2 molecule, or molecular mass, is 32.

3. CO: Atomic mass of C is 12, atomic mass of O is 16. Mass of the CO molecule, or molecular mass of $CO = 12 + 16 = 28$.

4. CO_2: Atomic mass of $C = 12$, atomic mass of O is 16. Mass of the CO_2 molecule, or molecular mass of $CO_2 = 12 + (16 \times 2) = 44$.

The gas with the **smallest molecular mass (lightest gas)** will **diffuse the fastest**. Answer *1*, H_2, is correct. The molecular mass of $H_2 = 2$ (**lightest mass**), therefore H_2 **diffuses the fastest** (at the fastest rate).

Try Sample Question #3, on page 26, then do Homework Questions, #1-4, pages 27-28, and #14-17, page 28.

REMEMBER: When you answer Regents questions on MOLES/ STOICHIOMETRY, use the Periodic Table and Table T.

SAMPLE REGENTS & REGENT–TYPE QUESTIONS AND SOLUTIONS

1. What is the percentage by mass of H_2O in $BaCl_2 \cdot 2H_2O$?

2. What is the approximate percent by mass of potassium in $KHCO_3$?
   (1) 19%   (2) 24%   (3) 61%   (4) 39%

3. Which gas would diffuse most rapidly under the same conditions of temperature and pressure?
   (1) gas *A*, molecular mass = 4
   (2) gas *B*, molecular mass = 16
   (3) gas *C*, molecular mass = 36
   (4) gas *D*, molecular mass = 49

SOLUTIONS

1. Look at Table T, below, or on page Reference Tables 25-26.

Table T

Percent Composition	% composition by mass = $\dfrac{mass\ of\ part}{mass\ of\ whole}$ x 100

Formula mass of H_2O = (2 X 1) + 16 = 18. Mass of **$2H_2O$ = 36.**
Formula mass of $BaCl_2 \cdot 2H_2O$:

Ba = 137	Mass of Ba			= 137
Cl = 35.5	Mass of Cl_2	=	2 X 35.5 =	71
H_2O = 18	Mass of $2H_2O$			= 36
	Formula mass of $BaCl_2 \cdot 2H_2O$			= **244**

In the formula above, put the mass of $2H_2O$ on top and the total mass of $BaCl_2 \cdot 2H_2O$ on bottom.
Percent composition by mass =

$\dfrac{mass\ of\ part}{mass\ of\ whole}$ x 100 = $\dfrac{mass\ of\ 2H_2O}{mass\ of\ BaCl_2 \cdot 2H_2O}$ x 100 = $\dfrac{36}{244}$ x 100 = 14.75%

Answer is *14.75%*.

2. Answer *4*. You want to find the percent by mass of potassium in $KHCO_3$. You might make a chart to find the mass of $KHCO_3$ (total mass).

Element	Atomic Mass	Number of Atoms	Molecular Mass
K	39	1	39
H	1	1	1
C	12	1	12
O	16	3	48
		Mass of $KHCO_3$	100 grams

Use the Percent Composition formula from Table T:

Table T

Percent Composition	% composition by mass = $\dfrac{mass\ of\ part}{mass\ of\ whole}$ x 100

The percentage composition of potassium in $KHCO_3$ =

$\dfrac{Mass\ of\ potassium}{Mass\ of\ KHCO_3\ (total\ mass)}$ = $\dfrac{39}{100}$ X 100 = 39 *percent*

3. Answer *1*. Graham's Law of Diffusion states that, under the same conditions of temperature and pressure, the heavier the gasses are, the slower the gases diffuse (spread out). The lighter the gases are, the faster the gases diffuse. Gas A has the smallest mass (the lightest) and therefore diffuses the fastest (most rapidly).

NOW LET'S TRY A FEW HOMEWORK QUESTIONS:

1. Given the balanced equation representing a reaction:
$$2H_2 + O_2 \longrightarrow 2H_2O$$
What is the total mass of water formed when 8 grams of hydrogen reacts completely with 64 grams of oxygen?
(1) 72 g   (2) 56 g   (3) 36 g   (4) 18 g

2A. At the same temperature and pressure, which gas will diffuse through air at the fastest rate?

  (1) CO   (2) O_2 (3) H_2 (4) CO_2

2B. Which gas diffuses most rapidly at STP?

  (1) Ar   (2) Kr   (3) N_2 (4) O_2

3. At STP, which gas will diffuse more rapidly than Ne?

  (1) He   (2) Ar   (3) Kr   (4) Xe

4. Given the same conditions of temperature and pressure, which noble gas will diffuse most rapidly?

  (1) He   (2) Ne   (3) Ar   (4) Kr

CONSTRUCTED RESPONSE QUESTIONS

5. What is the percent by mass of hydrogen in CH_3COOH (formula mass = 60)?

6. The percent by mass of Ca in $CaCl_2$ is equal to _____.

7. What is the percent by mass of water in the hydrate $Na_2CO_3 \cdot 10H_2O$ (formula mass = 286)?

8. What is the percent by mass of oxygen in Fe_2O_3 (formula mass = 160)?

9. How many moles is 39 grams of LiF equivalent to?

10. You have 162 grams of HBr. How many moles do you have?

11. What is the mass of 3 moles of SO_2?

12. What is the mass of 4 moles of CO_2?

13. What type of chemical reaction is shown by each equation:

  (1) $2 H_2 + O_2 \longrightarrow 2 H_2O$
  (2) $Zn + 2 HCl \dashrightarrow ZnCl_2 + H_2$
  (3) $2 Mg + O_2 \longrightarrow 2 MgO$
  (4) $2 H_2O \longrightarrow 2 H_2 + O_2$

14. Given the equation $FeAsS(s) \xrightarrow{\text{heat}} FeS(s) + As(g)$

  125.0 kg of FeAsS (arsenopyrite) is heated. Determine the total mass of As (arsenic) produced by the reaction.

15. A. Given the reaction:
  $$Ca(s) + 2H_2O(\ell) \rightarrow Ca(OH)_2(aq) + H_2(g).$$
  When 40.1 grams of Ca(s) reacts completely with the water, what is the total volume, at STP, of $H_2(g)$ produced?
  B. Given the equation:
  $$6CO_2(g) + 6H_2O(\ell) \rightarrow C_6H_{12}O_6(s) + 6O_2(g).$$
  What is the *minimum* number of liters of $CO_2(g)$, measured at STP, needed to produce 32.0 grams of oxygen?

16. Given the reaction:
  $$2C_2H_6(g) + 7O_2(g) \rightarrow 4CO_2(g) + 6H_2O(g).$$
  What is the total number of liters of $CO_2(g)$ produced by the complete combustion of 1 liter of $C_2H_6(g)$?

17. Given the reaction:
  $$2C_2H_6(g) + 7O_2(g) \rightarrow 4CO_2(g) + 6H_2O(g).$$
  What is the total number of liters of carbon dioxide formed by the complete combustion of 28.0 liters of $C_2H_6(g)$?

CHAPTER 6: SOLUTIONS

A **SOLUTION** is a homogeneous mixture (made up of two or more substances). You put salt in water and now you have a salt solution. Salt is the **solute** and water is the **solvent**.

Solute: A substance like salt or sugar that dissolves in the water.

```
        ┌──────┐ ←──── salt (solute)  ⎫ make up
        │      │                       ⎬ a solution
        └──────┘ ←──── H₂O (solvent)  ⎭
```

Solvent: Usually a liquid, for example, water. If water is the solvent, it is called an aqueous solution.

Sugar or salt dissolves faster (faster rate of dissolving) in water when you stir or shake the salt and water or sugar and water solution, or when you crush the sugar or raise the temperature of the water (as explained by the collision theory, in Chapter 7, page 7:1).

The more salt you have in a solution, the more **concentrated** is the solution. When there is very little salt in a solution, it is called **dilute**.

Question: Which statement describes KCl(aq)?
  (1) KCl is the solute in a homogeneous mixture.
  (2) KCl is the solute in a heterogeneous mixture.
  (3) KCl is the solvent in a homogeneous mixture.
  (4) KCl is the solvent in a heterogeneous mixture.
Solution: Answer *1*. In KCl(aq), "aq" means aqueous solution. (Aqueous means water is the solvent.) Solutions are homogeneous mixtures. Answer *1* says homogeneous mixture. The substance, for example KCl, before the "(aq)" is the solute (salt). Therefore, Answer *1*: KCl is the solute in a homogeneous mixture.

In a salt solution (examples LiCl(aq), KCl(aq)), you can separate the salt (LiCl) from the water (H₂O) by physical means (examples distillation, evaporation), see page 1:3.

Solubility shows the most salt that the water can hold, or the most salt that can dissolve in the water at a specific temperature.

TEMPERATURE, PRESSURE, AND THE NATURE OF SOLUTE AND SOLVENT AFFECT SOLUBILITY

TEMPERATURE: Look at Table G: Solubility Curves, on the next page. The horizontal line in the graph shows temperature. The vertical line shows how many grams of solute is **the most** that can dissolve in 100 grams of water .

TABLE G

Follow line SO_2 on Reference Table G. At 0°C, about 23 grams is the most that can dissolve. At 30°C, about 7 grams is the most that can dissolve, and at 60°C about 4 grams is the most that can dissolve. This shows that, as temperature increases (from 0°C to 30°C to 60°C), less grams (23, then 7, then 4) can dissolve in the water, which means the **solubility decreases**.

As you can see in Table G, in gases SO_2, NH_3 and HCl, as the temperature goes up, less grams can dissolve in the water, which means the solubility decreases.

Let's look at the salt $KClO_3$. Follow this line on **Reference Table G**. At 0°C, about 6 grams of $KClO_3$ is the most that can dissolve in 100 grams of water. At 30°C, about 12 grams of $KClO_3$ is the most that can dissolve in 100 grams of water. Obviously, in 200 grams of water, 24 grams can dissolve at 30°C. At 90°C, about 50 grams is the most $KClO_3$ that can dissolve in 100

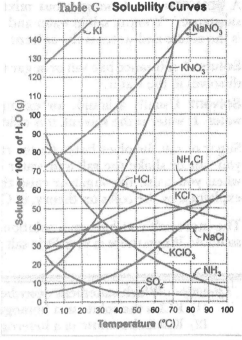

Table G **Solubility Curves**

grams of water. This shows that, as the temperature increases (from 0°C to 30°C to 90°C), more grams (6, then 12, then 50) can dissolve in the water, which means the solubility increases. As you see in Reference Table G, with salts like KNO_3, $NaNO_3$ and KCl, as temperature increases, solubility increases.

In short, temperature has an effect on solubility. When temperature increases:

    1. **Solubility** of a **solid** (example: KNO_3, $NaNO_3$) **increases**.
    2. **Solubility** of a **gas** (example: NH_3, HCl) **decreases**.

Question: Which compound's solubility decreases most rapidly as the temperature increases from 50°C to 70°C? (See Table G).
    (1) NH_3 (2) HCl   (3) SO_2 (4) KNO_3

Solution: Answer *1*. Follow the line for each substance in the table:

Substance	NH_3	HCl	SO_2	KNO_3
Solubility at 50°C, grams	29	58	6	83
(most that can dissolve)				
Solubility at 70°C, grams	18	51	4	134
(most that can dissolve)				
Decrease in solubility, grams	11	7	2	increased

Solubility decreased most rapidly for NH_3, by 11 grams.

TYPES OF SOLUTIONS

There are three types of solutions, saturated, unsaturated, and supersaturated.

Saturated solution: It contains the **most solute** (salt) that **can dissolve** at a given temperature. If you add another table-spoon of salt, no more can dissolve. If some of the added salt dissolves, an equal amount from the solution will fall to the bottom (crystallize). (In a saturated solution, the rate of crystallizing equals the rate of dissolving.)

At **70°C**, 134 grams of KNO_3 is the most that can dissolve in 100 mL of water. This is a **saturated** solution. Look at the solubility curve above. At 70°C, 134 grams of KNO_3 falls on the line. **Any point on the line is a saturated solution.**

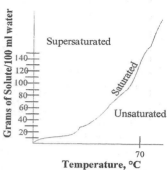

Solubility of KNO_3 in H_2O

Unsaturated solution: It contains **less solute** (salt) than a saturated solution. You can still add more salt and it will dissolve. On the solubility curve, it is below the line. Example: at 70°C, 133 grams, 100 grams or 40 grams of KNO_3 in 100 mL is an unsaturated solution (which is **below the line**).

Supersaturated solution: This is **only** a **temporary** situation, caused by slowly cooling a saturated solution. It has **more solute than** in a **saturated solution**. Example: At 70°C, 135 grams, 140 grams, and 145 grams of KNO_3 are supersaturated solutions (**above the line**).

Similarly, for $KClO_3$ at 50°C (see solubility curve on the previous page), 22 grams is a saturated solution (on the line), 15 grams or 10 grams is an unsaturated solution (below the line), and 23 grams or 40 grams is a supersaturated solution (above the line).

Question: How many grams of $NaNO_3$ would be needed to saturate 200 grams of H_2O at 40°C?

Solution: Needed to saturate means to make a saturated solution, which means it falls on the line in the solubility curve. Look at Table G (Solubility Curve). The x axis has temperature, the y axis has solute per 100 grams of H_2O (means solute in 100 grams of water). Look at 40°C. 105 grams of $NaNO_3$ is the most that can dissolve in 100 grams of water, which means 105 grams is needed to saturate 100 grams of water. The question asks for 200 grams of water, twice as much water, so twice as much salt, 2 x 105 grams = 210 grams, is needed.

A saturated solution of $KClO_3$ in 100 grams of water is cooled from 90°C to 20°C. How many grams of $KClO_3$ settle out? Look at Table G. At 90°C, 53 grams of $KClO_3$ dissolves and at 20°C, only 8 grams of $KClO_3$ dissolves, therefore, 53 grams – 8 grams, which equals 45 grams of $KClO_3$, falls out of the solution (settles out) when it is cooled.

CONSTRUCTING A SOLUBILITY CURVE

Let's draw a solubility graph from data given.

Solubility of KNO$_3$

Temperature	Grams Dissolved
0	14
20	34
40	62
60	105
70	134

1. On the x axis, write Temperature, ° Celsius. **Always** include units of measure (in this example, °C) on the axis. The lines must be evenly spaced, and there must be an equal number of degrees between each two lines. See graph below.

2. On the y axis, write Grams of Solute/100 grams water. Always include units of measure (in this example, grams). Again, the lines must be evenly spaced, and there must be an equal number of grams between each two lines. See graph below.

3. Plot the following **experimental data for KNO$_3$** on the graph. Draw a circle around each point. Connect the points with a best fit curve. Do **not** continue the line past the last point.

4. Put a title on the graph which shows what the graph is about. Example: "Effect of Temperature on Solubility."

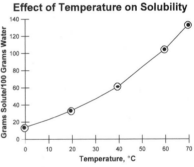

Effect of Temperature on Solubility

As you can see, **temperature** has an **effect** on **solubility.**

Pressure also has an **effect** on **solubility**.

Pressure makes **gases more soluble** and has almost **no effect on liquids and solids.** Example: High pressure forces carbon dioxide gas into water to make soda. (It makes carbon dioxide more soluble.) When you open the cap of a soda bottle, there is **less pressure** on the soda, the soda fizzes and gas escapes. (The gas becomes less soluble under low pressure.)

NATURE OF SOLUTE AND SOLVENT also **affects solubility**:

A **polar** molecule or ionic substance has 1 end positive and 1 end negative, + ▭ -.

A **nonpolar** molecule or substance does **not** have a positive end and a negative end, ▭.

Molecular polarity (whether the two molecules are both polar, both nonpolar, or one polar and one nonpolar) affects solubility.

In a solution, a **polar** (positive end + ▭- negative end) **solute** molecule (examples: alcohol, sugar) **dissolves** in a **polar solvent** (water).

The positive end of one molecule is attracted is to the negative end of the water (the oxygen atom). The negative end of the molecule is attracted to the positive end of the water (hydrogen). Polar substances dissolve in polar substances. **Like dissolves like**.

In a solution, a **nonpolar solute** (example: oil paint) **dissolves** in a **nonpolar solvent** (example: turpentine). An oil paint stain can be removed with turpentine. Oil paint dissolves in turpentine. **Like dissolves like**. But oil paint **(nonpolar)** does **not dissolve** in **polar substances** (example: water); oil paint and water (even when mixed) form two separate layers.

In short, **temperature**, **pressure**, and **nature of solute and solvent affect solubility**.

TABLE F: DETERMINING IF COMPOUNDS ARE SOLUBLE OR INSOLUBLE

Look at Table F, below or on page Reference Tables 6-7. Table F shows the solubilities of different compounds (which compounds are soluble, which are insoluble).

-soluble compounds dissolve nicely-a lot of a compound dissolves in water.

-insoluble compounds dissolve very little or not at all.

The **left box** of Table F lists ions that form **soluble compounds.** A few exceptions are given. The **right box** of Table F lists ions that form **insoluble compounds**. A few exceptions are given.

Part of Table F
Solubility Guidelines for Aqueous Solutions

Ions That Form *Soluble* Compounds	Exceptions	Ions That Form *Insoluble* Compounds	Exceptions
Group 1 Ions (Li^+, Na^+, etc)		carbonate(CO_3^{2-})	when combined with Group 1 ions or ammonium (NH_4^+)
ammonium (NH_4^+)		sulfide (S^{2-})	when combined with Group 1 ions or ammonium (NH_4^+)
halides (Cl^-, Br^-, I^-)	when combined with Ag^+, Pb^{2+}, or Hg_2^{2+}		

Question: Is NaCl soluble?

Solution: Look at Table F. The left box of Table F lists Group 1 ions (Li^+, Na^+, etc) as forming **soluble** compounds. Therefore you know **NaCl** is soluble. Or, you can look at halides in Table F. **Halides** form **soluble** compounds; therefore, you know NaCl is **soluble.**

Question: Is AgBr soluble?

Solution: Look at Table F. The left box lists halides (Cl^-, Br^-, I^-) as soluble, **but,** when combined with Ag^+, it is **insoluble.** Therefore, you know **Ag**Br, AgCl, and AgI are **insoluble.**

High Marks: Regents Chemistry Made Easy

Question: Is $CaCO_3$ soluble?

Solution: Look at Table F. The right box of Table F lists **carbonate** $(CO_3{}^{2-})$ ions as forming **insoluble compounds**, therefore you know $CaCO_3$ is **insoluble** (very low concentration of dissolved ions).

Question: Is $(NH_4)_2S$ soluble?

Solution: Look at Table F. The right box lists sulfide ions as forming **insoluble** compounds, **but** when combined with **ammonium** $(NH_4{}^+)$, it is **soluble**. Therefore you know $(NH_4)_2S$ is **soluble.**

Question: Based on Reference Table F, which of these saturated solutions has the *lowest* concentration of dissolved ions?
   (1) NaCl(aq)   (2) $MgCl_2$(aq) (3) $NiCl_2$(aq) (4) AgCl(aq)

Solution: Look at Table F. AgCl is insoluble. Insoluble means it has the lowest concentration of dissolved ions. Answer *4*

TABLE F: DETERMINING ELECTRICAL CONDUCTIVITY (ELECTROLYTES)
By looking at Table F, you saw that NaCl and $(NH_4)_2S$ are soluble. This means that NaCl and $(NH_4)_2S$ have **ions** that are **soluble** (in solution) and can **conduct electricity(are electrolytes)**. These soluble solutions can carry electricity and light a bulb.

You saw AgBr, AgCl and $CaCO_3$ are insoluble. They have ions that are **insoluble** (ions do not go into solution, have the **lowest concentration** of dissolved **ions**), and **cannot conduct electricity** (are nonelectrolytes or poor electrolytes.)   Electrical conductivity decreases when the concentration of ions decreases.

ENTROPY: When a **solid** (salts-NaCl, NH_4Cl: or sugar) **dissolves** in water, **entropy (disorder)** increases. A solid has a regular organized pattern. When the solid dissolves, the particles have more entropy (more disorder, more randomness).

$$H_2O$$

Example: NaCl(s) $\longrightarrow$ Na^+(aq) + Cl^-(aq)

NaCl **dissolves** in water,   **entropy increases**

Try Sample Questions #1-4, on page 12, then do Homework Questions, #1-6, page 13, & #16-18, page 14.

CONCENTRATION OF SOLUTION

There are different ways to describe the concentration of a solution. You can describe the **concentration** of a solution by **molarity, percent by volume, percent by mass,** or **parts per million.**

1. MOLARITY (M).
Molarity can be used to describe how concentrated the solution is (how much salt is in the solution). The molarity of a solution is the number

of **moles** of solute (e.g., salt) in **one liter** of **solution**. The formula for molarity is given in Table T.

Table T

CONCENTRATION	molarity = $\dfrac{moles\ of\ solute}{liters\ of\ solution}$

1 molar (1M) salt solution means there is **1 mole** (1 gram formula mass) of salt in one (1) liter of solution.

5 molar (5M) salt solution means there are **5 moles** (5 gram formula mass) of salt in one (1) liter of solution.

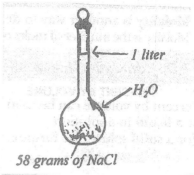

To make a **1M** solution of NaCl, take **1 gram formula mass** or 58 grams of NaCl and add water until the salt NaCl and water together fill up 1 liter.

To make a **5M** solution of NaCl, take 5 moles (**5 gram formula mass**) = 5 x 58 grams = 290 grams of NaCl and add water until the salt NaCl together with the water fills up 1 liter.

58 grams of NaCl

1 M NaCl Solution

Obviously, **.05M** solution of NaCl has .05 moles (**.05 gram formula mass**) = .05 x 58 grams = 2.90 grams of salt in 1 liter of solution.

Question: What is the concentration of a solution of **10 moles** of copper (II) nitrate in **5 liters** of solution?
(1) 0.5 M   (2) 2.0 M   (3) 5.0 M   (4) 10 M

Solution: Answer *2*. Use the molarity formula, given in Table T:

$$molarity = \frac{moles\ of\ solute\ (like\ salt)}{liters\ of\ solution}$$

$$= \frac{10\ moles}{5\ liters} = 2\ \frac{moles}{liter} = 2M\ solution = 2\ molar\ solution$$

If you have 50 moles (mol) of solute (salt) in 25 liters (L) of solution and you want to know the molarity (concentration), look at Table T for the molarity formula:

$$molarity = \frac{moles\ of\ solute}{liters\ of\ solution} = \frac{50\ moles}{25\ liters} = \frac{2\ moles}{liter}\ or\ \frac{2\ mol}{L} = 2\ M\ solution$$

Question: How many moles of HCl are in 4 liters of a 0.50 M(aq) solution?

Solution: M means molar (molarity). 0.50 M means 0.50 **molar** (**molarity** of 0.50 M).

$$molarity = \frac{moles\ of\ solute}{liters\ of\ solution} \qquad 0.50\ M = \frac{x\ moles}{4\ liters} \qquad x\ moles = 0.50\ M\ (4\ L) = 2\ mole$$

Question: How many moles of solute are contained in 200 mL of a 1M solution?

Solution: 1M solution means 1 mole of solute (salt) in 1 liter of solution. (One liter = 1000 mL). The question asks for 200 mL;

therefore, set up a proportion. Let x = number of moles of solute in 200 mL.

$$\frac{1 \ mole}{1000 \ mL} = \frac{x \ moles}{200 \ mL} \qquad 1000 \ x = 200, \quad x = \frac{200}{1000}, \quad x = 0.2 \ mole.$$

OR you can solve it like you did in the question right above, but change mL to liters (200 mL = 0.2 L), because the molarity equation uses liters of solution.

Molality is another way to describe the concentration of a solute. Molality is the number of moles of solute dissolved in a kilogram of solvent.

Do Homework Questions, #7-9, page 13, and #19-21, page 14.

2. PERCENT BY VOLUME
Percent by volume can be used to describe the concentration of a solid or a liquid in a solution.
For a **solid** solute, the formula for

$$percent \ by \ volume = \frac{grams \ of \ solute \ (salt)}{100 \ mL \ of \ solution} \ x \ 100$$

$10\% \ salt \ solution = \dfrac{10 \ grams \ of \ salt}{100 \ mL \ of \ solution} \ x \ 100.$ This means that you have 10 grams of salt. Add water to the 10 grams of salt until you have 100 mL of solution (salt together with water).
For a **liquid** solute, the formula for

$$percent \ by \ volume = \frac{mL \ of \ solute}{100 \ mL \ of \ solution} \ x \ 100$$

Orange drink has 10% orange juice (solute). Percent by volume of orange juice in orange drink $= \dfrac{10 \ mL \ orange \ juice \ (solute)}{100 \ mL \ of \ solution} \ x \ 100 = 10\%.$

3. PERCENT BY MASS
Percent by mass also describes the concentration of a solution.

$$percent \ by \ mass = \frac{grams \ of \ solute}{100 \ grams \ of \ solution} \ x \ 100$$

$$2\% \ salt \ solution = \frac{2 \ grams \ of \ salt}{100 \ grams \ of \ solution} \ x \ 100$$

2% salt solution means that you have 2 grams of salt. Add water to the 2 grams of salt until the solution (salt together with water) weighs 100 grams.

Question: What is the percent by mass of sodium chloride if:
    A. 8 grams of sodium chloride is dissolved in 100 grams of solution (salt and water).
    B. 8 grams of sodium chloride is dissolved in 200 grams of solution (salt and water).

Solution:
$$percent \ by \ mass = \frac{grams \ of \ solute}{100 \ grams \ of \ solution} \ x \ 100$$

A. $\qquad percent \ by \ mass = \dfrac{8 \ grams \ of \ NaCl}{100 \ grams \ of \ solution} \ x \ 100 = 8\%$

B. $$\text{percent by mass} = \frac{8 \text{ grams of NaCl}}{200 \text{ grams of solution}} \times 100 = 4\%$$

SUMMARY: Formulas

$$\% \text{ by volume} = \frac{\text{grams or mL of solute}}{100 \text{ mL of solution}} \times 100$$ ←This is in volume, mL

$$\% \text{ by mass} = \frac{\text{grams of solute}}{100 \text{ grams of solution}} \times 100$$ ←This is in mass, grams

4. PARTS PER MILLION (ppm)

Formula given in Table T:

Table T

CONCENTRATION	$\text{parts per million} = \dfrac{\text{mass of solute}}{\text{mass of solution}} \times 1000000$

A $CuSO_4$ solution contains .050 grams (g) $CuSO_4$ in 1000 grams (g) of solution (homogeneous mixture). Find ppm.

$$\text{parts per million (ppm)} = \frac{.050 \text{ g } CuSO_4}{1000 \text{ grams (g)}} \times 1000000 = 50 \text{ parts per million}$$

A solution has 300 parts per million of KOH. Find the number of grams (g) of KOH in 1000 grams (g) of this solution. (Use the equation above.)

$$300 = \frac{x \text{ grams}}{1000} \times 1000000, \ 300 = \frac{x \text{ grams}}{1000} \times 1000000, \ 300 = 1000 \, x, x = 0.3$$

Another solution has 0.040 g of nitrogen dissolved in 1000 g of water. Find the N_2 concentration of the solution in ppm.(Use equation above.)

$$ppm = \frac{0.0040g \ N_2}{0.004g \ N_2 + 1000g \ H_2O} \times 1000000 = \frac{0.004g \ N_2}{1000.004g \ solution} \times 1000000 = 4.0 \ ppm$$

In the ppm formula, both solute and solution must be in the same units (both in g or both in kilograms etc.). In % by mass formula, both also must be in g or kg, etc. In short, **molarity**, **percent by volume**, **percent by mass** and **parts per million (ppm)** can **describe** the **concentration of** a **solution**.

Do Homework Questions #10, page 13, and #22-24, page 14.

CONCENTRATION OF IONS AFFECTS HOW WELL SOLUTIONS CONDUCT ELECTRICITY

The higher the concentration of ions (the more ions), the better the solution conducts electricity. 1M NaCl breaks up into 1 mole of Na^+ ions and 1 mole of Cl^- ions in 1 liter of solution. In comparison, 5M NaCl breaks up into 5 moles of Na^+ ions and 5 moles of Cl^- ions in 1 liter of solution, which is a **higher concentration** of **ions,** and therefore is a **better conductor** of electricity.

0.1 M NaCl (or NaOH) only breaks up into 0.1 mole of Na^+ and 0.1 mole of Cl^- (or OH^-), a **low concentration** of **ions**, and therefore is a **poor conductor** of electricity.

Note: You will learn later that carbon (C) compounds such as CH_3OH do not break up into ions and do not conduct electricity.

BOILING POINT ELEVATION

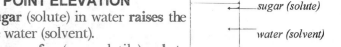
sugar (solute)
water (solvent)

The presence of **sugar** (solute) in water **raises the boiling point** of the water (solvent).

RULE: The presence of a (nonvolatile) **solute (salt or sugar) raises the boiling point of the solvent.** The greater the concentration of solute (more particles per liter of solution), the more it raises the boiling point. The more sugar (the more concentration of particles), the more it would raise the boiling point.

The boiling point of water is 100°C. One mole of dissolved particles in a kilogram of water raises the boiling point of water 0.52°C. If you add 1 mole of $C_6H_{12}O_6$ (glucose), which gives only one mole of dissolved particles ($C_6H_{12}O_6$ does not break up into two or more particles), to a kilogram of water, it raises the boiling point of water 0.52°C.

Molal boiling point elevation (for water) = 0.52°C. (Boiling point **elevates**, rises, goes up, by 0.52°C for each mole of particles in a kilogram of water.) 1 mole of particles in a kilogram of water elevates or raises the boiling point by 0.52°C.

If you add 1 mole of KCl, an ionic substance, it dissociates or breaks up into 1 mole of K^+ and 1 mole of Cl^-. Now you have 2 moles of dissolved particles (ions). Since KCl now has 2 moles of dissolved particles (K^+ and Cl^-), it raises the boiling point by 2 x 0.52°C = 1.04°C.

One mole of K_2SO_3 breaks up into 2 moles of K^+ and 1 mole of SO_3^{2-} (see Table E, polyatomic ions). You have 3 moles of dissolved particles (ions); the boiling point rises (goes up, increases) by 3 x 0.52°C = 1.56°C.

If 1 mole of K_2SO_3 produces 3 moles of dissolved particles, then 2 moles of K_2SO_3 produces 2 x 3 moles of dissolved particles = 6 moles of dissolved particles; it raises the boiling point by 6 x 0.52°C = 3.12 °C.

How many degrees the boiling point is raised is NOT part of the NYS Regents curriculum, but you must know that the more moles of dissolved particles, the higher the boiling point.

FREEZING POINT DEPRESSION

The presence of **sugar** (solute) in water **lowers the freezing point** of water (solvent).

RULE: The presence of any solute (salt or sugar) lowers the freezing point of the solvent. The greater the concentration of solute (more particles per liter of solution) the more it lowers the freezing point. The more sugar (the more concentration of particles) the more it lowers the freezing point.

The freezing point of water is 0°C. One mole of dissolved particles in one kilogram of water lowers the freezing point by 1.86°C. If you add one mole of $C_6H_{12}O_6$ (glucose), which gives only one mole of dissolved particles, to a kilogram of water, it lowers the freezing point by 1.86°C.

The molal freezing point depression (for water) is 1.86°C. The word "depression" means depress, go down. The freezing point goes down 1.86°C for each mole of dissolved particles in a kilogram of water.

If you add one mole of KCl, an ionic substance, it dissolves and breaks up into one mole of K^+ and one mole of Cl^-. Now you have two moles of dissolved particles. Since KCl has two moles of dissolved particles, K^+

and Cl⁻, it lowers the freezing point by 2 x 1.86°C = 3.72°C. Since the freezing point of water is 0°C, the freezing point of the KCl solution (KCl in water) is -3.72°C.

Three moles of dissolved particles (from 1 mole of K_2SO_3) lower the freezing point by 3 x 1.86°C = 5.58°C. Five moles of dissolved particles in a kilogram of water lower the freezing point by 5 x 1.86°C = 9.30°C. How many degrees the freezing point is lowered is NOT part of the NYS Regents curriculum, but you must know that the more moles of dissolved particles, the lower the freezing point.

Question: How are the boiling and freezing points of a sample of water affected when a salt is dissolved in the water?

    (1) boiling point decreases and freezing point decreases
    (2) boiling point decreases and freezing point increases
    (3) boiling point increases and freezing point decreases
    (4) boiling point increases and freezing point increases

Solution: Answer 3. When a solute (example: salt) is added to a solvent (example: H_2O), it causes the boiling point of the solvent to increase and the freezing point of the solvent to decrease.

Question: The freezing point of a 1.0 molal solution of $C_2H_4(OH)_2$ is closest to
    (1) +1.86°C   (2) -1.86°C   (3) -3.72°C   (4) +3.72°C

Solution: Answer 2. A **1 molal solution** means **1 mole of solute in a kilogram** of solvent. In this example, 1 mole of $C_2H_4(OH)_2$ has only **1 mole of particles**. Compounds with carbon generally do not break up into ions. The molal freezing point depression is 1.86°C. The freezing point of water is 0°C. One mole of solute $C_2H_4(OH)_2$ lowers the freezing point by 1.86°C. Therefore, the freezing point of the $C_2H_4(OH)_2$ solution is -1.86°C.

Similarly, if a nonvolatile substance, solute (example salt) is added to ethanol (solvent), it raises the boiling point and lowers the freezing point of ethanol (solvent) or the solution (ethanol and salt).

salt (solute)
Ethanol (solvent)
make up a solution

SUMMARY: The **addition** of a nonvolatile **solute** (sugar, salt) **raises** the **boiling point** and **lowers** the **freezing point** of the solvent or you can say the solution. The more solute (salt, sugar) added, the higher the boiling point and the lower the freezing point.

Try Sample Questions #5-7, on page 12, and then do Homework Questions, #11-15, pages 13-14.

REMEMBER: When you answer Regents questions on SOLUTIONS. use Table F. Table G. and Table T.

SAMPLE REGENTS & REGENT-TYPE QUESTIONS AND SOLUTIONS

1. A solution in which the crystallizing rate of the solute equals the dissolving rate of the solute must be
   (1) saturated   (2) unsaturated   (3) concentrated   (4) dilute

2. When sodium chloride is dissolved in water, the resulting solution is classified as a
   (1) heterogeneous compound   (3) heterogeneous mixture
   (2) homogeneous compound   (4) homogeneous mixture

3. According to Reference Table G, which compound's solubility decreases most rapidly as the temperature changes from 10°C to 70°C?
   (1) NH_4Cl (2) NH_3 (3) HCl   (4) KCl

4. Based on the full Reference Table F, on page Reference Tables 6, which of the following saturated solutions would be the *least* concentrated?
   (1) sodium sulfate   (3) copper (II) sulfate
   (2) potassium sulfate   (4) barium sulfate

5. When ethylene glycol (an antifreeze) is added to water, the boiling point of the water
   (1) decreases, and the freezing point decreases
   (2) decreases, and the freezing point increases
   (3) increases, and the freezing point decreases
   (4) increases, and the freezing point increases

6. How many moles of a nonvolatile, nonelectrolyte solute are required to lower the freezing point of 1,000 grams of water by 5.58°C?
   (1) 1   (2) 2   (3) 3   (4) 4

7. A 0.100-molal aqueous solution of which compound has the *lowest* freezing point?
   (1) $C_6H_{12}O_6$ (2) CH_3OH (3) $C_{12}H_{22}O_{11}$ (4) NaOH

SOLUTIONS:

1. Answer *1*. In a saturated solution, the rate of crystallizing equals the rate of dissolving.

2. Answer *4*. You learned solutions (example sodium chloride solution, salt solutions) are homogeneous mixtures.

3. Answer *2*. Look at Table G, on page Reference Tables 8:

Substance	NH_3	HCl	NH_4Cl	KCl
Solubility at 10°C, grams	70	77	33	30
Solubility at 70°C, grams	18	52	62	48
Decrease in solubility, grams	52	25	increases	increases

   Solubility decreases most rapidly for NH_3.

4. Answer *4*. Look at Table F on page Reference Tables 6. The left box lists sulfates, SO_4^{2-}, as soluble. Choices 1,2, and 3, sodium sulfate, potassium sulfate, and copper sulfate are all soluble: salt dissolves nicely (a lot of salt) in water. Now look at barium sulfate. The left box lists sulfates as soluble, except, when combined with Ba is **insoluble**. Almost nothing dissolves. The solution is least concentrated.

5. Answer 3. A solute such as ethylene glycol raises (increases) the boiling point and lowers (decreases) the freezing point of the solvent it is in.

6. Answer 3. The molal freezing point depression (for H_2O) = 1.86°C. (The word "depression" means depress, go down, the temperature goes down.)1 mole of particles in a kilogram of H_2O lowers the freezing point by 1.86°C. Since the water was lowered by 5.58°C, divide 5.58°C by 1.86°C to know the number of moles. Number of moles = $\frac{5.58}{1.86}$ = 3 moles.

7. Answer 4. 1 mole of dissolved particles in a kilogram of H_2O lowers the freezing point. 2 moles of dissolved particles lower the freezing point by twice as much. Choice 1, Choice 2, and Choice 3 each has 1 mole of dissolved particles. But Choice 4, NaOH (an ionic substance) breaks up into two particles, Na^+ and OH^-, and lowers the freezing point even more (twice as much). NaOH solution has the lowest freezing point.

NOW LET'S TRY A FEW HOMEWORK QUESTIONS:

1. Which solution is the most concentrated?
   (1) 1 mole of solute dissolved in 1 liter of solution
   (2) 2 moles of solute dissolved in 3 liters of solution
   (3) 5 moles of solute dissolved in 4 liters of solution
   (4) 4 moles of solute dissolved in 8 liters of solution

2. A solution in which the crystallizing rate of the solute equals the dissolving rate of the solute must be
   (1) saturated   (2) unsaturated (3) concentrated (4) dilute

3. According to Reference Table G, approximately how many grams of $KClO_3$ are needed to saturate 100 grams of H_2O at 40°C?
   (1) 6   (2) 16   (3) 38   (4) 47

4. According to Reference Table G, a temperature change from 60°C to 90°C has the *least* effect on the solubility of
   (1) SO_2 (2) NH_3 (3) KCl   (4) $KClO_3$

5. According to Reference Table G, which compound's solubility decreases most rapidly when the temperature increases from 50°C to 70°C?
   (1) NH_3 (2) HCl   (3) SO_2 (4) KNO_3

6. Based on full Reference Table F, on page Reference Tables 6-7, which of the following saturated solutions would be the *least* concentrated?
   (1) sodium sulfate   (3) copper (II) sulfate
   (2) potassium sulfate   (4) barium sulfate

7. What is the concentration of a solution of 10 moles of copper (II) nitrate in 5.0 liters of solution?
   (1) 0.50 M   (2) 2.0 M   (3) 5.0 M   (4) 10 M

8. What is the total number of moles of H_2SO_4 needed to prepare 5.0 liters of a 2.0 M solution of H_2SO_4?
   (1) 2.5   (2) 5.0   (3) 10   (4) 20

9. What is the molarity of a KF(aq) solution containing 116 grams of KF in 1.00 liter of solution?
   (1) 1.00 M   (2) 2.00 M   (3) 3.00 M   (4) 4.00 M

10. The formula for percent by volume of alcohol in water is
    (1) $\dfrac{mL \ solute}{100 \ mL \ of \ solution}$ (3) $\dfrac{mL \ solvent}{100 \ mL \ of \ solution}$
    (2) $\dfrac{mL \ solution}{100 \ mL \ of \ solvent}$ (4) $\dfrac{mL \ solute}{100 \ mL \ of \ solvent}$

11. The freezing point of a 1.00-molal solution of $C_2H_4(OH)_2$ is closest to
    (1) +1.86°C   (2) -1.86°C   (3) -3.72°C   (4) +3.72°C

---

12. Why is salt (NaCl) put on icy roads and sidewalks in the winter?
    (1) It is ionic and lowers the freezing point of water.
    (2) It is ionic and raises the freezing point of water.
    (3) It is covalent and lowers the freezing point of water.
    (4) It is covalent and raises the freezing point of water.

13. How are the boiling and freezing points of a sample of water affected when salt is dissolved in the water?
    (1) The boiling point decreases and the freezing point decreases.
    (2) The boiling point decreases and the freezing point increases.
    (3) The boiling point increases and the freezing point decreases.
    (4) The boiling point increases and the freezing point increases.

14. Compared to the normal freezing point and boiling point of water, a 1-molal solution of sugar in water will have a
    (1) higher freezing point and a lower boiling point
    (2) higher freezing point and a higher boiling point
    (3) lower freezing point and a lower boiling point
    (4) lower freezing point and a higher boiling point

15. Which solution containing 1 mole of solute dissolved in 1,000 grams of water has the lowest freezing point?
    (1) $KOH(aq)$ (2) $C_6H_{12}O_6(aq)$ (3) $C_2H_5OH(aq)$ (4) $C_{12}H_{12}O_{11}(aq)$

CONSTRUCTED RESPONSE QUESTIONS: Parts B-2 and C of NYS Regents Exam

Questions 3, 7, 8, 9, 12, 13, without giving the 4 choices, can also be used here.

16. Explain the difference between a saturated, unsaturated, and supersaturated solution.

17. Draw two solubility graphs, based on the data in the two experiments:

Temp, °C	$KClO_3$, g, in 100 g H_2O	Temp, °C	$KClO_3$, g, in 100 g H_2O	Temp, °C	NH_3, g, in 100 g H_2O	Temp, °C	NH_3, g, in 100 g H_2O
0	5	30	12	0	90	30	45
10	7	40	16	10	70	40	36
20	9	50	22	20	56	50	28

18. How do temperature, pressure, and nature of solute and solvent affect solubility?

19. How do you make a 3M solution of KBr? How many moles of KBr are in 1 liter of a .03 M solution?

20. What is the concentration of 20 moles of NaCl in 10 liters of solution?

21. What is the molarity of a $CaCl_2$ solution containing 330 grams of $CaCl_2$ in 1 liter of solution?

22. A solution contains 5 grams of KNO_3 in 100 grams of solution. What is the percent by mass of KNO_3?

23. What is the concentration of a solution in parts per million if 10 mg of LiF are dissolved in 5000 grams of solution?

24. A. What is the concentration of a solution in parts per million if 20 grams of Na_2S is dissolved in 40000 grams of solution?

    B. An aqueous solution has 0.0070 gram of oxygen dissolved in 1000. grams of water. Calculate the dissolved oxygen concentration of this solution in ppm.

CHAPTER QUESTION: Parts B-2 and C of NYS Regents Exam

25. A student observed the following reaction:
    $$AlCl_3(aq) + 3NaOH(aq) \longrightarrow Al(OH)_3(s) + 3NaCl(aq)$$
    After the products were filtered, which substance remained on the filter paper and why?

26. Which two solutions, when mixed together, will undergo a double replacement reaction and form a white, solid substance?
    Hint: Review double replacement reactions, page 5:22
    (1) $NaCl(aq)$ and $LiNO_3(aq)$ (3) $KCl(aq)$ and $LiCl(aq)$
    (2) $KCl(aq)$ and $AgNO_3(aq)$ (4) $NaNO_3(aq)$ and $AgNO_3(aq)$

CHAPTER 7:
KINETICS / EQUILIBRIUM

SECTION A

KINETICS is concerned with **rates** of reaction (**how fast**, how many moles are consumed or produced in a unit of time) and **mechanism** (**steps** in a chemical reaction).

COLLISION THEORY

For a **reaction** to **happen**, the (reactant) **particles** must have **effective collisions**, which means particles must have **enough energy** and **collide** at the correct angles (**proper orientation**). The collision theory explains the factors that affect the rate of reaction (see below).

FACTORS THAT AFFECT RATE OF REACTION

1. **CONCENTRATION**: When the concentration of one or more of the reactants increases, the rate of reaction increases. (More concentration of reactant means more frequent collisions.) With gases, an increase in pressure increases concentration (more frequent collisions) and increases the rate of reaction.

2. **TEMPERATURE**: Increase in temperature increases the rate of reaction. Increase in temperature means more kinetic energy, particles move faster, and therefore more collisions (with more force) and more effective collisions occur.

3. **SURFACE AREA**: Increasing the surface area increases the rate of reaction. The larger the surface area, the more exposed particles that can react, the more collisions and the faster the rate of reaction. Powdered Mg (magnesium) in HCl (hydrochloric acid) will react faster than a solid piece of Mg will, because powdered Mg has more surface area.

   Note: Similarly, higher temperature (more energy) or increasing surface area (powdered sugar instead of a sugar cube) increases or speeds up the rate of dissolving sugar in water (the sugar dissolves faster).

4. **NATURE OF REACTANTS**: A reaction that involves the smallest (least) amount of bond rearrangement (breaking and making new bonds) is fast. Ionic substances react very fast (ions just move close to each other and can collide at any angle). Hydrogen + oxygen → water is a slower reaction because bonds are broken and new ones are formed. (The particles must collide at the correct angle.)

5. **CATALYST**: A catalyst lowers the activation energy (amount of energy needed to start a reaction) and gives a faster rate of reaction. Particles need less energy to react (to have effective collisions). Therefore, more of the particles can now react.

SUMMARY:

Ionic compounds
More concentration
More pressure (gas)
High temperature ————→ Faster rate of reaction
More surface area
Catalyst

Question: At room temperature, which reaction would be expected to have the fastest reaction rate?

(1) $Pb^{2+} + S^= \rightarrow PbS$ (3) $N_2 + 2O_2 \rightarrow 2NO_2$

(2) $2H_2 + O_2 \rightarrow 2H_2O$ (4) $2KClO_3 \rightarrow 2KCl + 3O_2$

Solution: Answer *1*. Ionic reactions are the fastest. Pb^{2+} and $S^=$ just come very close together.

Question: Under which conditions will the rate of a chemical reaction always decrease?

(1) The concentration of the reactants decreases and the temperature decreases
(2) The concentration of the reactants decreases, and the temperature increases
(3) The concentration of the reactants increases, and the temperature decreases
(4) The concentration of the reactants increases, and the temperature increases

Solution: Answer *1*. If concentration decreases, rate of reaction decreases. If temperature decreases, rate of reaction decreases.

Try Sample Questions #1-2, on page 15, and then do Homework Questions, #1-8, pages 18-19, and #42-43, page 23.

HEAT OF REACTION:

Hydrogen + Oxygen → Water.

Each substance (hydrogen, oxygen, water) has its own potential energy (stored energy) or heat content.

Reactants	*Product*		
$2H_2 + O_2$ →	$2H_2O$	+	571.6 kJ
	Less heat content		*Heat is given off or*
	(than $H_2 + O_2$)		*energy is given off*

Equation to produce 1 H_2O: $H_2 + \frac{1}{2}O_2 \rightarrow H_2O + 285.8$ kJ.

To produce 2 (moles of) H_2O, 571.6 kJ are given off; therefore, to produce 1 (mole of) H_2O = ½ x 571.6 = 285.8 kJ are given off.

HEAT OF REACTION is the amount of heat given off or absorbed in a chemical reaction. Heat of reaction is the difference in heat content (potential energy) of the products and reactants.

$$\Delta H \quad = \quad H_{products} \quad - \quad H_{reactants}$$

Heat of	*Heat content*	*Heat content*
reaction	*of products*	*of reactants*

EXOTHERMIC REACTIONS

Exothermic reactions give off heat (energy) like in the example:
$$2H_2 + O_2 \longrightarrow 2H_2O + 571.6 \text{ kJ}.$$

If the products have less potential energy (PE) or heat content than the reactants (like in the example above), heat is given off and it is an **EXOTHERMIC REACTION. ΔH = negative**.

The energy diagram below has arbitrary numbers to help you understand the meaning of ΔH.

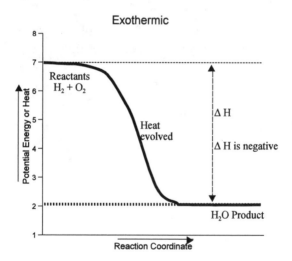

Look at the energy diagram.

$$\text{Heat of products, } H_{products} = 2$$
$$\text{Heat of reactants, } H_{reactants} = 7$$
$$\Delta H = H_{products} - H_{reactants}$$
$$\Delta H = 2 - 7 = -5$$

$\Delta H = -5$. **ΔH = -** (negative sign) means that it is an **EXOTHERMIC** reaction and **heat is given off**.

ACTIVATION ENERGY (spark, flame, high temperature, or in some cases, rubbing or striking two objects covered with chemicals together) is the smallest amount of energy needed to start a reaction.

Let's add **activation energy** to the previous example:

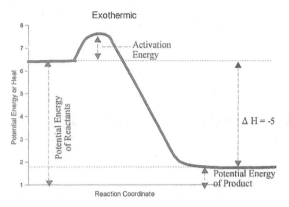

Exothermic

In this diagram, the activation energy is from the reactants to the top (the highest point) of the curve. In our example, $H_{REACTANTS}$, or potential energy of the reactants, is 7. Activation energy is therefore from 7 to the highest point of the curve.

CATALYST: Lowers the activation energy and makes a faster reaction.

Question: The potential energy diagram shown below represents the reaction A + B → AB. Which statement correctly describes this reaction?

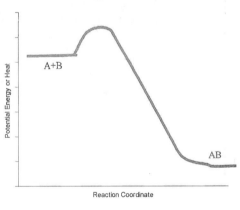

(1) It is endothermic and energy is absorbed.
(2) It is endothermic and energy is released.
(3) It is exothermic and energy is absorbed.
(4) It is exothermic and energy is released.

Solution: Answer *4*. In the diagram, when the **product is lower than the reactant** (product has less potential energy than the reactant), it is **an exothermic reaction** and energy is released.

Exothermic reaction: $2H_2 + O_2 \longrightarrow 2H_2O + 571.6\,kJ$. 571.6 kJ of heat after the arrow means 571.6 kJ of heat or energy is the product (produced) given off. By looking at the equation, you can tell it is an exothermic reaction because 571.6 kJ of **heat** or **energy after** the **arrow means energy** is **given off; exothermic reactions** give off energy.

HINT: Sometimes it is easier for the student to learn **endothermic** reactions after they know and remember exothermic reactions. If you need more time to study and know *exothermic* reactions, go on to the section on spontaneous reaction, and/or equilibrium, beginning on page **Chap 7**:9.

Try Sample Question #3, on page 15, and then do Homework Questions, #9-12, page 19, and #44, page 23.

ENDOTHERMIC REACTIONS

ENDOTHERMIC REACTIONS **take in** heat, like in the example:
$$N_2 + O_2 + 182.6 \text{ kJ} \longrightarrow 2NO.$$
Look at the potential energy diagram. In endothermic reactions, the reactants take in heat and form the products. In endothermic reactions, the products are higher up than the reactants, which mean the products have more heat or potential energy than the reactants.

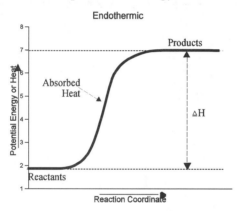

Look at the energy diagram above. As you can see in the diagram, the products are higher up (#7 for potential energy), than the reactants (#2), and therefore it is an endothermic reaction.

Let's add the activation energy to the diagram. Look at the diagram on the next page. The numbers represent:

① Potential energy of the reactants.
② Potential energy of the products.
③ $\Delta H = H_{\text{PRODUCTS}} - H_{\text{REACTANTS}}$.
④ Activation energy from the reactants to the highest point of the curve. The very top of the curve is called the **activated complex**.
⑤ Activation energy with catalyst. There is now an alternate pathway (another path) that has a lower activation energy that

the reactant can take to become product. You can see that the activation energy of the **catalyzed reaction** (with the catalyst) is **lower** than for the **uncatalyzed reaction**. The **catalyst lowers** the **activation energy**.

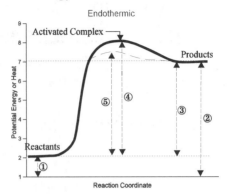

$\mathbf{\Delta H} = \mathbf{H}_{\text{products}} - \mathbf{H}_{\text{reactants}};$
$\mathbf{\Delta H} = 7 \quad -2 = +5;$ **ΔH is positive** in an **endothermic reaction**.

Questions:

1. Which interval represents the heat of reaction (ΔH)?
   (1) E   (2) F
   (3) C   (4) G

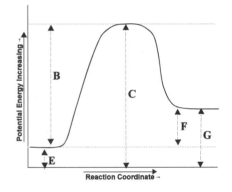

2. Interval B represents the
   (1) potential energy of the products
   (2) potential energy of the reactants
   (3) activation energy
   (4) activated complex

Solutions:

Question 1: Answer **2**. ΔH = H$_{\text{PRODUCTS}}$ − H$_{\text{REACTANTS}}$
    ΔH = G (in diagram) – E (in diagram) = F (in diagram).
Question 2: Answer **3**. Activation energy. From reactants to the highest point on the curve.

Endothermic reaction: $N_2 + O_2 + 182.6 \text{ kJ} \longrightarrow 2NO$.   182.6 kJ of heat before the arrow means 182.6 kJ of heat or energy is taken in; endothermic reactions take in energy.

Try Sample Questions #4-5, on page 15, and then do Homework Questions, #13-18, page 20 and #45, page 23.

Energies of Formation

This table is NOT part of the New York State Regents curriculum		
Compound	Heat of Formation ΔH_f *(kJ)*	
Aluminum oxide, Al_2O_3	-1675.7	
Calcium hydroxide, $Ca(OH)_2$	-985.2	
Hydrogen iodide, HI	26.5	
* minus sign indicates an exothermic reaction		

In the table above, ΔH_f tells you how much heat (how many kilojoules) is given off or absorbed when one mole of the compound is formed from its elements. Example: $2Al + 3/2O \longrightarrow_2 \quad Al_2O_3$, ΔH_f = -1675.7 kJ.

Look at the ΔH_f column in the table. As you know, ΔH_f = – (**minus, or negative, sign**) means it is an **exothermic reaction**. Forming aluminum oxide has ΔH_f = -1675.7, therefore it is exothermic. Forming calcium hydroxide has ΔH_f = -985.2, therefore it is also exothermic. The bottom of the table reminds you that a minus sign indicates an exothermic reaction.

As you learned, ΔH_f = + (**positive sign**) is an **endothermic reaction**. Forming hydrogen iodide has ΔH_f = 26.5, therefore it is endothermic.

REMINDER: ΔH_f = 26.5 means ΔH_f = **+26.5** (no sign in front of a number means "+").

TABLE I: HEATS OF REACTION

Now look at Table I below, and on **page Reference Tables 10.**

Part of Table I

Heat of Reaction at 101.3 kPa and 298 K	
Reaction	ΔH *(kJ)**
$CH_4(g) + 2O_2(g) \rightarrow CO_2(g) + 2H_2O(\ell)$	-890.4
$2CH_3OH(\ell) + 3O_2(g) \rightarrow 2CO_2(g) + 4H_2O(\ell)$	-1452
$NaOH(s) \xrightarrow{H_2O} Na^+(aq) + OH^-(aq)$	-44.51
$NH_4Cl(s) \xrightarrow{H_2O} NH_4^+(aq) + Cl^-(aq)$	+14.78
$N_2(g) + O_2(g) \rightarrow 2NO(g)$	+182.6

*The ΔH values are based on molar quantities represented in the equations.
A minus sign indicates an exothermic reaction.

In Table I, ΔH tells you how much heat (how many kilojoules) is given off or absorbed in the reaction.

Look at the ΔH column in Table I. As you know, ΔH = –(**minus or negative** sign) means it is an **exothermic reaction**. The reaction CH_4 + 2 $\longrightarrow O_2$ CO_2 + 2 H_2O has ΔH = -890.4kJ, therefore it is exothermic (gives off heat).

The reaction $2CH_3OH + 3O_2 \longrightarrow 2 CO_2 + 4H_2O$ has ΔH = -1452 kJ, therefore it is also an exothermic reaction.

TABLE I

The reaction of CH_3OH (see above) has **ΔH** a **bigger negative number** (-1452 kJ), therefore it **gives off more heat**.

The bottom of Table I reminds you that a minus sign indicates an exothermic reaction.

Look at Table I. You know ΔH = + (positive sign) is an endothermic reaction. The reaction $N_2(g) + O_2(g) \rightarrow 2NO(g)$ has ΔH = +182.6 kJ, therefore it is an endothermic reaction (heat is absorbed). Dissolving NH_4Cl has ΔH = + 14.78 kJ. The reaction $N_2 + O_2$ has **ΔH** a **bigger positive number** (+182.6 kJ), therefore it **absorbs more heat**.

SUMMARY:

ΔH negative means an exothermic reaction. Heat is given off, products have less energy than reactants.

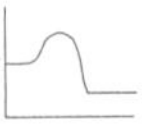

ΔH positive means an endothermic reaction. Heat is absorbed, products have more energy than reactants.

Question: According to Reference Table **I**, in which reaction do the products have a higher energy content than the reactants?

    (1) $CH_4(g) + 2O_2(g) \rightarrow CO_2(g) + 2H_2O(\ell)$
    (2) $2CH_3OH(\ell) + 3O_2(g) \rightarrow 2CO_2(g) + 4H_2O(\ell)$
    (3) $NH_4Cl(s) \xrightarrow{H_2O} NH_4^+(aq) + Cl^-(aq)$
    (4) $NaOH(s) \xrightarrow{H_2O} Na^+(aq) + OH^-(aq)$

Solution: Answer **3**. The question states that the products have higher energy than the reactants, therefore, it is an endothermic reaction and ΔH is positive. In Table I, Answer 3, $NH_4Cl \rightarrow NH_4^+$ + Cl^- has ΔH = +14.78. For all the other choices, ΔH is negative. Remember the diagram of an endothermic reaction. The products have more energy than the reactants, and ΔH is positive.

Find heat released when 1 mole of Al_2O_3 is formed from its elements. The compound that is formed (product, example Al_2O_3) is written after the arrow in an equation. Look for aluminum oxide (Al_2O_3) after the arrow on **Table I, page reference tables 10 (back of book)**.

        $4Al(s) + 3O_2(g) \rightarrow 2Al_2O_3(s)$. ΔH = - 3351 kJ.

ΔH = - means an exothermic reaction. Exothermic reactions give off heat. To produce (form) only 1 mole of Al_2O_3, ΔH = ½(-3351 kJ) = -1676 kJ.

Answer: To form 1 mole of Al_2O_3, 1676 kJ is given off (released).

You learned in chapter 1 that heat travels from higher temperature to lower temperature until both temperatures are the same. If a piece of hot metal at 100°C is put into cool distilled water at 25°C, the metal loses heat to the water. Heat lost by the metal equals heat gained by the water. **Do chapter question #52 on page 7:23.**

TABLE I *(side label)*

Exothermic Reaction: When NaOH dissolves in water, NaOH (or some salts, such as LiBr) gives off heat (exothermic reaction). Heat goes from NaOH (or from LiBr) to water, which causes the water to get hotter.

Endothermic Reaction: When NH_4Cl, ammonium chloride (or some other salts) dissolves in water, heat goes from the water to the NH_4Cl. NH_4Cl takes in heat from the water (endothermic reaction) and causes the water to get colder. This reaction can be used in a cold pack; mixing the chemical and water makes the pack cold.

Note: If your hand holds a test tube of NH_4Cl in water or a cold pack, heat goes from your hand (from higher temperature) to the test tube (lower temperature).

Try Sample Question #6, on page 16, then do Homework Questions, #19-20, pages 20-21, and #46, page 23.

SPONTANEOUS REACTION

A **SPONTANEOUS REACTION** is one that takes place under a specific set of conditions.

Spontaneous reactions occur in the direction of:

1. **Less energy (lower enthalpy)**: This favors exothermic reactions and
2. **Greater entropy** (randomness, disorder).

   A. Solids have the least entropy, liquids have more, and gases have the most entropy (disorder).

   B. When a solid (examples: salts-NaCl, NH_4Cl; or sugar) dissolves in water, entropy increases.

At low temperature, energy is important, and at high temperature, entropy is important.

FREE ENERGY CHANGE (ΔG)

To determine if a reaction is spontaneous, use the free energy change equation. **MEMORIZE this equation**:

$$\Delta G = \Delta H - T\Delta S.$$

If **ΔG is negative**, the reaction is **spontaneous**.

ΔG = free energy change   ΔH = heat of reaction
ΔS = change of entropy   T = Kelvin temperature.

REMEMBER: ΔG = negative means the reaction is **spontaneous**. (Obviously, if ΔG is positive, the reaction is not spontaneous.)

Clearly, if T is high (a large number), TΔS is a big number, and ΔS, entropy is important in determining ΔG.

If temperature is a small number, TΔS is a small number and ΔS, entropy, is not important in determining ΔG. Then, ΔH would be more important in determining ΔG.

REMEMBER: ΔG = zero at equilibrium.

Question: Which equation correctly represents the free energy change in a chemical reaction?

(1) $\Delta G = \Delta H + T\Delta S$ (3) $\Delta G = \Delta T - \Delta H \, \Delta S$

(2) $\Delta G = \Delta H - T\Delta S$ (4) $\Delta G = \Delta S - T\Delta H$

Solution: Answer *2*. $\Delta G = \Delta H - T\Delta S$.

Look at the ΔG column in the Table of Standard Energies of Formation of Compounds, below or on page Additional Tables 1:

Standard Energies of Formation of Compounds

Compound	ΔH *(kJ)*	ΔG *(kJ)*
Aluminum oxide, Al_2O_3		−1582.3
Calcium hydroxide, $Ca(OH_2)$		−897.5
Hydrogen iodide, HI		1.7

You know that, **if ΔG is negative, the reaction is spontaneous**. In forming aluminum oxide, $\Delta G = -1582.3$; therefore the reaction is spontaneous. In forming calcium hydroxide, $\Delta G = -897.5$; therefore the reaction is spontaneous.

You know that, **if ΔG is positive, the reaction is not spontaneous**. In forming hydrogen iodide $\Delta G = 1.7$ (meaning $+1.7$); therefore the reaction is not spontaneous.

Question: According to the Table of Standard Energies of Formation of Compounds, which compound will form spontaneously from its elements?

(1) ethene   (3) nitrogen (II) oxide

(2) hydrogen iodide   (4) magnesium oxide

Solution: **Answer *4*.** Look at the ΔG column in the table.

If ΔG is negative, the reaction is spontaneous.

MgO, has $\Delta G = -569.3$ and therefore the reaction is spontaneous.

Standard Energies of Formation of Compounds

Compound	ΔH	ΔG
Ethene		68.4
HI		1.7
Nitrogen (II) oxide		87.6
MgO		−569.3

Try Sample Questions #7-11, on page 16, and then do Homework Questions, #21-30, page 21, and #44, page 23.

EQUILIBRIUM

Equilibrium is when the forward and reverse reactions occur at the **same rate**. The **rates are equal** but the quantities (**amounts**) of reactants and products are **not necessarily** equal.

I. PHYSICAL EQUILIBRIUM: Rates of forward and reverse reactions are equal for a physical change. (You learned that in a physical change, there is only a change in appearance, but the substance is not changed; example: ice ⟶ water.)

1. **Phase Equilibrium**

   A. **Liquid – Gas**

   In a covered beaker of water, some water becomes water vapor (evaporation) and some water vapor becomes water (condensation). When water evaporates (1) (goes up) at the **same rate** as it condenses (2) (goes down), equilibrium exists.

   Phase equilibrium (dynamic equilibrium): Rate of liquid to gas equals rate of gas to liquid.
   Equilibrium vapor pressure is the pressure caused by the water vapor at equilibrium.

   B. **Solid – Liquid**

   Phase equilibrium is also when the rate of solid (example: ice) to liquid (example: water) equals rate of liquid to solid.

   Do Sample Question #14 on page 7:16

2. **SOLUTION EQUILIBRIUM**

   A. **Solids in Liquids**

   Some sugar on the bottom (crystallized sugar) dissolves at the **same rate** as some dissolved sugar crystallizes (goes to bottom). At equilibrium, the rate at which sugar crystallizes equals the rate at which sugar dissolves. When these rates are equal, it is called a **saturated solution**.

   Solubility of a solute (sugar, salt) is the mass of solute (how much solute) that dissolves in a given volume of solvent (H_2O, etc.) at equilibrium. Or, solubility is the concentration of solute in a saturated solution (the most it can hold).

   B. **Gas in Liquid**

   Equilibrium is when the **rate** that (1) gas dissolved in liquid goes up into the space above the liquid **equals** the **rate** that (2) undissolved gas on top goes into the liquid. **Equilibrium is disturbed** by **changes** in **temperature** or **pressure**. Increase in temperature decreases the solubility of gas. (When water is heated, gas bubbles form on the top of the water, as dissolved air goes out. High temperature decreases solubility.) Increase in pressure (high pressure) makes gases more soluble.

II. CHEMICAL EQUILIBRIUM is when the rates of the forward and reverse reactions are equal in a chemical reaction (chemical change). (In a chemical change, a new substance is produced). At equilibrium, the concentration of the reactants and products of the reaction remains constant.

$$2CO(g) + O_2(g) \underset{reverse}{\overset{forward}{\rightleftharpoons}} 2CO_2(g) \quad \text{(g means gas)}$$

To have equilibrium, we need a **closed container** so nothing can enter or leave (examples O_2 gas, CO_2 gas, water vapor, Cl_2 gas, see above.)

Question: Which is a property of a reaction that has reached equilibrium?
(1) The amount of products is greater than the amount of reactants.
(2) The amount of products equals the amount of the reactants.
(3) The rate of the forward reaction is greater than the rate of the reverse reaction.
(4) The rate of the forward reaction is equal to the rate of the reverse reaction and the reactions continue going on.

Solution: Answer **4**. At equilibrium, the rates of the forward and reverse reactions are equal; the forward reaction and the reverse reaction continue going on.

Try Sample Questions #12-14, on page 16, and then do Homework Questions, #31-33, pages 21-22, and #48, page 23.

LECHATELIER'S PRINCIPLE

If a system at equilibrium is **subjected to a stress**, the equilibrium will shift in the direction that relieves the stress.

1. **CHANGE IN CONCENTRATION**

$$\underset{\substack{\text{(Reactants)} \\ A + B}}{\text{Left Side}} \rightleftharpoons \underset{\substack{\text{(Products)} \\ C + D}}{\text{Right Side}}$$

   a. An **increase** in the concentration of *A* **causes** the reaction to **go** to the **right** (shifts the equilibrium to the right) in order to reduce the stress caused by too much *A*. *A* and *B* are used up faster, and more *C* and *D* are formed. At equilibrium, there a higher concentration (more) of *C* and *D* and a **lower** concentration of *B*.
   b. An **increase** in the concentration of *B* drives the **reaction** to the **right**.
   c. An **increase** in the concentration of *C* or *D* shifts equilibrium (causes reaction to **go**) to the **left** to reduce the stress. An increase in the concentration of *C* shifts the equilibrium to the left, decreases the concentration of *D*, and produces more *A* and *B*.

SUMMARY: An **increase in concentration of anything on the left side** (*A* or *B*) **causes the reaction to go to the right side**. An increase in concentration of anything on the right side (*C* or *D*) causes the reaction to go to the left side.

You learned, increase in concentration of A or B causes the reaction to go to the right, therefore obviously, decrease in concentration of A or B causes the reaction to go to the left (opposite way), producing more A and B and less (lower concentration of) C and D.

Note: An increase in concentration of A or B causes more collisions (see page 7:1) and more C and D are formed.

Question: : In the reaction $2SO_2 + O_2 \longrightarrow 2SO_3$ + energy, state how the equilibrium shifts when SO_3 is removed from the system.

Solution: When SO_3 (right side of the equation) is removed from the system, there is no stress (nothing) on the right side of the system, while there is stress on the left side, therefore the reaction goes to the right (shifts the equilibrium to the right, or you can say, equilibrium shifts to the right) to remove the stress.

---

2. **CHANGE OF PRESSURE** has an effect only on **gases**. If the **pressure** on an equilibrium system is **increased**, the reaction is driven in the direction which relieves the pressure (stress), in the **direction** of less gas molecules (**smaller number of gas molecules**).

In the **Haber process** for making ammonia ($NH_3(g)$, (g) means gas):

Left Side		Right Side
Reactant		Product
$N_2(g) + 3H_2(g)$	$\rightleftarrows$	$2NH_3(g)$
1 molecule N_2 + 3 molecules H_2	$\rightarrow$	2 molecules NH_3
4 molecules on left side		2 molecules on right side
		Smaller number of gas molecules is on right side (2 is less than 4)

Therefore, when there is **more pressure**, the reaction **goes** to the **right** (**smaller number of gas molecules**; right side only has two gas molecules). More NH_3 is produced and there is less N_2 and H_2 (because N_2 and H_2 are used to make NH_3).

In the reaction

Left Side		Right Side
$H_2(g) + Cl_2(g)$	$\rightleftarrows$	$2HCl(g)$
2 gas molecules	$\rightarrow$	2 gas molecules

Pressure has **no effect**, because there is the **same number of gas molecules on both sides**.

3. **CHANGE IN TEMPERATURE:** The **addition of heat**, according to LeChatelier's Principle, **shifts** equilibrium so that **heat is absorbed** (to relieve the stress). This **favors** the **endothermic reaction** (takes in heat). Removal of heat (cooling) favors the exothermic reaction.

In this reaction,

Reactant (left side)		Product (right side)
$H_2(g) + I_2(g)$ + heat (or energy)	$\rightleftarrows$	$2HI(g)$

Heat is a reactant (left side). Increase in temperature (heat, energy, or number of kJ) is like increasing the amount of reactant (left side), which causes reaction to go to the right, producing more HI. Therefore, obviously, decrease in temperature (opposite of increase in temperature) causes reaction to go to the left (opposite way), producing more H_2 and I_2.

In the reaction,

Reactant (left side)   Product (right side)
$$N_2 + 3H_2 \rightleftharpoons 2NH_3 + 91.8 \text{ kJ}$$

heat (91.8 kJ) is a product (right side). Increasing temperature (heat or number of kJ) is like increasing the amount of product (right side), which causes the reaction to go to the left, producing more N_2 and H_2. Therefore, obviously, decreasing temperature (cooling) (opposite of increasing temperature) causes reaction to go to the right (opposite way) producing more NH_3.

4. **CATALYSTS**: A catalyst **speeds up** the **reaction** (speeds up the forward and reverse reactions equally), but produces **no change** in the (equilibrium) **concentration**.

COMMON ION EFFECT

$AgCl(s) \rightleftharpoons Ag^+ + Cl^-$. According to LeChatelier, if you **add** more Ag^+ or Cl^-, the **reaction** would be shifted **to the left**. If you **add** more NaCl (Na^+Cl^-) or KCl (K^+Cl^-), add the Cl^- ions to the right side (after the arrow); the Cl^- ions unite with the Ag^+, forming more AgCl. An **increase** in the concentration of Cl^- **causes** a **decrease** in Ag^+ and an **increase** in **AgCl**.

In general, when an ionic compound like AgCl breaks up into ions:

Reactant (left side)   Products (right side)

Equation: AgCl(s) $\rightleftharpoons$ $Ag^+ + Cl^-$

an **increase** in one of the products, (such as Cl^-) causes the other product (this example: Ag^+) to **decrease** and the reactant (this example: AgCl) to **increase**.

REACTIONS GO TO COMPLETION

Some reactions go to completion (reaction in one direction only: Reactants form products, products do not form reactants).

Reactions go to completion because **(1)** a **GAS**, **(2)** an essentially un-ionized product (such as **WATER**), or **(3)** a **PRECIPITATE** (something insoluble) is one of the products:

EXAMPLES:

1. **Gas:** $H_2CO_3(aq) \rightarrow H_2O(\ell) + CO_2(g)$ (g means gas, ℓ means liquid)
(aq means dissolved in water)

2. **Water:** $HCl + NaOH \rightarrow NaCl + H_2O(\ell)$

3. **Precipitate** (means something insoluble):

$$NaCl(aq) + AgNO_3(aq) \rightarrow NaNO_3(aq) + AgCl(s) \quad \text{(s means solid)}$$

Look at **Table F,** page Reference Tables 6. The left box lists halides (Cl^- Br^-, I^-) as soluble, **but** when combined with Ag^+, it is **insoluble.** Therefore, you know **AgCl is insoluble.** (Chap. 6:5 explains how to use Table F.)

Try Sample Questions #15-17, on page 16, and then do Homework Questions, #34-41, pages 22-23, and #49-50, page 23.

REMEMBER: When you answer Regents questions on KINETICS/EQUILIBRIUM, use Table I and Table F.

SAMPLE REGENTS & REGENT-TYPE QUESTIONS AND SOLUTIONS

---

1. Given the reaction:
$$Zn(s) + 2HCl(aq) \rightarrow ZnCl_2 + H_2(g).$$
The reaction occurs more slowly when a single piece of zinc is used than when the same mass of powdered zinc is used. Why does this occur?
(1) The powdered zinc is more concentrated.
(2) The powdered zinc has a greater surface area.
(3) The powdered zinc requires less activation energy.
(4) The powdered zinc generates more heat energy.

2. As the number of effective collisions between the reactant particles in a chemical reaction decreases, the rate of the reaction
(1) decreases　　(2) increases　　(3) remains the same

3. In a chemical reaction, a catalyst changes the
(1) potential energy of the products
(2) potential energy of the reactants
(3) heat of reaction
(4) activation energy

4. In the diagram at right, which arrow represents the activation energy for the forward reaction?
(1) A　(2) B　(3) C　(4) D

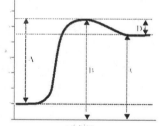

5. Which statement describes characteristics of an endothermic reaction?
(1) The sign of ΔH is positive, and the products have less potential energy than the reactants.
(2) The sign of ΔH is positive, and the products have more potential energy than the reactants.
(3) The sign of ΔH is negative, and the products have less potential energy than the reactants.
(4) The sign of ΔH is negative, and the products have more potential energy than the reactants.

---

6. According to the Table of Standard Energies of Formation of Compounds, which compound forms exothermically?
   (1) ethane, C_2H_6 (3) ethyne, C_2H_2
   (2) ethene, C_2H_4 (4) hydrogen iodide, HI

7. The change of reactants into products will always be spontaneous if the products, compared to the reactants, have
   (1) lower enthalpy and lower entropy
   (2) lower enthalpy and higher entropy
   (3) higher enthalpy and lower entropy
   (4) higher enthalpy and higher entropy

8. Based on the Table of Standard Energies of Formation of Compounds, which compound forms spontaneously, even though the ΔH for its formation is positive?
   (1) C_2H_4(g)   (2) C_2H_2(g)   (3) ICl(g)   (4) HI(g)

9. What is the value of ΔG for any chemical reaction at equilibrium?
   (1) one (2) zero (3) greater than one (4) less than one but not zero

10. The expression $\Delta H - T\Delta S$ is equal to the change in
    (1) binding energy (3)   free energy
    (2) ionization energy   (4) activation energy

11. Which reaction system tends to become *less* random as reactants form products?
    (1) $C(s) + O_2(g) \rightarrow CO_2(g)$ (3) $I_2(s) + Cl_2(g) \rightarrow 2ICl(g)$
    (2) $S(s) + O_2(g) \rightarrow SO_2(g)$ (4) $2Mg(s) + O_2(g) \rightarrow 2MgO(s)$

12. Which statement is true for a saturated solution?
    (1) It must be a concentrated solution.
    (2) It must be a dilute solution.
    (3) Neither dissolving nor crystallizing is occurring.
    (4) The rate of dissolving equals the rate of crystallizing.

13. Under which conditions are gases most soluble in water?
    (1) high pressure and high temperature
    (2) high pressure and low temperature
    (3) low pressure and high temperature
    (4) low pressure and low temperature

14. At which temperature will Hg(ℓ) and Hg(s) reach equilibrium?

Q 14
June 18 Q17

15. In which reaction will the point of equilibrium shift to the left when the pressure on the system is increased?
    (1) $C(s) + O_2(g) \rightleftharpoons CO_2(g)$ (3) $2Mg(s) + O_2(g) \rightleftharpoons 2MgO(s)$
    (2) $CaCO_3(s) \rightleftharpoons CaO(s) + CO_2(g)$ (4) $2H_2(g) + O_2(g) \rightleftharpoons 2H_2O(g)$

16. Given the equilibrium reaction at constant pressure:
    $$2HBr(g) + 72.7 \text{ kjoule} \rightleftharpoons H_2(g) + Br_2(g)$$
    When the temperature is increased, the equilibrium will shift to the
    (1) right, and the concentration of HBr(g) will decrease
    (2) right, and the concentration of HBr(g) will increase
    (3) left, and the concentration of HBr(g) will decrease
    (4) left, and the concentration of HBr(g) will increase

17. Given the reaction at equilibrium:
    $$N_2(g) + O_2(g) \rightleftharpoons 2NO(g).$$
    As the concentration of N_2(g) increases, the concentration of O_2(g) will
    (1) decrease   (2) increase   (3) remain the same.

SOLUTIONS

1. Answer 2. Powdered zinc has more surface area than a single piece of zinc and therefore a faster rate of reaction.

2. Answer 1. As temperature decreases, particles move slower (less kinetic energy). Therefore, there are fewer collisions and fewer effective collisions, and the rate of reaction decreases.

3. Answer 4. You learned that a catalyst lowers the activation energy and causes a faster rate of reaction.

4. Answer 1. Activation energy is the smallest amount of energy needed to start the reaction. On the diagram in Question 2, Arrow A is the activation energy from the reactants to the highest point of the curve.

5. Answer 2. Look at the diagram to the right. In an endothermic reaction, ΔH is positive, and the products have more energy than the reactants.

6. Answer 1. Look at the ΔH column in the Table of Standard Energies of Formation, on page Additional Tables 1. Ethane has ΔH = -20.2; therefore, it is an exothermic reaction. In all other choices, ΔH is positive and the reactions are endothermic.

7. Answer 2. You learned that spontaneous reactions occur in the direction of less energy (lower enthalpy) and greater entropy.

8. Answer 3. A compound that forms spontaneously has ΔG negative. Look at the Table of Standard Energies of Formation of Compounds. ICl has ΔG = -5.5 (negative). All the other choices have ΔG positive. In addition, ICl has ΔH = positive. No sign in front of ΔH or ΔG means it is positive.

Standard Energies of Formation of Compounds
in kJ

Compound	ΔH	ΔG
ICl	17.8	-5.5
C₂H₄	52.4	68.4
C₂H₂	227.4	209.9
HI	26.5	1.7

9. Answer 2. At equilibrium, ΔG = 0. (The forward and reverse reactions are not spontaneous.)

10. Answer 3. Memorize the formula for free energy change: ΔG = ΔH - TΔS. If ΔG is negative, the reaction is spontaneous.

11. Answer 4. You learned that solids have the least entropy, or randomness. Liquids have more entropy (randomness). Gases have the most entropy (randomness). The symbols in parentheses mean (s), solid, and (g), gas.
    In Answer 4, **gas** in the reactant **becomes solid** (no gas) in the product. A solid (product) is less random.

12. Answer 4. You learned that, in a saturated solution, the rate that sugar crystallizes equals the rate sugar dissolves. Both rates are equal.

---

13. Answer **2**. Gases are more soluble under high pressure and low temperature. (High temperature decreases solubility.)

14. Solid(s) mercury (Hg) changes to liquid(ℓ) mercury (Hg) and liquid mercury Hg(ℓ) changes to solid mercury Hg(s) at the same rate (equilibrium) at the melting point (equals freezing point). Look at Table S on page Reference Tables 24-25. The melting point of Hg (atomic number 80) is 234 K, which means, at 234 K, liquid and solid mercury (Hg) reach equilibrium.

15. Answer **2**. According to LeChatelier, when pressure is increased, reaction goes in the direction of less gas molecules (smaller number of gas molecules). The question states, when pressure is increased, the equilibrium shifts to the left, which means there are less gas molecules on the left side. This is Answer 2,

$$CaCO_3(s) \quad \rightarrow \quad CaO(s) + CO_2(g)$$

| No gas molecules | 1 gas molecule |
| Left Side | Right Side |

16. Answer **1**. According to LeChatelier, if there is an increase in temperature, the reaction goes in the direction to absorb heat, to remove the stress.

    Here, there is an increase in temperature. Therefore, HBr absorbs the heat (favors endothermic reaction), and the reaction goes to the right (equilibrium displaced or shifted to the right), producing **more** products, and therefore decreasing the reactant HBr.

17. Answer **1**. As more N_2 is added (increases), it takes along the O_2 (causes oxygen to decrease) to form more NO.

NOW LET'S TRY A FEW HOMEWORK QUESTIONS:

1a. Which conditions will increase the rate of a chemical reaction?
   (1) decreased temperature and decreased concentration of reactants
   (2) decreased temperature and increased concentration of reactants
   (3) increased temperature and decreased concentration of reactants
   (4) increased temperature and increased concentration of reactants

b. Given the reaction:

                             forward

$$N_2(g) + O_2(g) + 182.6kJ \quad \rightleftharpoons \quad 2NO(g)$$

                             reverse

Which change would cause an immediate increase in the rate of the forward reaction?
   (1) increasing the concentration of NO(g)
   (2) decreasing the reaction temperature
   (3) decreasing the reaction pressure
   (4) increasing the concentration of N_2(g)

2. As the number of moles per liter of a reactant in a chemical reaction increases, the number of collisions between the reacting particles
   (1) decreases   (2) increases   (3) remains the same

3. Which statement explains why the speed of some chemical reactions is increased when the surface area of the reactant is increased?
   (1) This change increases the density of the reactant particles.
   (2) This change increases the concentration of the reactant.
   (3) This change exposes more reactant particles to a possible collision.
   (4) This change alters the electrical conductivity of the reactant particles.

4. Given the reaction:
$$Mg(s) + 2HCl(aq) \rightarrow MgCl_2(aq) + H_2(g)$$
The reaction occurs more rapidly when a 10-gram sample of Mg is powdered, rather than in one piece, because powdered Mg has
(1) less surface area   (3) a lower potential energy
(2) more surface area   (4) a higher potential energy

5. Under which conditions will the rate of a chemical reaction always decrease?
(1) the concentration of the reactants decreases, and the temperature decreases
(2) the concentration of the reactants decreases, and the temperature increases
(3) the concentration of the reactants increases, and the temperature decreases
(4) the concentration of the reactants increases, and the temperature increases

6. As the number of effective collisions between the reactant particles in a chemical reaction decreases, the rate of the reaction
(1) decreases   (2) increases   (3) remains the same

7. Adding a catalyst to a chemical reaction will
(1) lower the activation energy needed
(2) lower the potential energy of the reaction
(3) increase the activation energy needed
(4) increase the potential energy of the reactants

8. If a catalyst is added to a system at equilibrium, and the temperature and pressure remain constant, there will be no effect on the
(1) rate of the forward reaction   (3) activation energy of the reaction
(2) rate of the reverse reaction   (4) heat of the reaction

9. Which potential energy diagram represents the reaction $A + B \rightarrow C + energy$?

(1)
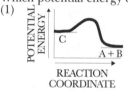

(3)
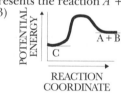

(2)
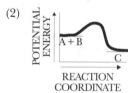

(4)

10. The potential energy diagram shown below represents the reaction A+B → AB. Which statement correctly describes this reaction?
(1) It is endothermic and energy is absorbed.
(2) It is endothermic and energy is released.
(3) It is exothermic and energy is absorbed.
(4) it is exothermic and energy is released.

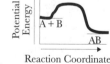

11. The potential energy diagram below shows the reaction $X + Y \rightleftharpoons Z$.
When a catalyst is added to the reaction, it will change the value of
(1) 1 and 2   (2) 1 and 3
(3) 2 and 3   (4) 3 and 4

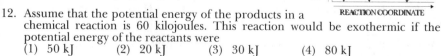

12. Assume that the potential energy of the products in a chemical reaction is 60 kilojoules. This reaction would be exothermic if the potential energy of the reactants were
(1) 50 kJ   (2) 20 kJ   (3) 30 kJ   (4) 80 kJ

13. The potential energy diagram shown below represents the reaction: $R + S + energy \rightarrow T$.

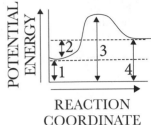

REACTION COORDINATE

Which numbered interval represents the potential energy of the product T?
  (1) 1   (2) 2   (3) 3   (4) 4

Base your answers to Questions 14 and 15 on the potential energy diagram of a chemical reaction shown below:

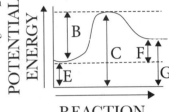

REACTION COORDINATE

14. Which interval represents the heat of reaction?
  (1) E   (2) F   (3) C   (4) G

15. Interval B represents the
  (1) potential energy of the products
  (2) potential energy of the reactants(4)
  (3) activation energy
  (4) activated complex

Base your answers to Questions 16 and 17 on the potential energy diagram of a chemical reaction shown below;

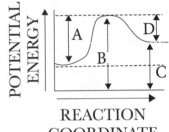

REACTION COORDINATE

16. Which arrow represents the activation energy for the forward reaction?
  (1) A   (2) B   (3) C   (4) D

17. The forward reaction is best described as an
  (1) exothermic reaction in which energy is released
  (2) exothermic reaction in which energy is absorbed
  (3) endothermic reaction in which energy is released
  (4) endothermic reaction in which energy is absorbed

---

18. Which statement describes characteristics of an endothermic reaction?
  (1) The sign of ΔH is positive, and the products have less potential energy than the reactants.
  (2) The sign of ΔH is positive, and the products have more potential energy than the reactants.
  (3) The sign of ΔH is negative, and the products have less potential energy than the reactants.
  (4) The sign of ΔH is negative, and the products have more potential energy than the reactants.

19A. $2C(s) + H_2(g) + 227.4$ kJ $\rightarrow C_2H_2(g)$ is an endothermic reaction.
  If 682.2 kilojoules are absorbed, how many moles of $C_2H_2(g)$ are produced?
  (1) 3   (2) 6   (3) 2   (4) 4

19B. At 101.3 kPa and 298 K, what is the total amount of heat released when one mole of water, $H_2O(g)$, is formed from its elements?
  (1) 241.8 kJ   (2) 393.5 kJ   (3) 483.6 kJ   (4) 227.4 kJ

---

20. A student observed that the temperature of water increased when a salt was dissolved in it. The student should conclude that dissolving the salt caused
    (1) an exothermic reaction   (2) an endothermic reaction
    (3) formation of an acidic solution (4) formation of a basic solution

21. Which equation is used to determine the free energy change during a chemical reaction?
    (1) $\Delta G = \Delta H - \Delta S$ (2) $\Delta G = \Delta H - T\Delta S$
    (3) $\Delta G = \Delta H + \Delta S$ (4) $\Delta G = \Delta H + T\Delta S$

22. A chemical reaction will always occur spontaneously if the reaction has a negative
    (1) ΔG (2) ΔH (3) ΔS (4) T

23. What is the free energy change for a system at equilibrium?
    (1) one   (3) zero
    (2) greater than one   (4) less than zero

24. Based on the Table of Standard Energies of Formation of Compounds, which compound will form spontaneously from its elements?
    (1) carbon dioxide (g)   (3) ethene (g)
    (2) nitrogen (II) oxide (g) (4) ethyne (g)

25. Which series of physical changes represents an entropy increase during each change?
    (1) gas → liquid → solid   (3) solid → gas → solid
    (2) liquid → gas → solid   (4) solid → liquid → gas

26. Based on the Table of Standard Energies of Formation of Compounds, which reaction occurs spontaneously?
    (1) $2C(s) + 3H_2(g) \rightarrow C_2H_6(g)$ (3) $N_2(g) + 2O_2(g) \rightarrow 2NO_2(g)$
    (2) $2C(s) + 2H_2(g) \rightarrow C_2H_4(g)$ (4) $N_2(g) + O_2(g) \rightarrow 2NO(g)$

27. Which change represents an increase of entropy?
    (1) $I_2(s) \rightarrow I_2(g)$ (3) $H_2O(g) \rightarrow H_2O(\ell)$
    (2) $I_2(g) \rightarrow I_2(\ell)$ (4) $H_2O(\ell) \rightarrow H_2O(s)$

28. Which reaction will occur spontaneously? (Refer to the Table of Standard Energies of Formation of Compounds.)
    (1) $\frac{1}{2}N_2(g) + \frac{1}{2}O_2(g) \rightarrow NO(g)$ (3) $2C(s) + 3H_2(g) \rightarrow C_2H_6(g)$
    (2) $\frac{1}{2}N_2(g) + O_2(g) \rightarrow NO_2(g)$ (4) $2C(s) + 2H_2(g) \rightarrow C_2H_4(g)$

29. According to the Table of Standard Energies of Formation of Compounds, ICl(g) is formed from its elements in a reaction that is
    (1) exothermic and spontaneous
    (2) exothermic and not spontaneous
    (3) endothermic and spontaneous
    (4) endothermic and not spontaneous

30. According to the Table of Standard Energies of Formation of Compounds, which compound is spontaneously formed, even though the reaction is endothermic?
    (1) ICl(g)   (2) $CO_2(g)$ (3) $H_2O(\ell)$ (4) $Al_2O_3(s)$

31. A system is said to be in a state of dynamic equilibrium when the
    (1) concentration of products is greater than the concentration of reactants
    (2) concentration of products is the same as the concentration of reactants
    (3) rate at which products are formed is greater than the rate at which reactants are formed
    (4) rate at which products are formed is the same as the rate at which reactants are formed

32. Given the reaction at equilibrum:
$$2CO(g) + O_2(g) \rightleftharpoons 2CO_2(g)$$
Which statement regarding this reaction is always true?
(1) The rates of the forward and reverse reactions are equal.
(2) The reaction occurs in an open system.
(3) The masses of the reactants and the products are equal.
(4) The concentrations of the reactants and the products are equal.

33. A closed system is shown in the diagram at right:
The rate of vapor formation at equilibrium is
(1) less than the rate of liquid formation
(2) greater than the rate of liquid formation
(3) equal to the rate of liquid formation

Vapor
Liquid

34. Given the reaction at equilibrium:
$$N_2(g) + O_2(g) = 2NO(g)$$
If the temperature remains constant and the pressure increases, the number of moles of $NO(g)$ will
(1) decrease   (2) increase   (3) remain the same

35. Given the reaction at equilibrium:
$$2H_2(g) + O_2(g) \rightleftharpoons 2H_2O(g) + heat$$
Which concentration changes occur when the temperature of the system is increased?
(1) The $[H_2]$ decreases and the $[O_2]$ decreases.
(2) The $[H_2]$ decreases and the $[O_2]$ increases.
(3) The $[H_2]$ increases and the $[O_2]$ decreases.
(4) The $[H_2]$ increases and the $[O_2]$ increases.

36. Given the reaction at equilibrium:
$$N_2(g) + 3H_2(g) \rightleftharpoons 2NH_3(g) + 91.8 \text{ KJ}$$
Which stress would cause the equilibrium to shift to the left?
(1) increasing the temperature   (3) adding $N_2(g)$ to the system
(2) increasing the pressure   (4) adding $H_2(g)$ to the system

37. Given a saturated solution of silver chloride at constant temperature:
$$AgCl(s) \rightleftharpoons Ag^+(aq) + Cl^-(aq)$$
As $NaCl(s)$ is dissolved in the solution, the concentration of the Ag^+ ions in the solution
(1) decreases, and the concentration of Cl^- ions increases
(2) decreases, and the concentration of Cl^- ions remains the same
(3) increases, and the concentration of Cl^- ions increases
(4) increases, and the concentration of Cl^- ions remains the same

38. Given the reaction at equilibrium:
$$PbCl_2(s) \rightleftharpoons Pb^{2+}(aq) + 2Cl^-(aq)$$
When $KCl(s)$ is added to the system, the equilibrium shifts to the
(1) right, and the concentration of $Pb^{2+}(aq)$ ions decreases
(2) right, and the concentration of $Pb^{2+}(aq)$ ions increases
(3) left, and the concentration of $Pb^{2+}(aq)$ ions decreases
(4) left, and the concentration of $Pb^{2+}(aq)$ ions increases

39. Given the reaction at equilibrium:
$$AgCl(s) \rightleftharpoons Ag^+(aq) + Cl^-$$
The addition of Cl^- ions will cause the concentration of $Ag^+(aq)$ to
(1) decrease as the amount of $AgCl(s)$ decreases
(2) decrease as the amount of $AgCl(s)$ increases
(3) increase as the amount of $AgCl(s)$ decreases
(4) increase as the amount of $AgCl(s)$ increases

40. When $AgNO_3(aq)$ is mixed with $NaCl(aq)$, a reaction occurs which tends to go to completion because
(1) a gas is formed   (3) a weak acid is formed
(2) water is formed   (4) a precipitate is formed

41. Given the reaction:
$$2Na(s) + 2H_2O(\ell) \rightarrow 2Na^+(aq) + 2OH^-(aq) + H_2(g)$$
This reaction goes to completion because one of the products formed is
(1) an insoluble base   (3) a precipitate
(2) a soluble base   (4) a gas

CONSTRUCTED RESPONSE QUESTIONS: Parts B-2 and C of NYS Regents Exam

42. Explain the collision theory.
    Give 5 factors that affect rate of reaction.

43. Explain why, if there is more concentration, more surface area, or higher temperature, the reaction is faster.

44. Draw an exothermic reaction diagram. Label potential energy of reactants, potential energy of products, activation energy, and ΔH. Give the formula for heat of reaction.

45. Draw an endothermic reaction diagram. Label potential energy of reactants, potential energy of products, activation energy, activation energy with catalyst, and ΔH. Give the formula for heat of reaction.

46. By looking at the Table of Heats of Reaction, what do you know about the reaction if ΔH is negative? What do you know if ΔH is positive?

47. How does entropy vary (differ) in solids, liquids, and gases?

48. Compare and contrast (similarities and differences) in phase equilibria, solution equilibria, and chemical equilibria.

49. Explain how a change in concentration affects equilibrium.

50. How does pressure affect equilibrium?

CHAPTER QUESTION: Parts B-2 and C of NYS Regents Exam

51. A hot pack contains chemicals that can be activated to produce heat. A cold pack contains chemicals that feel cold when activated.
    A. Based on energy flow, state the type of chemical change that occurs in a hot pack.
    B. A cold pack is placed on an injured leg. Indicate the direction of the flow of energy between the leg and the cold pack.
    C. Identify a reactant in Table I that, when mixed with water, could be used in a cold pack.

52. A student using a Styrofoam cup as a calorimeter added a piece of metal to distilled water and stirred the mixture as shown in the diagram below. The student's data is shown in the table below.

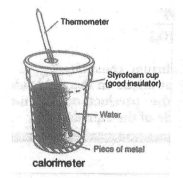

Thermometer

Styrofoam cup (good insulator)

Water

Piece of metal

calorimeter

DATA TABLE

Mass of H_2O	50.0 g
Initial temperature of H_2O	25.0°C
Mass of metal	20.0 g
Initial temperature of metal	100.0°C
Final temperature of H_2O + metal .	32.0°C

How much heat is lost or gained by the water? By the metal?
    Hint: Use $q = mC\Delta T$ (Chapter 1, Table T).
    m = mass of water ΔT = final temperature - initial temperature

SECTION B

THE EQUILIBRIUM CONSTANT: K_{EQ}

The **EQUILIBRIUM CONSTANT**: K_{eq} describes the ratio of **products** to **reactants**.

HOW TO FIND THE EQUILIBRIUM CONSTANT, K_{EQ}

	Reactants (left side)		Products (right side)

Equation : $\quad 2H_2(g) + O_2(g) \quad \rightarrow \quad 2\ H_2O(g)$

$\qquad\qquad\quad \uparrow \qquad\qquad\qquad\qquad\qquad \uparrow$

$\qquad\qquad\ $ coefficient $\qquad\qquad\qquad$ coefficient

Step 1: Put **products on top** and **reactants on bottom**:

$$\frac{Products\ on\ top}{Reactants\ on\ bottom} = \frac{[H_2O]}{[H_2][O_2]} \quad \begin{array}{l} \textit{compounds} \\ \textit{after the arrow} \\ \textit{compounds} \\ \textit{before the arrow} \end{array}$$

Use brackets [ ] to show concentration.

No sign (nothing) between the sets of brackets $[H_2]\ [O_2]$ means $[H_2]$ **times** $[O_2]$.

AND

Step 2: **Coefficient in equation becomes (equals) exponent in K_{eq}.**

In the equation, H_2O has a coefficient of 2, $2H_2O$; therefore, in K_{eq}, H_2O has an **exponent of 2**.

In the equation, H_2 has a coefficient of 2, $2H_2$,; therefore, in K_{eq}, H_2 has an **exponent of 2**.

$$K_{eq} = \frac{[H_2O]^2}{[H_2]^2\ [O_2]}.$$

As you can see, the equilibrium constant, K_{eq}, is the product of the molar concentrations on the right side of the equation divided by the product of the molar concentrations on the left side of the equation.

SUMMARY: K_{eq}

Step 1:

$$\frac{[\ ]\ [\ ] \quad \leftarrow Products}{[\ ]\ [\ ] \quad \leftarrow Reactants}$$

(right side of equation, **AFTER the arrow)**

(left side of equation, **BEFORE the arrow)**

Step 2: Any coefficient in the equation becomes an **exponent** in K_{eq}.

Rule: **NEVER** put a + sign in K_{eq}.

Question: Which is the correct equilibrium expression for the reaction

$$2A(g) + 3B(g) \rightleftharpoons C(g) + 3D(g)?$$

(1) $K = \dfrac{[2A]+[3B]}{[C]+[3D]}$ (2) $K = \dfrac{[C]+[3D]}{[2A]+[3B]}$ (3) $K = \dfrac{[A]^2\ [B]^3}{[C]\ [D]^3}$ (4) $K = \dfrac{[C]\ [D]^3}{[A]^2\ [B]^3}$

Solution: Answer *4*.

Reactants (left side)		Products (right side)
2A + 3B	→	C + 3D
↑ ↑ coefficient		↑ coefficient

K_{eq}:
Step 1:

$$\frac{Products\ on\ top\ (right\ side\ of\ the\ equation)}{Reactants\ on\ bottom\ (left\ side\ of\ the\ equation)} = \frac{[C]\ [D]}{[A]\ [B]}$$

Use [ ] to show concentration.

AND

Step 2: Any coefficient in the equation becomes a power or exponent in K_{eq}.

$$K_{eq} = \frac{[C]\ [D]^3 \ \leftarrow exponent\ or\ power}{[A]^2\ [B]^3 \ \leftarrow exponent\ or\ power}$$

Answer *4* is correct. Choices *1* and *2* are wrong answers. They have a "+" sign in K or K_{eq}. *NEVER* put a "+" sign in K. Choice *3* is also wrong. In K, the **products** (right side of the equation) **should be on top**; in Choice 3, the products are on the bottom.

Meaning of K_{eq}, the Equilibrium Constant:

1. Products are in the *top* of K_{eq}; therefore, a **large K_{eq}** means **a lot of products**.
2. A small K_{eq} means less products

RULE: K_{eq}, the equilibrium constant, **changes only with temperature.** Concentration and pressure have no effect on the equilibrium constant. Only **temperature** has an **effect on the equilibrium constant, K_{eq}.**

IONIZATION CONSTANT

Ionic substances break up into ions.

$$HC_2H_3O_2 + H_2O \rightleftharpoons H_3O^+ + C_2H_3O_2^-$$

acetic acid   water   hydronium ion   acetate ion

The **ionization constant** is really K_{eq} (the equilibrium constant) for ions. The ionization constant for acids is K_a (a is for acid); the ionization constant for bases is K_b (b is for base). It follows all the rules for K_{eq}.

HOW TO FIND THE IONIZATION CONSTANT

Reactant (left side)   Products (right side)

Equation: $HC_2H_3O_2 + H_2O \rightleftharpoons H_3O^+ + C_2H_3O_2^-$

IONIZATION CONSTANT: K_a

Step 1:

$$\frac{\text{Products } \text{(right side of equation, } \textit{AFTER} \text{ the arrow)}}{\text{Reactant } \text{(left side of equation, } \textit{BEFORE} \text{ the arrow)}} = \frac{[H_3O^+]\,[C_2H_3O_2^-]}{[HC_2H_3O_2]}$$

Do not put H_2O in the denominator. K_a already includes the H_2O. $(K_a = K_{eq} \times [H_2O])$.

$$K_a = \frac{[H_3O^+]\,[C_2H_3O_2^-]}{[HC_2H_3O_2]}$$

Step 2: If, in another equation, there is a coefficient of 2, 3, etc., make it a power of 2, 3, etc., in K_a.

Rule: **NEVER** put a **+ sig** in K_a.

Similarly, to calculate K_b, do not put H_2O in the denominator and, if an equation has a coefficient of 2, 3, etc., make it a power of 2, 3, etc. Never put a + sign in K_b.

Meaning of Ionization Constant:

The **bigger** the **ionization constant,** K_a or K_b, the **more ions** there are and the greater the degree of ionization.

Try Sample Questions #1-2, on page 28, and then do Homework Questions, #1-6, pages 30-31.

SOLUBILITY PRODUCT CONSTANT: K_{sp}

The **solubility product constant** (K_{sp}) indicates how much of a slightly soluble salt dissolves in a solution. To find the **SOLUBILITY PRODUCT CONSTANT** (K_{sp}) **multiply** the concentrations of the **ions together** (in a saturated solution). Any **coefficient** in the equation becomes an **exponent** in K_{sp}. Concentration of ions is shown by brackets [ ].

For silver chloride, $AgCl \rightarrow Ag^+ + Cl^-$,

$$K_{sp} = [Ag^+][Cl^-]$$

Question: What is the K_{sp} expression for the salt PbI_2?
  (1) $[Pb^{2+}][I^-]^2$ (3) $[Pb^{2+}][I_2]^2$
  (2) $[Pb^{2+}][2I^-]$ (4) $[Pb^{2+}][2I_2]^2$

Solution: Answer *1*. The equation is
$$PbI_2 \rightarrow Pb^{2+} + 2I^-$$
$$K_{sp} = [Pb^{2+}][I^-]^2$$
Explanation: K_{sp} = product of ions $[Pb^{2+}]$ times $[I^-]^2$ and, as you learned in K, the coefficient of 2 becomes an exponent of 2.

REMEMBER: The **coefficient**, 2 in any equation, becomes a **power or exponent** of 2 in K_{sp}.

Meaning of K_{sp}, the Solubility Product Constant:

A **bigger** K_{SP} means a **more** soluble compound. A **smaller** value of K_{sp} is **less** soluble. Insoluble compounds have a very small K_{sp}.

Look at the **Table of Equilibrium Constants**, below or at the bottom of the table of equilibrium constants on page Additional Tables 5 in Appendix II.

Table of Equilibrium Constants (partial)

Compound	K_{sp}	Compound	K_{sp}
AgBr	5×10^{-13}	Li_2CO_3	2.5×10^{-2}
AgCl	1.8×10^{-10}	$PbCl_2$	1.6×10^{-5}
$BaSO_4$	1.1×10^{-10}	$PbCO_3$	7.4×10^{-14}
$CaSO_4$	9.1×10^{-6}	AgI	8.3×10^{-17}

A **bigger value of K_{sp}** means a **more soluble** compound:

Li_2CO_3: $K_{sp} = 2.5 \times 10^{-2}$. (Remember: A smaller negative exponent = a bigger value.) This is the largest value of K_{sp} in the Table of Equilibrium Constants; therefore, it is the most soluble in this table.

AgI: $K_{sp} = 8.3 \times 10^{-17}$. (Remember: A bigger negative exponent = less value.) This is the smallest value of K_{sp} in the Table of Equilibrium Constants; therefore it is the least soluble (insoluble).

By the way, look also at Table F, on page Reference Tables 6, for AgI. The left box lists halides (Cl⁻, Br⁻, I⁻) as soluble, **but** when combined with Ag^+, it is **insoluble**. Therefore, you know **AgI** is **insoluble.** (**Chap 6**:5 explains how to use Table F.)

Question: Based on the Table of Equilibrium Constants, which compound is less soluble in water than $PbCO_3$ at 298 K and 1 atmosphere?

(1) AgI   (2) AgCl   (3) $CaSO_4$ (4) $BaSO_4$

Solution: Answer *1*. Look at the K_{sp}'s in the Table of Equilibrium Constants, on the previous page:

(1) AgI = 8.3×10^{-17} (3) AgCl = 1.8×10^{-10}
(2) $CaSO_4$ = 9.1×10^{-6} (4) $BaSO_4$ = 1.1×10^{-10}

$PbCO_3$ **has a K_{sp} = 7.4×10^{-14}.** The question is "Which compound is less soluble than $PbCO_3$?" Take the compound that has a smaller K_{sp} than $PbCO_3$ (which has 7.4×10^{-14}). AgI, **Answer 1**, with 8.3×10^{-17}, has the smallest value of K_{sp} (biggest negative exponent.)

SUMMARY: K_{sp} = the **product** of the concentrations **of its ions,** each raised to its appropriate power. **Bigger K_{sp} means more soluble. Smaller K_{sp}** means **less soluble.**

Try Sample Questions #3-5, on pages 28-29, then do Homework Questions, #7-14, page 31.

SAMPLE REGENTS & REGENT-TYPE QUESTIONS AND SOLUTIONS

1. The value of the equilibrium constant of a chemical reaction will change when there is an increase in the
   1. temperature   (3) concentration of the reactants
   2. pressure   (4) concentration of the products

2. Which is the equilibrium expression for the reaction:
$$3A(g) + B(g) \rightarrow 2C(g)?$$

1. $K = \dfrac{[3A][B]}{[2C]}$ (2) $K = \dfrac{[A]^3[B]}{[C]^2}$ (3) $K = \dfrac{[2C]}{[3A][B]}$ (4) $K = \dfrac{[C]^2}{[A]^3[B]}$

3. Given the equilibrium reaction:
$$AgCl(s) \leftrightarrows Ag^+(aq) + Cl^-(aq)$$

At 25°C, the K_{sp} is equal to (Hint: See Table of Equilibrium Constants, page Additional Tables 5)

1. 6.0×10^{-23} (2) 1.8×10^{-10} (3) 1.0×10^{-7} (4) 9.6×10^{-4}

4. Given the reaction at equilibrium:
$$Mg(OH)_2(s) \rightleftharpoons Mg^{2+}(aq) + 2OH^-(aq)$$
The solubility product constant for this reaction is correctly written as

1. $K_{sp} = [Mg^{2+}][2OH^-]$ (3) $K_{sp} = [Mg^{2+}][OH^-]^2$
2. $K_{sp} = [Mg^{2+}] + [2OH^-]$ (4) $K_{sp} = [Mg^{2+}] + [OH^-]^2$

5. Based on the Table of Equilibrium Constants, which of the following compounds is *least* soluble at 298 K? (See page Additional Tables 5)

1. $PbCl_2$ (2) PbI_2 (3) $AgCl$ (4) AgI

SOLUTIONS

1. Answer *1*. You learned that the equilibrium constant only changes with temperature.

2. Answer *4*.

	Reactants (left side)	Products (right side)
Equation:	$3A(g) + B(g)$ →	$2C(g)$
	↑ coefficient	↑ coefficient

$K_{eq} =$
Step 1:

$$\underbrace{\frac{\text{Products on top}}{\text{Reactants on bottom}}}_{} \begin{matrix} \textit{(right side of the equation,} \\ \textit{after the arrow)} \\ \textit{(left side of the equation,} \\ \textit{before the arrow)} \end{matrix} = \frac{[C]}{[A][B]}$$

and

Step 2: Make each coefficient in the equation into a power, or exponent, in K_{eq}:

$$K_{eq} = \frac{[C]^2}{[A]^3[B]}$$

3. Answer *2*. Look at the Table of Equilibrium Constants, at right, or at the bottom of the Table of Equilibrium Constants on page Additional Tables 5. The K_{sp} of AgCl $= 1.8 \times 10^{-10}$.

Table of Equilibrium Constants

Compound	K_{sp}
AgBr	5×10^{-13}
AgCl	1.8×10^{-10}

4. Answer *3*. Remember, you learned that, to find the solubility product, multiply the concentrations of its **ions** (in a saturated solution) together. The **coefficient** in the equation becomes an **exponent** in K_{sp}.

$$K_{sp} = [Mg^{2+}] \underset{times}{[OH^-]^2}$$

Remember: The coefficient 2 in the equation becomes a power, or exponent, of 2 in K_{sp}.

Table of Equilibrium Constants

Compound	K_{sp}
AgCl	1.8×10^{-10}
AgI	8.3×10^{-17}
$PbCl_2$	1.6×10^{-5}
PbI_2	7.1×10^{-9}

5. Answer **4**. Look at the bottom of the Table of Equilibrium Constants, K_{sp}. The biggest negative exponent is the smallest, or least, value of K_{sp}, or least soluble. AgI's K_{sp}, 8.3×10^{-17} is the least soluble.

NOW LET'S TRY A FEW HOMEWORK QUESTIONS:

See Table of Relative Strengths of Acids, page Additional Tables 4 in Appendix II. See Table of Equilibria, pages Additional Tables 5 .

1. Which is the correct equilibrium constant for
$$N_2(g) + 3H_2(g) \rightleftharpoons 2NH_3(g)?$$

   1. $K_{eq} = \dfrac{[NH_3]^2}{[N_2][H_2]^3}$ (3) $K_{eq} = \dfrac{[N_2][H_2]^3}{[NH_3]^2}$

   2. $K_{eq} = \dfrac{[2NH_3]}{[2N_2][3H_2]}$ (4) $K_{eq} = \dfrac{[2N][3H_2]}{[NH_3]^2}$

2. The value of the equilibrium constant of a chemical reaction will change when there is an increase in the
   1. temperature   (3) concentration of the reactants
   2. pressure   (4) concentration of the products

3. Which is the correct equilibrium expression for the reaction
$$4NH_3(g) + 5O_2(g) \rightleftharpoons 4NO(g) + 6H_2O(g)?$$

   1. $K_{eq} = \dfrac{[NO]^4[H_2O]^6}{[NH_3]^4[O_2]^5}$ (3) $K_{eq} = \dfrac{[NO]^4 + [H_2O]^6}{[NH_3]^4 + [O_2]^5}$

   2. $K_{eq} = \dfrac{[4NO][6H_2O]}{[4NH_3][5O_2]}$ (4) $K_{eq} = \dfrac{[4NO] + [6H_2O]}{[4NH_3] + [5O_2]}$

4. Which is the correct equilibrium expression for the reaction below?
$$4NH_3(g) + 7O_2(g) \rightleftharpoons 4NO_2(g) + 6H_2O(g)$$

   1. $K_{eq} = \dfrac{[NO_2][H_2O]}{[NH_3][O_2]}$ (3) $K_{eq} = \dfrac{[NH_3][O_2]}{[NO_2][H_2]}$

   2. $K_{eq} = \dfrac{[NO_2]^4[H_2O]^6}{[NH_3]^4[O_2]^7}$ (4) $K_{eq} = \dfrac{[NH_3]^4[O_2]^7}{[NO_2]^4[H_2O]^6}$

5. The ionization constants (K_a's) of four acids are shown below. Which K_a represents the *weakest* of these acids? (*Hint: Least ionized*)
   1. $K_a = 1.0 \times 10^{-5}$ (3) $K_a = 7.1 \times 10^{-3}$
   2. $K_a = 1.0 \times 10^{-4}$ (4) $K_a = 1.7 \times 10^{-2}$

6. Given the reaction,
$$HNO_2(aq) \rightleftharpoons H^+(aq) + NO_2^-(aq),$$
the ionization constant, K_a, is equal to

1. $\dfrac{[HNO_2]}{[H^+][NO_2^-]}$ (2) $\dfrac{[H^+][NO_2^-]}{[HNO_2]}$ (3) $\dfrac{[NO_2^-]}{[H^+][HNO_2]}$ (4) $\dfrac{[H^+][HNO_2]}{[NO_2^-]}$

7. Given the equilibrium reaction:
$$AgCl(s) \rightleftharpoons Ag^+(aq) + Cl^-(aq)$$
at 25°C, the K_{sp} is equal to

1. 6.0×10^{-23} (2) 1.8×10^{-10} (3) 1.0×10^{-7} (4) 9.6×10^{-4}

8. Based on the Table of Equilibrium Constants, which of the following is the *least* soluble compound?
1. AgBr   (2) AgCl   (3) Ag_2CrO_4 (4) $ZnCO_3$

9. Based on the Table of Equilibrium Constants, which of the following is the *least* soluble compound?
1. $ZnCO_3$ (2) $CaSO_4$ (3) AgBr   (4) AgI

10. Based on the Table of Equilibrium Constants, which of the following salts is the *most* soluble in water?
1. $PbCrO_4$ (2) AgBr   (3) $BaSO_4$ (4) $ZnCO_3$

11. Based on the Table of Equilibrium Constants, which of these is the *most* soluble in water?
1. AgCl   (2) $CaSO_4$ (3) PbI_2 (4) $PbCl_2$

12. Given the solution at equilibrium:
$$CaF_2(s) \rightleftharpoons Ca^{2+}(aq) + 2F^-(aq)$$
What is the solubility product expression (K_{sp})?
1. $K_{sp} = [Ca^{2+}][F^-]$ (3) $K_{sp} = [Ca^{2+}][F^-]^2$
2. $K_{sp} = [Ca^{2+}][2F^-]$ (4) $K_{sp} = [Ca^{2+}]^2[F^-]$

13. According to the *Table of Equilibrium Constants*, which salt would have the smallest K_{sp} value?
1. $AlBr_3$ (2) $PbBr_2$ (3) NaBr   (4) AgBr

14. What is the correct equilibrium expression (K_{sp}) for the reaction below?
$$Ca_3(PO_4)_2(s) \rightleftharpoons 3Ca^{2+}(aq) + 2PO_4^{3-}(aq)$$
1. $K_{sp} = [3Ca^{2+}][2PO_4^{3-}]$ (3) $K_{sp} = [Ca^{2+}]^3[PO_4^{3-}]^2$
2. $K_{sp} = [3Ca^{2+}] + [2PO_4^{3-}]$ (4) $K_{sp} = [Ca^{2+}]^3 + [PO_4^{3-}]^2$

CHAPTER 8:
ACIDS, BASES AND SALTS

SECTION A

An **ELECTROLYTE** is any substance (examples of electrolytes NaCl(s), KCl(s), s means solid, HCl) that when put into water dissolves, forming a solution (example NaCl(aq), aq means in water) that conducts electricity. The ability of a solution to conduct electric current is due to the presence of **ions**. Ionic compounds and many polar covalent compounds (HCl, HBr) dissolve in water and conduct electricity.

Lab test for electrolytes: If NaCl(s) or KCl(s) or any substance dissolved in water conducts electricity, you know NaCl(s) or KCl(s) etc. is an electrolyte.

ACIDS

1. Aqueous (H_2O) solutions of acids **conduct electricity** (acids are **electrolytes.**)

2. Acids (example, HCl) react with certain metals to liberate (produce) hydrogen gas (H_2).

   Look at **Table J** below or on page Reference Tables 12:

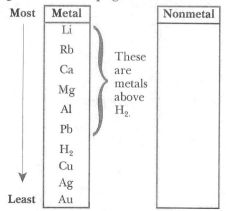

The word **most** is written at the top of Table J. The **most active** metals are near the top. The arrow shows that the activity decreases (the metals become less active) as you go down Table J.

Metals above hydrogen (H_2) are more active than hydrogen and **react** with **HCl (acid)** to **produce hydrogen.**

$$\text{Ca} \quad + \quad 2\text{HCl} \quad \longrightarrow \quad \text{CaCl}_2 \quad + \quad \text{H}_2$$

Metal   **Produces**

above H$_2$ **H$_2$**

3. Acids cause color changes in acid-base indicators:

         Blue litmus paper turns **red** in an **acid**.

         Phenolphthalein is **colorless** in an **acid**.

4. **Acids react** with **bases** (hydroxides, –OH) to form a **salt** and **water-neutralization** reaction:

$$\text{HCl} + \text{NaOH} \rightarrow \text{NaCl} + \text{H}_2\text{O}$$

     Acid   Base   Salt   Water

5. Acids have a **sour taste**.

Look at **Table K**, Common **Acids**, on page reference tables 13; **names** and **formulas** of acids are given.

TABLE K

BASES

1. Aqueous solutions of bases **conduct electricity** (bases are **electrolytes**).

2. Bases cause color changes in acid-base indicators:

         Red litmus paper turns <u>b</u>lue in a <u>b</u>ase.

         (*Hint*: *"Blue" begins with the letter "b," and "base" begins with the letter "b": **blue** in **base**).*

         Phenolphthalein is **pink** in a **base**.

3. **Bases react** with **acid** to form **salt** and **water-neutralization** reactions:

$$\text{NaOH} + \text{HCl} \rightarrow \text{NaCl} + \text{H}_2\text{O}$$

     Base   Acid   Salt   Water

4. Bases feel **slippery**.

Look at **Table L**, Common **Bases**, on page reference tables 14; **names** and **formulas** of bases are given.

TABLE L

Question: Given the reaction:

$$\text{KOH} + \text{HNO}_3 \rightarrow \text{KNO}_3 + \text{H}_2\text{O},$$

Which process is taking place?

(1) neutralization   (3) substitution

(2) esterification   (4) addition

Solution: Answer *1*.

$$\textbf{KOH} + \text{HNO}_3 \rightarrow \text{KNO}_3 + \text{H}_2\text{O}$$

     Base   Acid   Salt   Water

         = neutralization reaction

Question: An aqueous solution of an ionic compound turns red litmus blue, conducts electricity, and reacts with an acid to form a salt and water. This compound could be

1. HCl   (2) NaI   (3) KNO_3 (4) LiOH

Solution: If a solution of an ionic compound turns red litmus blue or if it reacts with an acid to form salt and water, it must be a base. Bases end in OH. Answer 4, LiOH.

Try Sample Questions #1-2, on page 11, and then do Homework Questions, #1-7, page 13, and #31-32, page 15.

ARRHENIUS THEORY

An **ARRHENIUS ACID** has **H** and **yields (releases or gives off) H⁺** (hydrogen ions) in aqueous solutions: HCl, HBr and H_2SO_4 are **Arrhenius acids** because they have **H** and give off H^+ ions in solutions. The H^+ ion is the only positive ion in these solutions. The H^+ (hydrogen ions) are always attached to H_2O, forming H_3O^+ (hydronium ions).
Note: You will learn in Chapter 10 that organic **acids** have the group COOH and give off H^+ ions in solution.

An **ARRHENIUS BASE** has **OH** and **yields (releases or gives off) OH⁻** (hydroxide ions) in aqueous solutions. NaOH, KOH and $Ca(OH)_2$ are **Arrhenius bases** because they have **OH** and give off OH^- ions in solution. The OH^- ion is the only negative ion in these solutions.

ARRHENIUS THEORY EXPLAINS ACIDS AND BASES

The properties of an **acid** (**acidic** properties) are because of H^+ **ions**, and the properties of a **base** (**basic** properties) are because of OH^- **ions**.

1. Acids and bases conduct electricity (are electrolytes) because they form ions. Example: HCl $\longrightarrow$ H^+ + Cl^- ions. (H^+ + $H_2O \longrightarrow H_3O^+$ ions.) NaOH $\longrightarrow Na^+$ + OH^- ions.

2. Acids and bases form salt and water. The H^+ from the acid combines or unites with the OH^- from the base to form water.

3. In the reaction

$$Ca + 2HCl \longrightarrow CaCl_2 + H_2,$$

the H^+ (hydrogen ion) gains an electron to form hydrogen.

Question: Which substance can be classified as an Arrhenius acid?

(1) HCl   (2) NaCl   (3) LiOH   (4) KOH

Solution: Answer *1*. HCl is an Arrhenius acid. It has H and gives off H^+ ions in solution.

ALTERNATE ACID-BASE THEORIES

There are alternate (other) acid-base theories. According to one **alternate acid-base theory** (called Bronsted-Lowry theory–the name Bronsted-Lowry is not on the Regents):

An **ACID** is an H^+ **donor (proton donor).** It gives away H^+(a proton).

A **BASE** is an H^+ **acceptor (proton acceptor).** It accepts H^+(a proton).

In the reaction below, according to the **alternate acid-base theory,** what is the **acid**?

donates (gives away)H^+(a proton)

$$HCl(g) + H_2O(\ell) \rightarrow H_3O^+(aq) + Cl^-(aq)$$
Acid

According to the **alternate acid-base theory, HCl** is an **acid** because it is an H^+ **donor (proton donor).** It **gives away** H^+ to H_2O and Cl^- is left.

Now, according to the **alternate acid-base theory,** what is the **base**?

↓ accepts H^+ (a proton)

$$HCl(g) \quad + \quad H_2O(\ell) \quad \rightarrow \quad H_3O^+(aq) + Cl^-(aq)$$
Base

According to the **alternate acid-base theory, H_2O** is a **base** because it is an H^+ **acceptor** (accepts an H^+ or proton). H_2O accepts H^+ (a proton) which HCl gives away.

You can substitute "Bronsted-Lowry theory" for "alternate acid-base theory". In the reaction above, HCl is the Bronsted-Lowry acid and H_2O is the Bronsted-Lowry base.

Question: In the reaction:
$$HNO_3 + H_2O \rightleftharpoons H_3O^+ + NO_3^-$$
according to the alternate acid-base theory, the two acids are
(1) H_2O and HNO_3 (3) H_2O and H_3O^+
(2) H_2O and NO_3^- (4) HNO_3 and H_3O^+

Solution: Answer *4*.

forward
$$HNO_3 + H_2O \quad \rightleftharpoons \quad H_3O^+ + NO_3^-$$
reverse

This is a reversible reaction:
Forward reaction:

donates (gives away) H^+

$$HNO_3 + H_2O \quad \rightarrow \quad H_3O^+ + NO_3^-$$
Acid

According to the **alternate acid-base theory, HNO_3** is the **acid** because it is an H^+ **donor (proton donor).** It gives away H^+ to the H_2O, and NO_3^- is left.

Reverse reaction:

donates H⁺ (gives away)

$$H_3O^+ + NO_3^- \rightarrow HNO_3 + H_2O$$
Acid

According to the **alternate acid-base theory, H_3O^+** is the **acid** because it is an **H^+ donor (proton donor)**; it gives away H^+ to the NO_3^-, and H_2O is left.

The correct answer is **4**. According to the alternate acid-base theory, the two acids are HNO_3 and H_3O^+.

For the answer to this question, you can say that the two Bronsted-Lowry acids are HNO_3 and H_3O^+.

Hint: To find 2 acids, etc. (see question on the previous page) **compare** HNO_3 **before** the arrow with NO_3^- **after** the arrow; HNO_3 and NO_3^- **differ only** by **H^+.** The one with 1 **more** H^+ (or more H^+) is the **acid, HNO_3.** (The one with 1 less H^+ is the base NO_3^-).

Similarly, compare H_3O^+ and H_2O (differ only by H^+). **H_3O^+** has 1 **more** H^+, the **acid.** (H_2O has 1 less H^+, the base).

Try Sample Questions #3-7, on page 11, then do Homework Questions, #8-18, pages 13-14, and #33-34, page 15.

Acidic solutions (acid in water (H_2O)) have more H^+ (H_3O^+) than OH^-. **Basic solutions have** more OH^- than H^+ (H_3O^+). But H_2O (water alone) has the same number of H^+ as OH^- ions.

Amphoteric substances can **sometimes** act like an **acid** and **sometimes** act like a **base**. Examples are **water and HSO_4^-**.

$$\begin{array}{cc} (1) & (2) \\ H_2O + H_2O & \rightarrow \quad H_3O^+ + OH^- \\ \text{Acid} & \text{Base} \end{array}$$

One water is an acid and one water is a base:

H_2O (1) acid, H^+ DONOR (proton donor), GIVES AWAY H^+ (a proton), and OH^- is left.

H_2O (2) BASE, H^+ ACCEPTOR (proton acceptor), ACCEPTS H^+(a proton) and becomes H_3O^+.

Water is amphoteric (amphiprotic or amphoterism) because water can be an acid or a base.

A **SALT** (example, NaCl) is an ionic compound that has positive ions other than hydrogen and negative ions other than hydroxide. For example, the salt NaCl is made up of Na^+ (positive ions) and Cl^- (negative ions); the salt KBr is made up of K^+ (positive ions) and Br^- (negative ions) and the salt KNO_3 is made of K^+ (positive ions) and NO_3^- (negative polyatomic ions, see Table E, page Reference Tables 5). Aqueous solutions of salts conduct electricity (salts are **electrolytes**).

Remember: Aqueous (water) solutions of acids, bases, and salts **conduct electricity** (acids, bases, and salts are **electrolytes**).

NEUTRALIZATION

NEUTRALIZATION occurs when there is the same number of H^+ **ions** as OH^- **ions** (equivalent or **equal amounts**):

$$acid + base \longrightarrow salt + water$$
$$HCl + NaOH \longrightarrow NaCl + H_2O$$

Neutralization can also be written as:

$$\underset{\text{from acid} \quad \text{from base}}{H^+ + OH^-} \longrightarrow H_2O$$

Acid **neutralizes** base or base **neutralizes** acid because there are **equal amounts** of H^+ (from the acid) and OH^- (from the base).

Question: Complete the neutralization reaction:

$$HBr + KOH \longrightarrow \underline{\hspace{2cm}} + \underline{\hspace{2cm}}$$

Solution:

Step 1: $H^+ + OH^-$ unite or combine to form H_2O. Write H_2O after the arrow.

$$HBr + KOH \longrightarrow \underline{\hspace{2cm}} + H_2O$$

Step 2: Look again at the equation. What is left over, (K and Br) combine or unite to form the salt.

$$K + Br \longrightarrow KBr$$

In a salt, write the metal(K) first and the nonmetal (Br) last. K is the metal (left of the zigzag line on the Periodic Table). Br is the nonmetal (right of the zigzag line on the Periodic Table). Write KBr. Put KBr after the arrow.

$$HBr + KOH \longrightarrow KBr + H_2O$$

For Practice: Write six neutralization reactions as we did above. For each reaction, select an acid from Table K and a base from Table L

TITRATION is used to find the molarity of an acid or base. In the example below, you want to find the molarity of an NaOH solution:

Red litmus paper turns blue in NaOH, a base.

When enough HCl is added, the blue litmus turns back to red.

Titration is adding measured volumes of an acid or base of known molarity to a base or acid of unknown molarity until neutralization occurs. In the picture to the right, the student added 15 mL of 1M HCl (20 mL - 5 mL = 15 mL) to NaOH in the flask until neutralization occurred (when the indicator changed color).

Endpoint of titration is the point when the indicator changes color. At

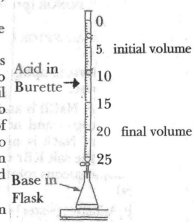

Acid in Burette →

0

5. initial volume

10

15

20 final volume

25

Base in Flask

neutralization (acid neutralizes base or base neutralizes acid), moles H^+ = moles OH^-. This is the basis for the formula used to solve titration problems. This formula is given in Table T.

Table T

Titration:	$M_A V_A = M_B V_B$	M_A = Molarity of H^+ V_A = Volume of Acid	M_B= Molarity of OH^- V_B= Volume of Base

Look at Table T above: $M_A V_A = M_B V_B$

M_A means moles of H^+/liter.

1M HCl yields or gives off 1 mole H^+/liter.

2M HCl yields or gives off 2 moles H^+/liter.

1M H_2SO_4 yields or gives off 2 moles H^+/liter. (H_2 has $2H^+$ in it).

2M H_2SO_4 yields or gives off 2 x 2 = 4 moles H^+/liter.

In short, M_A means **moles of H^+**/liter (or molarity of H^+ in Table T).

M_B means moles of OH^-/liter.

1M NaOH yields or gives off 1 mole OH^-/liter.

2M NaOH yields or gives off 2 moles OH^-/liter.

1M $Ca(OH)_2$ yields or gives off 2 moles OH^-/liter.

In short, M_B means **moles of OH^-**/liter (or molarity of OH^- in Table T).

V_A means volume of acid (in mL or liters).

V_B means volume of base (in mL or liters).

Question: How many milliliters of 4.00 M NaOH are required to exactly neutralize 50.0 milliliters of a 2.00 M solution of HNO_3?

   (1) 25.0 mL   (2) 50.0 mL   (3) 100. mL   (4) 200. mL

Solution: Answer *1*.

Use the titration formula: $M_A V_A = M_B V_B$ in Table T

2M HNO_3 yields or gives off 2 moles H^+/liter. Put **2** under M_A.

V_A = 50 mL. Put 50 under V_A.

4M NaOH yields or gives off 4 moles OH^-/liter. Put **4** under M_B.

V_B = X. Put X under V_B.

$$
\begin{array}{ccc}
M_A \, V_A & = & M_B \, V_B \\
2 \times 50 & = & 4 \times X \\
100 & = & 4X
\end{array}
$$

Divide by 4: $\dfrac{4x}{4} = \dfrac{100}{4}$

    X = 25 mL, Volume of NaOH = 25 mL Answer *1*.

Question: What is the molarity of an HCl solution if 20 mL of the acid is needed to neutralize 10 mL of a 0.5M NaOH solution?

Note: The volume of 20 mL of HCl in this problem was measured by subtracting the **final volume** of HCl (in a burette), 25 mL, minus the initial volume of HCl, 5 mL = 20 mL.

(1)1.0 M   (2).75 M   (3) .50 M   (4) .25M

Solution: Answer **4**.

Use the titration formula: $M_A V_A = M_B V_B$ in Table T.

M_A (molarity or moles H$^+$/liter) = X. Put X under M_A.
V_A = 20 mL. Put 20 under V_A.
M_B (molarity or moles OH$^-$/liter).
The problem has .5M NaOH. Put 0.5 under M_B.
V_B = 10 mL. Put 10 under V_B.

$$M_A \ V_A \quad = \quad M_B \ V_B$$
$$X \times 20 \quad = \quad 0.5 \times 10$$
$$20 \ X \quad = \quad 5$$

Divide by 20:

$$\frac{20x}{20} = \frac{5}{20}$$

X = .25 M_A = .25   Molarity of HCl solution = .25

Note: To find the molarity of H_2SO_4, an acid with **2** hydrogen atoms, divide M_A by 2 (because the acid has 2 hydrogen atoms). If M_A = .50, molarity of H_2SO_4 = .50/2 = .25 .
To find the molarity of $Ca(OH)_2$, a base with 2 OH$^-$, divide M_B by 2.

Try Sample Questions #8-9, on page 11, then do Homework Questions, #19-27, pages 14-15, and #35-37, page 15.

PH

pH indicates the strength of the acid or base (how strong the acid or base is). A **pH less than 7** is **acidic** (acid); the lower the number, the more acidic. **pH 7 = neutral**. **pH more than 7** is **basic** (base); the higher the number, the more basic.

pH Scale

| 0 | 1 | 2 | 3 | 4 | 5 | 6 | 7 | 8 | 9 | 10 | 11 | 12 | 13 | 14 |

strongest acid   weakest acid   neutral   weakest base   strongest base

As pH goes to lower numbers,
6, 5, 4, 3, 2, 1, 0 acid gets stronger

As pH goes to higher numbers,
8, 9, 10, 11, 12, 13, 14,
base gets stronger

0	1	2	3	4	5	6	7 ·	8	9	10	11	12	13	14
Strong Acid			Medium Strong Acid			Weak Acid	Neu-tral	Weak Base	Medium Strong Base			Strong Base		

Different pH's are neutral or strong, medium strong, or weak acids or bases.

Question: Identify each solution as acid, base, or neutral:

pH = 5, pH = 9, pH = 7

Solution: pH = 5 is an acid. pH less than 7 is an acid.
pH = 9 is a base. pH more than 7 is a base.
pH = 7 is neutral.

INDICATORS

INDICATORS change color when **pH changes**. They show whether a solution is an acid or base and how strong.

Let's understand how indicators work:

Table M
Common Acid-Base Indicators

Indicator	Approximate pH Range for Color Change	Color Change
methyl orange	3.2 - 4.4	red to yellow
bromthymol blue	6.0 - 7.6	yellow to blue
phenolphthalein	8.2-10	colorless to pink
litmus	5.5-8.2	red to blue
bromcresol green	3.8 - 5.4	yellow to blue
thymol blue	8.0 - 9.6	yellow to blue

Look at Table M. Table M shows the color of indicators at different pH's.

Methyl orange turns red in a solution with pH 3.2 or less. It turns yellow in a solution with pH 4.4 or more. A pH between 3.2 and 4.4 would be a blending of the colors (example: orange).

Bromthymol blue turns yellow in a solution with pH 6.0 or less (acid). It turns blue in a solution with a pH of 7.6 or more (base).

Bromcresol green turns yellow in a solution with pH 3.8 or less. It turns blue in a solution with pH 5.4 or more.

Let's now **use more than one indicator** to figure out the pH of a solution. Look at Table M again. If you put **bromcresol green** in a solution and the solution turns blue, the pH is 5.4 or more. If you now put **bromthymol blue** in another test tube with the same solution and it turns yellow, it shows the pH is 6.0 or less. **Conclusion:** The solution has a pH between 5.4 and 6.0.

Question: A solution turns yellow with thymol blue and blue with bromthymol blue. What is its pH?

Solution: Look at Table M. A solution that is **yellow with thymol blue** has a **pH of 8.0 or less**. A solution that is **blue** with

TABLE M

bromothymol blue has a **pH of 7.6 or more**. Therefore, you know that the pH is between **7.6** and **8.0**.

MEANING OF PH

pH is the **negative log** (exponent) of the H^+ concentration. In water, all H^+ ions are attached to water molecules, forming H_3O^+ (hydronium ions). If the H^+, hydrogen ion, (or H_3O^+, hydronium ion) concentration is $1 \times 10^{-4 \text{ -exponent}}$, pH is 4. Just **take the number 4** from the exponent (negative 4) and **make the number 4 = pH; pH is 4**. If H^+ concentration is 10^{-3}, just take the number 3, and pH = 3.

Remember: Look at H^+ concentration, then write only the number of the exponent for pH. If $H^+ = 1 \times 10^{-11}$, **pH = 11**.

At **pH 4**, the concentration of $H^+ = 1 \times 10^{-4} M$ = 0.0001 M (M = moles per liter), which means **1/10000 moles of H^+ in 1 liter of solution**.

At **pH 3**, the concentration of $H^+ = 1 \times 10^{-3} M$ = 0.001 M, which means **1/1000 moles of H^+ in 1 liter of solution**.

As you can see, a solution at pH 3 has 10 x as much H^+ as a solution at pH 4. When pH decreases by 1 unit (like from pH 4 to pH 3) the concentration of H^+, hydrogen ion, (or H_3O^+, hydronium ion) increases ten-fold, 10 times as much, therefore 10 times more acidic.

When pH increases by one unit (like from pH 5 to pH 6, which is less acidic), the H^+, hydrogen ion (or H_3O^+, hydronium ion) concentration decreases 10-fold (has only 1/10 as much H^+ ion as pH 5). When pH increases by two units (like from pH 5 to pH 7), the H^+ (or H_3O^+) concentration decreases by 10 x 10 = 100-fold (has only 1/100 as much H^+ ion (as pH 5)). When pH increases by 3 units (pH 5 to pH 8), the H^+ (or H_3O^+) concentration decreases by 10 x 10 x 10 = 1000-fold, (has only 1/1000 as much H^+ ion (as pH 5)).

Question: What is the pH of a 0.00001 molar HCl solution?

   (1)1   (2)9   (3) 5   (4) 4

Solution: Answer *3*. $HCl \longrightarrow H^+ + Cl^-$. 0.00001 molar HCl gives **0.00001 moles H^+** in a liter of solution which equals = 1×10^{-5} **moles H^+**. Write only the number of the exponent: **pH = 5**.

Try Sample Questions #10-12, page 11, and
Homework Questions #28-30, page 15, and #38, page 15.

REMEMBER: When you answer Regents questions on ACIDS, BASES and SALTS, use Table J, Table K, Table L, Table T and Table M.

SAMPLE REGENTS & REGENT-TYPE QUESTIONS
AND SOLUTIONS

1. Which 0.1 M solution will turn phenolphthalein pink?
   (2) HBr(aq)   (2) CO_2(aq)   (3) LiOH(aq)   (4) CH_3OH(aq)

2. Which metal will react with hydrochloric acid and produce H_2(g)?
   (1) Au   (2) Cu   (3) Mg   (4) Hg

3. Which substance is classified as an Arrhenius base?
   (1) HCl   (2) NaOH   (3) $LiNO_3$ (4) $KHCO_3$

4. According to the Arrhenius theory, a substance that is classified as an acid
   will always yield
   (1) H^+(aq)   (2) NH_4^+(aq)   (3) OH^-(aq)   (4) CO_3^{2-}(aq)

5. According to the alternate acid-base theory, a chloride ion, $\left[:\overset{..}{\underset{..}{Cl}}: \right]^-$, acts
   as a base when it combines with
   (1) an OH^- ion   (3) an H^- ion
   (2) a K^+ ion   (4) an H^+ ion

6. According to the alternate acid-base theory, an acid is
   (1) a proton donor, only
   (2) a proton acceptor, only
   (3) a proton donor and a proton acceptor
   (4) neither a proton donor nor a proton acceptor

7. Given the reaction at equilibrium:
   $$HSO_4^- + H_2O \rightleftharpoons H_3O^+ + SO_4^{2-}$$
   According to the alternate acid-base theory, the two bases are
   (1) H_2O and H_3O^+ (3) H_3O^+ and HSO_4^-
   (2) H_2O and SO_4^{2-} (4) H_3O^+ and SO_4^{2-}

8. What is the net ionic equation for a neutralization reaction?
   (1) $H^+ + H_2O \rightarrow H_3O^+$ (3) $2H^+ + 2O^{2-} \rightarrow 2OH^-$
   (2) $H^+ + NH_3 \rightarrow NH_4^+$ (4) $H^+ + OH^- \rightarrow H_2O$

9. If 20.0 milliliters of 1.0 M HCl was used to completely neutralize 40.0
   milliliters of an NaOH solution, what was the molarity of the NaOH
   solution?
   (1) 0.50 M   (2) 2.0 M   (3) 1.5 M   (4) 4.0 M

10. If an aqueous solution turns blue litmus red, which relationship exists
    between the hydronium ion and hydroxide ion concentrations?
    (1) $[H_3O^+] > [OH^-]$ (3) $[H_3O^+] = [OH^-] = 10^{-7}$
    (2) $[H_3O^+] < [OH^-]$ (4) $[H_3O^+] = [OH^-] = 10^{-14}$

11. An acidic solution could have a pH of
    (1) 7   (2) 10   (3) 3   (4) 14

12. An aqueous solution that has a hydrogen ion concentration of 1.0×10^{-8}
    mole per liter has a pH of
    (1) 6, which is basic   (3) 8, which is basic
    (2) 6, which is acidic   (4) 8, which is acidic

1. Answer *3*. Phenolphthalein turns pink in a base. Bases release OH^- (hydroxide) ions in solution. $NaOH$ and $LiOH$ are bases. Choice 4 is wrong because this choice contains C attached to OH, which is an alcohol. The OH is not a hydroxide ion (OH^-), but a hydroxyl group. You will learn more about this in Chapter 10. Alcohols have no effect on phenolphthalein or litmus paper.

2. Answer *3*. Look at Table J. All metals above H_2 (hydrogen) (example, Mg) react with hydrochloric acid (HCl) to liberate (produce) hydrogen. Choices 1, 2 and 4 are wrong: Cu, Hg and Au are metals below H_2 and do not react.

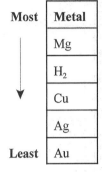

Most	Metal
	Mg
	H_2
	Cu
	Ag
Least	Au

3. Answer *2*. You learned that an Arrhenius base has OH and releases OH^- ions in solution.

4. Answer *1*. An Arrhenius acid releases (yields) H^+ ions in solution.

5. Answer *4*. According to alternate acid-base theory, a base is an H^+ (proton) acceptor, therefore Cl^- is a base when it accepts (combines with) H^+ (a proton).

6. Answer *1*. You learned that, according to alternate acid-base theory, an acid is a proton donor.

7. Answer *2*. This is a reversible reaction. According to the alternate acid-base theory, in the forward direction, $HSO_4^- + H_2O \leftrightarrows H_3O^+ + SO_4^{2-}$, H_2O is a **base** because the H_2O accepts H^+ (a proton) and becomes H_3O^+. In the reverse direction, $H_3O^+ + SO_4^{2-} \leftrightarrows HSO_4^- + H_2O$, SO_4^{2-} is a **base** because it accepts a H^+ (a proton) and becomes HSO_4^-. The two bases are H_2O and SO_4^{2-}.

8. Answer *4*. Neutralization = $H^+ + OH^- \rightarrow H_2O$.

9. Answer *1*. Use the titration formula, $M_A V_a = M_B V_B$ in Table T
    M_A (molarity or moles of H^+/liter).
    The problem has 1M HCl. Put 1 under M_A.
    V_A = 20 mL. Put 20 under V_A.
    M_B (molarity or moles of OH^-/liter = X. Put X under M_B.
    V_B = 40 mL
    $$M_A V_A = M_B V_B$$
    $$1 \times 20 = X \times 40$$
    $$20 = 40X$$
    Divide by 40:
    $$\frac{20}{40} = \frac{40X}{40}$$
    $$0.5 = X. \text{The molarity of NaOH} = 0.5$$

10. Answer *1*. It is an **acid** because blue litmus turns red in an acid. Acid solutions have more H^+ or H_3O^+ ($H^+ + H_2O$) than OH^-. Answer 1, H_3O^+ is greater than OH^-: $[H_3O^+] > [OH^-]$.

11. Answer *3*. An acidic solution has a pH less than 7. A pH of 3 is less than a pH of 7.

12. Answer *3*. H^+ concentration = 1.0×10^{-8}. As you learned, just take the number 8 and make that number = pH: pH = 8. You learned that a pH of more than 7 is basic, therefore pH 8 is basic.

NOW LET'S TRY A FEW HOMEWORK QUESTIONS:

1. Which species can conduct an electric current?
   (1) NaOH(s)　(2) CH_3OH(aq)　(3) H_2O(s)　(4) HCl(aq)

2. An aqueous solution of an ionic compound turns red litmus blue, conducts electricity, and reacts with an acid to form a salt and water. This compound could be
   (1) HCl　(2) NaI　(3) KNO_3　(4) LiOH

3. According to *Reference Table J*, which metal will react with 0.1 M HCl?
   (1) Au(s)　(2) Ag(s)　(3) Hg(ℓ)　(4) Mg(s)

4. Which 0.1 M solution turns phenolphthalein pink and red litmus blue?
   (1) HCl(aq)　(2) CO_2(aq)　(3) NaOH(aq)　(4) CH_3OH(aq)

5. At 298 K, which metal will release H_2(g) when reacted with HCl(aq)?
   (1) Au(s)　(2) Zn(s)　(3) Hg(ℓ)　(4) Ag(s)

6. Which compound reacts with an acid to form a salt and water?
   (1) CH_3Cl　(2) CH_3COOH　(3) KCl　(4) KOH

7. What color is phenolphthalein in a basic solution?
   (1) blue　(2) pink　(3) yellow　(4) colorless

8. Which substance can be classified as an Arrhenius acid?
   (1) HCl　(20 NaCl　(3) LiOH　(4) KOH

9. A. In an aqueous solution, which substance yields hydrogen ions as the only positive ions?
   (1) C_2H_5OH　(2) CH_3COOH　(3) KH　(4) KOH
   B. Which two formulas represent Arrhenius acids?
   (1) CH_3COOH and CH_3CH_2OH　(3) $KHCO_3$ and $KHSO_4$
   (2) $HC_2H_3O_2$ and H_3PO_4　(4) NaSCN and $Na_2S_2O_3$

10. According to the Arrhenius theory, the acidic property of an aqueous solution is due to an excess of
    (1) H_2　(2) H^+　(3) H_2O　(4) OH^-

11. Which substance, when dissolved in water, is an acid according to alternate acid-base theory?
    (1) CH_3OH　(2) NaOH　(3) C_2H_5COOH　(4) CH_3COO^-

12. According to the alternate acid-base theory, an acid is any species that can
    (1) donate a proton　(3) accept a proton
    (2) donate an electron　(4) accept an electron

13. Given the equation:
    $$H_2O + HF \rightleftharpoons H_3O^+ + F^-$$
    According to the alternate acid-base theory, which pair represents acids?
    (1) HF and F^-　(2) HF and H_3O^+　(3) H_2O and F^-　(4) H_2O and H_3O^+

14. Given the reaction:
    $$HCl(g) + H_2O(ℓ) \rightarrow H_3O^+(aq) + Cl^-(aq)$$
    According to the alternate acid-base theory, which reactant acted as an acid?
    (1) HCl(g), because it reacted with chloride ions
    (2) H_2O(ℓ), because it produced hydronium ions
    (3) HCl(g), because it donated protons
    (4) H_2O(ℓ), because it accepted protons

15. Given the reaction at equilibrium:
$$NH_4^+ + OH^- \rightleftharpoons H_2O + NH_3$$
According to the alternate acid-base theory, which species is the acid in the forward reaction?
(1) NH_3 (2) H_2O (3) OH^- (4) NH_4^+

16. In the reaction:
$$H_2O + CO_3^{2-} \rightleftharpoons OH^- + HCO_3^-$$
According to the alternate acid-base theory, the two acids are
(1) H_2O and OH^-
(2) H_2O and HCO_3^-
(3) CO_3^{2-} and OH^-
(4) CO_3^{2-} and HCO_3^-

17. In the reaction:
$$NH_3(g) + H_2O(\ell) \rightarrow NH_4^+(aq) + OH^-(aq)$$
According to the alternate acid-base theory, the $NH_3(g)$ acts as
(1) an acid, only
(2) a base, only
(3) both an acid and a base
(4) neither an acid nor a base

18. Given the reaction at equilibrium:
$$HSO_4^- + H_2O \rightleftharpoons H_3O^+ + SO_4^{2-}$$
According to the alternate acid-base theory, the two bases are
(1) H_2O and H_3O^+
(2) H_2O and SO_4^{2-}
(3) H_3O^+ and HSO_4^-
(4) H_3O^+ and SO_4^{2-}

19. Which reaction best illustrates amphoterism?
(1) $H_2O + HCl \rightarrow H_3O^+ + Cl^-$
(2) $NH_3 + H_2O \rightarrow NH_4^+ + OH^-$
(3) $H_2O + H_2SO_4 \rightarrow H_3O^+ + HSO_4^-$
(4) $H_2O + H_2O \rightarrow H_3O^+ + OH^-$

20. Given the reactions X and Y below:
$$X: H_2O + NH_3 \rightarrow NH_4^+ + OH^-$$
$$Y: H_2O + HSO_4^- \rightarrow H_3O^+ + SO_4^{2-}$$
Which statement describes the behavior of the H_2O in these reactions?
(1) Water acts as an acid in both reactions.
(2) Water acts as a base in both reactions.
(3) Water acts as an acid in reaction X and as a base in reaction Y.
(4) Water acts as a base in reaction X and as an acid in reaction Y.

21. Given the equation:
$$H^+ + OH^- \rightarrow H_2O$$
Which type of reaction does the equation represent?
(1) esterification
(2) decomposition
(3) hydrolysis
(4) neutralization

22. Which equation represents a neutralization reaction?
(1) $Ca(OH)_2 \rightarrow Ca^{2+} + 2OH^-$
(2) $CaCl_2 \rightarrow Ca^{2+} + 2Cl^-$
(3) $H^+ + OH^- \rightarrow HOH$
(4) $H^+ + F^- \rightarrow HF$

23. How many milliliters of 0.20 M HCl are needed to exactly neutralize 40.0 milliliters of 0.40 M KOH?
(1) 2.0 mL   (2) 40.0 mL   (3) 80.0 mL   (4) 160 mL

24. How many mL of 4M HCl are needed to exactly neutralize 20 mL of 2M $Ca(OH)_2$?
Hint: Look at the number of OH^- in $Ca(OH)_2$.
(1) 2 mL   (2) 4 mL   (3) 10 mL   (4) 20 mL

25. A 3.0 milliliter sample of HNO_3 solution is exactly neutralized by 6.0 milliliters of 0.50 M KOH. What is the molarity of the HNO_3 sample?
(1) 1.0 M   (2) 0.50 M   (3) 3.0 M   (4) 1.5 M

26. Which type of reaction occurs when 50 milliliter quantities of $Ba(OH_2)(aq)$ and $H_2SO_4(aq)$ are combined?
(1) hydrolysis (2) ionization (3) hydrogenation (4) neutralization

27. If 50.0 milliliters of 3.0 M HNO_3 completely neutralized 150.0 milliliters of KOH, what was the molarity of the KOH solution?
(1) 1.0 M   (2) 4.5 M   (3) 3.0 M   (4) 6.0 M

28. What are the relative ion concentrations in an acid solution?
(1) more H^+ ions than OH^- ions
(2) fewer H^+ ions than OH^- ions
(3) an equal number of H^+ ions and OH^- ions
(4) H^+ ions, but no OH^- ions

29. What is the pH of a solution with a hydronium ion concentration of 0.01 mole per liter?
(1) 1   (2) 2   (3) 10   (4) 14

30. Which pH value indicates the most basic solution?
(1) 7   (2) 8   (3) 3   (4) 11

CONSTRUCTED RESPONSE QUESTIONS: Parts B-2 and C of NYS Regents Exam

Questions 23, 24, 25, 27, 29, without giving the 4 choices, can also be used here.

31. Compare and contrast acids and bases.

32. How can you determine whether a solution is an acid or a base?

33. Give two examples of Arrhenius acids and bases; explain why they are Arrhenius acids and bases.

34. In the reaction $HF + H_2O \longrightarrow H_3O^+ + F^-$, according to the alternate acid-base theory, explain which is the acid and which is the base. Explain why it is the acid and why it is the base.

35. A. Select acids from Table K and bases from Table L; write 4 neutralization reactions
B. Write the formula for *one* substance that can neutralize lake water affected by **acid** rain.

36. How many mL of 2M NaOH are required to exactly neutralize 100 mL of 3 M solution of HBr?

37. How many mL of 2M HBr are needed to exactly neutralize 20 mL of 4M KOH?

38. You have an unknown solution. Make up the results of using 2 indicators and show how you can figure out the small range of pH of the solution.

CHAPTER QUESTION: Parts B-2 and C of NYS Regents Exam

39. A titration setup was used to determine the unknown molar concentration of a solution of NaOH. A 1.2 M HCl solution was used as the titration standard. The following data were collected.

	Trial 1	Trial 2	Trial 3	Trial 4
Amount of HCl standard used	10.0 mL	10.0 mL	10.0 mL	10.0 mL
Initial NaOH burette reading	0.0 mL	12.2 mL	23.3 mL	35.2 mL
Final NaOH burette reading	12.2 mL	23.2 mL	35.2 mL	47.7 mL

A. Calculate the volume of NaOH solution used to neutralize 10.0 mL of the standard HCl solution in trial 3. Show your work.
B. According to Reference Table M, what indicator would be most appropriate in determining the end point of the titration? Give one reason for choosing this indicator.
C. Calculate the average molarity of the unknown NaOH solution for all four trials. Your answer must include the correct number of significant figures and correct units.
Hint: Review significant numbers, pages 12:8-12:12.

THIS SECTION IS NOT IN THE NY STATE REGENTS CURRICULUM

HYDROLYSIS

Hydrolysis is when **salt reacts** with **water** to form **solutions** that are **acidic or basic**.

RELATIVE STRENGTHS OF ACIDS

ACID		BASE	K_a
HI	=	$H^+ + I^-$	very large
HBr	=	$H^+ + Br^-$	very large
HCl	=	$H^+ + Cl^-$	very large
HNO_3	=	$H^+ + NO_3^-$	very large
H_2SO_4	=	$H^+ + HSO_4^-$	large
HSO_4^-	=	$H^+ + SO_4^{2-}$	1.2×10^{-2}
H_3PO_4	=	$H^+ + H_2PO_4^-$	**7.5×10^{-3}**
H_2S	=	$H^+ + HS^-$	9.5×10^{-8}
NH_4^+	=	$H^+ + NH_3$	5.7×10^{-10}
HS^-	=	$H^+ + S^{2-}$	**1.3×10^{-14}**
H_2O	=	$H^+ + OH^-$	1×10^{-14}
NH_3	=	$H^+ + NH_2^-$	very small

Hydrolysis:

$$Na_2CO_3 + 2H_2O \longrightarrow 2NaOH + H_2CO_3$$

Salt   Water   **Strong base**: Metals of Group 1 form strong bases. Na is a metal in Group 1.   **Weak acid**: The top 5 on this table are strong acids: HI, HBr, HCl, HNO_3 and H_2SO_4. The acids below the top 5 are called weak acids.

Strong base (more OH^-) and *weak acid* makes a **solution that is basic**.

Hydrolysis:

$$NH_4I + H_2O \longrightarrow HI + NH_4OH$$

Salt   Water   Top of table, strong acid   Weak base

Strong acid (more H^+) and *weak base* makes a **solution that is acidic**.

NOT Hydrolysis:

$$KCl + H_2O \longrightarrow KOH + HCl$$

Salt   Water   strong base   strong acid

A *strong acid* plus a *strong base* is **not hydrolysis** (no extra H^+ or OH^-), solution is *not* acidic and *not* basic.

Table of Relative Strengths of Acids

CONJUGATE ACID-BASE PAIRS

CONJUGATE BASE:

$$HSO_4^- + H_2O \rightleftharpoons H_3O^+ + SO_4^{2-}$$

Acid 1   Conjugate Base 1

When a Bronsted-Lowry acid (such as HSO_4^-) gives up a proton, the remainder of the acid (SO_4^{2-}) becomes capable of accepting a proton. SO_4^{2-} can be considered a base. It is called a **conjugate base**. A **conjugate base** is what remains after the acid gives up a proton.

Question: If HSO_4^- is the acid, what is the conjugate base?

Solution: Cover the H^+ in HSO_4^- with your finger. Take away H^+ from the acid. What is left is the **conjugate base**, SO_4^{2-}.

Question: What is the conjugate base of HBr?

Solution: Cover the H^+ in HBr with your finger. Take away H^+ from the acid. The **conjugate base** is Br^-.

RULE: When the question is, "**What is the conjugate base?**", take away H^+ from the acid and what is left is the conjugate base.

CONJUGATE ACID:

$$HSO_4^- + H_2O \rightleftharpoons H_3O^+ + SO_4^{2-}$$

Acid 1   Base 2   Conjugate   Conjugate
             (accepts   **Acid 2**   Base 1
             protons
)

H_2O is a base (proton acceptor). It accepts a proton and becomes H_3O^+. H_3O^+ is the **conjugate acid**. A **conjugate acid** is what is formed when a base accepts a proton.

HINT: When the question asks, "**What is a conjugate acid?**", **add H^+ to the base.** ("Acid" begins with "a," "add" begins with "a;" add H^+ to the base.)

Question: What is the conjugate acid of H_2O?

Solution: Add H^+ to H_2O (conjugate **base**) and it **becomes H_3O^+ (conjugate acid)**.

FINDING A CONJUGATE ACID-BASE PAIR

In the previous example, Acid 1 and Conjugate Base 1 are a conjugate acid-base pair, HSO_4^- and SO_4^{2-}. Base 2 and Conjugate Acid 2, H_2O and H_3O^+, are also a conjugate acid-base pair.

As you can see, the two parts of a conjugate acid-base pair differ only by one H^+.

RULE: When the question asks for **conjugate acid-base pair**, choose the answer where the two things **differ only by one H^+ and nothing else.**

Question: In the reaction
$$H_2PO_4^- + H_2O \rightleftharpoons H_3PO_4 + OH^-,$$
which pair represents an acid and its conjugate base?

    (1) H_2O and $H_2PO_4^-$ (3) H_3PO_4 and OH^-

    (2) H_2O and H_3PO_4 (4) H_3PO_4 and $H_2PO_4^-$

Solution: Answer *4*. The question asks for an acid and its conjugate base, a **conjugate acid-base pair.** You learned that a conjugate acid-base pair **DIFFERS ONLY BY ONE H^+, NOTHING ELSE.**

Answer *4* is correct: H_3PO_4 and $H_2PO_4^-$ **DIFFER ONLY BY ONE H^+ ION.** Remember the conjugate acid-base pair differs only by one H^+.

Choice 1 is wrong: H_2O and $H_2PO_4^-$ differ by the number of oxygen atoms and the presence of phosphorus.

Choice 2 is wrong: H_2O and H_3PO_4 differ by the number of H and O atoms and the presence of P.

Choice 3 is wrong: H_3PO_4 and OH^- differ by the presence of P and the number of H and O atoms.

Try Sample Questions #1-2, on page 23, then do Homework Questions, #1-6, page 25.

IONIZATION CONSTANT FOR ACIDS, K_a
(The "a" in K_a stands for acids.)
You learned about the ionization constant in Chapter 6:

$$HCl \rightarrow H^+ + Cl^- \qquad K_a = \frac{[H^+][Cl^-]}{[HCl]} \qquad \frac{products}{reactants}$$

The **bigger** the **ionization constant**, K_{ion}, the **more ions** there are and the **greater** the degree of **ionization**. A **strong** acid has a LARGE IONIZATION CONSTANT, K_a, MORE IONS; a **weak** acid has a SMALL IONIZATION CONSTANT, K_a, LESS IONS.

HOW TO USE TABLE OF RELATIVE STRENGTHS OF ACIDS
This table is NOT part of the New York State Regents curriculum.
Look at the table called **Relative Strengths of Acids**, below or on page Additional Tables 4. The **strongest acids** are at the **top** of this table. HI, HBr, HCl, HNO_3 and H_2SO_4 are **strong** acids. Under K_a (the ionization constant) for these acids is written "very large" or "large;" this means that there are a lot of ions, a lot of H^+, and they are strong acids.

RELATIVE STRENGTHS OF ACIDS

Conjugate Pairs:		K_a
ACID	BASE	
HI $=$	$H^+ + I^-$	very large
HBr $=$	$H^+ + Br^-$	very large
HCl $=$	$H^+ + Cl^-$	very large
HNO_3 $=$	$H^+ + NO_3^-$	very large
H_2SO_4 $=$	$H^+ + HSO_4^-$	large
HSO_4^- $=$	$H^+ + SO_4^{2-}$	1.2×10^{-2}
H_3PO_4 $=$	$H^+ + H_2PO_4^-$	7.5×10^{-3}
H_2S $=$	$H^+ + HS^-$	9.5×10^{-8}
NH_4^+ $=$	$H^+ + NH_3$	5.7×10^{-10}
HS^- $=$	$H^+ + S^{2-}$	1.3×10^{-14}
H_2O $=$	$H^+ + OH^-$	1×10^{-14}
NH_3 $=$	$H^+ + NH_2^-$	very small

As you go down this table, the **acids get weaker**, and the **value of K_a** (the Ionization Constant) **gets smaller**. HS^- has $K_a = 1.3 \times 10^{-14}$ (very small), near the bottom of the table, therefore it is a very weak acid.

REMEMBER: A **bigger negative exponent** is a **smaller K_a**.

By looking at the table above, you can see that HS^- is a very weak acid, because it is near the **bottom** of the table, with a K_a of 1.3×10^{-14} (big negative exponent is a small K_a).

By looking at the table, you can see that HS^- is weaker than H_3PO_4. HS^- is lower down on the table than H_3PO_4, which shows that HS^- is a **weaker** acid than H_3PO_4.

The K_a of HS^- (1.3×10^{-14}) is smaller than the K_a of H_3PO_4 (7.5×10^{-3}), which shows that HS^- is a weaker acid than H_3PO_4.

TABLE OF RELATIVE STRENGTHS OF ACIDS:
FINDING CONJUGATE ACID-BASE PAIRS

By looking at the Table of Relative Strengths of Acids, on the previous page or on page Additional Tables 4, you can also **find** the **conjugate acid** of a base. **Find** the **conjugate acid** of HS^- (base). Find HS^- under base, go across, and it

RELATIVE STRENGTHS OF ACIDS

Conjugate Pair:		
ACID	BASE	K_a
HI	= $H^+ + I^-$	very large
H_2S	= $H^+ + HS^-$	9.5×10^{-8}
HS^-	= $H^+ + S^{2-}$	1.3×10^{-14}

has H_2S under acid. See the table, at right, and find it on the full table, on the first page of Section B.

The **top** of this table has the **strongest conjugate acids** and **weakest conjugate bases**. HI (strongest acid) gives I^- (weakest conjugate base). As you go down the table, the conjugate acids get weaker and the conjugate bases get stronger.

TABLE OF RELATIVE STRENGTHS OF ACIDS:
FINDING AMPHOTERIC SUBSTANCES

Amphoteric substances can sometimes be an acid and sometimes be a base. On the Table of Relative Strengths of Acids, **HS⁻** is in the **acid column** and also in the **base column**. Therefore, you know HS^- is **amphoteric**. Similarly, NH_3, or **any substance** in **both** the **acid and base columns** is **amphoteric**.

RELATIVE STRENGTHS OF ACIDS

Conjugate Pair:		
ACID	BASE	K_a
H_2S	= $H^+ + HS^-$	
HS^-	= $H^+ + S^{2-}$	

See the table at right, on the previous page, or on page Additional Tables 4.

TABLE OF RELATIVE STRENGTHS OF ACIDS:
CONDUCTORS OF ELECTRICITY

As you know, the **top** of the Table of Relative Strengths of Acids is the strongest acid, ionizes the most (biggest K_a, forms the most ions) and therefore is the **best conductor of electricity** (strongest electrolyte, bulb lights brightest).

As you go down this table, the acid gets weaker, ionizes less, and conducts electricity less (weaker electrolyte).

Question: Based on the Table of Relative Strengths of Acids, which of the following 0.1 M aqueous solutions is the best conductor of electricity?

    (1) HF   (2) H_2S (3) HNO_3 (4) CH_3COOH

Solution: Answer 3. Look at the Table of Relative Strengths of Acids. Answer 3, HNO_3, is highest on the table, therefore ionizes the most and is the best conductor of electricity.

Try Sample Questions #3-5, on page 23, and then do Homework Questions, #7-17, pages 25-26.

IONIZATION CONSTANT OF WATER, K_w

(The "w" in K_w stands for water.)

The **IONIZATION CONSTANT**, K_w, of water = the product of the **hydrogen (H^+)** ion concentration **times** the **hydroxide (OH^-)** ion concentration = 1×10^{-14}.

$$K_w = [H^+] \quad [OH^-] = 1 \times 10^{-14}.$$
$$K_w \text{ (pure water)} = [1 \times 10^{-7}][1 \times 10^{-7}] = 1 \times 10^{-14}.$$

In pure water, $H^+ = OH^-$ ($H^+ = 1 \times 10^{-7}$, $OH^- = 1 \times 10^{-7}$).

If one forgets the ionization constant of water, just look on the **Table of Relative Strengths of Acids**, in the beginning of Section B, or on the **Table of Constants for Various Equilibria**, on page Additional Tables 5.

At the bottom of the **Table of Relative Strengths of Acids** is written:

$$H_2O = H^+ + OH^- \quad K = 1 \times 10^{-14}.$$

At the top of the **Table of Constants for Various Equilibria**:

$$H_2O = H^+ + OH^- \quad K_w = 1 \times 10^{-14}.$$

Both of these entries mean that the ionization constant of H_2O is 1×10^{-14}.

In **acid and base solutions**, $H^+ \times OH^- = 1 \times 10^{-14}$, because the acids and bases are in water.

In **acid solutions** (acid in water), there is **more H^+** (more H_3O^+, because H^+ and H_2O yields H_3O^+) and **less OH^-**.

$$H^+ \times OH^- = 1 \times 10^{-14}.$$
$$K_w = [H^+][OH^-] = 1 \times 10^{-14}.$$

Examples of an acid solution:

  (1) $[H^+] = 1 \times 10^{-1}$ $[OH^-] = 1 \times 10^{-13}$
$$K_w = [H^+][OH^-] = 1 \times 10^{-14}$$
$$K_w = [1 \times 10^{-1}][1 \times 10^{-13}] = 1 \times 10^{-14}.$$

---

(2) $[H^+] = 1 \times 10^{-4}$ $[OH^-] = 1 \times 10^{-10}$

$$K_w = [H^+][OH^-] = 1 \times 10^{-14}$$
$$K_w = [1 \times 10^{-4}][1 \times 10^{-10}] = 1 \times 10^{-14}.$$

In **acid solutions**, H^+ has a smaller negative exponent, which means there is more H^+.

A **basic solution** (base in water) has more OH^- and less H^+ (less H_3O^+, because H^+ and $H_2O \rightarrow H_3O^+$): $H^+ \times OH^- = 1 \times 10^{-14}$.

Examples of a basic solution:

(1) $[H^+] = 1 \times 10^{-8}$ $[OH^-] = 1 \times 10^{-6}$

$$K_w = [H^+][OH^-] = 1 \times 10^{-14}$$
$$K_w = [1 \times 10^{-8}][1 \times 10^{-6}] = 1 \times 10^{-14}.$$

(2) $[H^+] = 1 \times 10^{-11}$ $[OH^-] = 1 \times 10^{-3}$

$$K_w = [H^+][OH^-] = 1 \times 10^{-14}$$
$$K_w = [1 \times 10^{-11}][1 \times 10^{-3}] = 1 \times 10^{-14}.$$

Question: If H^+ concentration is 1×10^{-2}, what is the OH^- (hydroxide ion) concentration?

Solution:
$$K_w = [H^+][OH^-] = 1 \times 10^{-14}.$$
$$K_w = [1 \times 10^{-2}][\mathbf{1 \times 10^{-12}}] = 1 \times 10^{-14}$$
$$(10^{-2} \times \mathbf{10^{-12}} = 10^{-14})$$

As you can see, $[OH^-] = 1 \times 10^{-12}$.

Reminder: In multiplication, you add exponents.

Or, $[H^+][OH^-] = 1 \times 10^{-14}$.

$$OH^- = \frac{1 \times 10^{-14}}{H^+} = \frac{1 \times 10^{-14}}{1 \times 10^{-2}} = 1 \times 10^{-12} \; OH^- \; Concentration.$$

Reminder: In division, subtract exponents.

Question: If $[OH^-] = 1 \times 10^{-12}$, what is H^+?

Solution:
$$K_w = [H^+][OH^-] = 1 \times 10^{-14}.$$
$$K_w = [\mathbf{1 \times 10^{-2}}][1 \times 10^{-12}] = 1 \times 10^{-14}$$
$$(\mathbf{10^{-2}} \times 10^{-12} = 10^{-14})$$

As you can see, $[H^+] = 1 \times 10^{-2}$.

Reminder: In multiplication, you add exponents.

Or, $[H^+][OH^-] = 1 \times 10^{-14}$.

$$H^+ = \frac{1 \times 10^{-14}}{OH^-} = \frac{1 \times 10^{-14}}{1 \times 10^{-12}} = 1 \times 10^{-2} \; H^+ \; Concentration.$$

Reminder: In division, subtract exponents.

Try Sample Questions #6-8, on page 23, then do
Homework Questions, #18-24, page 26.

SAMPLE REGENTS & REGENT-TYPE QUESTIONS
AND SOLUTIONS

1. Which compound is a salt?
    (1) CH_3OH (2) $C_6H_{12}O_6$ (3) $H_2C_2O_4$ (4) $KC_2H_3O_2$

2. The conjugate acid of the HS^- ion is
    (1) H^+ (2) S^{2-} (3) H_2O (4) H_2S

3. According to the alternate acid-base theory, which of the following is the *weakest* acid?
    (1) NH_4^+ (2) HSO_4^- (3) H_2SO_4 (4) HNO_3

4. According to the *Table of Relative Strengths of Acids*, what is the conjugate acid of the hydroxide ion (OH^-)?
    (1) O^{2-} (2) H^+ (3) H_2O (4) H_3O^+

5. According to the *Table of Relative Strengths of Acids*, which species is amphoteric (amphiprotic)?
    (1) HCl (2) HNO_2 (3) HSO_4^- (4) H_2SO_4

6. The value of the ionization constant of water, K_{ion}, will change when there is a change in
    (1) temperature   (3) hydrogen ion concentration
    (2) pressure   (4) hydroxide ion concentration

7. The $[OH^-]$ of a solution is 1×10^{-6}. At 298 K and 1 atmosphere, the product $[H_3O^+][OH^-]$ is
    (1) 1×10^{-2} (2) 1×10^{-6} (3) 1×10^{-8} (4) 1×10^{-14}

8. The concentration of hydrogen ions in a solution is 1.0×10^{-5} M at 298 K. What is the concentration of hydroxide ions in the same solution?
    (1) 1.0×10^{-14} M   (3) 1.0×10^{-7} M
    (2) 1.0×10^{-9} M   (4) 1.0×10^{-5} M

SOLUTIONS

1. Answer **4**. You learned that salt (example, $NaCl$) is an ionic compound that has positive ions other than hydrogen and negative ions other than hydroxide.

    Look at $_{19}K$ on the Periodic Table. In the upper right hand corner of the box in the Periodic Table is written "+1." K is a positive ion.

+1
K
19

Table E	
Polyatomic Ions	
NH_4^+	ammonium
$C_2H_3O_2^-$	acetate
CO_3^{2-}	carbonate

Look at the oxidation number of $C_2H_3O_2$ on Table E (for polyatomic ions). Table E has $C_2H_3O_2^-$ (negative one); therefore, $C_2H_3O_2^-$ is a negative ion. $K^+C_2H_3O_2^-$ is a salt made up of a positive and a negative ion.

2. Answer **4**. You learned that, to find the conjugate acid, add H^+ (to the base). (Acid begins with the letter "a;" add begins with the letter "a.)

$$H^+ \text{ and } HS^- \rightarrow H_2S$$
$$\text{base} \quad \text{conjugate acid}$$

Or, look at the Table of Relative Strengths of Acids. Find HS^- under base; go across, and it has H_2S under acid.

Table of Relative Strengths of Acids

Conjugate Pairs		
Acid	*Base*	K_a
$H_2S = H^+ + HS^-$		9.5×10^{-8}

3. Answer **1**. According to the alternate acid-base theory, the acid lowest down on the Table of Relative Strengths of Acids is the weakest acid. (In this theory, an acid is a proton donor.) Among the four choices, NH_4^+ is the lowest on this table, and therefore is the weakest acid.

Table of Relative Strengths of Acids

Conjugate Pairs		
Acid	*Base*	K_a
$HNO_3 = H^+ + NO_3^-$		very large
$H_2SO_4 = H^+ + HSO_4^-$		large
$HSO_4^- = H^+ + SO_4^{2-}$		1.2×10^{-2}
$NH_4^+ = H^+ + NH_3$		5.7×10^{-10}

4. Answer **3**. You learned that, to find the conjugate acid, add H^+ (to the base):

$$H^+ + OH^- \rightarrow H_2O$$
$$\text{base} \quad \text{conjugate acid}$$

Or, look at the Table of Relative Strengths of Acids. Find OH^- under base; go across, and it has H_2O under acid.

Table of Relative Strengths of Acids

Conjugate Pairs		
Acid	*Base*	K_a
$H_2O = H^+ + OH^-$		1.0×10^{-14}

5. Answer **3**. HSO_4^- is in the acid column and the base column of the Table of Relative Strengths of Acids. This means that HSO_4^- is amphoteric; it can sometimes be an acid and sometimes a base.

Rule: Any substance in both the acid and the base column is amphoteric (amphiprotic).

Table of Relative Strengths of Acids

Conjugate Pairs		
Acid	*Base*	K_a
$H_2SO_4 = H^+ + HSO_4^-$		Large
$HSO_4^- = H^+ + SO_4^-$		1.2×10^{-2}

6. Answer **1**. K_w, the ionization constant for water, is one type of equilibrium constant. The equilibrium constant only changes with temperature.

7. Answer **4**. $K_w = [H^+][OH^-] = 1 \times 10^{-14}$. K_w = product of H^+ and OH^- (in solution) $= 1 \times 10^{-14}$. As you learned, you can see that $K_w = 1 \times 10^{-14}$ on the bottom of the Table of Relative Strengths of Acids or the top of the Table of Constants for Various Equilibria.

8. Answer **2**. Hydroxide ions are OH^-, as shown in Table E, on page Reference Tables 5. $K_w = [H^+][OH^-] = 1 \times 10^{-14}$
$$[1 \times 10^{-5}][1 \times 10^{-9}] = 1 \times 10^{-14}$$
$$10^{-5} \times 10^{-9} = 10^{-14}$$

Remember, in **multiplying**, you **add** exponents.

Or, $[H^+][OH^-] = 1 \times 10^{-14}$

$$OH^- = \frac{1 \times 10^{-14}}{H^+} = \frac{1 \times 10^{-14}}{1 \times 10^{-5}} = 1 \times 10^{-9} \text{ hydroxide ions } (OH^-)$$

Remember, in **division**, **subtract** exponents.

NOW LET'S TRY A FEW HOMEWORK QUESTIONS:

1. Which compound is a salt?
   (1) C_2H_5OH (2) $C_{12}H_{22}O_{11}$ (3) CO_2 (4) $NaC_2H_3O_2$

2. Potassium chloride, KCl, is a salt derived from the neutralization of
   (1) a weak acid and a weak base
   (2) a weak acid and a strong base
   (3) a strong acid and a weak base
   (4) a strong acid and a strong base

3. Which type of reaction is represented by the following equation?
   $$Al_2S_3 + 6H_2O \rightarrow 2Al(OH)_3 + 3H_2S$$
   (1) neutralization   (3) electrolysis
   (2) dehydration   (4) hydrolysis

4. What is the conjugate base of NH_3?
   (1) NH_4^+ (2) NH_2^- (3) NO_3^- (4) NO_2^-

5. Given the reaction:
   $$HSO_4^- + H_2O \rightleftharpoons H_3O^+ + SO_4^{2-}$$
   Which is a Bronsted-Lowry conjugate acid-base pair?
   (1) HSO_4^- and H_3O^+ (3) H_2O and SO_4^{2-}
   (2) HSO_4^- and SO_4^{2-} (4) H_2O and HSO_4^-

6. In the reaction
   $$HSO_4^- + H_2O \rightleftharpoons H_3O^+ + SO_4^{2-},$$
   an acid-base conjugate pair is
   (1) HSO_4^- and SO_4^{2-} (3) SO_4^{2-} and H_3O^+
   (2) HSO_4^- and H_2O (4) SO_4^{2-} and H_2O

7. Which of the following acids is the weakest?
   (1) H_2S (2) HF (3) H_3PO_4 (4) HNO_2

8. According to the *Table of Relative Strengths of Acids*, what is the conjugate acid of the hydroxide ion (OH^-)?
   (1) O^{2-} (2) H^+ (3) H_2O (4) H_3O^+

9. Which of the following aqueous solutions is the *poorest* conductor of electricity? [Refer to the *Table of Relative Strengths of Acids*.]
   (1) 0.1 M H_2S (3) 0.1 M HNO_2
   (2) 0.1 M HF (4) 0.1 M HNO_3

10. According to the *Table of Relative Strengths of Acids*, which of the following is the strongest Bronsted-Lowry acid?
    (1) HS^- (2) H_2S (3) HNO_2 (4) HNO_3

11. Based on the *Table of Relative Strengths of Acids*, which species is amphoteric?
    (1) NH_2^- (2) NH_3 (3) I^- (4) HI

12. According to the *Table of Relative Strengths of Acids*, which species is amphoteric (amphiprotic)?
    (1) HS^- (2) HCl (3) NH_4^+ (4) HBr

13. According to the *Table of Relative Strengths of Acids*, which molecule is amphiprotic?
    (1) HCl (2) H_2SO_4 (3) NH_3 (4) H_2S

14. Based on the *Table of Relative Strengths of Acids*, which solution best conducts electricity?
    (1) 0.1 M HCl   (3) 0.1 M H_2S
    (2) 0.1 M CH_3COOH (4) 0.1 M H_3PO_4

15. Which of the following compounds is the strongest electrolyte?
    (1) NH_3 (2) H_2O (3) H_3PO_4 (4) H_2SO_4

16. Based on the *Table of Relative Strengths of Acids*, which of the following aqueous solutions is the best conductor of electricity?
    (1) 0.1 M HF   (3) 0.1 M H_2SO_4
    (2) 0.1 M H_2S (4) 0.1 M H_3PO_4

17. Which of the following is the *weakest* Bronsted-Lowry acid?
    (1) NH_4^+ (2) HSO_4^- (3) H_2SO_4 (4) HNO_3

18. A solution of a base differs from a solution of an acid in that the solution of a base
    (1) is able to conduct electricity
    (2) is able to cause an indicator color change
    (3) has a greater $[H_3O^+]$
    (4) has a greater $[OH^-]$

19. What is the hydroxide ion concentration of a solution that has a hydronium ion concentration of 1×10^{-9} mole per liter at 298K?
    (1) 1×10^{-5} mole per liter (3) 1×10^{-9} mole per liter
    (2) 1×10^{-7} mole per liter (4) 1×10^{-14} mole per liter

20. What is the H_3O^+ ion concentration of a solution whose OH^- ion concentration is 1×10^{-3} M?
    (1) 1×10^{-4} M (2) 1×10^{-7} M (3) 1×10^{-11} M (4) 1×10^{-14} M

21. An aqueous solution with a pH of 4 would have a hydroxide ion concentration of
    (1) 1×10^{-4} mol/L (3) 1×10^{-10} mol/L
    (2) 1×10^{-7} mol/L (4) 1×10^{-14} mol/L

22. Which 0.1 M solution has a pH greater than 7?
    (1) $C_6H_{12}O_6$ (2) CH_3COOH (3) KCl (4) KOH

23. When the salt Na_2CO_3 undergoes hydrolysis, the resulting solution will be
    (1) acidic, with a pH less than 7
    (2) acidic, with a pH greater than 7
    (3) basic, with a pH less than 7
    (4) basic, with a pH greater than 7

24. What is the pH of a solution that has an OH^- ion concentration of 1×10^{-5} mole per liter ($K_W = 1 \times 10^{-14}$)?
    (1) 1 (2) 5 (3) 7 (4) 9

CHAPTER 9:
OXIDATION / REDUCTION

SECTION A

OXIDATION AND REDUCTION

OXIDATION means LOSS of electrons. Hint: **O**xidation has the letter "**O**," **l**oss of electrons has the letter "**o**;" **o**xidation is **l**oss of electrons.

You learned metals lose electrons to have a complete outer shell (more stable, lower energy). $^{23}_{11}$Na (sodium) is a metal. Na **loses** an electron. (An electron can be shown by the e⁻).

An 11^{Na} 2 8 1 atom **breaks up** into a 11^{Na} 2 8 ion and **1 electron**.

$$Na \xrightarrow{\text{(breaks up into)}} Na^+ + e^-$$

Na loses the one electron (and becomes Na⁺ ion).

Sodium is **oxidized**. It is a **reducing agent**. (Na loses an electron and causes something else to gain the electron and be reduced.)

Look at the **Periodic Table**, on page Reference Tables 20-21. In every box of the Periodic Table, in the upper right hand corner, is the **oxidation number**.

Metals in Group 1 (for example, Na) have +1 oxidation number. Metals in Group 1 lose one electron (are oxidized) when they form a compound and have +1 oxidation number.

A 20^{Ca} 2 8 8 2 atom **breaks up** into a 20^{Ca} 2 8 8 ion and **2 electrons**.

$$Ca \xrightarrow{\text{(breaks up into)}} Ca^{2+} + 2e^-$$

Ca loses the two electrons (see above) and becomes Ca^{2+} ion.

Remember: Electrons (e⁻ or 2e⁻, etc.) after the arrow means Ca (before the arrow) loses electrons. Calcium loses **two electrons.** Calcium is **oxidized**. It is a **reducing agent**. Calcium loses two electrons and causes something else to gain electrons or be reduced.

Look at the **Periodic Table**. In every box of the Periodic Table, in the upper right hand corner, is the oxidation number. Metals in Group 2 (for example, Ca) have +2 oxidation number. Metals in Group 2 lose two electrons (are oxidized) when they form a compound and have a +2 oxidation number.

Any **free element** (not combined with any other element) has an oxidation number of **zero**. Example: Na = Na°.

High Marks: Regents Chemistry Made Easy

When any atom **loses** electrons, the oxidation number increases.

$Na° \rightarrow Na^+ + e^-$: Oxidation number increases from 0 to +1.
(Na^+ means Na^{1+}, but the 1 is understood and is not written.)

$Ca° \rightarrow Ca^{2+} + 2e-$: Oxidation number increases from 0 to +2.

Question: What happens to reducing agents in chemical reactions?
  (1) Reducing agents gain protons.   (3) Reducing agents are oxidized.
  (2) Reducing agents gain electrons.   (4) Reducing agents are reduced.

Solution: Answer **3**. You learned that Na loses an electron. Na is **oxidized**, Na is a **reducing agent**, because Na loses an electron and causes something else to be reduced, or gain the electron. **Oxidized = reducing agent.**

Question: Given the reaction:
$$2\,Na + 2\,H_2O \rightarrow 2\,Na^+ + 2\,OH^- + H_2,$$
Which substance is oxidized?
  (1) H_2 (2) H^+ (3) Na   (4) Na^+

Solution: Answer **3**. Remember, you learned that Na **loses an electron**, Na is **oxidized**. Elements of Group 1 **lose electrons** and are **oxidized**: $Na \rightarrow Na^+ + e^-$.

Question: Find the missing reactant in the balanced equation:
$$\underline{} \longrightarrow Ca^{2+} + 2e^-$$

Solution: Ca loses $2e^-$ (see previous page). Ca breaks up into $Ca^{2+} + 2e^-$. Therefore, the missing reactant is **Ca**.

REMEMBER: When a question asks for a **reducing agent**, find the substance that is oxidized:

OXIDIZED = REDUCING AGENT
REDUCING AGENT = OXIDIZED

Try Sample Questions #1-3, on page 7, then do Homework Questions, #1-8, pages 9-10, and #31-32, page 11.

Fluorine $^{19}_{9}F$ $\left(\frac{9p}{10n}\right)$)) 2 7

F gains one electron.
$F° + e \rightarrow F^-$
gains

REDUCTION means **GAIN** of electrons. Electrons are negative.

Fluorine gains one electron to have 8 electrons in the outer shell, a complete outer shell.

F **gains** an electron. F is **reduced**. It is an **oxidizing** agent. (F gains an electron and causes something else to be oxidized (lose an electron).

Oxygen $^{16}_{8}O$ $\left(\frac{8p}{8n}\right)$)) 2 6

O gains two electrons.
$O° + 2e \rightarrow O^{2-}$
gains

Look at the **Periodic Table**.

Oxygen gains two electrons to have a complete outer shell. O **gains** 2 electrons. O is **reduced**. O is an **oxidizing agent** (causes something else to be oxidized).

Look at the Periodic Table on page Reference Tables 20-21. Elements (nonmetals) in Group 16 (for example, O) have -2 oxidation number. Nonmetals in Group 16 gain 2 electrons when they form a compound and have -2 oxidation number.

Any free element (not combined with any other element) has an oxidation number of zero. When any atom **gains** electrons (negative particles), the oxidation number decreases.

$F° + e → F^-$; its oxidation number decreases from zero to negative 1.

$O° + 2e → O^{2-}$; its oxidation number decreases from zero to negative 2.

Question: In the reaction

$$2H_2(g) + O_2(g) → 2H_2O(g),$$

the oxidizing agent is

   (1) H_2 (2) O_2 (3) H^+ (4) O^{2-}

Solution: Answer 2. You learned that oxygen (O_2) gains electrons. Oxygen is reduced; it is an oxidizing agent (causes something else to be oxidized).

Oxygen is **reduced** = **oxidizing agent**.

$$2H_2 + O_2 → 2H_2O$$
$$\quad\quad O° \quad\quad\quad O^{2-}$$

oxidation number of   *oxidation number of oxygen*
free oxygen = 0   *in a compound = -2*

Question: Find the missing product in the balanced equation:

$$Cl_2 + 2e^- \longrightarrow \underline{\quad\quad}$$

Solution: Cl gains one electron, forming Cl^-. Cl_2 gains two electrons, forming 2 Cl^-. The missing product is **2 Cl^-**.

Remember: When a question asks for an **oxidizing agent**, find the substance that is reduced:

      REDUCED = OXIDIZING AGENT
      OXIDIZING AGENT = REDUCED

Try Sample Questions #4-6, page 7, then do Homework Questions #9-15, page 10, and #33-35, page 11.

REDOX REACTIONS

REDOX REACTIONS have **ox**idation and **red**uction reactions. In the word *redox*, "red" means reduction and "ox" means oxidation. One atom loses electrons (is oxidized) and the other atom gains electrons (is reduced). There is a **transfer of electrons** from one atom

$^{23}_{11}Na \left(\frac{11p}{12n}\right) \,2\,8\,1$

Oxidation
Na loses 1 electron.
$Na → Na^+ + e-$

$7\,2 \left(\frac{9p}{10n}\right) ^{19}_9 F$

Reduction
F gains 1 electron.
$F + e- → F^-$

(example Na) to another atom (example F) in a redox reaction. In redox reactions, the total number of electrons lost is equal to the total number of electrons gained.

Redox reactions (see equation in drawing on previous page), like all reactions, have **conservation of matter** (Na atom or Na ion before and after the arrow) and **conservation of charge** (Na°, **zero charge, before** the **arrow**; Na$^+$ (1+ charge) and e$^-$ (1− charge) = **zero charge, after** the **arrow**). The charges are the same on both sides of the equation.

Question: Which equation represents a redox reaction?
- (1) $2Na^+ + S^{2-} \rightarrow Na_2S$
- (2) $H^+ + C_2H_3O_2^- \rightarrow HC_2H_3O_2$
- (3) $NH_3 + H^+ + Cl^- \rightarrow NH_4^+ + Cl^-$
- (4) $Cu + 2Ag^+ + 2NO_3^- \rightarrow 2Ag + Cu^{2+} + 2NO_3^-$

Solution: Answer 4. In a redox reaction, the oxidation numbers change. $Cu \rightarrow Cu^{2+} + 2e^-$. The oxidation number of Cu changes from zero to +2. $Ag^+ + e^- \rightarrow Ag^0$; $2Ag^+ + 2e^- \rightarrow 2Ag^0$. The oxidation number of Ag changes from +1 to zero. Any element alone has an oxidation number of zero, for example Ag^0.

You can easily recognize a redox reaction when you have a **free element** by **itself**, example Cu, on one side of the arrow (equation) and an ion, Cu^{2+} (or a compound, $CuCl_2$) on the other side of the arrow.

Read again the question and solution above. In a redox reaction, the **metal**, such as Cu, **loses electrons** and **gives** (transfers) the **electrons** to the **ion** such as Ag$^+$, which **gains an electron.**

Question: Which is a redox reaction?
- (1) $H^+ + Cl^- \rightarrow HCl$
- (3) $NaOH + HCl \rightarrow NaCl + H_2O$
- (2) $Fe + 2 HCl \rightarrow FeCl_2 + H_2$
- (4) $MgO + H_2SO_4 \rightarrow MgSO_4 + H_2O$

Solution: Answer 2. You can easily recognize a redox reaction when you have a **free element** by **itself**, example Fe, on one side of the arrow (equation) and Fe in a compound, example $FeCl_2$ (or Fe ion, Fe^{2+}) on the other side of the arrow.

You can recognize $PCl_5 \rightarrow PCl_3 + Cl_2$ is also a redox reaction. The free element Cl_2 is on one side of the arrow (after the arrow) and Cl in PCl_5 is on the other side of the arrow (before the arrow). In a redox reaction, the oxidation numbers of the elements change before and after the arrow.

A half-reaction shows only the oxidation or reduction part of a redox reaction. See the two examples (questions) below.

Question: In the (redox) reaction $2Mg + O_2 \longrightarrow 2MgO$, which half-reaction correctly represents oxidation?
- (1) $Mg + 2e^- \rightarrow Mg^{2+}$
- (3) $Mg^{2+} \rightarrow Mg + 2e^-$
- (2) $Mg^{2+} + 2e^- \rightarrow Mg$
- (4) $Mg \rightarrow Mg^{2+} + 2e^-$

Solution: Answer 4 is an oxidation reaction, because Mg loses two electrons and goes from Mg° to Mg^{2+}, which is oxidation.

Question: In the redox reaction in the question above, which half reaction correctly represents reduction?

(1) $O_2 + 4e^- \rightarrow 2O^{2-}$

(2) $2O^{2-} + 4e^- \rightarrow O_2$

(3) $2O^{2-} \longrightarrow O_2 + 4e^-$

(4) $O_2 \longrightarrow 2O^{2-} + 4e^-$

Solution: Answer *1* is a reduction reaction, because O_2 gains four electrons and goes from $O_2{}^0$ to $2O^{2-}$, which is reduction.

DETERMINING OXIDATION AND REDUCTION

$Na \longrightarrow Na^+ + e^-$. The oxidation number of sodium increases from zero (Na^0) before the arrow to +1 (Na^+) after the arrow. When the oxidation number increases or goes to the right on the number line (example: 0 to +1), it shows oxidation.

$F + e^- \longrightarrow F^-$. The oxidation number of fluorine decreases from zero (F^0) before the arrow to -1 (F^-) after the arrow. When oxidation number decreases or goes to the left on the number line, (example: 0 to -1), it shows reduction. This is the opposite of oxidation.

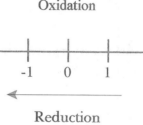

Question: Find the missing product in the balanced equation:

$$2Na + F_2 \longrightarrow \underline{\hspace{3cm}}$$

Solution: Two Na atoms unite with F_2 to form 2 NaF. The coefficient 2 in front of 2 NaF means 2 Na and 2 F atoms.

Question: Find the missing reactant in the balanced equation:

$$2Mg + \underline{\hspace{3cm}} \longrightarrow 2\ MgO$$

Solution:

2 Mg + <u>missing</u> $\longrightarrow$ 2 MgO

2 MgO = 2 Mg + 2 O

2 Mg + <u>missing</u> $\longrightarrow$ 2 Mg + 2 O

2 Mg + 2 O (oxygen) $\longrightarrow$ 2 Mg + 2 O

You see 2 O is missing; therefore, the missing reactant must be O_2.

Try Sample Questions #7-8, pages 7-8, then do Homework Questions, #16-21, pages 10-11, and #36-37, page 11.

RULES FOR OXIDATION NUMBERS

Let's review the first five rules and learn a few more:

1. Free elements (not combined with any other element) have an oxidation number of zero.

2. All metals in Group 1 have an oxidation number of +1. Look on the Periodic Table.

3. All metals in Group 2 have an oxidation number of +2. Look on the Periodic Table.

4. F always has an oxidation number of -1 in all compounds.

5. You learned in Chapter 3 that, in a compound, **the sum of the oxidation numbers must equal zero.**

   Examples:

   Sodium Chloride: Look at the oxidation numbers of $_{11}$Na (sodium) and $_{17}$Cl (chlorine) in the upper right hand corners of the boxes on the Periodic Table. Na^+, Cl^-: +1 and -1 = 0. *The sum of the oxidation numbers in a compound must equal zero.*

   Magnesium Chloride: *The sum of the oxidation numbers in a compound must equal zero.* Look at the oxidation numbers of $_{12}$Mg (magnesium) and $_{17}$Cl (chlorine) in the upper right hand corners of the boxes on the Periodic Table. The oxidation number of Mg is +2, and Cl is -1; therefore, you need 1 Mg (+2) and 2 Cl (2 times -1 = -2) to equal zero. +2 and -2 = 0. *The sum of the oxidation numbers in a compound must equal zero.* The formula for magnesium chloride is therefore $MgCl_2$.

   Note: You can use the criss-cross method on page 4:25 to write formulas.

6. In **ions (charged particles)**, the **sum of the oxidation numbers** of all the atoms **must equal** the **charge of the ion.**

   Example:

   Sulfate ion has a 2- charge, SO_4^{2-}. See Table E on page Reference Tables 5. (By convention, a -2 oxidation number is written as a charge of "2-"; -2 = 2-.) Oxygen has an oxidation number of -2 (upper right hand corner of of the box on the Periodic Table). Four oxygen atoms have 4 x (-2) = -8. The oxidation number of S (sulfur) must be +6 to give the SO_4^{2-} ion a charge of 2-. The sum of oxidation numbers, (oxygen) -8 and (sulfur) +6 = 2- (the charge of the ion). **Do Sample Question #12, P. 9:8**

7. Oxygen has an oxidation number of -2 in all its compounds *except* in **peroxides** (example, H_2O_2), when oxygen has an oxidation number of -1, and in compounds with F (OF_2), when oxygen has an oxidation number of +2.

8. Hydrogen has an oxidation number of +1 in all compounds *except* in **metal hydrides** (metal and hydrogen, examples: LiH and CaH_2), when hydrogen has an oxidation number of -1.

Question: What is the oxidation number of chlorine in $HClO_4$?

   (1) +1   (2) +5   (3) +3   (4) +7

Solution: Answer 4. You learned that, in a compound, the sum of the oxidation numbers must equal zero. You need to find the

PERIODIC TABLE

oxidation numbers of the atoms in $HClO_4$. Look at the oxidation number of oxygen in the upper right hand corner of the box on the Periodic Table. O is -2, so **4 atoms of oxygen = 4 x -2 = -8.** **Hydrogen has +1 as its oxidation number.**

$$H \quad Cl \quad O_4$$
$$+1 \ (H) \ \_\_ \ -8 \ (O_4) \ = -7,$$

therefore Cl must be +7, because +7 and -7 = 0. The sum of oxidation numbers in a compound must equal zero.

Try Sample Questions #9-11, on page 7, then do Homework Questions, #22-30, pages 10-11, and #38-39, page 11.

SAMPLE REGENTS & REGENT-TYPE QUESTIONS
AND SOLUTIONS

1. Given the redox reaction: $Fe^{2+}(aq) + Zn(s) \rightarrow Zn^{2+}(aq) + Fe(s)$
    Which species acts as a reducing agent?
    (1) $Fe(s)$ (2) $Fe^{2+}(aq)$ (3) $Zn(s)$ (4) $Zn^{2+}(aq)$

2. Given the reaction: $Zn(s) + Cu^{2+}(aq) \rightarrow Zn^{2+}(aq) + Cu(s)$
    Which particles must be transferred from one reactant to the other reactant?
    (1) ions   (2) neutrons   (3) protons   (4) electrons

3. Given the balanced equation: $2Al(s) + 6H^+(aq) \rightarrow 2Al^{3+}(aq) + 3H_2(g)$
    When 2 moles of $Al(s)$ completely reacts, what is the total number of moles of electrons transferred from $Al(s)$ to $H^+(aq)$?
    (1) 5   (2) 6   (3) 3   (4) 4

4. In the reaction $Cu + 2Ag^+ \rightarrow Cu^{2+} + 2Ag$, the oxidizing agent is
    (1) Cu (2) Cu^{2+} (3) Ag^+ (4) Ag

5. Which half -reaction correctly represents reduction?
    (1) $Fe^{2+} + 2e^- \rightarrow Fe$ (3) $Fe + 2e^- \rightarrow Fe^{2+}$
    (2) $Fe^{2+} + e^- \rightarrow Fe^{3+}$ (4) $Fe + e^- \rightarrow Fe^{3+}$

6. Given the reaction: $2Li(s) + Cl_2(g) \rightarrow 2LiCl(s)$
    As the reaction takes place, the $Cl_2(g)$ will
    (1) gain electrons (2) lose electrons (3) gain protons (4) lose protons

7. Which statement correctly describes a redox reaction?
    (1) The oxidation half-reaction and the reduction half-reaction occur simultaneously.
    (2) The oxidation half-reaction occurs before the reduction half-reaction.
    (3) The oxidation half-reaction occurs after the reduction half-reaction.
    (4) The oxidation half-reaction occurs spontaneously, but the reduction half-reaction does not.

8. Which equation represents a redox reaction?
    (1) $2Li^+ + S^{2-} \rightarrow Li_2S$
    (2) $H^+ + C_2H_3O_2^- \rightarrow HC_2H_3O_2$
    (3) $NH_3 + H^+ + Cl^- \rightarrow NH_4^+ + Cl^-$
    (4) $Zn + 2Ag^+ + 2NO_3^- \rightarrow 2Ag + Zn^{2+} + 2NO_3^-$

9. Oxygen has an oxidation number of –2 in
   (1) O_2　　　(2) NO_2　　　(3) Na_2O_2　　　(4) OF_2

10. In which species does hydrogen have an oxidation number of – 1?
   (1) H_2O　　　(2) H_2　　　(3) NaH　　　(4) NaOH

11. The oxidation number of nitrogen in N_2O is
   (1) +1　　　(2) +2　　　(3) –1　　　(4) –2

12. What are the two oxidation states of nitrogen in NH_4NO_2?

SOLUTIONS

1. Answer *3*. A substance that is **oxidized** is a **reducing agent**. $Zn^0 \rightarrow Zn^{2+} + 2e^-$. Zn is oxidized; Zn is a reducing agent.

2. Answer *4*. Electrons are transferred in the reaction. Zn loses two electrons and forms zinc ions: $Zn \rightarrow Zn^{2+} + 2e^-$. The two electrons are transferred to the Cu^{2+} ions.

3. Answer *2*. Aluminum goes from oxidation number of zero (free element has an oxidation number of zero) to oxidation number of +3. $Al^0 \rightarrow Al^{3+} + 3e^-$. 1 mole of Al loses 3 moles of electrons, or transfers 3 moles of electrons to the hydrogen ions (H^+) (which are reduced). Obviously, 2 moles of Al loses (2 x 3) moles of electrons, or transfers 6 moles of electrons to the H^+.

4. Answer *3*. Oxidizing agent = reduced = gain of electrons. $Ag^+ + e^- \rightarrow Ag^0$. Ag^+ (silver ion) gains an electron and becomes Ag (silver). Ag^+ is reduced; Ag^+ is an oxidizing agent.

5. Answer *1*. Reduction is a gain of electrons. Fe^{2+} gains 2 electrons and becomes Fe^0. The oxidation number goes from +2 to zero (any free element has an oxidation number of zero). Gain of electrons (negative particles) causes the oxidation number to decrease.
   Choices 2, 3 and 4 are incorrect equations. In these choices, the oxidation number of the product went up.

6. Answer *1*. $2Li + Cl_2^0 \rightarrow 2LiCl$. Cl_2, or any free element (not combined with any other element), has an oxidation number of zero. Look in the upper right hand corner of the box on the Periodic Table, page Reference Tables 14, for the oxidation number of chlorine: $_{17}Cl$; the oxidation number of $_{17}Cl$ is – 1. Or, in LiCl, Li has + 1 oxidation number, therefore Cl has – 1 oxidation number. In the equation, $2Li + Cl_2 \rightarrow 2LiCl$, the oxidation number of Cl goes from zero in Cl_2 to – 1 (more negative) in LiCl by gaining an electron.

7. Answer *1*. In a redox reaction, oxidation and reduction occur simultaneously.

8. Answer *4*. In a redox reaction, the oxidation numbers change.
   $Zn \rightarrow Zn^{2+} + 2e^-$. The oxidation number of Zn changes from zero to +2. $Ag^+ + e^- \rightarrow Ag^0$; $2Ag^+ + 2e^- \rightarrow 2Ag^0$. The oxidation number of Ag changes from +1 to zero.
   You can easily recognize a redox reaction when you have a free element by itself, example Zn, on one side of the arrow (equation), and an ion, Zn^{2+}, on the other side of the arrow.

9. Answer *2*. You learned that oxygen has an oxidation number of –2 in all compounds, except in peroxides (examples: H_2O_2 and Na_2O_2) and in compounds with F (OF_2). Therefore, the oxidation number of oxygen in NO_2 is –2.

In Answer 2, NO_2 is a compound; oxygen has an oxidation number of -2 in compounds.

In Choice 3, Na_2O_2, sodium peroxide, the oxidation number of oxygen is -1. In Choice 4, OF_2, the oxidation number of oxygen is +2. Choice 1 is O_2 (a free element). The oxidation number of oxygen, or any free element (an element all by itself), is zero.

10. Answer 3. Hydrogen has an oxidation number of -1 in metal hydrides (metal and hydrogen) like NaH. Na (sodium) is a metal (left of the zigzag line on the Periodic Table).

11. Answer 1. In N_2O, or any compound, the sum of the oxidation numbers must equal zero. Look in the upper right hand corner of the box on the Periodic Table, on page Reference Tables 20-21, for the oxidation number of oxygen. Oxygen has an oxidation number of -2; therefore, in N_2O, N_2, or two nitrogen atoms, must equal +2; therefore, one nitrogen atom has an oxidation number of +1.

12. The two ions in NH_4NO_2 are NH_4^+ and NO_2^-, see Table E on page Reference Tables 5. In NH_4^+, H has an oxidation number of +1; 4 H has 4 x (+1) = +4, therefore N must have -3 so the NH_4^+ ion has a charge of 1+. In NO_2^-, O has an oxidation number of -2; 2 O has 2 x (-2) = -4, so N must have +3, so the NO_2^- ion has a charge of 1-.

NOW LET'S TRY A FEW HOMEWORK QUESTIONS:

1. In the reaction $2Al(s) + 3Cu^{2+}(aq) \rightarrow 2Al^{3+}(aq) + 3Cu(s)$, the Al(s)
   (1) gains protons   (3) gains electrons
   (2) loses protons   (4) loses electrons

2. Which change occurs when an Sn^{2+} ion is oxidized?
   (1) Two electrons are lost   (3) Two protons are lost
   (2) Two electrons are gained   (4) Two protons are gained

3. Given the reaction:
$$Zn + 2HCl \rightarrow ZnCl_2 + H_2$$
   Which statement best describes what happens to the zinc?
   (1) The oxidation number changes from +2 to 0, and the zinc is reduced.
   (2) The oxidation number changes from 0 to +2, and the zinc is reduced.
   (3) The oxidation number changes from +2 to 0, and the zinc is oxidized.
   (4) The oxidation number changes from 0 to +2, and the zinc is oxidized.

4. When a substance is oxidized, it
   (1) loses protons   (3) acts as an oxidizing agent
   (2) gains protons   (4) acts as a reducing agent

5. Given the reaction:
$$2Na + 2H_2O \rightarrow 2Na^+ + 2OH^- + H_2$$
   Which substance is oxidized?
   (1) H_2 (2) H^+ (3) Na   (4) Na^+

6. Given the reaction:
$$Zn(s) + 2HCl(aq) \rightarrow ZnCl_2(aq) + H_2(g)$$
   Which equation represents the correct oxidation half-reaction?
   (1) $Zn(s) \rightarrow Zn^{2+} + 2e^-$ (3) $Zn^{2+} + 2e^- \rightarrow Zn(s)$
   (2) $2H^+ + 2e^- \rightarrow H_2(g)$ (4) $2Cl^- \rightarrow Cl_2(g) + 2e^-$

7. In the reaction $4Zn + 10HNO_3 \rightarrow 4Zn(NO_3)_2 + NH_4NO_3 + 3H_2O$, the zinc is
   (1) reduced, and the oxidation number changes from 0 to +2
   (2) oxidized, and the oxidation number changes from 0 to +2
   (3) reduced, and the oxidation number changes from +2 to 0
   (4) oxidized, and the oxidation number changes from +2 to 0

8. In the reaction $2Al + 3Ni(NO_3)_2 \rightarrow 2Al(NO_3)_3 + 3Ni$, the aluminum is
   (1) reduced, and its oxidation number increases
   (2) reduced, and its oxidation number decreases
   (3) oxidized, and its oxidation number increases
   (4) oxidized, and its oxidation number decreases

9. In the reaction $Pb + 2Ag^+ \rightarrow Pb^{2+} + 2Ag$, the Ag^+ is
   (1) reduced, and the oxidation number changes from +1 to 0
   (2) reduced, and the oxidation number changes from +2 to 0
   (3) oxidized, and the oxidation number changes from 0 to +1
   (4) oxidized, and the oxidation number changes from +1 to 0

10. Which half-reaction correctly represents a reduction reaction?
    (1) $Sn^0 + 2e^- \rightarrow Sn^{2+}$ (3) $Li^0 + e^- \rightarrow Li^+$
    (2) $Na^0 + e^- \rightarrow Na^+$ (4) $Br_2^0 + 2e^- \rightarrow 2Br^-$

11. Which is the oxidizing agent in the reaction $2Fe^{2+} + Cl_2 \rightarrow 2Fe^{3+} + 2Cl^-$?
    (1) Fe^{2+} (2) Cl_2 (3) Fe^{3+} (4) Cl^-

12. In the reaction $2H_2(g) + O_2(g) \rightarrow 2H_2O(g)$, the oxidizing agent is
    (1) H_2 (2) O_2 (3) H^+ (4) O^{2-}

13. In the reaction $Pb + 2Ag^+ \rightarrow Pb^{2+} + 2Ag$, the oxidizing agent is
    (1) Ag^+ (2) Ag (3) Pb (4) Pb^{2+}

14. In the reaction $2Mg + O_2 \rightarrow 2MgO$, the magnesium is the
    (1) oxidizing agent and is reduced (3) reducing agent and is reduced
    (2) oxidizing agent and is oxidized (4) reducing agent and is oxidized

15. Given the reaction:
    $$2KCl(\ell) \rightarrow 2K(s) + Cl_2(g)$$
    In this reaction, the K^+ ions are
    (1) reduced by losing electrons   (3) oxidized by losing electrons
    (2) reduced by gaining electrons   (4) oxidized by gaining electrons

16. Given the redox reaction:
    $$2I^-(aq) + Br_2(\ell) \rightarrow 2Br^-(aq) + I_2(s)$$
    What occurs during this reaction?
    (1) The I^- ion is oxidized, and its oxidation number increases.
    (2) The I^- ion is oxidized, and its oxidation number decreases.
    (3) The I^- ion is reduced, and its oxidation number increases.
    (4) The I^- ion is reduced, and its oxidation number decreases.

17. For a redox reaction to occur, there must be a transfer of
    (1) protons   (2) neutrons   (3) electrons   (4) ions

18. All redox reactions involve
    (1) the gain of elect ons, only   (3) Both the gain and the loss of electrons
    (2) the loss of electrons, only   (4) neither the gain nor the loss of electrons

19. Given the redox reaction:
    $$Fe^{2+}(aq) + Zn(s) \rightarrow Zn^{2+}(aq) + Fe(s)$$
    Which species acts as a reducing agent?
    (1) $Fe(s)$ (2) $Fe^{2+}(aq)$ (3) $Zn(s)$ (4) $Zn^{2+}(aq)$

20. In any oxidation-reduction reaction, the total number of electrons gained is
    (1) less than the total number of electrons lost
    (2) greater than the total number of electrons lost
    (3) equal to the total number of electrons lost
    (4) unrelated to the total number of electrons lost

21. A. A redox reaction is a reaction in which
    (1) only reduction occurs (3) reduction and oxidation occur at the same time
    (2) only oxidation occurs (4) reduction occurs first, and then oxidation occurs

    B. Given the balanced equation representing a reaction
    $$2Fe + 3Cu^{2+} \rightarrow 2Fe^{3+} + 3Cu$$
    When the iron atoms lose six moles of electrons, how many moles of electrons are gained by the copper ions?
    (1) 12 moles   (2) 2 moles   (3) 6 moles   (4) 3 moles

---

22. In which substance is the oxidation number of nitrogen zero?
    (1) N_2 (2) NH_3 (3) NO_2 (4) N_2O

23. Oxygen will have a positive oxidation number when combined with
    (1) fluorine   (2) chlorine   (3) bromine   (4) iodine

24. In which compound does chlorine have the highest oxidation number?
    (1) $KClO$ (2) $KClO_2$ (3) $KClO_3$ (4) $KClO_4$

25. What is the oxidation number of oxygen in HSO_4^-?
    (1) $+1$ (2) -2 (3) $+6$ (4) -4

26. What is the oxidation number of carbon in $NaHCO_3$?
    (1) $+6$ (2) $+2$ (3) -4 (4) $+4$

27. Oxygen has an oxidation number of -2 in
    (1) O_2 (2) NO_2 (3) Na_2O_2 (4) OF_2

28. The oxidation number of nitrogen in N_2O is
    (1) $+1$ (2) $+2$ (3) -1 (4) -2

29. What is the oxidation number of chlorine in $HClO_4$?
    (1) $+1$ (2) $+5$ (3) $+3$ (4) $+7$

30. The oxidation number of nitrogen in N_2 is
    (1) $+1$ (2) 0 (3) $+3$ (4) -3

CONSTRUCTED RESPONSE QUESTIONS: Parts B-2 and C of NYS Regents Exam

31. Write equations showing (see previous page):
    a. K losing 1 electron
    b. Mg losing 2 electrons
    c. Al losing 3 electrons

32. Find the missing reactant or product in each balanced equation:
    a. $Fe \rightarrow$ ___ $+ 2e^-$ b. ___ $\rightarrow Na^+ + 1e^-$ c. $Au \rightarrow$ ___ $+ 1e^-$

33. Write an equation showing:
    a. Cl gaining 1 electron b. S gaining 2 electrons c. N gaining 3 electrons

34. A-Complete the balanced equation:
    a. $Br_2 + 2e^- \rightarrow$ ___   b. ___ $+ 4e^- \rightarrow 2O^{2-}$ c. $I_2 + 2e^- \rightarrow$ ___
    B-Why do elements in Group 2 (example Ca) have a $+2$ oxidation number and elements in Group 17 (example Cl) have a -1 oxidation number?

35. A- Compare and contrast oxidation and reduction.
    B- What is the difference between the substance oxidized and the oxidizing agent?

36. In the reaction $Zn + Cr^{3+} \rightarrow Zn^{2+} + Cr$
    A- Write the half-reaction for the reduction that occurs.
    B. Write the half-reaction for the oxidation that occurs.

37. Fill in the missing reactant or product in the redox reaction:
    a. $2 Cr + 3 O_2 \rightarrow$ ___ b. $2 Rb +$ ___ $\rightarrow 2RbI$ c. $2 Li + Br_2 \rightarrow$ ___

38. How does the oxidation number of oxygen vary in different compounds?

39. What is the oxidation number of Cl in HCl, HClO, $HClO_2$ and why are the oxidation numbers different?

SECTION B

ELECTROCHEMICAL CELL

There are two types of electrochemical cells, voltaic and electrolytic. First you will learn about the voltaic cell, and then the electrolytic .

Voltaic Cell

A voltaic cell uses redox reactions (oxidation-reduction reactions) that are spontaneous (reactions take place by themselves) to produce electricity. A battery is an example of a voltaic cell.

Description of the Voltaic Cell

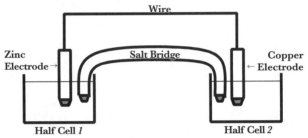

Voltaic Cell

There are **two half cells**. In each half cell is a **metal strip** called an **electrode**. There is a **wire** connecting the two electrodes. ELECTRONS travel through the wire. There is a **salt bridge** connecting the two half cells to permit IONS to flow between the two half cells.

Let's see what happens in the voltaic cell:

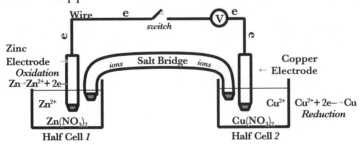

In half cell *1*, the **zinc electrode loses two electrons** and forms Zn^{2+}, which goes into solution: $Zn \rightarrow Zn^{2+} + 2e^-$. **Remember: Electrons ($2e^-$ or e^-, etc.) after** the **arrow means Zn (before the arrow) loses electrons.**

When the switch is closed, the **electrons (e) flow from the Zn through the wire to the Cu** in half cell 2. In half cell 2, there are Cu^{2+} ions from the $Cu(NO_3)_2$ solution. The Cu^{2+} in the solution **gains 2 electrons** (reduction) to form Cu: $Cu^{2+} + 2e^- \rightarrow Cu°$.

Ions travel through the salt bridge connecting the two half cells. (Now you have a **complete circuit**: Oxidation (Zn loses electrons) in half cell *1*, electrons traveling through the wire, reduction (Cu^{2+} gains electrons) in half cell *2*, and ions flowing through the salt bridge. In a voltaic cell, oxidation and reduction reactions produce electricity.

Comparison of the Two Half Cells of the Voltaic Cell

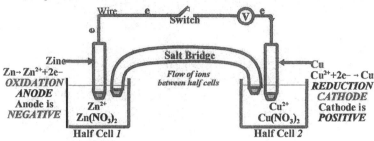

Half Cell 1	Half Cell 2
The name of the electrode where **oxidation** takes place is the **anode**.	The name of the electrode where **reduction** takes place is the **cathode**.
Hint: Look at the first three letters of the word "anode." "An" stands for anode; "o" stands for oxidation.	*Hint: Think of the words, "Cat eats red meat." Cat = cathode; red = reduction.*
ANODE-OXIDATION	**CATHODE-REDUCTION**
Oxidation: Zn loses two electrons and forms Zn^{2+}:	*Reduction*: Cu^{2+} gains two electrons and forms Cu:
$$Zn \rightarrow Zn^{2+} + 2e^-$$	$$Cu^{2+} + 2e^- \rightarrow Cu$$
ANODE IS NEGATIVE.	**CATHODE IS POSITIVE.**

The **direction of electron (e) flow** in a voltaic cell is **from the anode** (where **oxidation** takes place) **to the cathode** (where **reduction** takes place.

Answer Questions 1, 2, and 3 based on the following diagram. This is the *same* diagram as those you just had.

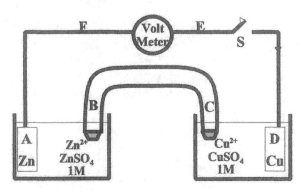

Question 1: When the switch is closed, which group of letters correctly represents the direction of electron flow?
(1) $A \rightarrow B \rightarrow C \rightarrow D$ (2) $A \rightarrow F \rightarrow E \rightarrow D$ (3) $D \rightarrow C \rightarrow B \rightarrow A$ (4) $D \rightarrow E \rightarrow F \rightarrow A$

Solution: Answer 2. As you learned, electrons go from the zinc anode through the wire to the copper cathode. Look at the diagram. This path is $A \rightarrow F \rightarrow E \rightarrow D$. Electrons go from the more active metal (Zn) to the less active metal (Cu). Table J shows which metals are more active. (You will learn later about Table J.) Zn is higher up on Table J, therefore Zn is more active than Cu and electrons go from Zn to Cu.

Question 2: Which statement correctly describes the direction of flow for the ions in this cell when the switch is closed?
(1) Ions move through the salt bridge from B to C, only.
(2) Ions move through the salt bridge from C to B, only.
(3) Ions move through the salt bridge in both directions.
(4) Ions do not move through the salt bridge in either direction.

Solution: Answer 3. Ions move in both directions.

Question 3: As the voltaic cell operates, why does the mass of the Zn (zinc) electrode decrease?

Solution: Zn atoms lose electrons, forming Zn^{2+}, which goes into the solution. $Zn \rightarrow Zn^{2+} + 2e^-$. (See diagram on page 9:12.)

Question: Which species acts as the anode when the reaction
$$Zn(s) + Pb^{2+}(aq) \rightarrow Zn^{2+}(aq) + Pb(s)$$
occurs in a voltaic cell?
(1) $Zn(s)$ (2) $Zn^{2+}(aq)$ (3) $Pb^{2+}(aq)$ (4) $Pb(s)$

Solution: Answer 1. The anode is oxidized. Zinc is oxidized as $Zn \rightarrow Zn^{2+} + 2e^-$; therefore, zinc is the anode.

Try Sample Question #1, on page 30, then do Homework Questions, #1-3, pages 31-32, and #26, page 34.

TABLE J:
ACTIVITY OF ELEMENTS AND SPONTANEOUS REACTIONS

Table J compares how active each metal and nonmetal is. Look at Table J on the next page or on page Reference Tables 11-12. The word **most** (next to **metals** and next to **nonmetals**) is written on the **top** of the table and the word **least** (next to **metals** and next to **nonmetals**) is written on the **bottom**.

The **metal** at the **top** of Table J is **most active**. Li is at the top of Table J and is the most active metal on the table. As you go down Table J, the metals become less active, as shown by the arrow going down from most to least. The **least active** metal is at the **bottom**. **Metals higher up** are **more active** and **replace metals below** them **from compounds**.

TABLE J (side label)

Examples:
1: **Li** is a **metal above K** and **replaces K** from its compounds.

Li + KCl ⟶ LiCl + K.

This **reaction** is **spontaneous** (the reaction takes place).

2: **Ca** is a **metal above Mg** and **replaces Mg** from its compound.

$Ca + MgCO_3$ ⟶ $CaCO_3 + Mg$

This **reaction** is **spontaneous**.

Table J
Activity Series

Most	Metals	Nonmetals	Most
	Li	F_2	
	Rb	Cl_2	
	K	Br_2	
	Cs	I_2	
	Ba		
	Sr		
	Ca		
	Na		
	Mg		
	Al		
	Zn		
	Cr		
	Fe		
	Ni		
	Sn		
	Pb		
	H_2		
	Cu		
	Ag		
Least	Au		Least

Metals lower down are **less active** and do **not** replace metals above them from compounds.

Example: K is below Li and does not replace Li from its compounds. **No reaction takes place.**

K + LiCl ⟶ No reaction

Reaction is **not spontaneous** (no reaction takes place).

The right column of Table J shows the activity of nonmetals. Look at Table J again. As you saw before, the word **most** (next to nonmetals) is written at the **top** and the word **least** is written at the **bottom**. The **nonmetal** at the **top** is the **most active**. F is at the top of Table J and is the most active nonmetal. As you go down Table J, the nonmetals become less active, as shown by the arrow going down from most to least. The **least active** nonmetal is at the **bottom**.

Nonmetals higher up in Table J are more active and **replace nonmetals below them from compounds**.

Example:

F_2 is a **nonmetal above Cl** and **replaces Cl_2** from its compounds.

$$F_2 + 2\ NaCl \longrightarrow 2\ NaF + Cl_2$$

The **reaction** is **spontaneous** (reaction takes place).

Nonmetals **lower down** are less active and do **not** replace nonmetals above them from compounds. Reaction is **not spontaneous**.

Example: $Cl_2 + 2\ NaF \longrightarrow$ No reaction

The **reaction** is **not** spontaneous (no reaction takes place).

Question: Which metal is less active than Mg?

(1) Al   (2) Ba   (3) Ca   (4) Rb

Solution: Answer 1. Al is below Mg on Table J, therefore it is less active.

Question: Which metal has lower activity than H_2?

(1) Al   (2) Na   (3) Cu   (4) Fe

Solution: Answer 3. Cu is below H_2 in Table J.

Question: Which reaction will take place spontaneously?

(1) $Ni^{2+} + Pb \longrightarrow Ni + Pb^{2+}$ (3) $Sr^{2+} + Sn \longrightarrow Sr + Sn^{2+}$

(2) $Au^{3+} + Al \longrightarrow Au + Al^{3+}$ (4) $Fe^{2+} + Cu \longrightarrow Fe + Cu^{2+}$

Solution: Answer 2. The metal (Al) must be higher in Table J than the ion or metal in the compound. In Choice 2, Al is higher in Table J than Au. Therefore, Al replaces Au from the compound. The reaction is spontaneous.

Question: Which metal will react spontaneously with Ag^+ ions, but not with Zn^{2+} ions?

(1) Cu   (2) Au   (3) Al   (4) Mg

Solution: Answer 1 Cu will react spontaneously with Ag^+, but not with Zn^{2+}. Cu is above Ag and therefore replaces Ag^+ from a compound. Cu is below Zn and **cannot** replace Zn^{2+} ions from a compound.

Note: Copper is a better choice than iron to make jewelry (bracelets, necklaces, earrings, etc.). Copper (Cu) is below iron (Fe) on table J, therefore copper is less active and reacts less with its surroundings (air, water, etc.), then copper jewelry keeps its shine (tarnishes less).

TABLE J: OXIDATION AND REDUCTION

You learned **metals lose electrons** (oxidized) to form compounds. **Metals** that are **more active easily lose electrons** (are **easily oxidized**) and form compounds. Na is active (see Table J). Na reacts vigorously with cold water, easily loses electrons and is easily oxidized. Au (gold) is not active. Au does not react with water and it is not easily oxidized.

Nonmetals gain electrons (reduced) to form compounds. **Nonmetals** that are **more active easily gain electrons** (are **easily reduced**) and form compounds. Nonmetals gain electrons because there is a tendency to have a complete outer shell, octet (more stable, lower energy). F is the most active nonmetal. F is the most easy to gain electrons and be reduced.

Part of Table J
Activity Series

Metals		**Metals**	**Nonmetals**		Nonmetals
Activity decreases as you **go down Table J.** Metals on top are most easily oxidized.	Most	*Most easily oxidized* ↓ Li Rb K	*Most easily reduced* ↓ F_2 Cl_2 Br_2	Most	**Activity decreases** as you **go down Table J.** Nonmetals on top are most easily reduced.
		Ca	I_2		
As you go down, less easily oxidized		Na Mg			As you go down, less easily reduced
		Fe H_2 Cu			
Bottom is least easily oxidized	Least	Ag Au		Least	Bottom is least easily reduced
		↑ *Least easily oxidized*	↑ *Least easily reduced*		

(Look at Table J, above.)

RULE:

Most active metal: Top of Table J. Most easily loses an electron. **Most easily oxidized.**

RULE (continued):

As you go down Table J, metals are less active and less easily oxidized, as shown by the arrow going down, from most to least.

Least active metal: Bottom of Table J. Hardest to lose electrons, **least easily oxidized** (hardest to be oxidized).

Most active nonmetal: Top of Table J: Most easily gains an electron, **most easily reduced**.

As you go down Table J, nonmetals are less active and less easily reduced, as shown by the arrow going down, from most to least.

Least active nonmetal: Bottom of Table J: Hardest to gain electrons, **least easily reduced**. (Table J lists only the nonmetals in Group 17).

Question: Which is most easily oxidized?

    (1) Li   (2) K   (3) Ca   (4) Na

Solution: Answer 1. Look at these elements in Table J. They are all metals. Metals at the top of Table J are most active and are most easily oxidized. Lithium is at the top and is most easily oxidized.

Question: Which is least easily oxidized (hardest to be oxidized)?

    (1) Ca   (2) Ni   (3) Cu   (4) Au

Solution: Answer 4. Look at these elements in Table J. They are all metals. Metals at the bottom of Table J are least active and are least easily oxidized. Au is at the bottom and least easily oxidized.

Question: When the switch is closed, electrons flow **in both diagrams** from

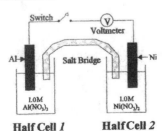 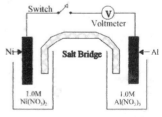

    (1) Al(s) to Ni (s)   (3) Al^{3+}(aq) to Ni^{2+}(aq)

    (2) Ni(s) to Al(s)   (4) Ni^{2+}(aq) to Al^{3+}(aq)

Solution: Answer 1. Al is higher up on Table J than Ni (see Table J, on page Reference Tables 11-12. Therefore, Al is more active and more **easily oxidized** (more easily loses an electron). Therefore,

---

electrons flow from Al to Ni. It does not matter if the Al is on the left side or the right side of the diagram. Al is more active than Ni, therefore electrons flow from Al to Ni in both diagrams. Answer 1.

To find the properties of the ion (least or most easily oxidized, least or most easily reduced), take the properties of the element and reverse them:
1. Reverse most to least or least to most.
2. Reverse reduced to oxidized or oxidized to reduced.
F is most easily reduced. For F^-, change most to least and reduced to oxidized: F^- (ion) is least easily oxidized.

Question: Which of the following ions is most easily oxidized?
(1) F^- (2) Cl^- (3) Br^- (4) I^-

Solution: Answer 4. The ion that is most easily oxidized has its atom that is (reverse way) least easily reduced. Find the atom that is least easily reduced. All the choices are nonmetals. The least easily reduced nonmetal is at the bottom of Table J. I is at the bottom of Table J, least easily reduced.

CALCULATING CELL VOLTAGE (E^0)

Let's find out how much voltage (E^0) is produced in a voltaic cell. Use the reduction potential table to calculate $E°$. Look at the Table of Standard Reduction Potentials on the next page or on page Additional Tables 6. The table lists reduction reactions that take place in a half cell. (In the diagram below, reduction takes place in half cell 2.) The second column of the table is labeled "E^0". E^0, or half cell potential, is how many volts of energy is given off or absorbed in that reaction.

The E^0 number is figured out for a specific reaction by comparing it to hydrogen (reference point), which is given an $E^0 = 0$.

Question: How much voltage (E^0) is produced in the voltaic cell below, and is the reaction spontaneous? (For simplicity, the voltaic cell diagram is printed again on this page:)

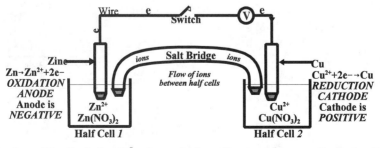

Solution: To find the E^0 of a voltaic cell, you have to find the $E°$'s of half cell 2 and half cell 1, and add the $E°$'s together.

In **half cell 2**, you have $Cu^{2+} + 2e^- \rightarrow Cu$. Look for $Cu^{2+} + 2e^- \rightarrow Cu$ in the Table of Reduction Potentials, to the right or on page Additional Tables 6. Go across to the E^0 column and read the E^0 for that reaction:

$Cu^{2+} + 2e^- \rightarrow Cu$: $E^0 = +0.34V$ (volts).

The Table of Reduction Potentials and this example are exactly the same: $Cu^{2+} + 2e \rightarrow Cu$.

Table of Reduction Potentials

Half Reaction	E^0 (*volts*)
$F_2 + 2e^- \rightarrow 2F^-$	+2.87
$Au^{3+} + 3e^- \rightarrow Au$	+1.50
$Cu^{2+} + 2e^- \rightarrow Cu$	+0.34
$2H^+ + 2e^- \rightarrow H_2$	0.00
$Zn^{2+} + 2e^- \rightarrow Zn$	-0.76

In **half cell 1**, you have $Zn \rightarrow Zn^{2+} + 2e$. This **example starts with zinc**. The Table of Reduction Potentials has

$$Zn^{2+} + 2e^- \rightarrow Zn: E^0 = -0.76V.$$

which is the reverse reaction. The table starts with Zn^{2+} (reverse, opposite way), therefore, in this example, **change** the **sign** of E^0.

$$Zn \rightarrow Zn^{2+} + 2e^- : E^0 = +0.76V.$$

Add together the E^0 from each of the two half cells:

$Cu^{2+} + 2e^- \rightarrow Cu: E^0 = +0.34V.$
$Zn \rightarrow Zn^{2+} + 2e^- : E^0 = \underline{+0.76V.}$
$E^0 = \mathbf{+1.10}V$ (volts)

This voltaic cell produces $E^0 = +1.10$ **volts**. E^0 **is positive, +1.10V**. Whenever E^0 is positive, the **reaction is spontaneous** (the reaction takes place).

REMEMBER: If E^0 is **positive**, the **reaction** is **spontaneous**, the reaction takes place. If E^0 is **negative**, **no reaction takes place**. $E^0 = 0$ at **equilibrium**.

Question: Given the reaction:

$$Mg + Zn^{2+} \rightarrow Mg^{2+} + Zn,$$

What is the **cell voltage (E^0)** for the overall reaction and is the reaction spontaneous?

Solution: This **example starts with Mg**. (Mg is left of the arrow.) The Table of Reduction Potentials starts with Mg^{2+} (reverse or opposite reaction), therefore, in this example, **change** the **sign** of E^0: $E^0 = +2.37V$.

Table of Reduction Potentials

Half Reaction	E^0 (*volts*)
$2H^+ + 2e^- \rightarrow H_2$	0.00
$Zn^{2+} + 2e^- \rightarrow Zn$	-0.76
$Mg^{2+} + 2e^- \rightarrow Mg$	-2.37

This example and the table start with Zn^{2+}; leave the sign of E^0 the way it is: $E^0 = -0.76V$.

Add the two E^0's:

$Mg \rightarrow Mg^{2+}:$ $E^0 = +2.37V$
$Zn^{2+} \rightarrow Zn:$ $E^0 = \underline{-0.76V}$
$E^0 = \mathbf{+1.61V}$ *(volts)*

This voltaic cell produces $E^0 = +1.61$ volts. E^0 is positive, +1.61V. Whenever E^0 is positive, the **reaction is spontaneous** (the reaction takes place).

Question: Given the reaction:

$$Al(s) + 3 Ag^+ \rightarrow Al^{3+} + 3 Ag(s).$$

Based on the Table of Reduction Potentials, what is the potential ($E°$) for the overall reaction and is the reaction spontaneous?

(1) +0.74 V   (2) +1.66 V   (3) +2.46 V   (4) +4.06 V

Solution: Answer *3*. This **example starts with Al**. The Table of Reduction Potentials starts with Al^{3+} (reverse or opposite reaction); therefore, in this example, **change** the **sign** of E^0: $E^0 = +1.66V$.

Table of Reduction Potentials

Half Reaction	E^0 (*volts*)
$Ag^+ + e^- \rightarrow Ag$	+0.80
$Al^{3+} + 3e^- \rightarrow Al$	−1.66

This example and the table start with Ag^+; leave the sign of E^0 the way it is: $E^0 = +0.80V$.

Add the two E^0's:

$$Al \rightarrow Al^{3+}: \quad E^0 = +1.66V$$
$$Ag^+ \rightarrow Ag: \quad E^0 = \underline{+0.80V}$$
$$E^0 = \textbf{+2.46 V } \textit{(volts)}$$

This voltaic cell produces $E° = +2.46$ volts. $E°$ is positive, +2.46 **volts.** Whenever $E°$ is positive, the **reaction is spontaneous** (the reaction takes place).

Question: Given the reaction

$$Pb + Cu^{2+} \longrightarrow Pb^{2+} + Cu,$$

is the reaction spontaneous?

Solution:

Step 1: Find $E°$. This **example starts with Pb**. (Pb is left of the arrow.) The Table of Reduction Potentials starts with Pb^{2+} (reverse or opposite reaction), therefore, in this example, **change** the **sign** of $E°$: $E° = +0.13V$.

Table of Reduction Potentials

Half Reaction	E^0 Volts
$Cu^{2+} + 2e^- \longrightarrow Cu$	+.34
$2H^+ + 2e^- \longrightarrow H_2$	0
$Pb^{2+} + 2e^- \longrightarrow Pb$	−0.13

This example and the table start with Cu^{2+}; leave the sign of $E°$ the way it is: $E° = +0.34V$

Add the two $E°$'s:

$$Pb \rightarrow Pb^{2+} \quad E° = +0.13V$$
$$Cu^{2+} \rightarrow Cu \quad E° = \underline{+0.34V}$$
$$E° = \textbf{+0.47V } \textit{(volts)}$$

Step 2: **Rule:**   If $E°$ is positive, reaction is spontaneous (reaction takes place).

If $E°$ is negative, reaction is not spontaneous (no reaction takes place).

In this example, $E° = 0.47$, $E°$ is positive, $+0.47$, therefore the reaction is spontaneous.

Question: Given the reaction:
$$Cu + Zn^{2+} \rightarrow Cu^{2+} + Zn$$
is this reaction spontaneous?

Solution:

Step 1: Find the $E°$'s and add the two $E°$'s:

$$Cu \rightarrow Cu^{2+} \qquad E° = -0.34$$
$$Zn^{2+} \rightarrow Zn \qquad E° = \underline{-0.76}$$
$$E° = \textbf{\textit{-1.10V}} \textit{ (volts)}$$

Step 2: **Rule:** If $E°$ is positive, reaction is spontaneous.

If $E°$ is negative, reaction is not spontaneous.

In this example, $E° = -1.10$, $E°$ is negative, therefore the reaction is not spontaneous. (The reaction does not take place.)

Question: Which reduction half reaction has a standard electrode potential (E^0) of 1.50 volts?

(1) $Au^{3+} + 3e \rightarrow Au$ (3) $Co^{2+} + 2e \rightarrow Co$
(2) $Al^{3+} + 3e \rightarrow Al$ (4) $Ca^{2+} + 2e \rightarrow Ca$

Solution: Answer *1*. Read the Table of Reduction Potentials:

$$Au^{3+} + 3e \rightarrow Au: E° = +1.50V.$$

Answer 1 and Table of Reduction Potentials start with Au^{3+}. Leave the sign of E^0 the way it is, $E^0 = +1.50V$.

Table of Reduction Potentials

Half Reaction	E^0 (volts)
$Au^{3+} + 3e^- \rightarrow Au$	+1.50

Question: What is the standard electrode potential (E^0) assigned to the half reaction $Cu^{2+} + 2e \rightarrow Cu$, when compared to the standard hydrogen half reaction?

(1) +0.34 V   (2) -0.34 V   (3) +0.52 V   (4) -0.52 V

Solution: Answer *1*. Read the Table of Reduction Potentials. This example is exactly the one on the table. This example and table start with Cu^{2+}. Leave the sign of E^0 the way it is: $E^0 = +0.34V$.

Table of Reduction Potentials

Half Reaction	E^0 (volts)
$Cu^{2+} + 2e \rightarrow Cu$	+0.34
$2H^+ + 2e^- \rightarrow H_2(g)$	0.00

TABLE OF REDUCTION POTENTIALS:
RELATIONSHIP OF E⁰ AND LOCATION

Look at the E^0 column in the Table of Reduction Potentials. The **biggest electrode potential** is at the **top**: E^0 = +2.87V. As you go down the table, the potential (E^0) gets smaller and smaller, until the bottom of the table has the smallest electrode potential, $E^0 = -3.04$V.

Table of Reduction Potentials

Half Reaction	E° (volts)	
$F_2(g) + 2e^- \rightarrow 2F^-$	+2.87	(top)
$2H^+ + 2e^- \rightarrow H_2(g)$	0.00	
$Zn^{2+} + 2e^- \rightarrow Zn(s)$	-0.76	
$Li^+ + e^- \rightarrow Li(s)$	-3.04	(bottom)

In short, the **biggest E^0** is on **top** of the Table of Reduction Potentials, and the **smallest E^0** is on the **bottom**.

Question: Based on the Table of Reduction Potentials, which half cell has a lower electrode potential than the standard hydrogen half cell?

(1) $Au^{3+} + 3e \rightarrow Au$ (3) $Cu^+ + e \rightarrow Cu.$

(2) $Hg^{2+} + 2e \rightarrow Hg$ (4) $Pb^{2+} + 2e \rightarrow Pb$

Solution: Answer **4**. Look at the Table of Reduction Potentials. Lower electrode potential than hydrogen means looking lower down on the table (*under* hydrogen) (lower E^0).

$2H^+ + 2e \rightarrow H_2$: $E^0 = 0.00$V

$Pb^{2+} + 2e \rightarrow Pb$, $E^0 = -0.13$V is the only one *under* hydrogen (lower than hydrogen). All the other three choices, Au^{3+}, Hg^{2+} and Cu^+ are on top of (above) hydrogen.

Table of Reduction Potentials

Half Reaction	E^0 (volts)
$Au^{3+} + 3e^- \rightarrow Au$	+1.50
$Hg^{2+} + 2e^- \rightarrow Hg$	+0.85
$Cu^+ + e^- \rightarrow Cu$	+0.52
$2H^+ + 2e^- \rightarrow H_2(g)$	0.00
$Pb^{2+} + 2e^- \rightarrow Pb$	-0.13

TABLE OF REDUCTION POTENTIALS:
REDUCTION AND OXIDATION

REDUCTION: The **top** of the Table of Reduction Potentials is the **most easily reduced** (strongest oxidizing agent). As you go down the table, atoms or ions are less easily reduced.

OXIDATION: The **bottom** of the table (right side of the equation) is the **most easily oxidized** (strongest reducing agent).

Li is the most easily oxidized. As you go up the table, atoms or ions are less easily oxidized.

Table of Reduction Potentials

Half Reaction	E^0 (volts)
↓ *Most easily reduced*	
$F_2 + 2e^- \rightarrow 2F^-$	+2.87
$Cl_2 + 2e^- \rightarrow Cl^-$	+1.36
$Br_2 + 2e^- \rightarrow Br^-$	+1.09
$I_2 + 2e^- \rightarrow I^-$	+0.54
$Li^+ + e^- \rightarrow Li(s)$	-3.04
Most easily ↑ oxidized	

Question: Based on the Table of Reduction Potentials, which of the following ions is most easily oxidized?

    (1) F^- (2) Cl^- (3) Br^- (4) I^-

Solution: Answer *4*. Look at the Table of Reduction Potentials, above. The top of the table is most easily reduced. The bottom of the table is most easily oxidized. As you go up towards the top of the table, ions get less easily oxidized. I^- is lower down on the table than the other three, and therefore is the most easily oxidized.

TABLE OF REDUCTION POTENTIALS:
SPONTANEOUS REACTIONS

RULE: Substances to the left and those below them on the right react spontaneously with each other.

Anything on the left side will react spontaneously with anything below and to the right. F_2 reacts spontaneously with Cl^-.

Anything on the right side will react spontaneously with anything to the left and above it. Cl^- reacts spontaneously with F_2.

In short: F_2 reacts spontaneously with Cl^-, and Cl^- reacts spontaneously with F_2.

Table Of Reduction Potentials

Half Reaction		E^0 *(volts)*
Left	Right	
$F_2 + 2e^- \rightarrow 2F^-$		
$Au^{+3} + 3e^- \rightarrow Au$		
$Cl_2 + 2e^- \rightarrow 2Cl^-$		

The arrow goes from the upper left to the lower right.
REMEMBER the direction of the arrow: ◣

Question: According to the Table of Reduction Potentials, which metal will react spontaneously with Ag^+ ions, but not with Zn^{2+} ions?

    (1) Cu   (2) Au   (3) Al   (4) Mg

Solution: Answer *1*. Cu will react spontaneously with Ag^+. Cu reacts with anything above it and to the left.

REMEMBER: Ag^+ (upper left) reacts with anything below it and to the right. Ag^+ reacts with Cu.

Table of Reduction Potentials

Half Reaction	E^0 *volts)*
$Ag^+ + e^- \rightarrow Ag$	+0.80
$Cu^{2+} + 2e^- \rightarrow Cu$	+0.34
$Zn^{2+} + 2e^- \rightarrow Zn(s)$	−0.76

REMEMBER: Upper left ↔ lower right, or lower right ↔ upper left: ◣. This is the only direction for spontaneous reactions.

Cu
Zn^{2+}

Therefore, Cu cannot react with Zn^{2+} — wrong direction.

Question: According to the Table of Reduction Potentials, which reaction will take place spontaneously?

(1) $Ni^{2+} + Pb \rightarrow Ni + Pb^{2+}$ (3) $Sr^{2+} + Sn \rightarrow Sr + Sn^{2+}$
(2) $Au^{3+} + Al \rightarrow Au + Al^{3+}$ (4) $Fe^{2+} + Cu \rightarrow Fe + Cu^{2+}$

Solution: Answer **2**. Remember the direction of the arrow for spontaneous reactions: upper left ↔ lower right, or lower right ↔ upper left. Look at the table. Au^{3+} reacts spontaneously with anything below it and to the right. Au^{3+} reacts spontaneously with Al.

Additional Information: Al reacts spontaneously with Au^{3+}.

Table of Reduction Potentials

Half Reaction	E^0
Left Right	
$Au^{+3} + 3e^- \rightarrow Au$	+1.50
⋮	
$Al^{3+} + 3e^- \rightarrow Al$	−1.66

Try Sample Questions #2-6, on page 30, then do Homework Questions, #4-20, pages 32-33, and #27-29, page 34

ELECTROLYTIC CELL AND ELECTROLYSIS

When E^0 is negative, a reaction will not occur spontaneously. When NO reaction will take place spontaneously (not spontaneous or non-spontaneous chemical reaction), see Table J, an **electrolytic cell** uses (needs) a **power source** or battery (**sources of electricity**, electric current, or electrical energy) **to produce chemical reactions** (chemical change).

Example: *Electrolysis of Water*:
Electricity breaks down water into hydrogen and oxygen:

$$2\,H_2O \xrightarrow{\text{electricity}} 2\,H_2 + O_2$$

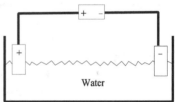

Water

Comparison of Voltaic and Electrolytic cells

SAME:

   Voltaic and electrolytic cells have ANODE-OXIDATION (oxidation takes place at the anode) and CATHODE-REDUCTION (reduction takes place at the cathode).

DIFFERENT:

   1. **Voltaic cell**: CHEMICAL REACTIONS PRODUCE ELECTRICITY. (Chemical energy is changed into electrical energy).

      Electrolytic cell: ELECTRICITY PRODUCES CHEMICAL REACTIONS. (Electrical energy is changed into chemical energy).

   2. **Voltaic cell**: The ANODE IS NEGATIVE (example, Zn) and the CATHODE IS POSITIVE.

      Electrolytic cell: **POLARITIES ARE REVERSED**: The ANODE IS *POSITIVE*, the CATHODE IS *NEGATIVE*.

ELECTROLYSIS OF KCl: Electrical energy (electricity) produces a **chemical reaction** (chemical change or chemical energy).

$$2KCl \longrightarrow 2K + Cl_2$$

Step 1: Electrons (negatively charged particles) go from the negative end of the electric source (battery) through the wire to the **electrode**, which is now negative.

Step 2: Positive ions go to the negative electrode (positive is attracted to negative — opposites attract) and gain an electron:

$$\underset{\text{reduction.}}{K^+ + e^- \rightarrow K,}$$

You learned that **reduction** takes place at the **cathode** (**cat** eats **red** meat = **cathode reduction**).

The movement of ions carries the electrical current through the liquid KCl.

Step 3: The other electrode obviously is the anode, oxidation,

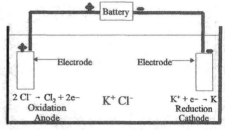

positive. The negative chloride ions are attracted to the positive electrode (opposites attract), loses electrons, and becomes chlorine, Cl_2.

$$2\,Cl^- \longrightarrow Cl_2 + 2e^-$$

The electrons go from the positive electrode through the wire back to the battery (positive end).

REMEMBER:

Voltaic Cell: The ANODE IS NEGATIVE (example, Zn).

Electrolytic Cell: The **polarities are reversed**. The CATHODE IS *NEGATIVE*.

Electroplating is not part of the NYS Regents curriculum, but applications are asked on the exam.

ELECTROPLATING

Electroplating is another example of an **electrolytic cell**. An electric current is used to produce chemical reactions. A layer of metal, such as silver or copper, coats or covers any object (spoon, fork) to be plated.

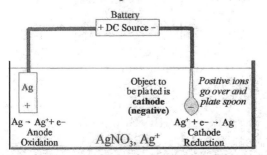

You want to cover the spoon with silver. Make the spoon the cathode (**negative electrode**). The cathode is negative in an electrolytic cell. Ag^+ (silver ions) goes over to the cathode (spoon), gains an electron and becomes silver, which covers the spoon.

$$Ag^+ + e^- \rightarrow Ag$$
Reduction.

You put the spoon in a solution of $AgNO_3$ to provide more Ag^+.

You make the anode a bar of silver. At the anode, the reaction $Ag \rightarrow Ag^+ + e^-$ takes place. Now you have more Ag^+.

Question: The diagram below shows the electrolysis of fused KCl:

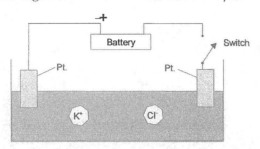

What occurs when the switch is closed?
  (1) Positive ions migrate toward the anode, where they lose electrons.
  (2) Positive ions migrate toward the anode, where they gain electrons.
  (3) Positive ions migrate toward the cathode, where they lose electrons.
  (4) Positive ions migrate toward the cathode, where they gain electrons.

Solution: Answer *4*. You know that, in an electrolytic cell, the cathode is negative (reverse of voltaic cell), and positive ions go over to the cathode and gain electrons.

Try Sample Questions #7-8, on page 30, then do Homework Questions, #21-23, pages 33-34, and #30-31, page 34.

BALANCING OXIDATION-REDUCTION EQUATIONS
Let's learn how to balance oxidation-reduction equations.

Question: Given the reaction:

$$\_Mg + \_Cr^{3+} \rightarrow \_Mg^{2+} + \_Cr$$

When the equation is properly balanced using the smallest whole numbers, the sum of the coefficients will be
  (1) 10   (2) 7   (3) 5   (4) 4

Solution: Answer *1*.

Step 1: Write the equations:

Oxidation: $Mg \rightarrow Mg^{2+} + 2e^-$ **Reduction**: $Cr^{3+} + 3e^- \rightarrow Cr$

(Any element alone has an oxidation number of zero.)

Step 2: Loss of electrons must = gain of electrons. Put coefficients in front of oxidation and reduction equations so that loss of electrons = gain of electrons:

$3(Mg \rightarrow Mg^{2+} + 2e^-)$ means $3\,Mg \rightarrow 3Mg^{2+} + 6e^-$ (3 Mg loses 6 electrons)

$2(Cr^{3+} + 3e^- \rightarrow Cr)$ means $2Cr^{3+} + 6e^- \rightarrow 2Cr$ (2 Cr^{3+} gains 6 electrons).

Add oxidation and reduction reactions in order to get **one net equation**. (The 6 electrons after the arrow in the first equation cancel the 6 electrons before the arrow in the second equation.)

$$3Mg + 2Cr^{3+} \rightarrow 3Mg^{2+} + 2Cr$$

The question asked for the **sum** of the coefficients. Add the coefficients: $3 + 2 + 3 + 2 = 10$. Answer 1 is correct.

Question: Write the balanced equation for the reaction of hydrochloric acid and potassium permanganate to produce water, potassium chloride, manganese (II) chloride, and chlorine gas,

__HCl + __KMnO$_4$ → __H$_2$O + __KCl + __MnCl$_2$ + __Cl$_2$.

Solution:

1. Look at the oxidation number for the elements on the upper right hand corner of the box in the Periodic Table, on page Reference Tables 20-21. The oxidation number of any free element (by itself) is zero. Write oxidation numbers as superscripts to the right of the symbols of the elements:

   $H^+ Cl^- + K^+ Mn^{7+} O_4^{2-} \rightarrow H_2^+ O^{2-} + K^+ Cl^- + Mn^{2+} Cl_2^- + Cl_2^0$
   + written to the right of the symbol of the element (example, H or K) means 1+; the 1 is understood and not written. Similarly, – to the right of the symbol of the element (example, Cl) means 1–; the 1 is understood and not written.

2. Write the two reactions for oxidation and reduction:
   $$2Cl^- \rightarrow Cl^\circ_2 + 2e^- \qquad Mn^{7+} + 5e^- \rightarrow Mn^{2+}$$

3. Number of electrons lost = number of electrons gained:
   $$5(2Cl^- \rightarrow Cl_2 + 2e^-) = 10Cl^- \rightarrow 5Cl_2 + 10e^-$$
   $$2(Mn^{7+} + 5e^- \rightarrow Mn^{2+}) = 2Mn^{7+} + 10e^- \rightarrow 2Mn^{2+}$$

4. Add the oxidation and reduction reactions. Net reaction =
   $$10Cl^- + 2Mn^{7+} \rightarrow 5Cl_2 + 2Mn^{2+}$$

5. Put coefficients in the (skeleton) equation:
   $$10HCl + 2KMnO_4 \rightarrow H_2O + KCl + 2MnCl_2 + 5Cl_2.$$

The K and O atoms in $2KMnO_4$ form $2KCl + 8H_2O$.
$$10HCl + 2KMnO_4 \rightarrow 8H_2O + 2KCl + 2MnCl_2 + 5Cl_2.$$

The $2KCl$ and $2MnCl_2$ in the product (right side of the equation) need 6 more HCl molecules, which must come from 6 more HCl.

The balanced equation:
$$16HCl + 2KMnO_4 \rightarrow 8H_2O + 2KCl + 2MnCl_2 + 5Cl_2.$$

Question: Given the reaction in a voltaic cell:
$$Mg(s) + Ni^{2+}(aq) \rightarrow Mg^{2+}(aq) + Ni(s)$$
What is the total number of moles of electrons needed to completely reduce 6 moles of $Ni^{2+}(aq)$ ions?

Solution: In the equation $Mg(s) + Ni^{2+}(aq) \rightarrow Mg^{2+}(aq) + Ni(s)$, Ni^{2+} gains electrons (Ni^{2+} is reduced) to form Ni.

   1 Ni^{2+} gains 2 electrons   to form 1 Ni.

or

1 mole of Ni^{2+} gains 2 moles of electrons to form 1 mole of Ni.

6 moles of Ni^{2+} gains 6 x 2 = 12 moles of electrons to form 6 moles of Ni.

Answer: 12 moles of electrons.

Question: Is the equation balanced:
$$3Fe^{3+} + Al \rightarrow 3Fe^{2+} + Al^{3+}$$

Solution: In a balanced equation, there is the same number of atoms of each element on both sides of the equation. Left side of the equation has $3Fe + 1Al$; right side of the equation has $3Fe + 1Al$. There is the same number of atoms of each element on both sides of the equation ($3Fe + 1Al$).

In a balanced equation, loss of electrons equals gain of electrons.

   Reduction: $Fe^{3+} + e^- \rightarrow Fe^{2+}$, therefore $3Fe^{3+} + 3e^- \rightarrow 3Fe^{2+}$
   Oxidation: $Al \rightarrow Al^{3+} + 3e^-$ $Al \rightarrow Al^{3+} + 3e^-$
   (Any element alone has an oxidation number of zero.)

Equation is balanced: loss of electrons equals gain of electrons.

   Al loses 3 electrons; $3 Fe^{3+}$ gains 3 electrons.

Try Sample Question #9, on page 30, then do
Homework Questions, #24-25, page 34.

REMEMBER: When you answer Regents questions on ELECTRO-CHEMISTRY, use Table J and the Periodic Table.

SAMPLE REGENTS & REGENT-TYPE QUESTIONS AND SOLUTIONS

Give your answers to Questions 1 and 2 based on the following reaction:
$$Mg(s) + 2Ag^+(aq) \rightarrow Mg^{2+}(aq) + 2Ag(s)$$

1. Which species undergoes a loss of electrons?
   (1) $Mg(s)$ (2) $Ag^+(aq)$ (3) $Mg^{2+}(aq)$ (4) $Ag(s)$

2. What is the cell voltage ($E°$) for the overall reaction?
   (1) $+1.57$ V   (2) $+2.37$ V   (3) $+3.17$ V   (4) $+3.97$ V

3. Which half-cell reaction serves as the arbitrary standard used to determine the standard electrode potential and activity series in Table J?
   (1) $Na^+ + e^- \rightarrow Na(s)$ (3) $F_2(g) + 2e^- \rightarrow 2F^-$
   (2) $Ag^+ + e^- \rightarrow Ag(s)$ (4) $2H^+ + 2e^- \rightarrow H_2(g)$

4. Which reaction will take place spontaneously?
   (1) $Cu + 2H^+ \rightarrow Cu^{2+} + H_2$ (3) $Pb + 2H^+ \rightarrow Pb^{2+} + H_2$
   (2) $2Au + 6H^+ \rightarrow 2Au^{3+} + 3H_2$ (4) $2Ag + 2H^+ \rightarrow 2Ag^+ + H_2$

5. When equilibrium is attained in a voltaic cell, the cell voltage is
   (1) between 0 and -1 (3) between 0 and $+1$
   (2) 0   (4) greater than $+1$

6. According to *Reference Table J*, which molecule is most easily reduced?
   (1) Br_2 (2) Cl_2 (3) F_2 (4) I_2

7. In which kind of cell are the redox reactions made to occur by an externally applied electrical current (external power source)?
   (1) voltaic cell   (3) battery
   (2) chemical cell   (4) electrolytic cell

8. Which atom forms an ion that would migrate toward the cathode in an electrolytic cell?
   (1) F   (2) I   (3) Na   (4) Cl

9. Which quantities are conserved in all oxidation-reduction reactions?
   (1) charge, only   (3) both charge and mass
   (2) mass, only   (4) neither charge nor mass

SOLUTIONS

1. Answer **1**. $Mg^0 \rightarrow Mg^{2+} + 2e^-$. Mg loses electrons. [Mg(s) means Mg solid.]

2. Answer **3**. $Mg + 2Ag^+ \rightarrow Mg^{2+} + 2Ag$. Look at the Table of Reduction Potentials. This example and the table start with Ag^+; leave the sign of E^0 the way it is: $E^0 = +0.80$V (volts). (It doesn't matter how many Ag^+ there are; the $E^0 = +0.80$.) This example starts with Mg. **The**

Table of Reduction Potentials

Half Reaction	E^0 (volts)
$Ag^+ + e^- \rightarrow Ag$	$+0.80$
$2H^+ + 2e^- \rightarrow H_2$	0
$Zn^{2+} + 2e^- \rightarrow Zn$	-0.76
$Mg^{2+} + 2e^- \rightarrow 2Mg$	-2.37

table starts with Mg^{2+} (reverse, or opposite reaction); therefore, **in this example, change the sign of E^0.** $E^0 = +2.37$ V.

Add the two E^0's:

$$Ag^+ + e^- \to Ag \qquad E^0 = +0.80V$$
$$Mg \to Mg^{2+} + 2e^- \qquad \underline{E^0 = +2.37V}$$
$$E^0 = +3.17V, \text{ Answer 3.}$$

3. Answer *4*. Half cell reaction $2H^+ + 2e^- \to H_2$ serves as the arbitrary standard.

4. Answer *3*. The **metal** at the **top** of Table J is **most active**(the word most is written at the top of Table J). Metals **higher up** are **more active** and **replace** metals **below** them **from compounds**.Pb is higher up on Table J than H. Therefore, Pb replaces H from its compounds and the reaction is spontaneous.

5. Answer *2*. At equilibrium, the voltage, E^0, is zero.

6. Answer *3*. Look at Table J. F_2 is on the **top** of Table J. $F_2$is the most active nonmetal, and is the **most easily reduced**.

7. Answer *4*. In an electrolytic cell, **electric current** (electricity) produces redox reactions.

8. Answer *3*. You learned that, in an electrolytic cell, the cathode is negative. Positive ions go to the negative cathode. Look at the oxidation numbers in the upper right hand corner of the boxes on the Periodic Table for F, I, Na and Cl. The only positive ion is Na, with a +1 oxidation number. The three other choices are all negative ions (F^-, Cl^- and I^-).

9. Answer *3*. In a redox reaction, there is conservation of charge (electrons lost = electrons gained) and conservation of mass (same number of atoms of each element on both sides of the equation).

NOW LET'S TRY A FEW HOMEWORK QUESTIONS:

1. Which species acts as the anode when the reaction
$$Zn(s) + Pb^{2+}(aq) \to Zn^{2+}(aq) + Pb(s)$$
occurs in an voltaic cell?
(1) $Zn(s)$ (2) $Zn^{2+}(aq)$ (3) $Pb^{2+}(aq)$ (4) $Pb(s)$

2. The purpose of a salt bridge in an electrochemical cell is to
(1) allow for the flow of molecules between the solutions
(2) allow for the flow of ions between the solutions
(3) prevent the flow of molecules between the solutions
(4) prevent the flow of ions between the solutions

Base your answers to Questions 3 and 4 on the diagram below of an electrochemical cell at 298K:

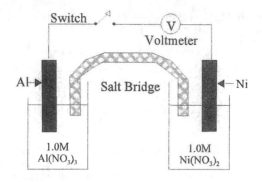

3. When the switch is closed, electrons flow from
    (1) Al(s) to Ni (s)   (3) Al^{3+}(aq) to Ni^{2+}(aq)
    (2) Ni(s) to Al(s)   (4) Ni^{2+}(aq) to Al^{3+}(aq)

4. What is the maximum cell voltage (E^0) when the switch is closed?
    (1) +1.40V   (2) −1.40V   (3) +1.92V   (4) −1.92V

5. Given the reaction:
    $$Zn(s) + Cu^{2+}(aq) \rightleftharpoons Zn^{2+}(aq) + Cu(s)$$
    When the chemical cell reaction reaches equilibrium, the measured voltage will be
    (1) 0.00 V   (2) 0.34 V   (3) 0.76 V.   (4) 1.10 V

6. Which overall reaction in a chemical cell has the highest net potential (E^0)?
    (1) $Zn(s) + 2H^+ \rightarrow Zn^{2+} + H_2(g)$ (3) $Mg(s) + 2H^+ \rightarrow Mg^{2+} + H_2(g)$
    (2) $Ni(s) + 2H^+ \rightarrow Ni^{2+} + H_2(g)$ (4) $Sn(s) + 2H^+ \rightarrow Sn^{2+} + H_2(g)6$

7. Given the reaction:
    $$Mg(s) + Zn^{2+}(aq) \rightarrow Mg^{2+}(aq) + Zn(s)$$
    What is the cell voltage (E^0) for the overall reaction?
    (1) +1.61 V   (2) −1.61 V   (3) +3.13 V   (4) −3.13 V

8. Given the reaction:
    $$Zn(s) + Br_2(aq) \rightarrow Zn^{2+}(aq) + 2Br^-(aq)$$
    What is the net cell potential (E^0) for the overall reaction?
    (1) +0.76 V   (2) −1.09 V   (3) +1.85 V   (4) 0.00 V

9. What is the standard electrode potential (E^0) assigned to the half-reaction Cu^{2+} + $2e^- \rightarrow Cu(s)$, when compared to the standard hydrogen half-reaction?
    (1) +0.34 V   (2) −0.34 V   (3) +0.52 V   (4) −0.52 V

Base your answers to Question 10 on *Reference Table J* and to Question 11 on the *Table of Reduction Potentials.* Use the diagram below for both questions.

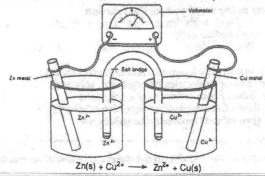

10. When this cell operates, the electrons flow from the
    (1) copper half-cell to the zinc half-cell through the wire
    (2) copper half-cell to the zinc half-cell through the salt bridge
    (3) zinc half-cell to the copper half-cell through the wire
    (4) zinc half-cell to the copper half-cell through the salt bridge

11. What is the potential (E^0) of this cell?
    (1) +1.10 V   (2) +0.42 V   (3) −1.10 V   (4) −0.42 V

12. Based on *Reference Table J*, which of the following ions is most easily oxidized?
    (1) F⁻   (2) Cl⁻   (3) Br⁻   (4) I⁻

13. According to *Reference Table J*, which is the strongest reducing agent?
    (1) Li(s)   (2) Na(s)   (3) $F_2(g)$ (4) $Br_2(\ell)$

14. Which ion can be both an oxidizing agent and a reducing agent?
    (1) Sn^{2+} (2) Cu^{2+} (3) Al^{3+} (4) Fe^{3+}

15. Based on *Reference Table J*, which of the following elements is the strongest reducing agent?
    (1) Fe   (2) Sr   (3) Cu   (4) Cr

16. According to *Reference Table J*, which will reduce Mg^{2+} to Mg(s)?
    (1) Fe(s)   (2) Ba(s)   (3) Pb(s)   (4) Ag(s)

17. Based on *Reference Table J*, which metal will react with H^+ ions to produce $H_2(g)$?
    (1) Au   (2) Ag   (3) Cu   (4) Mg

18. According to *Reference Table J*, which metal will react spontaneously with Al^{3+} at 298 K?
    (1) Cu   (2) Au   (3) Li   (4) Ni

19. According to *Reference Table J*, which reaction will take place spontaneously?
    (1) $Ni^{2+} + Pb(s) \rightarrow Ni(s) + Pb^{2+}$ (3) $Sr^{2+} + Sn(s) \rightarrow Sr(s) + Sn^{2+}$
    (2) $Au^{3+} + Al(s) \rightarrow Au(s) + Al^{3+}$ (4) $Fe^{2+} + Cu(s) \rightarrow Fe(s) + Cu^{2+}$

20. Based on *the Table of Reduction Potentials*, which ion will reduce Ag^+ to Ag?
    (1) F⁻   (2) I⁻   (3) Br⁻   (4) Cl⁻

21. Given the equation for the electrolysis of a fused salt:
    $$2LiCl(\ell) + electricity \rightarrow 2Li(\ell) + Cl_2(g)$$
    Which reaction occurs at the cathode?
    (1) $2Cl^- \rightarrow Cl_2(g) + 2e^-$ (3) $Li^+ + e^- \rightarrow Li(\ell)$
    (2) $2Cl^- + 2e^- \rightarrow Cl_2(g)$ (4) $Li^+ \rightarrow Li(\ell) + e^-$

Base your answers to Questions 22 A and B on the diagram below, which represents the electroplating of a metal fork with Ag(s).

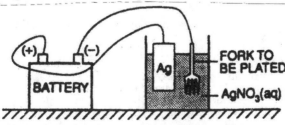

22. A. Which part of the electroplating system is provided by the fork?
    (1) the anode, which is the negative electrode
    (2) the cathode, which is the negative electrode
    (3) the anode, which is the positive electrode
    (4) the cathode, which is the positive electrode

B. Which equation represents the half-reaction that takes place at the fork?
(1) $Ag^+ + NO_3^- \rightarrow AgNO_3$ (3) $AgNO_3 \rightarrow Ag^+ + NO_3^-$
(2) $Ag^+ + e^- \rightarrow Ag(s)$ (4) $Ag(s) \rightarrow Ag^+ + e^-$

---

23. The diagram below shows a key being plated with copper in an electrolytic cell.
    Given the reduction
reaction for this cell:
$Cu^{2+}(aq) + 2e^- \quad Cu(s)$
This reduction occurs at
(1) A, which is the anode
(2) B, which is the anode
(3) A, which is the cathode
(4) B, which is the cathode

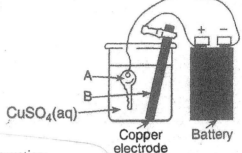

CuSO$_4$(aq)

Copper electrode   Battery

24a. Given the unbalanced equation:
    _MnO$_2$ + _HCl → _MnCl$_2$ + _H$_2$O + _Cl$_2$
    When the equation is correctly balanced, using the smallest whole-number
    coefficients, the coefficient of HCl is
    (1) 1   (2) 2   (3) 3   (4) 4

b. Given the reaction:
    _Cr + _Fe^{2+} → _Cr^{3+} + _Fe
    When the reaction is completely balanced, using the smallest whole-number
    coefficients, the sum of the coefficients is
    (1) 3   (2) 4   (3) 6   (4) 10

25. How many moles of electrons does zinc lose when 4 moles of Zn° react with HCl
    in the reaction
    $Zn + HCl \rightarrow ZnCl_2 + H_2$
    (1) 8   (2) 1   (3) 2   (4) 4

CONSTRUCTED RESPONSE QUESTIONS: Parts B-2 and C of NYS Regents Exam

26. Draw a voltaic cell. The 2 electrodes are zinc and copper. Label cathode and
    anode, positive and negative, oxidation and reduction, direction of flow of
    electrons.

27. Explain why the following reactions are or are not spontaneous:
        Hint: Look at the full Table J on pages Reference Tables 11-12.
    a. Zn + HCl   b. Cu + PbSO$_4$ c. Ba + NaCl   d. Co + FeSO$_4$
    e. Mn + PbSO$_4$ f. Rb + KCl   g. F$_2$ + HCl   h. I$_2$ + HBr
    i. Br$_2$ + HF

28. Explain how the activity of metals and nonmetals changes as you go down Table J.

29. Compare and contrast Rb and Ag and compare and contrast Cl$_2$ and Br$_2$ (include
    activity, oxidized, reduced).

30. Compare and contrast voltaic and electrolytic cells.

31. A. Draw and explain an electrolytic cell for $2 KCl \rightarrow 2K + Cl_2$.
    B. Explain why AgNO$_3$ is a better choice than AgCl for use in electroplating (an
       electrolytic cell). Hint: See complete Table F on page Reference Tables 6-7;
       review Table F: Solubility in Chapter 6.

32. Balance this redox reaction, using the smallest whole-number coefficients:
    $Cu(s) + AgNO_3(aq) \rightarrow Cu(NO_3)_2(aq) + Ag(s)$.

33. Balance the equation for the reaction, using the smallest whole-number
    coefficients:
    _Fe$_2$O$_3$ + _CO → _Fe + _CO$_2$.

CHAPTER 10: ORGANIC CHEMISTRY

SECTION A

ORGANIC CHEMISTRY is the study of carbon and carbon compounds. There are a large number of carbon compounds. The carbon atoms bond together (join together) to form chains C-C-C, branches $\begin{smallmatrix} C \\ | \\ C-C-C \end{smallmatrix}$, rings ⬡ , or networks of carbon atoms.

COMMON CHARACTERISTICS OF ORGANIC COMPOUNDS

1. Organic compounds are generally **nonpolar** (no end is positive and no end is negative).
2. Because they are nonpolar, they are **soluble** in nonpolar solvents (example: benzene).
3. Organic compounds are **non-electrolytes** (they don't conduct an electric current).
4. They have **low melting points**.
5. **Reactions** are generally **slower** than inorganic reactions.
6. Reactions have **high activation energy**.

BONDING

Carbon, $^{12}_{6}C$, $\frac{6p}{6n}\begin{smallmatrix}) &) \\ & \\ ) &) \end{smallmatrix}$ 2 4, has 4 valence electrons, and forms four covalent bonds. A bond is made up of 2 electrons or 1 pair of electrons. The bonds around the carbon atom are shaped like a **tetrahedron**. (The four bonds of the carbon atoms are directed toward the corners of a tetrahedron.)

Organic compounds can be classified (made, sorted) into groups having similar structures and characteristics. **Hydrocarbons** are organic compounds that contain only carbon and hydrogen atoms.

Organic compounds generally use the IUPAC system of naming (examples: ethane on the next page, ethene and propyne later in the chapter).

ALKANES

ALKANES are **hydrocarbons** with the formula C_nH_{2n+2}. This formula is given in **Table Q**, below, or on page Reference Table 19.

Table Q

		Example	
Name	General Formula	Name	Structural Formula
alkanes	C_nH_{2n+2}	ethane	H H | | H-C-C-H | | H H

Table P

Prefix	Number of Carbon Atoms
meth-	1

Methane is the first member of the alkane series. Methane ends in "-ane" because it is part of the alkane series. All members of the alkane series end in **-ane**.

Look at **Table P** above or on page Reference Table 18. If an organic compound begins (prefix) with "**meth-**" (or "m-" for short), it has one carbon atom. Therefore, you know that **meth**ane, which begins with "meth-," has one carbon atom.

To find the **molecular formula** of **methane**, use the alkane formula C_nH_{2n+2}, given in Table Q. Let "n" = number of carbon atoms. Methane has one carbon atom. The number of hydrogen atoms is 2 times the number of carbon atoms, plus 2 more:

$C_nH_{2n+2} = C_1H_{(2\times1)+2} = CH_4$, the molecular formula for methane.

The **structural formula** tells how the atoms are arranged:

There is a single bond (shown by one line) between C and H. A bond is made up of a pair of electrons (two electrons). Methane and all alkanes only have **single bonds** and are **saturated**.

$$H - \underset{\underset{H}{|}}{\overset{\overset{H}{|}}{C}} - H$$

←single bond

single bond

Ethane is the next member of the alkane series:

Ethane similarly ends in "-ane" because it is part of the alkane series. All members of the alkane series end in **-ane**.

Look at **Table P**; if the organic compound begins (prefix) with "eth-" (or "e" for short), it has **2 carbon atoms**. Therefore you know **eth**ane, which begins with "eth-," has 2 carbon atoms.

Table P

Prefix	Number of Carbon Atoms
eth-	2

You can figure out the **molecular formula** of **ethane**. Use the **alkane formula** C_nH_{2n+2}, given in **Table Q**. Let "n" = number of carbon atoms.

Ethane has **two carbon atoms**. The number of hydrogen atoms is 2 times the number of carbon atoms, plus 2 more:

$C_nH_{2n+2} = C_2H_{(2\times2)+2} = C_2H_6$, the molecular formula for ethane.

Look at the **structural formula** for ethane, $H-\overset{\overset{\displaystyle H}{|}}{C}-\overset{\overset{\displaystyle H}{|}}{C}-H$. Ethane has 2 carbon atoms, 6 hydrogen atoms, and single bonds between them.

Table Q

Name	General Formula	Examples	
		Name	Structural Formula
alkanes	C_nH_{2n+2}	ethane	$H-\overset{\overset{\displaystyle H}{|}}{C}-\overset{\overset{\displaystyle H}{|}}{C}-H$

By looking at Table Q, at the structural formula of ethane, which is an alkane, you can see that ethane (similarly, all alkanes) has single bonds between carbon atoms (single carbon-carbon bonds). $-C\underset{\underset{\textit{Single Bond}}{|}}{\top}C-$ Alkanes are saturated because they have single bonds between carbon atoms.

Propane is the next member of the alkane series:

The third member is propane. Prop**ane** ends in "**-ane**" as part of the alkane series. Look at Table P. If an organic compound begins (prefix) with "**prop-**" (or "**pro**" for short), it has 3 carbon atoms. Therefore you know **pro**pane, which begins with "prop-," has 3 carbon atoms.

Table P

Prefix	Number of Carbon Atoms
prop-	3

Again, you can figure out the **molecular formula** of **propane**. Use the alkane formula C_nH_{2n+2}, given in Table Q. Let "n" = number of carbon atoms. **Propane** has **three carbon atoms**. The number of hydrogen atoms is 2 times the number of carbon atoms, plus 2 more:

$C_nH_{2n+2} = C_3H_{(3\times2)+2} = C_3H_8$, the molecular formula for propane.

Look at the structural formula for propane, $H-\overset{\overset{\displaystyle H}{|}}{C}-\overset{\overset{\displaystyle H}{|}}{C}-\overset{\overset{\displaystyle H}{|}}{C}-H$. Propane has 3 carbon atoms, 8 hydrogen atoms and single bonds between them. Look at **Table Q**. Look at alkanes and the structural formula for ethane. You see ethane, and therefore **all alkanes**, have **single** bonds.

By looking at the formula, you can name this organic compound. Look at Table P again. 3 carbon atoms equals prop; 3 carbon atoms with **single bonds** is propane.

For larger molecules, it is easier to write a condensed structural formula. In a **condensed structural formula**, just draw lines between carbon atoms, but not between carbon and hydrogen atoms. The condensed structural formula for propane is written

$$CH_3 \text{—} CH_2 \text{—} CH_3$$

Line shows bond between
2 carbon atoms

PROPANE

Structural Formula:	$H - \overset{\displaystyle H}{\underset{\displaystyle H}{\overset{\displaystyle |}{\underset{\displaystyle |}{C}}}} - \overset{\displaystyle H}{\underset{\displaystyle H}{\overset{\displaystyle |}{\underset{\displaystyle |}{C}}}} - \overset{\displaystyle H}{\underset{\displaystyle H}{\overset{\displaystyle |}{\underset{\displaystyle |}{C}}}} - H$
Condensed Formula:	$CH_3 \text{-} CH_2 \text{-} CH_3$

In the condensed formula, the H_3 means the 3 hydrogen atoms surrounding the carbon atom, as shown by the formulas in the box above. The line in the formula between CH_3 and CH_2 means the bond between the two carbon atoms.

In **short, Alkanes** have **single bonds**, are **saturated**, and have the **formula C_nH_{2n+2}**. Each alkane differs from the next one by CH_2. For example, CH_4 differs from C_2H_6 by CH_2. The alkane series is a **homologous series** because each one differs from the next one by the same additional unit (CH_2). You will learn later about alkenes and alkynes, which are also homologous series.

As the members of a series (example: alkane series) **increase** in molecular **size** (increase in number of carbon atoms), dispersion forces (intermolecular forces) increase and the **boiling point** and **freezing point increase**.

Butane is the next member of the alkane series.

Butane ends in "-ane" as part of the alkane series. Look at **Table P**. If the organic compound begins (prefix) with "but-," it has 4 carbon atoms. Therefore you know that butane, which begins with "but-," has 4 carbon atoms.

Table P

Prefix	Number of Carbon Atoms
but-	4

To find the **molecular formula** of **butane**, use the alkane formula (see Table Q on the previous page), C_nH_{2n+2}. n = number of carbon atoms = 4:

$C_nH_{2n+2} = C_4H_{(4 \times 2)+2} = C_4H_{10}$, the molecular formula for butane.

By looking at the structural formula above, you know it is butane; "but-" because of 4 carbon atoms, "-ane" because it has single bonds.

$$H - \overset{\displaystyle H}{\underset{\displaystyle H}{\overset{\displaystyle |}{\underset{\displaystyle |}{C}}}} - \overset{\displaystyle H}{\underset{\displaystyle H}{\overset{\displaystyle |}{\underset{\displaystyle |}{C}}}} - \overset{\displaystyle H}{\underset{\displaystyle H}{\overset{\displaystyle |}{\underset{\displaystyle |}{C}}}} - \overset{\displaystyle H}{\underset{\displaystyle H}{\overset{\displaystyle |}{\underset{\displaystyle |}{C}}}} - H$$

Another structural formula for C_4H_{10} is at right, where the carbon and hydrogen atoms are arranged differently.

How do you **name** this compound?

$$
\begin{array}{c}
H \\
| \\
H-C-H \\
H \quad | \quad H \\
| \quad | \quad | \\
H-C_1-C_2-C_3-H \\
| \quad | \quad | \\
\text{carbon 1} \quad H \quad H \quad H \quad \text{carbon 3} \\
\text{carbon 2}
\end{array}
$$

2-methylpropane

RULE: Count the number of carbon atoms in the longest unbroken chain of carbon atoms. Three carbon atoms = "prop-" (Table P); three carbon atoms with **single bonds** is propane. The CH_3 is attached to C_2, or the second carbon atom. Since CH_4 is methane, **-CH₃** (1 hydrogen less than methane) is **a methyl group**. Since the CH_3 is attached to the second carbon atom, you call it 2-methyl. The compound is called **2-methylpropane**.

Study and understand the name with the structural formula. Later you will draw these formulas.

ISOMERS are compounds that have the **same molecular formula** but **different structural formulas**; therefore, isomers have **different physical** (example, boiling point) and **chemical properties**.

Example, **the isomers of C_4H_{10}**:

$$
\begin{array}{c}
H \quad H \quad H \quad H \\
| \quad | \quad | \quad | \\
H-C-C-C-C-H \\
| \quad | \quad | \quad | \\
H \quad H \quad H \quad H
\end{array}
$$

normal butane

and

$$
\begin{array}{c}
H \\
| \\
H-C-H \\
H \quad | \quad H \\
| \quad | \quad | \\
H-C--C--C-H \\
| \quad | \quad | \\
H \quad H \quad H
\end{array}
$$

2-methyl propane

Normal butane and 2-methyl propane are isomers because they have the same molecular formula, C_4H_{10}, but different structural formulas.

Pentane:

Pentane ends in "-ane" as part of the alkane series. Look on Table P. If an organic compound begins (prefix) with "pent-," it has 5 carbon atoms. Therefore you know that **pent**ane has 5 carbon atoms.

Table P

Prefix	Number of Carbon Atoms
pent-	5

To find the **molecular formula** of **pentane**, use the alkane formula, C_nH_{2n+2}, given in **Table Q**. n = number of carbon atoms = 5:

$C_nH_{2n+2} = C_5H_{(5\times2)+2} = C_5H_{12}$, the **molecular formula** for pentane.

Let's look at three **isomers of pentane**. They all have five C atoms and 12 H atoms:

A. **Normal, or n-pentane**: All the carbon atoms are in a straight chain (no branching).

$$\begin{array}{ccccc} H & H & H & H & H \\ | & | & | & | & | \\ H\text{-}C\text{-}C\text{-}C\text{-}C\text{-}C\text{-}H \\ | & | & | & | & | \\ H & H & H & H & H \end{array}$$

normal pentane

B. **2-methylbutane**:

Explanation of the name, 2-methylbutane:

Count carbon atoms in the longest unbroken chain of carbon atoms. **Four carbon** atoms = "**but**" (Table P); four carbon atoms with **single bonds** (Table Q) = bu**tane**.

2-methylbutane

It is called "**2-methyl**" because the **CH₃** (methyl) is attached to **C₂**, the second carbon, therefore "2-methylbutane."

Study and understand the structural formula. Later, you will draw these formulas.

C. **2, 2-dimethylpropane**:

Explanation of the name, 2,2-dimethylpropane

Rule: Count carbon atoms in the longest unbroken chain of carbon atoms. **Three carbon** atoms = "**prop**" (Table P); three carbon atoms with **single bonds** (Table Q) = propane. CH₃ (methyl) on top and CH₃ (methyl) on bottom = two methyls = **dimethyl**. "**2, 2**" dimethyl means **both methyls** are attached to **C₂**; therefore, it is called 2,2-dimethylpropane.

Study and understand the structural formula.

As the number of carbon atoms in the molecule increases, the number of isomers increases.

ALKANES WITH 6-10 CARBON ATOMS

For alkanes with 6-10 carbon atoms, continue to use Table P. Use the prefix given for the number of carbon atoms (Table P); follow the alkane formula C_nH_{2n+2} (Table Q). As you did previously, look at the structural formula for ethane. You see ethane and therefore all alkanes have single bonds (Table Q).

Table P

Prefix	Number of Carbon Atoms
hex	6
hept	7
oct	8
non	9
dec	10

Table Q			
Name	General Formula	Examples	
		Name	Structural Formula
alkanes	C_nH_{2n+2}	ethane	H H H-C-C-H H H

Examples:

Heptane: 7 carbon atoms = hept

$C_nH_{(2n + 2)} =$

$C_7H_{(2 \times 7) + 2} = C_7H_{16}$.

Nonane: 9 carbon atoms = non

$C_nH_{(2n + 2)} =$

$C_9H_{(2 \times 9) + 2} = C_9H_{20}$.

In short, **alkanes** have **single bonds**, are **saturated**, have the formula C_nH_{2n+2}, and are a homologous series in which members differ by CH_2.

Question: Determine the number of carbon atoms in one molecule of an alkane that has 22 hydrogen atoms in the molecule.

Solution: Alkane formula: CnH_{2n+2}.

Let n = # of C atoms.

Let # of H atoms = 2n + 2 (twice number of C atoms + 2).

# of H atoms = 22 given.

Set the two equations equal to each other: 2n + 2 = 22

n = 10

Note: $C_{10}H_{22}$ is decane.

Carbon always has 4 bonds: A carbon atom has 4 electrons in its outer shell, or 4 valence electrons: C o 2 4. A carbon atom will get a complete outer shell with 8 electrons. 8 electrons form 4 bonds, because a bond is made up of 2 electrons. Therefore, in any compound, **there are 4 bonds surrounding each carbon atom.** Look at Table Q. You will see that each carbon atom is surrounded by 4 bonds. When you **draw alkanes** or any other organic structure, make sure each **carbon atom has 4 bonds.**

ALKYL GROUPS

ALKYL groups have one less hydrogen than the corresponding alkane.

CH_3 is methyl. It has one less H than CH_4 (methane). Methyl begins with "meth-" (Table P); it has one carbon atom.

C_2H_5 is ethyl. It has one less H than C_2H_6 (ethane). Ethyl begins with "eth-" (Table P); it has two carbon atoms.

C_3H_7 is propyl. It has one less H than C_3H_8 (propane). Propyl begins with "prop-" (Table P); it has three carbon atoms.

Examples:

2-methylpropane:
CH₃ (methyl
group) on second
carbon of propane.

3-ethylpentane:
C_2H_5 (ethyl group) on third
carbon of pentane

*Try Sample Questions #1-3, on page 19, then do
Homework Questions, #1-13, page 21, and #22-23, page 22.*

HOW TO DRAW ALKANES

Table P

Prefix	Number of Carbon Atoms
hex	6
hept	7
oct	8
non	9
dec	10

Table Q
Homologous Series of Hydrocarbons

		Example	
Name	General Formula	Name	Structural Formula
alkanes	C_nH_{2n+2}	ethane	H H | | H-C-C-H | | H H

Draw hexane.

Step 1: Hexane begins with "hex-." Look at Table P above or on page Reference Tables 18. Hex means 6 carbon atoms. Draw a chain of 6 carbon atoms.

Step 2: Hexane ends in "ane," therefore you know it is an alkane. Look on Table Q at alkanes and the structural formula for ethane. You see ethane and therefore all alkanes have single bonds. Put single bonds between the carbon atoms: -C-C-C-C-C-C-.

Step 3: Each carbon atom must have 4 bonds or lines attaching it to other atoms. Put bonds or lines around all the carbon atoms. Attach hydrogens to the end of each bond:

The formula for alkanes, given on Table Q, is

$$C_nH_{2n+2} = C_6H_{(2x6)+2} = C_6H_{14}.$$

You can use the formula to check that you have put the correct number of hydrogen atoms, 14, into your drawing.

Draw heptane.

Step 1: Heptane begins with "hept." Look at Table P, on the previous page or page Reference Tables 18. Hept means 7 carbon atoms. Draw a chain of 7 carbon atoms.

Step 2: Heptane ends in "-ane," therefore you know it is an alkane. Look on Table Q at alkanes and the structural formula for ethane. You see ethane and therefore all alkanes have single bonds. Put single bonds between the carbon atoms in the chain: -C-C-C-C-C-C-C-.

Step 3 (last step): Each carbon atom must have 4 bonds or lines attaching it to other atoms. Put bonds or lines around all the carbon atoms. Attach hydrogen atoms to the end of each bond.

$$
\begin{array}{ccccccc}
H & H & H & H & H & H & H \\
| & | & | & | & | & | & | \\
H-C & -C & -C & -C & -C & -C & -C-H \\
| & | & | & | & | & | & | \\
H & H & H & H & H & H & H
\end{array}
$$

The formula for alkanes, given on Table Q, is

$$C_nH_{2n+2} = C_7H_{(2\times7)+2} = C_7H_{16.}$$

You can use the formula to check that you have put the correct number of hydrogen atoms, 16, into your drawing.

The condensed formula for heptane is

$$CH_3\text{-}CH_2\text{-}CH_2\text{-}CH_2\text{-}CH_2\text{-}CH_2\text{-}CH_3.$$

Only lines between carbon atoms are drawn, not those between carbon and hydrogen.

Heptane

Structural Formula:	$\begin{array}{ccccccc} H & H & H & H & H & H & H \\ \| & \| & \| & \| & \| & \| & \| \\ H-C & -C & -C & -C & -C & -C & -C-H \\ \| & \| & \| & \| & \| & \| & \| \\ H & H & H & H & H & H & H \end{array}$
Condensed Formula:	$CH_3\text{-}CH_2\text{-}CH_2\text{-}CH_2\text{-}CH_2\text{-}CH_2\text{-}CH_3$

Draw 2-methylbutane.

Step 1: Butane begins with "but." Look at Table P. "But" means 4 carbon atoms. Draw a chain of 4 carbon atoms.

Step 2: Butane ends in "-ane." Therefore, look on Table Q at alk**anes**, and the structural formula for ethane. You see ethane, and therefore all alkanes, have single bonds. Draw single bonds between the carbon atoms, C-C-C-C.

Step 3: Methyl begins with "meth-" (Table P). It has one carbon atom. Add the 1 carbon atom to the formula from Step 2. 2-methyl means the added C is on the second carbon atom.

$$
\begin{array}{c}
\quad\ -C- \\
\ \ \ \ | \\
-C_1-C_2-C_3-C_4-
\end{array}
$$

Step 4 (last step): Each **carbon** atom **must** have **4 bonds** or lines attaching it to other atoms: $-\overset{|}{\underset{|}{C}}-$. Put bonds or lines around all carbon atoms. Attach a hydrogen atom to the end of each bond.

2-methylbutane

The formula for alkanes given in Table Q is $C_nH_{2n+2} = C_5H_{(2\times5)+2} = C_5H_{12}$. You can use this formula to check that you've put the correct number of hydrogen atoms, 12, into your drawing.

For practice: Draw 2,2,4-trimethyl pentane. See question 27b, page 10:22.

MODEL of heptane, C_7H_{16}:

Use an organic chemistry model kit. Use 7 black balls to show 7 carbon atoms, and 16 yellow balls to show 16 hydrogen atoms. Use sticks to represent single bonds connecting carbon and carbon or carbon and hydrogen atoms. A model of heptane can also be made by using a space-filling model kit.

Do Homework Questions #24-27, page 22.

ALKENES

ALKENES are hydrocarbons with the formula C_nH_{2n} (also known as the **ethylene series**). Alkenes have one double bond.

Ethene is the first member of the alkene series:

Ethene ends in "**-ene**" because it is part of the alkene series. All members of the **alkene** series end in "**-ene**." Look at Table P, here, or on page Reference Tables 18 . If the compound begins (prefix) with "eth-" (or "e" for short), it

Table P

Prefix	Number of Carbon Atoms
eth-	2

has 2 carbon atoms. Therefore you know that **eth**ene, which begins with "eth-," has 2 carbon atoms.

There is no alkene corresponding to methane, because alkenes must have a double bond between two carbon atoms. Ethene is commonly called ethylene.

Ethene and **all alkenes** have the formula C_nH_{2n}. The formula is given on Table Q on the next page. Let "n" = number of carbon atoms. Ethene has two carbon atoms:

$C_nH_{2n} = C_2H_{(2\times2)} = \mathbf{C_2H_4}$, the **molecular formula for ethene**.

By looking at the structural formula of ethene in Table Q, on the next page, you know it is ethene: "**eth-**" because it has 2 carbon atoms, "**-ene**" because it has a double bond (shown in Table Q). On Table Q, the structural formula with the name, ethene, is given.

Name	General Formula	Example	
		Name	Structural Formula
Alkene	C_nH_{2n}	ethene	$\begin{array}{cc} H & H \\ \backslash & / \\ C & = C \\ / & \backslash \\ H & H \end{array}$

By looking at Table Q, at the structural formula of ethene (which is an example of alkenes), you see that ethene (similarly all alkenes) has a double bond between 2 carbon atoms: $-\overset{|}{C}=\overset{|}{\underset{\uparrow}{C}}-$

Double Bond

When you see a double bond in the structural formula, you know it ends in -ene.

In short, **ethene** and **all alkenes** have the formula $\mathbf{C_nH_{2n}}$, have **one double bond**, and are **unsaturated**, because two carbon atoms share more than one pair of electrons.

Propene is the next member of the alkene series:

Table P

Prefix	Number of Carbon Atoms
prop-	3

Propene ends in "-ene" because it is part of the alkene series. All members of the alkene series end in "-ene." Look at Table P. If the organic compound begins(prefix) with "prop-" (or "pro-" for short), it has 3 carbon atoms. Therefore you know that **pro**pene, which begins with "prop-," has 3 carbon atoms.

Propene and **all alkenes** have the formula $\mathbf{C_nH_{2n}}$, given in Table Q. Let "n" = number of carbon atoms. Propene has three carbon atoms:

$C_nH_{2n} = C_3H_{(3x2)} = \mathbf{C_3H_6}$, the **molecular formula** for **propene**.

Propene and **all alkenes** have **one double bond** and are **unsaturated**.

You learned, in the condensed formula (see below), to just draw the **lines between carbon atoms** and not between carbon and hydrogen atoms. The double bond (=) between CH_2 and CH means the bond between the two carbon atoms.

By looking at the structural or condensed formula , you know it is **propene**: "**prop-**" because it has 3 carbon

Structural Formula	$\begin{array}{c} H \quad\quad H \quad H \\ \backslash \quad\quad | \quad | \\ C = C - C - H \\ / \quad\quad\quad | \\ H \quad\quad\quad H \end{array}$
Condensed Formula	$CH_2 = CH\text{-}CH_3$

atoms (see Table P), "**-ene**" because it has a double bond (see Table Q). Look at alkenes and the structural formula for ethene. You see ethene, and therefore all **alkenes**, have a **double bond.**

2-Butene:

Butene and **all alkenes** have the formula C_nH_{2n}. Let "n" = number of carbon atoms. Butene has four carbon atoms:

$C_nH_{2n} = C_4H_{(4x2)} = \mathbf{C_4H_8}$, the **molecular** formula for **butene**.

2-butene means 4 C atoms and double bond after the second carbon atom.

Table P

Prefix	Number of Carbon Atoms
but-	4

H—C_1—C_2≡C_3—C_4—H

double bond after second carbon

$CH_3 - CH = CH - CH_3$
Condensed Formula

Table Q

		Example	
Name	General Formula	Name	Structural Formula
alkenes	C_nH_{2n}	ethene	C = C

By looking at the structural or condensed formula, you know it is **butene**: "**but-**" because it has 4 carbon atoms (Table P), "**-ene**" because it has a double bond (Table Q), 2-butene because the double bond is after the second carbon atom.

2-Pentene:

Pentene and **all alkenes** have the formula C_nH_{2n}. Let "n" = number of carbon atoms. Pentene has five carbon atoms:

$C_nH_{2n} = C_5H_{(5x2)} = \mathbf{C_5H_{10}}$, the **molecular** formula for **pentene**.

By looking at the structural or condensed formula, you see 5 carbon atoms, therefore "**pent**," and you see "=," **double bond**, therefore "**-ene**," **pentene**. 2-pentene because it has a double bond after the second carbon atom. See Tables P and Q. **Do sample question 4b, page 19.**

Table P

Prefix	Number of Carbon Atoms
pent-	5

H—C_1—C_2≡C_3—C_4—C_5—H

double bond after second carbon atom

$CH_3 - CH = CH - CH_2 - CH_3$
Condensed Formula

For alkenes with 6-10 carbon atoms, use the prefix given for the number of carbon atoms (Table P); follow the alkene formula C_nH_{2n} (Table Q). Look at the structural formula for ethene and you know all alkenes have double bonds (Table Q).
Examples:

Hexene: 6 carbon atoms= hex $C_nH_{2n} = C_6H_{2 \times 6} = C_6H_{12}$.

Decene: 10 carbon atoms = dec $C_nH_{2n} = C_{10}H_{2 \times 10} = C_{10}H_{20}$.

HOW TO DRAW ALKENES

Table P

Prefix	Number of Carbon Atoms
hex	6
hept	7
oct	8
non	9
dec	10

Table Q
Homologous Series of Hydrocarbons

Name	General Formula	Name	Example Structural Formula
alkenes	C_nH_{2n}	ethene	$\begin{array}{ccc} H & & H \\ \backslash & & / \\ & C = C & \\ / & & \backslash \\ H & & H \end{array}$

Draw 2-octene.

Step 1: 2-octene begins with "oct." Look at Table P. Oct means 8 carbon atoms. Draw a chain of 8 carbon atoms: C C C C C C C C

Step 2: Octene ends in "-ene," therefore you know it is an alkene. Look on Table Q at alkenes and the structural formula for ethene. You see ethene and therefore all alkenes have a double bond. 2-octene has the double bond after the second carbon atom. Put a double bond after the second carbon atom and single bonds between the other carbon atoms:
$C_1\text{-}C_2=C_3\text{-}C_4\text{-}C_5\text{-}C_6\text{-}C_7\text{-}C_8\text{-}$.

Step 3 (last step)**:** Draw lines or bonds around each carbon atom. Make sure each carbon atom has 4 bonds, not 3 or 5.

> **Important:** When an atom has a **double bond** on **one side** and a **single bond** on the **other side**, this carbon atom already has three bonds, and you may **only add one bond** to this carbon atom. Carbon atom 2 and carbon atom 3 each have two bonds on one side and one bond on the other side, therefore you may only add one bond to each of these carbon atoms.

$$- \overset{|}{C_1} - \overset{|}{C_2} = \overset{|}{C_3} - \overset{|}{C_4} - \overset{|}{C_5} - \overset{|}{C_6} - \overset{|}{C_7} - \overset{|}{C_8} -$$

Attach a hydrogen atom to the end of each bond:

$$H - \overset{\overset{H}{|}}{\underset{\underset{H}{|}}{C_1}} - \overset{\overset{H}{|}}{C_2} = \overset{\overset{H}{|}}{C_3} - \overset{\overset{H}{|}}{\underset{\underset{H}{|}}{C_4}} - \overset{\overset{H}{|}}{\underset{\underset{H}{|}}{C_5}} - \overset{\overset{H}{|}}{\underset{\underset{H}{|}}{C_6}} - \overset{\overset{H}{|}}{\underset{\underset{H}{|}}{C_7}} - \overset{\overset{H}{|}}{\underset{\underset{H}{|}}{C_8}} - H$$

The formula for alkenes, given on Table Q, is $C_nH_{2n} = C_8H_{2x8} = C_8H_{16}$. You can use this formula to check that you have put the proper number of hydrogen atoms, 16, into your drawing.

RULE: Each carbon atom must have 4 bonds, (4 lines touching each carbon atom), not 3 lines, not 5 lines, etc., as shown in Table Q.

Example: $-\overset{\displaystyle |}{C}=$

4 lines must touch each carbon atom.

DIENES are unsaturated hydrocarbons that have **two** double bonds. They are **not** alkenes.

Do Homework Questions #29-31, page 22.

A few common names, ethylene (page 10:10) and acetylene (page 10:14) are sometimes used, which are not IUPAC names.

ALKYNES

ALKYNES are hydrocarbons with the formula C_nH_{2n-2} (also known as the **acetylene series**).

Ethyne (also known as acetylene) is the first member of the alkyne series: $H-C\equiv C-H$. Ethyne ends in "**-yne**" because it is part of the alkyne series. All members of the alkyne series end in "**-yne**." Look at Table P. If the organic compound begins (prefix) with "eth-," it has 2

Table P

Prefix	Number of Carbon Atoms
eth-	2
prop-	3

carbon atoms. Therefore you know that **eth**yne, which begins with "**eth-**," has 2 carbon atoms.

Ethyne and **all alkynes** have the formula C_nH_{2n-2}. The formula is given in Table Q. Let "n" = number of carbon atoms. Ethyne has two carbon atoms:

$C_nH_{2n-2} = C_2H_{(2x2)-2} = \mathbf{C_2H_2}$, the **molecular formula for ethyne**.

By looking at the structural formula, you know it is ethyne: "**eth-**" because it has 2 carbon atoms; "**-yne**" because it has a triple bond (shown in Table Q). On Table Q, the structural formula with the name, ethyne, is given (see below).

Table Q

Name	General Formula	Example	
		Name	**Structural Formula**
Alkyne	C_nH_{2n-2}	ethyne	$H-C\equiv C-H$

By looking at Table Q , at the structural formula of ethyne (which is an example of alkynes), you can see that ethyne (similarly all alkynes) has a triple bond between 2 carbon atoms.

$-C\underset{\uparrow}{\equiv}C-$
Triple Bond

When you see a triple bond in the structural formula, you know it ends in -yne.

Do sample question 6, page 10:19.

In short, ethyne and **all alkynes** have the **formula C_nH_{2n-2}**, have **one triple bond** and are **unsaturated**, because two carbon atoms share more than one pair of electrons.

Propyne:

Propyne is the next member of the alkyne series:

Propyne and **all alkynes** have the formula C_nH_{2n-2}, given in Table Q. Let "n" = number of carbon atoms. Propyne has three carbon atoms:

$C_nH_{2n-2} = C_3H_{(3x2)-2} = C_3H_4$, the **molecular formula** **for propyne.**

$H-C\equiv C-\overset{\overset{\displaystyle H}{|}}{\underset{\underset{\displaystyle H}{|}}{C}}-H$

$CH\equiv C-CH_3$
Condensed
Formula

By looking at the structural or condensed formula, you know it is propyne: "**prop-**" because it has three carbon atoms (see Table P); "**-yne**" because it has a triple bond.

Look at **Table Q** on the previous page. Look at alkynes and the structural formula for ethyne. You see ethyne, and therefore all **alkynes**, has a **triple bond.**

2-Butyne:

Butyne and **all alkynes** have the formula C_nH_{2n-2}, given in Table Q. Let "n" = number of carbon atoms. Butyne has four carbon atoms:

$C_nH_{2n-2} = C_4H_{(4x2)-2} = C_4H_6$, the **molecular formula** for 2-butyne.

By looking at the structural or condensed formula, you know it is 2-butyne: "**but-**" because it has 4 carbon atoms (see Table P); "**-yne**" because it has a triple bond; 2-butyne because a triple bond is after the second carbon atom. Look at **Table Q** on the previous page. Look at alkynes and the structural formula for ethyne. You see ethyne, and therefore all **alkynes**, have a **triple bond.**

$H-\overset{\overset{\displaystyle H}{|}}{\underset{\underset{\displaystyle H}{|}}{C}}-C\equiv C-\overset{\overset{\displaystyle H}{|}}{\underset{\underset{\displaystyle H}{|}}{C}}-H$
triple bond after
second carbon atom

$CH_3-C\equiv C-CH_3$
Condensed
Formula

2-Pentyne:

Pentyne and **all alkynes** have the formula C_nH_{2n-2}. Let "n" = number of carbon atoms. Pentyne has five carbon atoms (see Table P):

$C_nH_{2n-2} = C_5H_{(5x2)-2} = C_5H_8$, the **molecular formula** for 2-pentyne.

2-pentyne means a triple bond after the second carbon atom:

$H-\overset{\overset{\displaystyle H}{|}}{\underset{\underset{\displaystyle H}{|}}{C_1}}-C_2\equiv C_3-\overset{\overset{\displaystyle H}{|}}{\underset{\underset{\displaystyle H}{|}}{C_4}}-\overset{\overset{\displaystyle H}{|}}{\underset{\underset{\displaystyle H}{|}}{C_5}}-H$.
triple bond after
second carbon atom

TABLE P AND TABLE Q

TABLE P AND TABLE Q

TABLE P AND TABLE Q

Look at **Table Q.** Look at alkynes and the structural formula for ethyne. You see ethyne, and therefore all **alkynes**, have a **triple bond.**

For alkynes with 6-10 carbon atoms, use the prefix given for the number of carbon atoms (Table P, below); follow the alkyne formula C_nH_{2n-2} (Table Q, below). Look at the structural formula for ethyne, and you know every alkyne has triple bonds (Table Q).

Examples:

Octyne: 8 carbon atoms= oct $\quad C_nH_{(2n-2)} = C_8H_{(2\times8)-2} = C_8H_{14}.$

Decyne: 10 carbon atoms = dec $\quad C_nH_{(2n-2)} = C_{10}H_{(2\times10)-2} = C_{10}H_{18}.$

Try Sample Questions #6-8, page 19, then do Homework Questions #20-21, page 22, and #32, page 22.

HOW TO DRAW ALKYNES

Table P

Prefix	Number of Carbon Atoms
hex	6
hept	7
oct	8
non	9
dec	10

Table Q
Homologous Series of Hydrocarbons

Name	General Formula	Name	Example Structural Formula
alkynes	C_nH_{2n-2}	ethyne	$H-C\equiv C-H$

Let's draw **3-decyne.**

Step 1: Decyne begins with dec-. Look at Table P. Dec means 10 carbon atoms. Draw a chain of 10 carbon atoms: C C C C C C C C C C

Step 2: Decyne ends in "-yne," therefore you know it is an alkyne. Look on Table Q at alkynes and the structural formula for ethyne. You see ethyne and therefore all alkynes have a triple bond. 3-decyne has the triple bond after the third carbon atom. Put a triple bond after the third carbon atom and single bonds between the other carbon atoms:

$$-C_1-C_2-C_3\equiv C_4-C_5-C_6-C_7-C_8-C_9-C_{10}-.$$

Step 3 (last step): Draw lines or bonds around each carbon atom. Make sure each carbon atom has 4 bonds, not 3 or 5.

Important: When an atom has a **triple bond** on **one side** and a **single bond** on the **other side**, this carbon atom already has four bonds, and **do not add** any **bonds** to this carbon atom. Carbon 3 and carbon 4 each have three bonds on one side and one bond on the other side (C_3 and C_4 each already has four bonds), therefore **do not add any bonds** to these carbon atoms.

Attach a hydrogen atom to the end of each bond:

$$H - \overset{\overset{\displaystyle H}{|}}{\underset{\underset{\displaystyle H}{|}}{C_1}} - \overset{\overset{\displaystyle H}{|}}{\underset{\underset{\displaystyle H}{|}}{C_2}} - C_3 \equiv C_4 - \overset{\overset{\displaystyle H}{|}}{\underset{\underset{\displaystyle H}{|}}{C_5}} - \overset{\overset{\displaystyle H}{|}}{\underset{\underset{\displaystyle H}{|}}{C_6}} - \overset{\overset{\displaystyle H}{|}}{\underset{\underset{\displaystyle H}{|}}{C_7}} - \overset{\overset{\displaystyle H}{|}}{\underset{\underset{\displaystyle H}{|}}{C_8}} - \overset{\overset{\displaystyle H}{|}}{\underset{\underset{\displaystyle H}{|}}{C_9}} - \overset{\overset{\displaystyle H}{|}}{\underset{\underset{\displaystyle H}{|}}{C_{10}}} - H$$

The formula for alkynes, given on Table Q, is

$$C_nH_{2n-2} = C_{10}H_{(2 \times 10)-2} = C_{10}H_{18}.$$

You can use this formula to check that you have put the proper number of hydrogen atoms, 18, into your drawing.

REMEMBER: Put hydrogen atoms around all the carbon atoms, **except** where you have a double or triple bond. Look at the structural formula above. **Carbon atoms 3 and 4** have a **triple** and **a single bond** (C_3 and C_4 each already has four bonds), therefore **do not add any more bonds** to carbon 3 and carbon 4.

SUMMARY: How to draw carbon compounds

Use the same four steps to draw **all** carbon compounds:

Step 1: Look at the prefix (Table P). This tells you the number of carbon atoms. Draw the chain of carbon atoms.

Step 2: Put bonds between the carbon atoms. Look at the ending (Table Q).

-ane means all single bonds.

-ene means a double bond.

- yne means a triple bond.

A number before -ene or -yne tells you that the double or triple bond is right after that carbon atom.

2-pentene: Double bond after second carbon atom.

1-butyne: Triple bond after first carbon atom.

Put in the double bond (in alkenes) or triple bond (in alkynes), and then put single bonds between all other carbon atoms.

Step 3 (when needed): If you have an alkyl group like methyl, CH_3, or a functional group (discussed later in this chapter), put it in. (The number before CH_3, like 2-methyl, tells which carbon atom to put the methyl on.)

Step 4 (last step): Draw lines or bonds around all the carbon atoms. Put hydrogen atoms at the ends of the bonds.

MODEL of propyne, C_3H_4:
Use 3 black balls to show 3 carbon atoms, and 4 yellow balls to show 4 hydrogen atoms. Use 3 springs to represent the carbon-carbon triple bond connecting the C_1 and C_2 atoms. Use sticks to represent the single bond between C_2 and C_3 and the bonds between carbon and hydrogen atoms. A model of propyne can also be made by using a space-filling model kit.

SUMMARY: Bonds in Alkanes, Alkenes and Alkynes

A single bond is made up of 2 electrons or 1 pair of electrons.

A single covalent bond is made up of 2 electrons or 1 pair of electrons shared between 2 atoms.

Alkanes have a single covalent bond made up of 2 electrons or 1 pair of electrons shared between 2 carbon atoms. See formula page 4:12

Alkenes and alkynes have multiple covalent bonds (more than 1 pair of electrons shared between 2 carbon atoms).

Alkenes-double bond-2 pairs of electrons shared. See formula p. 4:12

Alkynes-triple bond-3 pairs of electrons shared. See formula p. 4:12

BENZENE

Benzene has 6 carbon atoms in a ring.

Benzene, C_6H_6
First member and simplest member of the benzene series

The benzene family are cyclic, aromatic hydrocarbons, having the formula C_nH_{2n-6}.

In **benzene**, the bond between any two carbon atoms is the average (resonance) of single and double bonds (see below).

Toluene, C_7H_8
Second member of the benzene series

For simplicity, **benzene** can be shown as:

Try Sample Questions #9-10, page 20, then do Homework Questions, #33-36, page 22.

SAMPLE REGENTS & REGENT-TYPE QUESTIONS AND SOLUTIONS

1. Which property is generally characteristic of an organic compound?
   (1) low melting point
   (2) high melting point
   (3) soluble in polar solvents
   (4) insoluble in nonpolar solvents

2. The four single bonds of a carbon atom are spatially directed toward the corners of a regular
   (1) triangle   (2) rectangle   (3) square   (4) tetrahedron

3. Which is an isomer of

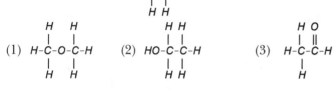

$$H-\underset{\underset{H}{|}}{\overset{\overset{H}{|}}{C}}-\underset{\underset{H}{|}}{\overset{\overset{H}{|}}{C}}-OH ?$$

(1) $H-\underset{\underset{H}{|}}{\overset{\overset{H}{|}}{C}}-O-\underset{\underset{H}{|}}{\overset{\overset{H}{|}}{C}}-H$ (2) $HO-\underset{\underset{H}{|}}{\overset{\overset{H}{|}}{C}}-\underset{\underset{H}{|}}{\overset{\overset{H}{|}}{C}}-H$ (3) $H-\underset{\underset{H}{|}}{\overset{\overset{H}{|}}{C}}-\overset{\overset{O}{\|}}{C}-H$

4a. Which is the correct name for the substance at right?   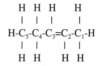
   (1) ethanol   (2) ethyne   (3) ethane   (4) ethene

4b. What is the chemical name of this compound

$$H-\underset{\underset{H}{|}}{\overset{\overset{H}{|}}{C_5}}-\underset{\underset{H}{|}}{\overset{\overset{H}{|}}{C_4}}-\underset{\underset{H}{|}}{\overset{\overset{H}{|}}{C_3}}=\underset{\underset{H}{|}}{\overset{\overset{H}{|}}{C_2}}-\underset{\underset{}{}}{\overset{\overset{H}{|}}{C_1}}-H$$

5. A molecule of ethane and a molecule of ethene both have the same
   (1) empirical formula
   (2) molecular formula
   (3) number of carbon atoms
   (4) number of hydrogen atoms

6. Given the structural formula for ethyne:
   $$H-C\equiv C-H.$$
   What is the total number of electrons shared between the carbon atoms?
   (1) 6   (2) 2   (3) 3   (4) 4

7. Which is the general formula for an alkyne?
   (1) C_nH_{2n-2} (2) C_nH_{2n+2} (3) C_nH_{2n} (4) C_nH_{2n-6}

8. Which structural formula represents a member of the series of hydrocarbons having the general formula C_nH_{2n-2}?

   (1) $H-\underset{\underset{H}{|}}{\overset{\overset{H}{|}}{C}}-\underset{\underset{H}{|}}{\overset{\overset{H}{|}}{C}}-H$ (3) $\overset{H}{\underset{H}{}}C=C\overset{H}{\underset{H}{}}$

   (3) $H-C\equiv C-H$ (4) $\overset{H}{\underset{H}{}}C=C\overset{H}{}-\overset{H}{\underset{H}{}}C-H$

9. Which compound is a member of the hydrocarbon series with the general formula C_nH_{2n-6}?

(1) C_3H_8 (2) C_4H_8 (3) C_5H_8 (4) C_7H_8

10. For simplicity, the structure of benzene is often represented as

(1) ⬠   (2) ▢   (3) △   (4) ⬡

SOLUTIONS

1. Answer *1*. Organic compounds have low melting points.

2. Answer *4*. You learned that the bonds of a carbon atom are directed towards the corners of a tetrahedron.

3. Answer *1*. Isomers have the same molecular formula but different structural formulas. Isomers must have the same number of carbon, hydrogen and oxygen atoms. The example has 2 carbon atoms, 6 hydrogen atoms, and 1 oxygen atom. Answer 1 also has 2 carbon atoms, 6 hydrogen atoms, and 1 oxygen atom.

4a. Answer *4*. The substance has two (2) carbon atoms, therefore the name begins with "eth" (see Table P). On Table Q, you see that compounds with C=C end in "ene"; therefore this substance is ethene.

4b. The name of this compound is 2-pentene (not 3-pentene). In the example, you need to count (number) the C atoms from the right end to the left end so **the number in front of pentene** is the **lowest possible number,** which is **2** (not 3).

5. Answer *3*. Ethane and ethene both begin with the prefix "eth." You learned that the prefix "eth" means two carbon atoms. Both ethane and ethene have the same number of carbon atoms.

6. Answer *1*. One bond, or 1 line between the two carbon atoms, has 1 pair of shared electrons, or 2 electrons. There are three lines, therefore three bonds, between the two carbon atoms. Each line or bond equals a pair of shared electrons. 3 lines or bonds = 3 x 2 = 6 electrons.

7. Answer *1*. Alkynes have the formula C_nH_{2n-2} (see Table Q).

8. Answer *2* belongs to the formula C_nH_{2n-2}. $C_2H_{(2x2)-2} = C_2H_2$. The other choices do not fit the formula C_nH_{2n-2}. You can also know that Choice 2 is the correct answer by realizing that C_nH_{2n-2} is in the alkyne series. The alkyne series has a triple bond (see Table Q). Choice 2 is the only one with a triple bond.

9. Answer *4*. The benzene series has the formula C_nH_{2n-6}. Choice 4 is toluene, and toluene follows the formula C_nH_{2n-6}. $C_7H_{(2x7)-6} = C_7H_{14-6} = C_7H_8$. The other choices do not fit the formula.

10. Answer *4*. You learned that, for simplicity, benzene, C_6H_6, can be represented as ⬡ .

1. A compound that is classified as organic must contain the element
   (1) carbon   (2) nitrogen   (3) oxygen   (4) hydrogen

2. What is the geometric shape of a methane molecule?
   (1) triangular   (2) rectangular (3) octahedral   (4) tetrahedral

3. In a molecule of CH_4, the hydrogen atoms are spatially oriented toward the corners of a regular
   (1) pyramid   (2) tetrahedron (3) square   (4) rectangle

4. Compared to the rate of inorganic reactions, the rate of organic reactions generally is
   (1) slower because organic particles are ions
   (2) slower because organic particles contain covalent bonds
   (3) faster because organic particles are ions
   (4) faster because organic particles ontain covalent bonds

5. Which is the general formula for the alkane series of hydrocarbons?
   (1) C_nH_{2n+2} (2) C_nH_{2n} (3) C_nH_{2n-2} (4) C_nH_{2n-6}

6. Which is a saturated hydrocarbon?
   (1) ethene   (2) ethyne   (3) propene   (4) propane

7. Which structural formula represents a molecule of butane?

   (1)
   ```
        H H H H
        | | | |
   H-C=C-C=C-H
   ```
   (3)
   ```
     H H H H
     | | | |
   H-C-C-C-C-H
     | | | |
     H H H H
   ```
   (2)
   ```
     H H H H
     | | | |
   H-C-C=C-C-H
     |     |
     H     H
   ```
   (4)
   ```
          H H
          | |
   H-C≡C-C-C-H
          | |
          H H
   ```

8. Which compounds are isomers?
   (1) CH_3Br and CH_2Br_2 (3) CH_3OH and CH_3CHO
   (2) CH_3OH and CH_3CH_2OH (4) CH_3OCH_3 and CH_3CH_2OH

9. Which pair of compounds are isomers?
   (1) C_6H_6 and C_6H_{12} (3) CH_3CH_2OH and CH_3COOH
   (2) C_2H_4 and C_2H_6 (4) CH_3CH_2OH and CH_3OCH_3

10. Which of the following hydrocarbons has the *lowest* normal boiling point?
    (1) ethane   (2) propane   (3) butane   (4) pentane

11. The compound C_4H_{10} belongs to the series of hydrocarbons with the general formula
    (1) C_nH_{2n} (2) C_nH_{2n+2} (3) C_nH_{2n-2} (4) C_nH_{2n-6}

12. Which compound is a saturated hydrocarbon?
    (1) ethane   (2) ethene   (3) ethyne   (4) ethanol

13. What is the correct name for the substance below?

    ```
              H
              |
        H H-C-H H  H
        | | | | |  |
      H-C---C---C-C-H
        | | | | |  |
        H   H   H  H
    ```

    (1) ethanol   (3) 2-ethylpentane
    (2) 2-methylbutane   (4) butane

14. Which is the correct name for the substance below?

$$H-C=C-H$$ with H H above the carbons

(1) ethanol (2) ethyne (3) ethane (4) ethene

15. Which structural formula correctly represents 2-butene?

(1) $H-C=C-C-C-H$ with H H H H above and H H below

(2) $H-C-C=C-C-H$ with H H H H above and H H below

(3) $H-C=C-C-C-H$ with H H above and H H below

(4) $H-C-C=C-C-H$ with H H above and H H below

16. What is the formula of pentene?
(1) C_4H_8 (2) C_4H_{10} (3) C_5H_{10} (4) C_5H_{12}

17. Which of the following hydrocarbons has the *highest* normal boiling point?
(1) butene (2) ethene (3) pentene (4) propene

18. Alkenes *differ* from alkanes in that alkanes
(1) are hydrocarbons   (3) have the general formula C_nH_{2n}
(2) are saturated compounds (4) undergo addition reactions

19. In which pair of hydrocarbons does each compound contain only one double bond per molecule?
(1) C_2H_2 and C_2H_6 (3) C_4H_8 and C_2H_4
(2) C_2H_2 and C_3H_6 (4) C_6H_6 and C_7H_8

20. Which hydrocarbon is a member of the series with the general formula C_nH_{2n-2}?
(1) ethyne (2) ethene (3) butane (4) benzene

21. If a hydrocarbon molecule contains a triple bond, its IUPAC name ends in
(1) "ane" (2) "ene" (3) "one" (4) "yne"

CONSTRUCTED RESPONSE QUESTIONS: Parts B-2 and C of NYS Regents Exam

22. Compare and contrast the number of carbon atoms, the types of bonds and saturation in methane, ethane, propane, butane, and pentane.

23. How do the isomers of pentane differ from one another?

24. Draw propane, hexane, and butane.

25. Draw pentane, octane, and decane.

26. Draw 2-methylheptane and 3-ethylpen ane

27. A. Draw 3-ethylnonane and 4-ethyldecane.

27. B. Draw 2,2,4-trimethyl pentane. *Hint:* Draw 5 C atoms. Trimethyl means **three** methyls (three CH_3). **2,2,4**-trimethyl means 1 methyl (1 C with 3 H's) on **C$_2$**, another methyl (1 C with 3 H's) on **C$_2$**, and 1 methyl (1 C with 3 H's) on **C$_4$**.

28. Compare and contrast the number of carbon atoms, the types of bonds, and saturation in ethene, butene, octene, nonene, and decene.

29. Draw 2-butene, 2-pentene, and 2-hexene.

30. Draw 3-decene, 4-nonene, and 3-heptene.

31. Draw 3-octene and 2-methyl-3-octene.

32. Compare and contrast the number of carbon atoms, the types of bonds, and saturation in ethyne, butyne, heptyne, and octyne.

33. Draw 2-butyne, 2-pentyne, and 2-hexyne.

34. Draw 3-decyne, 4-octyne, and 3-heptyne.

35. Draw 2-methyl-4-decyne and 3-nonyne.

36. Draw 2-methyl-3-octyne and 3-ethyl-4-nonyne.

OTHER ORGANIC COMPOUNDS

THESE ORGANIC COMPOUNDS AND THEIR FUNCTIONAL GROUPS (SHOWN IN THE GRAY BOXES BELOW) ARE WRITTEN IN TABLE R.

Each functional group causes the compound (example: alcohol, acid) to have specific physical and chemical properties. For example, the functional group OH causes the alcohol to have specific characteristics.

Table R

Class of Compound	Functional Group	
Alcohol	–OH	

TABLE R

ALCOHOLS are organic compounds in which one or more hydrogens of a hydrocarbon are replaced by an **OH group** (hydroxyl group). The **OH of alcohols** is a **hydroxyl** group, while the OH in bases (NaOH) is a hydroxide ion (OH⁻). **Alcohols** are not bases and have the **functional group OH**, as shown in Table R, above, or on page Reference Tables 22.

Organic compounds can be named by using the IUPAC system. **Alcohols** are **named** by **dropping** the "e" from the corresponding **alkane** and **adding** "ol." Example: Drop "-e" from methan"e" and add "-ol," methanol.

Examples of Alcohols

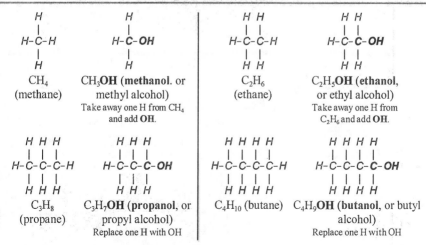

CH₄ (methane)	CH₃OH (methanol. or methyl alcohol) Take away one H from CH₄ and add OH.	C₂H₆ (ethane)	C₂H₅OH (ethanol, or ethyl alcohol) Take away one H from C₂H₆ and add OH.
C₃H₈ (propane)	C₃H₇OH (propanol, or propyl alcohol) Replace one H with OH	C₄H₁₀ (butane)	C₄H₉OH (butanol, or butyl alcohol) Replace one H with OH

$$H-\underset{\underset{H}{|}}{\overset{\overset{H}{|}}{C}}-\underset{\underset{H}{|}}{\overset{\overset{H}{|}}{C}}-\underset{\underset{H-C-H}{|}}{\overset{\overset{H}{|}}{C}}-OH$$

$$H-C-H$$
$$|$$
$$H$$

C_4H_9OH
2-butanol

$$H-\underset{\underset{H}{|}}{\overset{\overset{H}{|}}{C}}-\underset{\underset{H}{|}}{\overset{\overset{OH}{|}}{C}}-\underset{\underset{H}{|}}{\overset{\overset{H}{|}}{C}}-H$$

C_3H_7OH
2-propanol

$CH_3 - CHOH - CH_3$
Condensed Formula

$$H-\underset{\underset{OH}{|}}{\overset{\overset{H}{|}}{C_1}}-\underset{\underset{OH}{|}}{\overset{\overset{H}{|}}{C_2}}-H$$

1,2-ethanediol
Common Name:
Ethylene glycol

1,2-ethanediol: ethane = 2 carbon atoms with single bonds; diol = 2 OH.

It is called 1,2-ethanediol because OH is attached to the first and second carbon atoms, C_1 and C_2.

Commonly called **ethylene glycol**, antifreeze.

Ethylene glycol has 2 OH groups.

$$H-\underset{\underset{OH}{|}}{\overset{\overset{H}{|}}{C_1}}-\underset{\underset{OH}{|}}{\overset{\overset{H}{|}}{C_2}}-\underset{\underset{OH}{|}}{\overset{\overset{H}{|}}{C_3}}-H$$

1,2,3-propanetriol
Common Name:
Glycerol

3 carbons with single bonds = propane.

3 OH = triol

OH is attached to the first, second and third carbon atoms: C_1, C_2 and C_3, therefore: 1,2,3-propanetriol. Glycerol has 3 OH groups.

MONOHYDROXY ALCOHOLS have **only one OH** group.

Primary alcohols: The C that is attached to the OH group **is attached** only **to one carbon or to no carbons**.

Examples of primary alcohols:

$$H-\underset{\underset{H}{|}}{\overset{\overset{H}{|}}{C}}-\underset{\underset{H}{|}}{\overset{\overset{H}{|}}{C}}-OH$$

ETHANOL
Carbon attached to OH is
attached to **only one carbon**

$$H-\underset{\underset{H}{|}}{\overset{\overset{H}{|}}{C}}-OH$$

METHANOL
Carbon attached to OH is
attached to no other carbon
(only to H)

In the IUPAC system of nomenclature, primary alcohols are named by dropping the "e" from the corresponding alkane and adding "ol." Example: Drop "-e" from methan"e" and add "-ol," methan**ol**.

Primary Alcohols

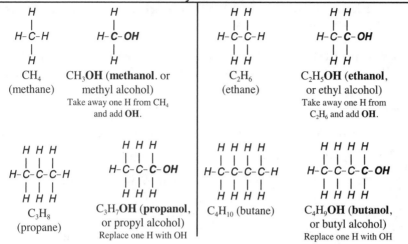

CH_4 (methane)	CH_3OH (**methanol.** or methyl alcohol) Take away one H from CH_4 and add **OH**.	C_2H_6 (ethane)	C_2H_5OH (**ethanol,** or ethyl alcohol) Take away one H from C_2H_6 and add **OH**.
C_3H_8 (propane)	C_3H_7OH (**propanol,** or propyl alcohol) Replace one H with OH	C_4H_{10} (butane)	C_4H_9OH (**butanol,** or butyl alcohol) Replace one H with OH

As you can see, methanol, ethanol, propanol and butanol are examples of primary alcohols because the C that is attached to the OH group is attached only to ONE CARBON, OR TO NO CARBON ATOMS.

SECONDARY ALCOHOLS: The C attached to OH is attached to **two other C atoms.**

R_1, R_2 = hydrocarbon radical, and has C in it. R_1, R_2 can be CH_3, C_2H_5, etc.

Example of secondary alcohol:

$$H-\overset{\underset{|}{H}}{\underset{H}{C}}-\overset{\underset{|}{H}}{\underset{H-C-H}{C}}-\overset{\underset{|}{H}}{\underset{H}{C}}-OH \qquad or \qquad H-\overset{\underset{|}{H}}{\underset{H}{C}}-\overset{\underset{|}{OH}}{\underset{H}{C}}-\overset{\underset{|}{H}}{\underset{H}{C}}-H$$

The C attached to the OH is attached to TWO OTHER C ATOMS.

TERTIARY ALCOHOLS: The C attached to OH is attached to **three** other carbon atoms.

R_1, R_2, R_3 = hydrocarbon radical, and has C in it.

Example of tertiary alcohol:

The C attached to the OH is attached to THREE OTHER C ATOMS.

DIHYDROXY ALCOHOLS: Compounds containing **two OH groups** are called **dihydroxy** (dihydric) **alcohols** or **glycols**. Example: Ethylene glycol (see structure on second page of Section B).

TRIHYDROXY ALCOHOLS: Compounds with **three OH groups** are known as **trihydroxy** (trihydric) **alcohols**. Example: Glycerol (see structure on second page of Section B).

Table R

Class of Compound	Functional Group
Aldehyde	O ∥ - C - H

An **ALDEHYDE** has the functional group

O ∥ - C - H . It has a single bond between C and H and a double bond between C and O, as shown in Table R, to the left, or on page Reference Tables 22.

$$ \begin{array}{c} O \\ \parallel \\ H\text{-}C\text{-}H \end{array} $$

methanal

"-al" at the end of the word **means an aldehyde.**
"M" means **1 C** atom.
Common Name: *Formaldehyde*

Aldehydes are **named** by dropping the final "e" from the corresponding alkane and adding **al**.

Aldehydes are oxidized to form acids.

SUMMARY: When you see O ∥ - C - H **at the end** of a chain of carbon atoms, you know it is an **aldehyde**.

Table R

Class of Compound	Functional Group	General Formula
Ketone	O ∥ - C -	O ∥ R-C-R¹

In a **KETONE**, the C=O bond is in the middle, and carbon atoms are on both sides of C=O, as shown in Table R, to the left, or on page Reference Tables 22. A ketone can be shown by

$$ \begin{array}{c} O \\ \parallel \\ R\text{-}C\text{-}R^1 \end{array} $$

R and R¹ are hydrocarbon groups which contain C. Ketones are named by dropping the final "e" from the corresponding alkane and adding one. Ket**one**, add **one**. ("o" has the sound of "o" in the word "own.")

By looking at the structural or condensed formula, you know it is a **ketone**, because it has the

$$ \begin{array}{c} H \quad O \quad H \\ | \quad \parallel \quad | \\ H\text{-}C\text{-}\,C\text{-}C\text{-}H \\ | \qquad\quad | \\ H \qquad\quad H \end{array} $$

Propanone

"pro" means   "-one" at the end of
3 C atoms   the word means a
                     ket**one**.

Common Name: *Acetone*

---

functional group $-\overset{\overset{\displaystyle O}{\|}}{C}-$ in the middle of the compound (see Table R).

The condensed formula can be written with or without the single line between the carbon atoms. (See formulas to the right.)

$$CH_3-\overset{\overset{\displaystyle O}{\|}}{C}-CH_3 \qquad CH_3\overset{\overset{\displaystyle O}{\|}}{C}CH_3$$

Condensed Formula   Condensed Formula

SUMMARY: When you see $-\overset{\overset{\displaystyle O}{\|}}{C}-$ **in the middle,** between 2 carbon atoms, you know it is a **ketone.**

Table R

Class of Compound	Functional Group	General Formula
Organic Acid	$-\overset{\overset{\displaystyle O}{\|}}{C}-OH$	$R-\overset{\overset{\displaystyle O}{\|}}{C}-OH$

TABLE R

ORGANIC ACIDS have the functional group $-\overset{\overset{\displaystyle O}{\|}}{C}-OH$, as shown in Table R, second column. Organic acids can be shown by $R-\overset{\overset{\displaystyle O}{\|}}{C}-OH$, given in the third column of Table R. "R" can be CH_3–methyl, C_2H_5–ethyl, etc. They are named by dropping the final "e" from the corresponding alkane and **adding "oic acid."**

The first two members of the organic acid series are formic acid and acetic acid:

$$H-\overset{\overset{\displaystyle O}{\|}}{C}-OH$$

Methanoic Acid
"meth" means 1 carbon atom
Common Name: *Formic Acid*

$$H-\overset{\overset{\displaystyle H}{|}}{C}-\overset{\overset{\displaystyle O}{\|}}{C}-OH$$
$$\underset{\displaystyle H}{|}$$

Ethanoic Acid
"eth" means 2 carbon atoms
Common Name: *Acetic Acid*

$$CH_3-\overset{\overset{\displaystyle O}{\|}}{C}-OH$$

Condensed Formula

Organic acids, like other acids (example: HCl) are **electrolytes.** When organic acids are dissolved in water, they produce hydrogen ions (see page 8:3) and **conduct** an **electric current.**

SUMMARY: When you see $-\overset{\overset{\displaystyle O}{\|}}{C}-OH$ **, you know it is an acid.**

Table R

Class of Compound	Functional Group	General Formula	Example
Ether	– O –	R –O –R´	$CH_3OCH_2CH_3$ Methyl ethyl ether

The functional group of an **E**THER is – O –, as shown in Table R, second column. Ethers can be shown by R – O – R´ (Table R, column 3). R and R´ are hydrocarbon groups which contain carbon.

Different functional groups, -O- in ether, -OH in alcohol and other functional groups cause the ether or alcohol, etc., to have different chemical and physical properties (see bottom of page 1:2).

You learned that C_2H_5 is ethyl. **Diethyl** = 2(C_2H_5) ethyls. In an **ether** there is **always an O in the middle** between two carbon atoms (or two hydrocarbon groups). Look at Table R. You see O is in the middle in column 2 (functional group), column 3 (general formula) and column 4 (example). By looking at the structural or condensed formula, you know it is an ether because – O – is in the middle of the compound.

H H   H H
| | | |
H-C-C-O-C-C-H
| | | |
H H   H H
Diethyl Ether
$CH_3CH_2OCH_2CH_3$
Condensed Formula

SUMMARY: When you see O in the middle between two carbon atoms, you know it is an ether.

Try Sample Questions #1-5, page 46, then do Homework Questions, #1-18, pages 49-50 and #40, page 51.

*Hint: It might be **easier** for the student to go to pages 10:31-10:35 and draw these compounds before learning halides, amines, etc.*

Table R

Class of Compound	Functional Group
Halide	– F (fluoro-) – Cl(chloro-) – Br (bromo-) – I (iodo-)

HALIDE (also called **halocarbon**) is one or more halogens (F, Cl, Br, I) attached to one carbon atom or a chain of carbon atoms. (There are also hydrogen atoms attached to the carbon atoms.)

H   Br H H
| | | |
H—C₁—C₂—C₃—C₄—H
| | | |
H   H   H H

2-bromobutane
$CH_3 – CHBr – CH_2 – CH_2$
Condensed Formula

H   H
\   /
C = C
/   \
H   Cl

Chloroethene double bond between two C atoms

H H Br H
| | | | right side
H-C₄-C₃- C₂-C₁ -H count
| | | |
H H H H

2-bromobutane

This is also 2-bromobutane. In this diagram,

you must count the carbon atoms from the right side so Br will be on the second carbon, C_2, and not on C_3. (If you count from the left side, Br would be on C_3). Br or any functional group must be on the lowest numbered C, in this example C_2.

Halides (also called halocarbons) are formed by substitution and addition reactions, which you will learn about later in the chapter.

SUMMARY: When you see, F, Cl, Br or I attached to carbon, you know it is a halide.

Table R

Class of Compound	Functional Group
Amine	$-\overset{\mid}{N}-$

An **AMINE** has (functional group) $-\overset{\mid}{N}-$, as shown in Table R.

An amine has $-\overset{\mid}{N}-$ attached to carbon or hydrogen atoms.

It is named by dropping the final "e" from the name of the hydrocarbon and adding "-amine.

$$H-\overset{\overset{\displaystyle H}{\mid}}{\underset{\underset{\displaystyle H}{\mid}}{C}}-NH_2$$

Methanamine
$CH_3 - NH_2$
Condensed Formula

SUMMARY: When you see $-\overset{\mid}{N}-$ as the only functional group, you know it is an amine.

Class of Compound	Functional Group
Amino Acid	$-\overset{\overset{\displaystyle H}{\mid}}{\underset{\underset{\displaystyle NH_2}{\mid}}{C}}-\overset{\overset{\displaystyle O}{\parallel}}{C}-OH$

An **AMINO ACID** has both an amine (NH_2) and an organic acid $-\overset{\overset{\displaystyle O}{\parallel}}{C}-OH$.

Table R gives the functional groups of amine and acid, which you learned about on the previous pages.

The general formula for amino acids is $R-\overset{\overset{\displaystyle H}{\mid}}{\underset{\underset{\displaystyle NH_2}{\mid}}{C}}-\overset{\overset{\displaystyle O}{\parallel}}{C}-OH$. R can be H, CH_3, C_2H_5. The NH_2 is usually on the carbon next to $-\overset{\overset{\displaystyle O}{\parallel}}{C}-OH$. (Amino acids are not given in Table R.)

SUMMARY: When you see NH_2 and COOH, you know it is an amino acid.

Table R

Class of Compound	Functional Group
Amide	O $\overset{\parallel}{\underset{}{} } $ $-C-NH$

An **AMIDE** has the functional group $-\overset{O}{\overset{\parallel}{C}}-NH$ (see Table R) at the end of a carbon chain. The carbon is attached to the oxygen by a double bond and to the $-\overset{|}{NH}$ (amine group) by a single bond.

Amides are named by dropping the final "e" from the name of the hydrocarbon and adding -amide.

$$H-\overset{\overset{\displaystyle H}{|}}{\underset{\underset{\displaystyle H}{|}}{C}}-\overset{\overset{\displaystyle H}{|}}{\underset{\underset{\displaystyle H}{|}}{C}}-\overset{O}{\overset{\parallel}{C}}-NH_2$$

Propanamide

SUMMARY: **When** you see $-\overset{O}{\overset{\parallel}{C}}-NH$, you know it is an amide.

$$CH_3 - CH_2 - \overset{O}{\overset{\parallel}{C}} - NH_2$$

Condensed Formula

Table R

Class of Compound	Functional Group
Ester	O $\overset{\parallel}{\underset{}{}}$ $-C-O-$

An **ESTER** has $-\overset{O}{\overset{\parallel}{C}}-O-$ between two carbon chains. The carbon is attached to one oxygen by a double bond and to the other oxygen by a single bond.

An ester is formed by the reaction of an acid and an alcohol.

To name an ester:

1. Look at the carbon chain after the $-\overset{O}{\overset{\parallel}{C}}-O-$; write its prefix (example: meth-, eth-, etc.) and add -yl to the end of the prefix.

   In this formula, prefix eth- and -yl equal ethyl.

$$H-\overset{\overset{\displaystyle H}{|}}{\underset{\underset{\displaystyle H}{|}}{C}}-\overset{\overset{\displaystyle H}{|}}{\underset{\underset{\displaystyle H}{|}}{C}}-\overset{O}{\overset{\parallel}{C}}-O-\overset{\overset{\displaystyle H}{|}}{\underset{\underset{\displaystyle H}{|}}{C}}-\overset{\overset{\displaystyle H}{|}}{\underset{\underset{\displaystyle H}{|}}{C}}-H$$

Ethyl Propanoate

2. Then give the name of the carbon chain that includes the C=O (propane); leave off the last letter and add -oate; in this case, propanoate. This example is ethyl propanoate.

Hint: Compare example of ester given in Table R.

Esters have fruity odors (examples: banana, pineapple, apple).

SUMMARY: When you see $-\overset{O}{\overset{\parallel}{C}}-O-$ between two carbon chains, you know it is an ester.

REMEMBER: By looking at the functional groups, you can tell if the compound is an **acid**, **aldehyde**, **alcohol**, etc.

Do Homework Questions #19-23, page 50, and #41, page 52.

DRAWING STRUCTURAL FORMULAS

HINT: Look at the examples on Table R (column 4) to see where to place each functional group, in the middle or at the end of the carbon chain.

ALCOHOLS

Draw 2-pentanol.

Step 1: "Pent" means 5 carbon atoms (Table P, page Reference Tables 18).

Step 2: Pentanol means single bonds. (-ane means single bonds: Table Q, page Reference Tables 19). Draw 5 carbon atoms with single bonds between them. ‎ ‎ ‎ ‎ ‎ ‎ -C-C-C-C-C-

Step 3: Look at and use Table R to put in the functional group.

2-Pentanol ends in **-ol**. Look at the examples in column 4 in Table R on page Reference Tables 22 or below . Which example also ends in -ol? The example 1-propanol ends in -ol. Go across and you see in column 1 that it is an alcohol.

TABLE R

Organic Functional Groups

Class of Compound	Functional Group	General Formula	Example
Alcohol	-OH	R-OH	$CH_3CH_2CH_2OH$ 1-propanol

In the second column, it has the -OH functional group. Put OH in your drawing. 2-pentanol means put OH on the second carbon atom (OH can be put on top of carbon 2 or below carbon 2). (If no number is given, put the OH on the first or last carbon atom.)

$$\overset{\displaystyle OH}{\underset{\displaystyle}{|}}$$
$$-C_1- C_2 -C_3 -C_4 -C_5-$$

The third column on the table shows that the OH is attached to a carbon atom.

Step 4: Draw lines or bonds around each carbon atom. Make sure each carbon has 4 bonds.

$$\overset{OH}{\underset{}{|}}$$
$$- \overset{|}{C} - \overset{|}{\underset{|}{C}} - \overset{|}{C} - \overset{|}{C} - \overset{|}{C} -$$

Attach a hydrogen atom to the end of each bond, except if the bond already has the OH.

$$\begin{array}{ccccc} H & OH & H & H & H \\ | & | & | & | & | \\ H-C- & C & -C-C-C-H \\ | & | & | & | & | \\ H & H & H & H & H \end{array}$$

TABLE R

Draw butanal.

Step 1: The prefix but means 4 carbon atoms (Table P, page Reference Tables 18).

Step 2: Butanal means single bonds. (-ane means single bonds: Table Q, page Reference Tables 19). Draw 4 carbon atoms with single bonds between them. -C-C-C-C-

Step 3: Look at and use Table R to put in the functional group. Butanal ends in **-al**. Look at the examples in column 4 in Table R on page Reference Tables 22 or below . Which example in Table R ends in -al? The example propanal ends in -al. Go across and you see in column 1 that it is an aldehyde.

TABLE R

Organic Functional Groups

Class of Compound	Functional Group	General Formula	Example
Aldehyde	$\overset{O}{\overset{\|}{-C-H}}$	$\overset{O}{\overset{\|}{R-C-H}}$	$CH_3CH_2-\overset{O}{\overset{\|}{C}}-H$ propanal

In the second column, it has $\overset{O}{\overset{\|}{-C-H}}$ (functional group).

Put $\overset{O}{\overset{\|}{-C-H}}$ in your drawing, but **REMEMBER:** The C of $\overset{O}{\overset{\|}{-C-H}}$ is the last (fourth) carbon, which you already drew. $C_1-C_2-C_3-\overset{O}{\overset{\|}{C_4}}-H$

Step 4: Draw lines or bonds around each carbon atom. Make sure each carbon atom has 4 bonds, not 5 or 3.

$-\overset{\|}{\underset{\|}{C}}-\overset{\|}{\underset{\|}{C}}-\overset{\|}{\underset{\|}{C}}-\overset{O}{\overset{\|}{\underset{\blacktriangle}{C}}}-H$ | (do NOT add a bond)

The **aldehyde** carbon, $\overset{O}{\overset{\|}{-C-H}}$ (at the end of this compound), already has **four bonds**. Do not add any bonds to it. Attach a hydrogen atom to each remaining bond on C_1, C_2, and C_3.

$H-\overset{H}{\underset{H}{C_1}}-\overset{H}{\underset{H}{C_2}}-\overset{H}{\underset{H}{C_3}}-\overset{O}{\overset{\|}{\underset{\blacktriangle}{C}}}-H$ | (do NOT add a bond)

In Table R, column 3 shows you the general formula: $\overset{O}{\overset{\|}{R-C-H}}$. R can be an atom or group of atoms.

KETONES

Draw 2-hexanone.

Step 1: The prefix hex means 6 carbons (Table P).

Step 2: Hexan**one** means single bonds. (-ane means single bonds: Table Q). Draw 6 carbon atoms with single bonds between them:

$$-C-C-C-C-C-C-$$

Step 3: Look at and use Table R to put in the functional group.

2-hexanone ends in -**one**. Look at the examples in column 4 in Table R on page Reference Tables 22 or below. Which example also ends in -one? The example 2-pentanone ends in -one. Go across and see in column 1 that it is a ketone.

TABLE R

Organic Functional Groups

Class of Compound	Functional Group	General Formula	Example
Ketone	$\overset{O}{\underset{}{\overset{\|\|}{-C-}}}$	$\overset{O}{\underset{}{\overset{\|\|}{R-C-R'}}}$	$\overset{O}{\underset{\text{2-pentanone}}{\overset{\|\|}{CH_3-C-CH_2-CH_2-CH_3}}}$

In the second column, it has $\overset{O}{\underset{}{\overset{\|\|}{-C-}}}$ (functional group).

Put $\overset{O}{\underset{}{\overset{\|\|}{-C-}}}$ in your drawing; 2-hexanone means put the $\overset{O}{\underset{}{\overset{\|\|}{-C-}}}$ on the second carbon atom.

$$C_1-\overset{O}{\overset{\|\|}{C_2}}-C_3-C_4-C_5-C_6$$

Step 4: Draw lines or bonds around each carbon atom. Each carbon atom must have 4 bonds, not 5 and not 3. The **ketone carbon**, $\overset{O}{\underset{}{\overset{\|\|}{-C-}}}$, already has **four bonds**. Do not add any

$$H-\overset{\overset{\displaystyle H}{|}}{C_1}-\overset{\overset{\displaystyle O}{\|\|}}{C_2}-\overset{\overset{\displaystyle H}{|}}{\underset{\underset{\displaystyle H}{|}}{C_3}}-\overset{\overset{\displaystyle H}{|}}{\underset{\underset{\displaystyle H}{|}}{C_4}}-\overset{\overset{\displaystyle H}{|}}{\underset{\underset{\displaystyle H}{|}}{C_5}}-\overset{\overset{\displaystyle H}{|}}{\underset{\underset{\displaystyle H}{|}}{C_6}}-H$$

bonds to it. Attach a hydrogen atom to the end of each bond on C_1, C_3, C_4, C_5, and C_6.

In Table R, column 3 gives the general formula $\overset{O}{\underset{R-C-R'}{\overset{\|\|}{}}}$. R can be any atom or group of atoms.

ACIDS

Draw hexanoic acid.

Step 1: The prefix hex means 6 carbon atoms (Table P).

Step 2: Hexan**oic** acid means single bonds. (-ane means single bonds: Table Q). Draw 6 carbon atoms with single bonds between them:

$$-C-C-C-C-C-C-$$

Step 3: Look at and use Table R to put in the functional group. Hexanoic acid ends in -**oic acid.** You can look at the example in column 4 in Table R on the next page and then go to columns 1 and 2 and put $\overset{O}{\underset{-C-OH}{\overset{\|\|}{}}}$ in your drawing (as you did in the previous examples).

REMEMBER: The C of $\overset{O}{\underset{-C-OH}{\|}}$ is the last (sixth) carbon atom that you already drew.

$$C_1 - C_2 - C_3 - C_4 - C_5 - \overset{O}{\underset{}{\overset{\|}{C_6}}} - OH$$

OR, EASY METHOD FOR ACIDS: You know hexanoic acid is an acid because it has the word acid. You see organic acid in column 1 and the functional group $\overset{O}{\underset{-C-OH}{\|}}$ in column 2. Put $\overset{O}{\underset{-C-OH}{\|}}$ in your drawing. See drawing (with 6 C atoms) above.

REMEMBER: The C of $\overset{O}{\underset{-C-OH}{\|}}$ is the last (sixth) carbon atom that you already drew.

TABLE R

Organic Functional Groups

Class of Compound	Functional Group	General Formula	Example
Organic Acid	$\overset{O}{\underset{-C-OH}{\|}}$	$\overset{O}{\underset{R-C-OH}{\|}}$	$\overset{O}{\underset{CH_3CH_2-C-OH}{\|}}$ propanoic acid

Step 4: Draw lines or bonds around each carbon atom.

Each carbon atom must have 4 bonds, not 3 or 5. The **acid carbon**, $\overset{O}{\underset{-C-OH}{\|}}$, already **has four bonds**. Do **not add** any **bonds** to it. Attach a hydrogen atom to each remaining bond on C_1, C_2, C_3, C_4, and C_5.

In Table R, column 3 gives the general formula for acids. "R" can be carbon or hydrogen atoms.

$$H-\overset{\overset{H}{|}}{C_1}-\overset{\overset{H}{|}}{\underset{\underset{H}{|}}{C_2}}-\overset{\overset{H}{|}}{\underset{\underset{H}{|}}{C_3}}-\overset{\overset{H}{|}}{\underset{\underset{H}{|}}{C_4}}-\overset{\overset{H}{|}}{\underset{\underset{H}{|}}{C_5}}-\overset{O}{\overset{\|}{C}}-OH$$

ETHERS

Draw ethyl propyl ether.

Ethers are easy to draw. Look on Table R, column 2 , below for the the functional group for ethers, -O-. Look at the **example**, methyl ethyl ether, in column 4 on **Table R**.

Table R

Class of Compound	Functional Group	General Formula	Example
Ether	**-O-**	**R-O-R'**	$CH_3OCH_2CH_3$ methyl ethyl ether

The formula of methyl ethyl ether begins with CH_3, a methyl group; the first word in the name is methyl. The functional group -O- goes after the methyl group. Obviously, $-CH_2CH_3$, the end of the example on Table R, an ethyl group, is the second word in the name.

Let's draw ethyl propyl ether (similar to above example):

- Begin with ethyl (eth- means 2 carbon atoms, see Table P), -C-C-.
- Put -O- for the ether (Table R); you now have -C-C-O-.
- Put propyl (prop- means 3 carbon atoms, see Table P) after the -O-: -C-C-O-C-C-C-.
- Draw lines or bonds around each carbon atom and put a hydrogen atom at the end of each bond.

$$H - \overset{\overset{\displaystyle H}{|}}{\underset{\underset{\displaystyle H}{|}}{C}} - \overset{\overset{\displaystyle H}{|}}{\underset{\underset{\displaystyle H}{|}}{C}} - O - \overset{\overset{\displaystyle H}{|}}{\underset{\underset{\displaystyle H}{|}}{C}} - \overset{\overset{\displaystyle H}{|}}{\underset{\underset{\displaystyle H}{|}}{C}} - \overset{\overset{\displaystyle H}{|}}{\underset{\underset{\displaystyle H}{|}}{C}} - H$$

Do Homework Questions #42-44, page 52.

First study halides, amines, etc., pages 10:28-10:31, then learn to draw these structures on the following pages.

HALIDES (HALOCARBONS) CONTAIN F, CL, BR OR I.

Draw 3-fluoroheptane.

Step 1: The prefix hept means 7 carbon atoms (Table P).

Step 2: 3-fluorohept**ane** means single bonds. (-ane means single bonds: Table Q). Draw 7 carbon atoms with single bonds between them:

-C-C-C-C-C-C-C-

Step 3: Look at and use Table R to put in the functional group. 3-fluoroheptane ends in -**ane** (and also has fluoro in it). Look at the example in column 4 in Table R below. The example 2-chloropropane ends in ane (and has Cl in it). Go across to columns 1 and 2. In column 2 it has the functional group -F (fluoro). Put F in your drawing. 3-fluoro- means fluorine is on the third carbon atom. Put F on the third carbon atom. (If no number is given, the fluorine is on the first or last carbon atom.)

$$C_1 - C_2 - \overset{\overset{\displaystyle F}{|}}{C_3} - C_4 - C_5 - C_6 - C_7$$

Or, Easy Method 2: You see in the top box in Table R on page Reference Tables 22 or below, under functional group (second column), it shows that, when you have the word "fluoro-," it means F. Put F in your drawing. See drawing above. 3-fluoro- means fluorine is on the third carbon atom. Put F on the third carbon atom. (If no number is given, the fluorine is on the first or last carbon atom.)

TABLE R
Organic Functional Groups

Class of Compound	Functional Group	General Formula	Example
Halide (halocarbon)	- F (fluoro-) - Cl (chloro-) - Br (bromo-) - I (iodo-)	R – X (X represents any halogen)	$CH_3CHClCH_3$ 2-chloropropane

---

Step 4: Draw lines or bonds around each carbon atom. Make sure each carbon atom has 4 bonds.

$$- \overset{|}{\underset{|}{C}} - \overset{|}{\underset{|}{C}} - \overset{|}{\underset{|}{C}} - \overset{F}{\underset{|}{C}} - \overset{|}{\underset{|}{C}} - \overset{|}{\underset{|}{C}} - \overset{|}{\underset{|}{C}} -$$

Attach a hydrogen atom to each bond, **except** where the fluorine is already attached.

$$H - \overset{H}{\underset{H}{C}} - \overset{H}{\underset{H}{C}} - \overset{F}{\underset{H}{C}} - \overset{H}{\underset{H}{C}} - \overset{H}{\underset{H}{C}} - \overset{H}{\underset{H}{C}} - \overset{H}{\underset{H}{C}} - H$$

See Table R: compounds that have carbon and halogen (F, Cl, Br, I) are called halides or halocarbons (column 1).

AMINES

Draw octanamine.

Step 1: The prefix oct means 8 carbon atoms (Table P).

Step 2: Octanamine means single bonds. (-ane means single bonds: Table Q). Draw 8 carbon atoms with single bonds between them:

-C-C-C-C-C-C-C-C-

Step 3: Look at and use Table R to put in the functional group. Octanamine ends in **-amine**. Look at the examples in column 4 in Table R on page Reference Tables 22 or below. Which example also ends in -amine? The example 1-propanamine ends in -amine. Go across and you see in column 1 that it is an amine.

Table R

Class of Compound	Functional Group	General Formula	Example
Amine	$- \overset{|}{N} -$	$R - \overset{\overset{R^1}{|}}{N} - R^{11}$	$CH_3CH_2CH_2NH_2$ 1-propanamine

In the second column, it has the $- \overset{|}{N} -$ functional group. Put $- \overset{|}{N} -$ in the structural formula. Since no number is given before the amine, example: octanamine, put the amine on the end carbon.

$$-C-C-C-C-C-C-C-C-\overset{|}{N}-$$

Or, Method 2: Octan**amine** has the word amine in it. You see the word amine in column 1 of Table R. Go across. In the second column there is the functional group $- \overset{|}{N} -$. Put $- \overset{|}{N} -$ in the structural formula. See drawing above. Since no number is given before the amine, example: octanamine, put the amine on the end carbon.

Step 4: Draw lines or bonds around all the carbon atoms. Make sure each carbon atom has 4 bonds.

$$- \overset{|}{\underset{|}{C}} - \overset{|}{\underset{|}{C}} - \overset{|}{\underset{|}{C}} - \overset{|}{\underset{|}{C}} - \overset{|}{\underset{|}{C}} - \overset{|}{\underset{|}{C}} - \overset{|}{\underset{|}{C}} - \overset{|}{\underset{|}{C}} - \overset{|}{N} -$$

Put hydrogen atoms on all the carbon and nitrogen bonds.

Draw 2-pentanamine.

Step 1: The prefix pent means 5 carbon atoms (Table P).

Step 2: Pentanamine means single bonds. (-ane means single bonds: Table Q). Draw 5 carbon atoms with single bonds between them:

-C-C-C-C-C-

Step 3: Since pentanamine ends in amine, go to the example in column 4 on Table R (on the previous page) that ends in amine. The example is 1-propanamine. Go across and you see in column one that it is an

amine; in the second column, $-\overset{|}{N}-$ is the functional group.

Amine	$-\overset{|}{N}-$

Put $-\overset{|}{N}-$ in the structural formula. 2-Pentanamine means put the amine on the 2nd carbon atom (the amine can be above or below the second carbon atom):

$$-C-\overset{|}{\underset{|}{C}}-C-C-C-$$
$$-\overset{|}{N}-$$

Step 4: Draw lines or bonds around all the carbon atoms. Make sure each carbon atom has 4 bonds. Put hydrogen atoms on all the carbon and nitrogen bonds.

$$H-\overset{H}{\underset{H}{C}}-\overset{H}{C}-\overset{H}{\underset{H}{C}}-\overset{H}{\underset{H}{C}}-\overset{H}{\underset{H}{C}}-H$$
$$H-N-H$$

AMIDES

Draw pentanamide.

Step 1: The prefix pent- means 5 carbon atoms (Table P).

Step 2: Pentanamide means single bonds (**-ane** means single bonds, Table Q). Draw 5 carbon atoms with single bonds between them: -C-C-C-C-C- .

Step 3: Look at and use Table R to put in the functional group. Pentanamide ends in **-amide**. Look at the examples in column 4 in Table R on page Reference Tables 22 or below. Which example also ends in -amide? The example propanamide ends in -amide. Go across and you see in column 1 that it is an amide.

Table R

Class of Compound	Functional Group	General Formula	Example
Amide	$\overset{O}{\underset{}{\overset{\|}{-C}}}-\overset{\|}{NH}$	$R-\overset{O}{\overset{\|}{C}}-\overset{R^1}{\underset{}{\overset{\|}{N}}}H$	$CH_3CH_2-\overset{O}{\overset{\|}{C}}-NH_2$ propanamide

In the second column, it has the $-\overset{\overset{\text{O}}{\|}}{\underset{}{C}}-\overset{|}{N}H$ functional group. Put $-\overset{\overset{\text{O}}{\|}}{\underset{}{C}}-\overset{|}{N}H$ in the structural formula.

REMEMBER: The C of $-\overset{\overset{\text{O}}{\|}}{\underset{}{C}}-\overset{|}{N}H$ is the last (fifth) carbon that you already drew.

$C-C-C-C-\overset{\overset{\text{O}}{\|}}{\underset{}{C}}-\overset{|}{N}H$

Method 2: Pentanamide has the word **amide** in it. You see the word amide in column 1 of Table R. Go across. In the second column there is the functional group $-\overset{\overset{\text{O}}{\|}}{\underset{}{C}}-\overset{|}{N}H$. Put $-\overset{\overset{\text{O}}{\|}}{\underset{}{C}}-\overset{|}{N}H$ in the structural formula. See drawing (with 5 C atoms) above.

REMEMBER: The C of $-\overset{\overset{\text{O}}{\|}}{\underset{}{C}}-\overset{|}{N}H$ is the last (fifth) carbon that you already drew.

Step 4: Draw lines or bonds around all the carbon atoms. Make sure each carbon has 4 bonds. The **amide carbon,** $-\overset{\overset{\text{O}}{\|}}{\underset{}{C}}-\overset{|}{N}H$, already **has four bonds.** Do **not add any bonds** to it.

$-\overset{|}{\underset{|}{C}}-\overset{|}{\underset{|}{C}}-\overset{|}{\underset{|}{C}}-\overset{|}{\underset{|}{C}}-\overset{\overset{\text{O}}{\|}}{\underset{}{C}}-\overset{|}{N}-H$

Put hydrogen atoms on all the available carbon and nitrogen bonds.

$H-\overset{\overset{H}{|}}{\underset{\underset{H}{|}}{C}}-\overset{\overset{H}{|}}{\underset{\underset{H}{|}}{C}}-\overset{\overset{H}{|}}{\underset{\underset{H}{|}}{C}}-\overset{\overset{H}{|}}{\underset{\underset{H}{|}}{C}}-\overset{\overset{\text{O}}{\|}}{\underset{}{C}}-\overset{|}{N}-H$

ESTERS

Draw methyl butanoate.

Step 1: The prefix but means 4 carbon atoms (Table P).

Step 2: Butanoate means single bonds. Draw a chain of 4 carbon atoms with single bonds: C-C-C-C .

Step 3: Our example, methyl butanoate, ends in -oate. Go to Table R, column 4 and find an example that ends in -oate. You see methyl propanoate (same ending). Look in column 1 and you see it is an ester, and in column 2 the functional group is $\overset{\overset{\text{O}}{\|}}{\underset{}{-C-O-}}$. See Table R below.

Table R

Class of Compound	Functional Group	General Formula	Example
Ester	$\overset{\overset{\text{O}}{\|}}{\underset{}{-C-O-}}$	$R-\overset{\overset{\text{O}}{\|}}{\underset{}{C}}-O-R^1$	$CH_3CH_2-\overset{\overset{\text{O}}{\|}}{\underset{}{C}}-O-CH_3$ **methyl propanoate**

(Hint: In the methyl propanoate example, you see $\overset{\overset{\text{O}}{\|}}{\underset{}{-C-O-}}$ was put on the last carbon of the 3-carbon chain prop.)

Put $\overset{O}{\underset{-C-O-}{\|}}$ on the **last carbon** of your **4-carbon** $\overset{\displaystyle O}{\underset{\displaystyle |\ \ \ |\ \ \ |}{-C-C-C-C-O}}$ with double bond O on last carbon

chain. But **REMEMBER**, the C of $\overset{O}{\underset{-C-O-}{\|}}$ is the last (fourth) carbon that you already drew.

Put the **methyl group after** the $\overset{O}{\underset{-C-O-}{\|}}$ functional group (just like in the example in column 4). See drawing at right.

$$\overset{\displaystyle O \qquad\qquad\quad H}{\underset{\displaystyle |\ \ \ |\ \ \ |\qquad\qquad |}{-C-C-C-C-O-C-H}}$$
with H below last C

Step 4: Draw lines or bonds around all the carbon atoms. Each carbon atom must have four bonds. The C in $\overset{O}{\underset{-C-O-}{\|}}$ already has four bonds. Do not add a bond to it.

$$\overset{\displaystyle H\ \ \ H\ \ \ H\ \ \ O\qquad\ \ H}{\underset{\displaystyle H\ \ \ H\ \ \ H\qquad\qquad H}{H-C-C-C-C-O-C-H}}$$

Attach a hydrogen atom to the end of each remaining bond.

Do Homework Questions #45-47, page 52.

EFFECT OF FUNCTIONAL GROUP ON PHYSICAL PROPERTIES
BOILING POINTS

1. ALCOHOLS AND ACIDS

Alcohols have the functional group **OH** and **acids** have the functional group **COOH**. You learned that water has hydrogen bonds. The positive part, H, of 1 water molecule is attracted to the negative part, O, of the next water molecule. Alcohols and acids similarly have **hydrogen bonds**. In alcohols, the positive part, H, of alcohol is attracted to the negative part, O, of the next alcohol molecule; in acids, the positive part, H, is attracted to the negative part, $\overset{O}{\underset{-C-O-}{\|}}$, of the next acid molecule.

Propanol, an alcohol, has a higher boiling point than propane, because propanol has the **extra hydrogen bond** or intermolecular forces holding the molecules tightly together.

In short, **hydrogen bonds** in **alcohols** and also in **acids** cause the **boiling point** to be **higher** than in the alkane with the same number of carbon atoms.

2. AMINES

Amines also have a hydrogen bond, but it is a **weaker hydrogen bond**. Therefore, the boiling point of amines is higher than in alkanes (with the same number of carbon atoms), because of the hydrogen bond, but lower than in alcohols and acids, because the hydrogen bond is weaker.

3A. ETHERS, ALDEHYDES AND KETONES

Ethers have -O-, oxygen between 2 carbon atoms. **Aldehydes** and **ketones** have a C=O group. All three have oxygen bonded to carbon and **not** bonded to hydrogen. Oxygen must be bonded to hydrogen in order to have hydrogen bonds.

Aldehydes and ketones have the C=O group; ethers have oxygen, −O−, between two carbon atoms. These groups, C=O and −O−, make the **molecules polar** (1 part of molecule positive, 1 part negative). There is an **intermolecular attraction** between the molecules in ethers, aldehydes and ketones, **between the positive end** of one molecule and the **negative end** of the **next molecule**. This attraction is weaker than a hydrogen bond but stronger than the attraction of molecules in hydrocarbons. Therefore, the **boiling point** of **ethers, aldehydes** and **ketones** is **lower than** an **acid** or **alcohol** (which have hydrogen bonds) but **higher than** an **alkane** (with the same number of carbon atoms).

3B. HALIDES (HALOCARBONS)

The halides have elements in group 17 of the periodic table, **F (fluorine), Cl (chlorine), Br (bromine),** and **I (iodine),** attached to **C (carbon),** forming C-F, C-Cl, C-Br, or C-I. Just like ethers, aldehydes, and ketones, these groups, C-F, C-Cl, C-Br, and C-I make the **molecules polar** (1 part of molecule positive (C), 1 part negative (F or Cl or Br or I)). There is an **intermolecular attraction** between the molecules in halides, **between the positive end** of **one molecule** and the **negative end** of the **next molecule**. This attraction is weaker than a hydrogen bond but stronger than the attraction of molecules in hydrocarbons. Therefore, the **boiling point** of **halides** is **lower than** an **acid** or **alcohol** (which have hydrogen bonds) but **higher than** an **alkane** (with the same number of carbon atoms).

4. HYDROCARBONS

Hydrocarbons (only have carbon and hydrogen) have **no hydrogen bonds** and are **not polar**. Therefore, hydrocarbons have a **lower boiling point than acids, alcohols, amines, ethers, aldehydes or ketones**.

MELTING POINTS

Similarly, because of the hydrogen bonds, acids and alcohols have the highest melting points. Just like in the case of boiling points, ethers, aldehydes and ketones have lower melting points than acids and alcohols but higher than alkanes with the same number of carbon atoms. Hydrocarbons have the lowest melting points.

SOLUBILITY

Polar substances (1 part positive, 1 part negative) dissolve in polar substances. Polar substances dissolve in water, which is also polar. **Like dissolves like,** which means polar dissolves in polar.

Small **alcohols** and **acids** (each containing up to 4 carbon atoms) are polar and dissolve easily in water. The OH of the alcohol and the COOH of the acid make the molecules polar.

When the carbon chains are large, the nonpolar hydrocarbons become more important and the molecule becomes less polar. It dissolves less in water or does not dissolve.

Ethers, aldehydes and **ketones** are slightly polar. When they have small carbon chains, they dissolve in water but less than acids and alcohols. Large chains do not dissolve in water.

Hydrocarbons are nonpolar (not polar) and do not dissolve in water.

Do Homework Questions #48-49, page 52.

ORGANIC REACTIONS

1. **SUBSTITUTION**: Replacement of one kind of atom or group by another kind of atom or group.

$$
\begin{array}{ccccccc}
& H\ H & & & & H\ H & \\
& |\ \ | & & & & |\ \ | & \\
H\text{-}C\text{-}C\text{-}H & + & F_2 & \rightarrow & H\text{-}C\text{-}C\text{-}F & + & HF \\
& |\ \ | & & & & |\ \ | & \\
& H\ H & & & & H\ H & \\
\text{Ethane} & & \text{Fluorine} & & \text{Fluoroethane} & & \text{Hydrogen} \\
& & & & & & \text{Fluoride}
\end{array}
$$

The hydrogen atom of ethane is replaced by F = substitution.

Hydrogen atoms can similarly be replaced by chlorine (Cl_2), bromine (Br_2), or iodine (I_2).

$$
\begin{array}{ccccccc}
& H\ H & & & H\ H & & \\
& |\ \ | & & & |\ \ | & & \\
H\text{-}C\text{-}C\text{-}H & + & Cl_2 & \rightarrow & H\text{-}C\text{-}C\text{-}Cl & + & HCl \\
& |\ \ | & & & |\ \ | & & \\
& H\ H & & & H\ H & & \\
& \text{Ethane Chlorine} & & & \text{Chloroethane} & & \text{Hydrogen} \\
& & & & & & \text{Chloride}
\end{array}
$$

SUMMARY: Substitution only happens in **alkanes** (alkanes have single bonds between carbon atoms).

2. **ADDITION**: **Adding** one or more atoms at a double or triple bond:

$$
\begin{array}{ccccc}
& H\ H & & & H\ H \\
& |\ \ | & & & |\ \ | \\
H\text{-}C\text{=}C\text{-}H & + & F_2 & \rightarrow & H\text{-}C\text{-}C\text{-}H \\
& & & & |\ \ | \\
& & & & F\ F \\
\text{Ethene} & & \text{Fluorine} & & \text{1,2-Difluoroethane}
\end{array}
$$

F_2 is **added** to ethene. As you can see, 1,2-difluoroethane has two **more** F than ethene.

$$H-\overset{\displaystyle H}{\underset{\displaystyle |}{C}}=\overset{\displaystyle H}{\underset{\displaystyle |}{C}}-H \ + \ Cl_2 \ \rightarrow \ H-\overset{\displaystyle H}{\underset{\displaystyle |}{\underset{\displaystyle Cl}{C}}}-\overset{\displaystyle H}{\underset{\displaystyle |}{\underset{\displaystyle Cl}{C}}}-H$$

Ethene   Chlorine 1,2-Dichloroethane

Cl_2 is **added** to ethene(see above). 1,2-dichloroethane has two **more** Cl than ethene.

SUMMARY: Addition happens in **alkenes and alkynes** (double or triple bond). Addition does **not** happen in alkanes (single bond).

---

3. **FERMENTATION**:

$$C_6H_{12}O_6 \xrightarrow{\ zymase\ } 2C_2H_5OH + 2CO_2$$

    Glucose   **Ethanol**   **Carbon Dioxide**

Glucose is broken down into **ethanol and carbon dioxide** by enzymes (zymase produced by yeast).

4. **ESTERIFICATION**: **Acid** and **alcohol** produce **ester** and **water**.

$$H-\overset{\displaystyle H}{\underset{\displaystyle |}{\underset{\displaystyle H}{C}}}-\overset{\displaystyle O}{\underset{\displaystyle \|}{C}}-\underline{OH} \ + \ \underline{H}O-CH_3 \ \longrightarrow \ H-\overset{\displaystyle H}{\underset{\displaystyle |}{\underset{\displaystyle H}{C}}}-\overset{\displaystyle O}{\underset{\displaystyle \|}{C}}-O-\overset{\displaystyle H}{\underset{\displaystyle |}{\underset{\displaystyle H}{C}}}-H \ + \ H_2O$$

ethanoic acid + methanol $\longrightarrow$ methyl ethanoate + water

Acid   +   *Alcohol* $\longrightarrow$ *Ester*   +   *H$_2$O*

The OH from the acid unites with the H from the alcohol to form H_2O (water). The rest of the acid unites with the rest of the alcohol to form the ester.

Fats are esters derived from glycerol (an alcohol that has 3 OH groups) and long chain fatty acids.

    Glycerol is a trihydroxy alcohol.

(In neutralization, you had

$$acid + base \rightarrow salt + water$$
$$HCl + NaOH \rightarrow NaCl + H_2O)$$

In esterification,

$$acid + alcohol \rightarrow ester + H_2O.$$

5. **SAPONIFICATION** (hydrolysis) is an **ester** breaking up into acid and alcohol (reverse of esterification).

Saponification: fat (+ strong base) $\rightarrow$ soap   + glycerol
            ester   *salt of acid*   *alcohol*

Saponification produces **soap**.

---

6. **COMBUSTION** means **burning** (reacting with oxygen). Hydrocarbons **burn (unite with oxygen)** to form carbon dioxide (CO_2) and water. In a limited supply of oxygen, carbon (C) and carbon monoxide (CO) are formed (not enough oxygen to form CO_2).

7. **POLYMERIZATION** involves **smaller molecules joining together** to form **one big molecule**. A **monomer** is a **small** molecule. A **polymer** is a **large** molecule. Monomers join together to form a polymer.

   Amino Acid + Amino Acid + Amino Acid $\longrightarrow$ Protein
   monomer   *monomer*   *monomer*   *polymer*

   There are two types of polymerization:
   (The words condensation and addition are not on the Regents, but you must know the ideas about polymerization.)

   a. **Condensation Polymerization** is monomers joining together by **dehydration synthesis (removing water)** to form a polymer. Examples are nylon, polyester and silicone. **Naturally occuring polymers** include starch, protein and cellulose.

   b. **Addition Polymerization** is monomers joining together by breaking a double or triple bond (adding together the monomers) to form a polymer. Ethene (commonly called ethylene) + ethene (ethylene) + ethene (ethylene) join together by **breaking the double bond** to form polyethylene:

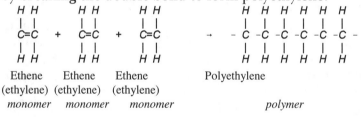

   $n=$number of ethene joining together $\longrightarrow$ Polyethylene

IUPAC NAME OF A COMPOUND

Question: What is the IUPAC name of

left side
$$H-\overset{\overset{\displaystyle H}{|}}{\underset{\underset{\displaystyle H}{|}}{C}}-\overset{\overset{\displaystyle H}{|}}{\underset{\underset{\displaystyle H}{|}}{C}}-\overset{\overset{\displaystyle H}{|}}{\underset{\underset{\displaystyle H}{|}}{C}}-\overset{\overset{\displaystyle H}{|}}{\underset{\underset{\displaystyle H}{|}}{C}}-\overset{\overset{\displaystyle H}{|}}{\underset{\underset{\displaystyle H}{|}}{C}}-\overset{\overset{\displaystyle Cl}{|}}{\underset{\underset{\displaystyle Cl}{|}}{C}}-\overset{\overset{\displaystyle H}{|}}{\underset{\underset{\displaystyle H}{|}}{C}}-H$$
right side

(1) 2-chloroheptane   (3) 2,2-dichloroheptane

(2) 6-chloroheptane   (4) 6,6-dichloroheptane

Solution: Review halides, bottom of page 10:28.

There are seven carbons with single bonds; it is an alkane, **heptane.**

There are **two chlorine** atoms; it is called **di**chloro. Therefore, you have **dichloro**heptane.

If you count the C atoms from left to right, the Cl atoms are on C-6.

If you count the C atoms from right to left, the Cl atoms are on C-2.

Count from right to left so the Cl atoms (or any functional group) is on the lower numbered carbon, C-2 and not C-6.

Since there is a Cl atom on C-2, write 2 because the Cl is on C-2.

There is another Cl on C-2; **both Cls are on C-2, so write 2,2.** You have **2,2-** dichloroheptane.

IUPAC name is 2,2-dichloroheptane.

Try Sample Questions #6-13, pages 46-47, then do Homework Questions, #24-39, pages 50-51 and #50-51, page 52.

FINDING MISSING REACTANTS AND PRODUCTS IN ORGANIC REACTIONS

In an equation, the number of atoms on the left side of the arrow must equal the number of atoms on the right side of the arrow.

After the elements or compounds are correctly written, you can only change the (coefficient) number in front of the element or compound. If there is no coefficient in front of C_2H_6, it means 1. $C_2H_6 = 1 \ C_2H_6$ molecule. If there is a coefficient of 3 in front of CO_2, it means 3 x 1 atoms of carbon = 3 carbon atoms and 3 x 2 atoms of oxygen = 6 oxygen atoms. A coefficient in front of the compound goes for all the elements in the compound.

Question: Find the missing product in this balanced equation:

$$C_2H_6 + Cl_2 \longrightarrow C_2H_5Cl + \underline{\hspace{2cm}}$$

Solution: A balanced equation must have equal numbers of atoms of each element on both sides of the equation.

Left Side	Right Side	Missing on Right Side
C: 2 atoms	C: 2 atoms	
H: 6 atoms	H: 5 atoms	H: 1 atom
Cl: 2 atoms	Cl: 1 atom	Cl: 1 atom

Therefore, the missing product must be HCl. This is a substitution reaction; a hydrogen atom of ethane is replaced by chlorine in the equation above.

Question: Find the missing reactant in this balanced equation:

$$C_2H_4 \ + \ \underline{\hspace{2cm}} \longrightarrow C_2H_4Br_2$$

Solution: A balanced equation must have equal numbers of atoms of each element on both sides of the equation.

Left Side	Missing on Left Side	Right Side
C: 2 atoms		C: 2 atoms
H: 4 atoms		H: 4 atoms
	Br: 2 atoms	Br: 2 atoms

Therefore, the missing reactant is Br_2. This is an addition reaction; Br_2 is added to C_2H_4 in the equation above.

Question: Find the missing reactant in the following balanced equation:

$$\underline{\hspace{2cm}} + 5O_2 \longrightarrow 3CO_2 + 4H_2O$$

Solution:

Left Side	Missing on Left Side:	Right Side
	3 C atoms	$3CO_2$ = 3 x 1 C = **3 C** atoms
	8 H atoms	$4H_2O$ = 4 x 2 H = **8 H** atoms
$5O_2$ = 5 x 2 = **10 O atoms**		$3CO_2$ = 3 x 2 O = 6 O atoms
		$4H_2O$ = 4 x 1 O = 4 O atoms
		Therefore, you have **10 O** atoms.

The missing reactant must be C_3H_8.

Question: Find the missing reactant in the balanced equation:

$$\underline{\hspace{2cm}} \longrightarrow 2C_2H_5OH \ + \ 2CO_2$$

Solution: A balanced equation must have equal numbers of atoms of each element on both sides of the equation.

Left Side	Missing on Left Side:	Right Side
	6 C atoms	$2C_2H_5OH$ means: 2 x 2 carbon atoms = 4 carbon atoms $2CO_2$ means: 2 x 1 carbon atoms = 2 carbon atoms Total: **6 C** atoms
	12 H atoms	$2C_2H_5OH$ means: 2 x (5+1) hydrogen atoms = **12 H** atoms
	6 O atoms	$2C_2H_5OH$ means: 2 x 1 oxygen atoms = 2 oxygen atoms $2CO_2$ means: 2 x 2 oxygen atoms = 4 oxygen atoms Total: **6 O** atoms

Therefore, the missing reactant must be $C_6H_{12}O_6$. This is a fermentation reaction; glucose is broken down to ethanol and carbon dioxide.

SAMPLE REGENTS & REGENT-TYPE QUESTIONS AND SOLUTIONS

1. Which structural formula represents a primary alcohol?

   (1)
   $$\begin{array}{ccc} H & H & O \\ | & | & // \\ H-C-C-C \\ | & | & \backslash \\ H & H & H \end{array}$$

   (2)
   $$\begin{array}{ccc} H & H & O \\ | & | & // \\ H-C-C-C \\ | & | & | \\ H & H & OH \end{array}$$

   (3)
   $$\begin{array}{ccc} H & H & H \\ | & | & | \\ H-C-C-C-OH \\ | & | & | \\ H & H & H \end{array}$$

   (4)
   $$\begin{array}{ccc} H & H & H \\ | & | & | \\ H-C- C -C-H \\ | & | & | \\ H & OH & H \end{array}$$

2. Which two compounds are monohydroxy alcohols?
   (1) ethylene glycol & ethanol   (3) methanol & ethanol
   (2) ethylene glycol & glycerol   (4) methanol & glycerol

3. What is the name of the compound with the following formula?

   $$\begin{array}{ccc} H & O & H \\ | & || & | \\ H-C-C-C-H \\ | & & | \\ H & & H \end{array}$$

   (1) propanone   (2) propanol   (3) propanal   (4)propanic acid

4. Which is the general formula for organic acids?

   (1)
   $$\begin{array}{c} O \\ // \\ R- C \\ \backslash \\ H \end{array}$$

   (2)
   $$\begin{array}{c} O \\ // \\ R- C \\ \backslash \\ OH \end{array}$$

   (3)
   $$\begin{array}{c} R_1 \\ \diagdown \\ C=O \\ \diagup \\ R_2 \end{array}$$

   (4) R_1-O-R_2

5. Which type of compound is represented by the structural formula below?

   $$\begin{array}{cccc} H & H & H & H \\ | & | & | & | \\ H-C-C-O-C-C-H \\ | & | & | & | \\ H & H & H & H \end{array}$$

   (1) a ketone   (2) an aldehyde   (3) an ester   (4) an ether

6. What type of reaction is
   $$CH_3CH_3 + Cl_2 \rightarrow CH_3CH_2Cl + HCl?$$
   (1) an addition reaction   (3) a saponification reaction
   (2) a substitution reaction   (4) an esterification reaction

7. Which is the structural formula for 2-chlorobutane?

   (1)
   $$\begin{array}{cccc} H & H & H & Cl \\ | & | & | & | \\ H-C-C-C-C-H \\ | & | & | & | \\ H & H & H & Cl \end{array}$$

   (3)
   $$\begin{array}{cccc} H & H & H & Cl \\ | & | & | & | \\ H-C-C-C-C-H \\ | & | & | & | \\ H & H & H & H \end{array}$$

   (2)
   $$\begin{array}{cccc} H & Cl & H & H \\ | & | & | & | \\ H-C-C-C-C-H \\ | & | & | & | \\ H & H & H & H \end{array}$$

   (4)
   $$\begin{array}{cccc} H & H & Cl & H \\ | & | & | & | \\ H-C-C-C-C-H \\ | & | & | & | \\ H & H & Cl & H \end{array}$$

8. Which structural formula represents the product formed from the reaction of Cl_2 and C_2H_4?

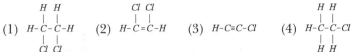

(1) $H-\overset{\overset{\displaystyle H}{|}}{\underset{\underset{\displaystyle Cl}{|}}{C}}-\overset{\overset{\displaystyle H}{|}}{\underset{\underset{\displaystyle Cl}{|}}{C}}-H$ (2) $H-\overset{\overset{\displaystyle Cl}{|}}{C}=\overset{\overset{\displaystyle Cl}{|}}{C}-H$ (3) $H-C\equiv C-Cl$ (4) $H-\overset{\overset{\displaystyle H}{|}}{\underset{\underset{\displaystyle H}{|}}{C}}-\overset{\overset{\displaystyle H}{|}}{\underset{\underset{\displaystyle H}{|}}{C}}-Cl$

9. Which type of reaction do ethane molecules and ethene molecules undergo when they react with chlorine?
   (1) Ethane and ethene both react by addition.
   (2) Ethane and ethene both react by substitution.
   (3) Ethane reacts by substitution and ethene reacts by addition.
   (4) Ethane reacts by addition and ethene reacts by substitution.

10. The fermentation of $C_6H_{12}O_6$ will produce CO_2 and
    (1) $C_3H_5(OH)_3$ (2) C_2H_5OH (3) $Ca(OH)_2$ (4) $Cr(OH)_3$

11. Which is a product of the hydrolysis of an animal fat by a strong base?
    (1) water (2) gasoline (3) soap (4) toluene

12. Which polymers occur naturally?
    (1) starch and nylon (3) protein and nylon
    (2) starch and cellulose (4) protein and plastic

13. A condensation polymerization reaction produces a polymer and
    (1) H_2 (2) O_2 (3) CO_2 (4) H_2O

SOLUTIONS

1. Answer **3**. Answer 3 is a primary alcohol, because the OH is attached to a carbon that is attached to one other carbon atom. Choice 1 is an aldehyde, Choice 2 is an (organic) acid, and Choice 4 is a secondary alcohol. In a secondary alcohol, the OH is attached to a carbon that is attached to *two other* carbon atoms.

2. Answer **3**. Monohydroxy alcohol has only one OH group. The correct answer is Answer 3, methanol and ethanol.

Methanol

Ethanol

   Methanol and ethanol have only one OH, and therefore they are monohydroxy alcohols. Choices 1 and 2 are wrong because ethylene glycol is a dihydroxy alcohol. Choice 4 and also, again, Choice 2, are wrong because glycerol is a trihydroxy alcohol.

3. Answer **1**. You learned that, in a ketone, the $\overset{O}{\overset{||}{C}}$ is in the **middle**, between two carbon atoms. You know that the compound is a ketone. Ketone ends in "-one," therefore the compound must end in "**-one**." Answer 1 is the only choice that ends in "**-one**."
   Additional information: There are three carbon atoms; therefore the compound begins with "pro-."

4. Answer **2**. $R-\overset{\displaystyle O}{\overset{\displaystyle \|}{\underset{\diagdown\ OH}{C}}}$ is the formula for an organic acid.

5. Answer **4**. An ether is represented by R_1-O-R_2. R_1 and R_2 are hydrocarbon groups. In an ether, there is always an O in the middle between two carbon atoms (or two hydrocarbon groups). The compound is diethyl ether.

6. Answer **2**. In CH_3CH_3, one Cl atom is replacing or substituting for an H and produces CH_3CH_2Cl. CH_3CH_3 = C_2H_6 = alkane. In alkanes, substitution takes place.

7. Answer **2**. 2-chloro**butane** means: "**bu**" = 4 carbonatoms," **ane**" means part of the alkane series, with single bonds between the carbon atoms. "Chloro" means 1 chlorine atom is attached.

$$\overset{\displaystyle H\ \ Cl\ \ H\ \ H}{\underset{\displaystyle H\ \ H\ \ H\ \ H}{H-\overset{|}{\underset{|}{C_1}}-\overset{|}{\underset{|}{C_2}}-\overset{|}{\underset{|}{C_3}}-\overset{|}{\underset{|}{C_4}}-H}}$$

"2-chloro" means the chlorine atom is attached to the second carbon atom.
Answer 2 has 4 carbon atoms (alkane series) and has 1 chlorine atom attached to the second C atom.

8. Answer **1**:
The double bond is broken and two chlorine atoms are added. Or, C_2H_4 is an alkene; addition takes place. The question is $C_2H_4 + Cl_2$, therefore 2 carbon atoms, 4 hydrogen and 2 Cl atoms. Answer 1 is the only choice with 2C, 4H and 2Cl atoms.

$$\overset{\displaystyle H\ \ H}{H-\overset{|}{C}=\overset{|}{C}-H} + Cl_2 \rightarrow \overset{\displaystyle H\ \ H}{\underset{\displaystyle Cl\ \ Cl}{H-\overset{|}{\underset{|}{C}}-\overset{|}{\underset{|}{C}}-H}}$$

C_2H_4

9. Answer **3**. Eth**ane** is part of the alk**ane** series, and alkanes react by substitution. Eth**ene** is part of the alk**ene** series, and alkenes react by addition.

10. Answer **2**. The process of fermentation is:

$$C_6H_{12}O_6 \rightarrow C_2H_5OH + CO_2$$
$$\text{glucose} \quad\quad \text{ethanol} \quad \text{carbon dioxide}$$

11. Answer **3**. Hydrolysis of fat by a strong base produces soap and glycerol.

12. Answer **2**. You learned that starch and cellulose are naturally occurring polymers.

13. Answer **4**. In condensation polymerization, monomers join together to produce a polymer and water (H_2O).

1. Which is an example of a monohydroxy alcohol?
   (1) methanal   (2) methanol   (3) glycol   (4) glycerol

2. Which organic compound is classified as a primary alcohol?
   (1) ethylene glycol   (3) glycerol
   (2) ethanol   (4) 2-butanol

3. To be classified as a tertiary alcohol, the functional –OH group is bonded to a carbon atom that must be bonded to a total of how many additional carbon atoms?
   (1) 1   (2) 2   (3) 3   (4) 4

4. Which is the structural formula for 2-propanol?

   (1) H-C-C-C-OH   (2) H-C-C-C-H   (3) H-C-C-C-C-OH   (4) H-C-C-C-C-H

5. What is the structural formula for 1,2-ethanediol?

   (1) H-C-C-OH   (2) H-C-C-C-OH   (3) H-C-C-H   (4) H-C-C-C-H

6. What is the total number of carbon atoms in a molecule of glycerol?
   (1) 1   (2) 2   (3) 3   (4) 4

7. Given the structural formulas for three alcohols:

   All are classified as
   (1) monohydroxy alcohols   (3) tertiary alcohols
   (2) secondary alcohols   (4) primary alcohols

8. Which structural formula represents a dihydroxy alcohol?

9. What is the total number of hydroxyl groups contained in one molecule of 1,2-ethanediol?
   (1) 1   (2) 2   (3) 3   (4) 4

10. Which class of organic compounds can be represented as R–OH?
    (1) acids   (2) alcohols   (3) esters   (4) ethers

11. What is the structural formula of an aldehyde?

12. Given the compound:

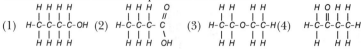

Which structural formula represents an isomer?

(1) H-C-C-C-C-OH (2) H-C-C-C- C (3) H-C-C-O-C-C-H(4) H-C-C-C-C-H

13. What is the name of the compound with the following formula?

H-C-C-C-H

(1) propanone (2) propanol (3) propanal (4) propanoic acid

14. Which organic compound is classified as an acid?
(1) CH_3CH_2COOH (3) $C_{12}H_{22}O_{11}$
(2) CH_3CH_2OH (4) $C_6H_{12}O_6$

15. Which structural formula represents an organic acid?

(1) H-C-C-C-OH (2) H-C- C -C-H (3) H-C-C- C (4) HO-C-C-C-OH

16. Which compound is an organic acid?
(1) CH_3OH (2) CH_3OCH_3 (3) CH_3COOH (4) CH_3COOCH_3

17. Which is the structural formula for diethyl ether?

(1) H-C-C-O-C-C-H (2) H-C-O-C-H (3) H-C-C-C-C-C-H (4) H-C-C-C-H

18. Which class of compounds has the general formula $R_1 - O - R_2$?
(1) esters   (2) alcohols   (3) ethers   (4) aldehydes

19. What is the name of the compound with the formula CH_3-NH_2?
(1) methanal   (3) amino acid
(2) methanamine   (4) methyl propanoate

20. Which compound is classified as an ester?
(1) CH_3CH_2COOH (3) $CH_3CH_2COOC_2H_5$
(2) CH_3CH_2OH (4) CH_4

21. Which class of organic compounds can be represented as R-C-C-OH with NH₂ ?

(1) ether   (2) amino acid   (3) ester   (4) amide

22. What is the structural formula for an amide?
(1) -C-   (2) -C-C-C-OH   (3) CH_3CH_2 - C - NH₂ (4) C-C-C-OH

23. Which organic compound is classified as a halocarbon?
(1) C_4H_9Cl (2) CH_3OH (3) C_2H_4 (4) $C_7H_{15}OH$

24. Which hydrocarbon will undergo a substitution reaction with chlorine?
(1) methane   (2) ethyne   (3) propene   (4) butene

25. Which compound can undergo an addition reaction?
    (1) CH_4 (2) C_2H_4 (3) C_3H_8 (4) C_4H_{10}

26. The type of reaction represented by
$$C_2H_4 + H_2 \rightarrow C_2H_6$$
    is called
    (1) substitution   (2) polymerization   (3) addition   (4) esterification

27. The products of the fermentation of a sugar are ethanol and
    (1) water   (2) oxygen   (3) carbon dioxide (4) sulfur dioxide
    (2) oxygen   (4) sulfur dioxide

28. The reaction
$$CH_2CH_2 + H_2 \rightarrow CH_3CH_3$$
    is an example of
    (1) substitution   (2) addition   (3) esterification (4) fermentation

29. Which substance is a product of a fermentation reaction?
    (1) glucose   (2) zymase   (3) ethanol   (4) water

30. Given the equation
$$C_6H_{12}O_6 \xrightarrow{\text{zymase}} 2C_2H_5OH + 2CO_2$$

    The reaction represented by this equation is called
    (1) esterification   (3) fermentation
    (2) saponification   (4) polymerization

31. Which alcohol reacts with C_2H_5COOH to produce the ester $C_2H_5COOC_2H_5$?
    (1) CH_3OH (2) C_2H_5OH (3) C_3H_7OH (4) C_4H_9OH

32. In which type of reaction are long-chain molecules formed from smaller molecules?
    (1) substitution   (3) fermentation
    (2) saponification   (4) polymerization

33. Which type of reaction is used in the production of nylon?
    (1) substitution   (3) esterification
    (2) saponification   (4) polymerization

34. Which is a product of a condensation reaction?
    (1) O_2 (2) CO_2 (3) H_2 (4) H_2O

35. Which substance is made up of monomers joined together in long chains?
    (1) ketone   (2) protein   (3) ester   (4) acid

36. Condensation polymerization is best described as
    (1) a dehydration reaction   (3) a reduction reaction
    (2) a cracking reaction   (4) an oxidation reaction

37. In which organic reaction is sugar converted to an alcohol and carbon dioxide?
    (1) esterification   (3) substitution
    (2) addition   (4) fermentation

38. Which reaction best represents the complete combustion of ethene?
    (1) $C_2H_4 + HCl \rightarrow C_2H_5Cl$ (3) $C_2H_4 + 3O_2 \rightarrow 2CO_2 + 2H_2O$
    (2) $C_2H_4 + Cl_2 \rightarrow C_2H_4Cl_2$ (4) $C_2H_4 + H_2O \rightarrow C_2H_5OH$

39. Which materials are naturally occurring polymers?
    (1) nylon and cellulose   (3) starch and cellulose
    (2) nylon and polyethylene   (4) starch and polyethylene

CONSTRUCTED RESPONSE QUESTIONS: Parts B-2 and C of NYS Regents Exam

40. By the chemical formula, how can you recognize alcohol, aldehyde, ketone, acid, and ether?

---

41. By the chemical formula, how can you recognize a halide, amine, amino acid, amide, and ester?

42. Draw 2-hexanol, 3-pentanol, hexanal, octanal.

43. Draw 3-heptanone, 2-decanone, decanoic acid, heptanoic acid.

44. Draw methyl ethyl ether, methyl propyl ether.

45. Draw 3-bromononane, 2-fluorodecane.

46. Draw decanamine, nonanamine, heptanamide, octanamide,.

47. Draw methyl propanoate, ethyl butanoate.

48. Explain how the boiling points of alcohols, acids, amines, ethers, aldehydes, ketones, and hydrocarbons differ and why.

49. Explain why small alcohols and acids dissolve easily in water.

50. Compare and contrast substitution and addition (which compounds and how it works.)

51. Explain the reaction between ethanoic acid and methanol.

52. Given the incomplete reaction:

$$CH_3CH_2\ CH_2\overset{\displaystyle O}{\hat{C}}\text{-OH} + x \rightarrow CH_3CH_2CH_2\overset{\displaystyle O}{\hat{C}}\text{-OCH}_2CH_3 + H_2O$$

Which compound is represented by x?

(1) CH_3CH_2OH (2) $CH_3\overset{\displaystyle O}{\hat{C}}\text{-H}$ (3) $CH_3OCH_2CH_3$ (4) $CH_3\overset{\displaystyle O}{\hat{C}}CH_3$

53. Find the missing product in this reaction:

$$2C_4H_{10} + 13O_2 \longrightarrow 8CO_2 + \underline{\hspace{1.5cm}}$$

CHAPTER QUESTION: Parts B-2 and C of NYS Regents Exam

54. Base your answers to this question on the article below and on your knowledge of chemistry.

For coffee beans to be labeled "decaffeinated," at least 97% of the caffeine must be removed. There are three primary methods for decaffeination: chemical extraction, the Swiss water process, and supercritical fluid extraction

Although all methods of decaffeinating coffee involve the use of "chemicals," one process has been traditionally referred to as "chemical extraction," probably because it uses organic solvents that are not typically part of our normal environment. The traditional method uses two slightly varied options using dichloromethane (CH_2Cl_2) or ethyl acetate ($CH_3COOC_2H_5$) as solvents. With both solvents, the beans are first soaked in water to soften them and speed the decaffeinating process. The beans are then soaked in one of the two solvents, which dissolves the caffeine in the bean. Once the solvent has removed the caffeine, the coffee beans are treated with steam. This evaporates the organic solvent along with the caffeine.

A. Draw the structural formula for ethyl acetate. (The correct IUPAC name is ethyl ethanoate).

B. To what class of organic compounds does dichloromethane belong?

C. Indicate whether you think the caffeine molecule is polar or nonpolar; then explain your answer in terms of the solubility of the caffeine molecule.

55. Methanethiol, CH_3SH, has the functional group (-SH) and has a structure similar to methanol. Draw the structural formula for methanethiol.

CHAPTER 11: NUCLEAR CHEMISTRY

STABLE AND UNSTABLE NUCLEI

Most nuclei are **stable (don't change)**, but some nuclei are **not stable (unstable)**. An unstable nucleus decays (breaks down) spontaneously (by itself), giving off rays and particles.

Whether a nucleus is stable or unstable (breaks up) depends on the ratio of neutrons to protons. (However, there are also other factors that affect stability.) In general, for elements of **small atomic number** (from 1 to about 20), when the **ratio** of $\frac{neutron}{proton}$ is **about 1**, the nucleus of the **isotope** is **stable**. Isotopes have the same atomic number but different mass numbers (different number of neutrons). $^{16}_{8}O$ and $^{18}_{8}O$ are isotopes of oxygen.

> Low atomic number (atomic number 1-20)
> Ratio of $\frac{neutron}{proton} = 1$
> Nucleus of isotope is stable.

Determine if isotope is stable: $^{16}_{8}O$, oxygen, low atomic number. Atomic number 8 = 8 protons. Mass # - atomic # = # of neutrons = 16 - 8 = 8 neutrons. Ratio of $\frac{neutron}{proton} = \frac{8n}{8p} = 1$.
Ratio of $\frac{neutron}{proton} = 1$. Isotope is stable.

Determine if isotope is stable: $^{12}_{6}C$, carbon, low atomic number. Atomic # 6 = 6 protons. Mass # - atomic # = 6 neutrons.
Ratio of $\frac{neutron}{proton} = \frac{6n}{6p} = 1$. Ratio of $\frac{neutron}{proton} = 1$. Isotope is stable.
For elements with **higher atomic numbers** (atomic number 21 and higher), the **nucleus** is generally **stable** when the **ratio** of $\frac{neutron}{proton}$ is **more than one**. For elements like $^{56}_{26}Fe$ with atomic number 26, stable isotopes have a ratio of about 1.15. Ratio of $\frac{neutron}{proton} = \frac{30n}{26p} = 1.15$.

The ratio of $\frac{neutron}{proton}$ for stable isotopes increases for heavier elements; the **ratio** for **stability**

> High atomic number (atomic number 82 and 83)
> Ratio of $\frac{neutron}{proton} = $ about 1.5
> Stable isotope.

for the **heaviest** (stable) **isotopes**, like $^{206}_{82}Pb$ and $^{209}_{83}Bi$, is **1.5**.

RADIOACTIVITY

In radioactivity, the nucleus of an unstable isotope or element decays (breaks down) spontaneously (by itself) and gives off rays and particles, or you can say, gives off radiation (rays and particles).

1. **ALPHA DECAY:** The nucleus (of a radioactive isotope) decays and gives off **alpha particles**. Look at **Table O**, at right or on page Reference Tables 17. The **alpha** particle is a helium **nucleus, $_2^4$He;** its **symbol** is **α**.

Example:

$$_{88}^{226}\text{Ra} \rightarrow \ _{86}^{222}\text{Rn} + \ _2^4\text{He}$$
$$\qquad\qquad\quad new \qquad\quad alpha$$

Table O

Symbols Used in Nuclear Chemistry		
Name	Notation	Symbol
alpha particle	$_2^4$He or $_2^4$α	α
beta particle	$_{-1}^{0}$e or $_{-1}^{0}$β	β⁻
gamma radiation	$_0^0$γ	γ
positron	$_{+1}^{0}$e	β⁺

The nucleus of radium decays and gives off (emits) alpha particles. The atomic number on the left side of the arrow, 88, is equal to all the atomic numbers on the right side of the arrow: 86 and 2. The mass number, 226, on the left side of the arrow is equal to all the mass numbers on the right side of the arrow: 222 and 4. (As you can see, the new element produced has atomic number 86 and mass number 222. Look at the Periodic Table. The symbol of the element with atomic number 86 is Rn.)

2. **BETA DECAY.** The nucleus decays and gives off (emits) a **beta particle**. Look at **Table O**. The beta particle is a (high speed) electron, $_{-1}^{0}$e; its **symbol** is **β⁻**. An example of beta decay is:

$$_{90}^{234}\text{Th} \rightarrow \ _{91}^{234}\text{Pa} + \ _{-1}^{0}\text{e}.$$

The atomic number on the left side of the arrow, 90, is equal to all the atomic numbers on the right side of the arrow: 91 and -1 = 90.

The mass number on the left side of the arrow, 234, is equal to all the mass numbers on the right side of the arrow, 234 and 0. As you can see, the new element produced has atomic number 91 and mass number 234. Look at the Periodic Table. The symbol of the element with atomic number 91 is Pa.

3. **POSITRON DECAY.** The nucleus decays and gives off a **positron.** Look at **Table O**. The positron, $_{+1}^{0}$e, has a mass of about zero like an electron and a positive charge, +1; its **symbol** is **β⁺**.

By the way, TRANSMUTATION is when the nucleus of an atom decays and one element changes into another element. In our example of alpha decay, $_{88}$Ra is changed into a new element, $_{86}$Rn.

In beta decay, $_{90}$Th is changed into a new element, $_{91}$Pa.

4. **GAMMA RAYS:** The nucleus decays and gives off **gamma rays.** (Gamma rays are high energy rays, similar to high-energy x-rays.) They have no charge and no mass.

Look at Table O on the previous page . The box next to the word gamma shows $_{0}^{0}\gamma$ (zero mass, zero charge). Gamma has no mass and no charge.

Each **radioactive isotope** has a specific **decay mode**, which means each radioactive isotope decays and gives off either **alpha, beta, gamma, or positron.** Example: **Alpha decay mode** means the radioactive isotope decays and **gives off alpha particles;** beta decay mode means the isotope gives off beta particles, etc.

Radioactive Particles

	Alpha	Beta	Gamma	Positron
Mass	4 u	1/1837 u	0	1/1837 u
Charge	+2	-1	0	+1
Penetrating power	low (lowest, least) (stopped by paper) 1	medium (stopped by aluminum) 100	very high (highest) (stopped by lead) 10000	medium 100
Ionizing power (how easily it removes electrons from atoms, forming ions)	very high (highest, greatest) ionizes gases	ionizes gases (much less than alpha)	very low (lowest) ionizes atoms in solids- damages cells	ionizes gases
When a positron hits an electron, they annihilate each other, forming two gamma rays (high penetration and high ionizing power).				

Alpha particle: **lowest penetrating** power, **highest ionizing** power.

Gamma: highest penetrating power, lowest ionizing power.

Beta, positron: medium penetrating power and ionizing power.

Question: What is the decay mode (does it give off alpha, beta, or gamma), and complete the equation:

$$_{90}^{234}\text{Th} \longrightarrow {}_{91}^{234}\text{Pa} + X$$

Solution: In $_{90}^{234}$Th, 90 is the atomic number and 234 is the mass number. The **atomic number** on the **left side** of the arrow must equal all the **atomic numbers** on the **right side** of the arrow:

$$_{90}\text{Th} \rightarrow {}_{91}\text{Pa} + X$$
left side = right side

Put -1 at the bottom of the X, $_{-1}X$.

Similarly, look at the mass numbers. The **mass number** on the **left side** of the arrow must **equal all** the **mass numbers** on the **right side** of the arrow:
$$^{234}\text{Th} \rightarrow {}^{234}\text{Pa} + X.$$
left side = right side

Put 0 at the top of X.

The answer is $_{-1}^{0}x$. Look at Table O. It shows that an electron ("e" means electron) has the atomic number of -1 and the mass number of 0. The correct answer is $_{-1}^{0}e$. Put this answer in the equation instead of X.

$$_{90}^{234}\text{Th} \longrightarrow {}_{91}^{234}\text{Pa} + {}_{-1}^{0}e$$

Look again at the table. Next to $_{-1}^{0}e$ is written beta. When a **beta** particle or electron is **given off**, the **decay mode** is **beta**.

Table O
Symbols Used in Nuclear Chemistry

Name	Notation	Symbol
alpha	$_{2}^{4}He$ or $_{2}^{4}\alpha$	α
beta particle	$_{-1}^{0}e$ or $_{-1}^{0}\beta$	β⁻
gamma radiation	γ	γ
positron	$_{+1}^{0}e$ or $_{+1}^{0}\beta$	β⁺

SEPARATING ALPHA, BETA, AND GAMMA PARTICLES

How can you separate radioactive emissions: alpha particles, beta particles and gamma rays? You can separate alpha, beta and gamma by using an electric or magnetic field. In an electric field, an *alpha* particle, which is POSITIVE ($_{2}^{4}He$ has 2 protons, +), IS DEFLECTED (BENDS) TOWARD THE NEGATIVE ELECTRODE. (**Positive attracts negative: Opposites attract.**) A *beta* particle (high speed electron), which is NEGATIVE ($_{-1}^{0}e$), is DEFLECTED TOWARD THE POSITIVE ELECTRODE; and gamma, with NO CHARGE, is NOT DEFLECTED. It doesn't bend and is not affected by the electric field. The diagram below shows how alpha, beta and gamma are separated by using an electric field (one end positive and one end negative).

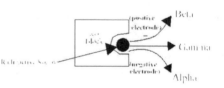

Separating Alpha, Beta and Gamma Rays

Try Sample Question #1-3, page 16, then do Homework Questions #1-9, page 19, and #35-36, page 22.

HALF-LIFE
We learned that unstable nuclei decay spontaneously and give off rays and particles. How fast does the unstable nucleus decay and give off

alpha, beta, positron, or gamma? Each radioactive isotope has its own rate of decay (half-life). **Half-life** is the time it takes **for half** of the sample to decay. Drawing a half-life graph is shown at the end of the chapter.

Look at **Table N**, below or on page Reference Tables 16. Table N lists radioisotopes (isotopes that are radioactive), such as ^{16}N and ^{226}Ra, which spontaneously decay and give off rays and particles. It lists the radioactive **nuclide (radioisotope)**, half-life, decay mode (does it give off alpha, beta, or positron) and name of the nuclide (isotope). The names for β^-, α, and β^+ (in the decay mode column below) are given in Table O. Look at the nuclide ^{16}N and look at its half-life. The **half-life is 7.2 seconds**. In 7.2 seconds, half of ^{16}N will decay. This is a **very fast** reaction.

Selected Radioisotopes			
Nuclide	Half-life	Decay Mode	Nuclide Name
^{16}N	7.13 s	β^-	nitrogen-16
^{226}Ra	1599 y	α	radium-226
^{198}Au	2.695 d	β^-	gold-198
^{37}K	1.23 s	β^+	potassium-37

How fast does radium decay? You learned that $^{226}_{88}Ra$ decays and gives off alpha particles (alpha decay). Look at Table N, above, at ^{226}Ra and its half-life.

The **half-life** of ^{226}Ra is **1,600 years**. In 1,600 years, half of ^{226}Ra will decay. This is a **very slow** reaction.

Meaning of Half-Life: Look at Table N, above. Look at ^{198}Au. It takes 2.69 days for half of the sample to decay. In 2.69 days, half decays or is no longer there. If you have 10 grams of Au, in 2.69 days, ½ x 10 grams, 5 grams will decay and you only have 5 grams left.

In another 2.69 days, half again decays. ½ x 5 = 2½ grams more have decayed, and you only have 2½ grams left.

Similarly, in another 2.69 days, half decays. ½ x 2½ = 1¼ grams more decay, and you only have 1¼ or 1.25 grams left (that **remains**).

In **8.07 days**, you have gone from **10 grams to 1.25 grams**. You see it took **3 half-lives** to go from 10 grams to 1.25 grams.

Amount remaining (left) after 8 days (see above) = 1.25 grams

$$\text{Fraction Remaining (left)} = \frac{\text{Amount Remaining (left)}}{\text{Original Amount}} = \frac{1.25 \text{ grams}}{10 \text{ grams}} = \frac{1}{8} \text{ left}$$

FRACTION REMAINING

Question: $^{131}_{53}I$ has a half-life of 8.07 days. A 10 gram sample was allowed to decay for 32 days. What fraction will remain?

Solution: Find fraction remaining (remains).

Use the formula: Fraction remaining = $(\frac{1}{2})^n$ n = **number of half-lives.**
In this formula, you have two unknowns, fraction remaining and n, therefore first find n (number of half-lives) and then substitute for n in the formula: Fraction remaining = $(\frac{1}{2})^n$

First find n (number of half-lives) $= \dfrac{total\ time}{half\text{-}life\ (means\ 1\ half\text{-}life)}$

$$= \frac{32}{8} = 4$$

or you can find n (the number of half-lives) because you know one half-life takes 8 days, 2 half-lives take about 16 days, 3 half-lives about 24 days, and 4 half-lives about 32 days. In 32 days, the number of half-lives is 4.

Then substitute 4 for n (number of half-lives) in the formula fraction remaining = $(\frac{1}{2})^n$:

Fraction remaining $= (\frac{1}{2})^n = (\frac{1}{2})^4 = (\frac{1}{2}) \times (\frac{1}{2}) \times (\frac{1}{2}) \times (\frac{1}{2}) = \dfrac{1}{16}$

HALF-LIFE

Question: 100 grams of a radioisotope decayed to 12½ grams after 90.7 years. What was the half-life?

Solution: Half-life means one half-life.

Step 1: Find how many half-lives it took to go from 100 grams to 12½ grams. Total time equals 90.7 years (given).

	half-life		half-life		half-life	
100 g	⟶	50 g	⟶	25 g	⟶	12½ g

Three half-lives passed.

Step 2: To **find half-life** take **total time and divide it by number of half-lives** (which you found in Step 1) $= \dfrac{90.7\ years}{3\ half\text{-}lives} = $ **30.2 years.**
Half-life $= 30.2$ years.

What you just did in Step 2 to find half-life can also be made into a formula.

Half-life (one half-life) $= \dfrac{total\ time}{number\ of\ half\text{-}lives}$

Half-life $= \dfrac{90.7}{3} = 30.2$ Half-life $= 30.2$ years

Do sample question #4, page 11:16

TIME ELAPSED (AGE OF A SAMPLE OR FOSSIL)

If 1/8 of a sample of ^{226}Ra remains, how much time elapsed? As you can see, it takes three half-lives to have 1/8 left:

$$1 \longrightarrow \tfrac{1}{2}, \quad \tfrac{1}{2} \longrightarrow 1/4, \quad 1/4 \longrightarrow 1/8.$$

One half-life of ^{226}Ra is 1600 years (see Table N), therefore three half-lives is 3 x 1600 = 4800 years.

INITIAL AMOUNT (ORIGINAL AMOUNT)

Question: A radioisotope has a half-life of 10 days. 1 gram remains after 40 days. What was the initial amount (original amount) of the radioisotope?

Solution: The question asks for initial amount (original amount) of the radioisotope.

Use the formula: Fraction remaining = $(\frac{1}{2})^n$ **n = number of half-lives.**

In this formula, you have two unknowns, fraction remaining and n, therefore, first find n (number of half-lives) and then substitute for n in the formula Fraction remaining = $(\frac{1}{2})^n$

Step 1: Find fraction remaining.

A. Find n (number of half-lives). You can use the formula

$$\text{number of half-lives} = \frac{total\ time}{half\text{-}life\ (one\ half\text{-}life)}$$

$$\text{number of half-lives} = \frac{40}{10} = 4 \text{ half-lives}$$

or you can find n (the number of half-lives) because you know one half-life = 10 days, 2 half-lives = 20 days, 4 half-lives = 40 days

In 40 days, number of half-lives = 4.

B. Then substitute 4 for n (number of half-lives) in the formula **fraction remaining** = $(\frac{1}{2})^n$

$$\text{Fraction remaining} = (\tfrac{1}{2})^n = (\tfrac{1}{2})^4 = (\tfrac{1}{2}) \times (\tfrac{1}{2}) \times (\tfrac{1}{2}) \times (\tfrac{1}{2}) = \tfrac{1}{16}$$

Step 2: Use the formula below to find **original amount** (or amount remaining). See formula below.

$$\text{Formula: fraction remains} = \frac{amount\ remains}{original\ amount}$$

$$\frac{1}{16} = \frac{1\ gram}{X}$$

Cross multiply

X (initial or original amount) = 16 grams.

• To find amount remains (remaining) in another example, use the same formula as in Step 2 right above, but write X for amount remains.

TRANSMUTATION

In **radioactivity**, the nucleus of an unstable isotope or element decays (breaks down) spontaneously (by itself) and gives off rays and particles.

Artificial Radioactivity: Elements can be made radioactive by bombarding their nuclei with high energy particles.

You learned before, in **(natural) transmutation**, the nucleus **spontaneously decays** and changes into a **new element.** You can easily **recognize** natural transmutation. There is only one element (example Ra) before the arrow, which changes into a new element (Rn), after the arrow.

Examples: $\underset{88}{\overset{226}{}}Ra \rightarrow \underset{\underset{new}{86}}{\overset{222}{}}Rn + \underset{\underset{alpha}{2}}{\overset{4}{}}He$ $\underset{90}{\overset{234}{}}Th \rightarrow \underset{\underset{new}{91}}{\overset{234}{}}Pa + \underset{\underset{beta}{-1}}{\overset{0}{}}e.$

In **artificial transmutation**, the nucleus first is **bombarded** with **high energy** particles, then decays and changes into a **new element.** ARTIFICIAL TRANSMUTATION is **transforming** one **element into another** element by **bombarding** the nucleus with high energy **particles**, such as protons, $\underset{1}{\overset{1}{}}H$, neutrons, $\underset{0}{\overset{1}{}}n$, and alpha particles, $\underset{2}{\overset{4}{}}He$. These particles are listed in Table O, below, or on page Reference Tables 17.

For **example**, an $\underset{13}{\overset{27}{}}Al$ nucleus is bombarded by a $\underset{2}{\overset{4}{}}He$ (alpha particle):

$$\underset{13}{\overset{27}{}}Al + \underset{2}{\overset{4}{}}He \rightarrow \underset{15}{\overset{30}{}}P + \underset{0}{\overset{1}{}}n$$

and aluminum is transformed or **changed into** $\underset{15}{\overset{30}{}}P$.

Table O

Symbols Used in Nuclear Chemistry		
Name	Notation	Symbol
alpha particle	$\underset{2}{\overset{4}{}}He$ or $\underset{2}{\overset{4}{}}\alpha$	α
neutron	$\underset{0}{\overset{1}{}}n$	γ
proton	$\underset{1}{\overset{1}{}}H$ or $\underset{1}{\overset{1}{}}p$	p

You can **recognize** artificial transmutation. You have on the left side, **before** the **arrow**, an **atom** and a **high energy particle** (for example: $\underset{2}{\overset{4}{}}He$, **alpha**; $\underset{1}{\overset{1}{}}H$, proton), combining to make a **new element**, on the right side, after the arrow.

TABLE O

	high energy	new		
atom	particle	element		

$$^{27}_{13}\text{Al} + ^{4}_{2}\text{He} \rightarrow ^{30}_{15}\text{P} + ^{1}_{0}\text{n}$$

Remember, you learned that the **atomic numbers** (bottom numbers, next to the symbols), on the **left side** of the equation (before the arrow), **equal** the **atomic numbers** on the **right** side of the equation (after the arrow):

Left side: atomic numbers 13 + 2 = Right side: atomic numbers 15 + 0

The **mass numbers** (top numbers next to the symbols) on the **left side** of the equation must **equal** the **mass numbers** on the **right** side of the equation.

Left side: mass numbers 27 + 4 = Right side: mass numbers 30 + 1.

Another **example** of artificial transmutation is:

$$^{9}_{4}\text{Be} + ^{1}_{1}\text{H} \rightarrow ^{6}_{3}\text{Li} + ^{4}_{2}\text{He}.$$

Be is bombarded by a $^{1}_{1}\text{H}$ (proton) and is transformed or changed into $^{6}_{3}\text{Li}$. The **proton** (bombarding the nucleus) and the beryllium **nucleus** both have **positive** charge and therefore **repel** each other. The **accelerator** (made up of electric and magnetic fields) gives the proton enough kinetic **energy** or speed (to **overcome** the **repulsion** and) to penetrate the beryllium nucleus. In general, any particle with a positive charge (such as $^{4}_{2}\text{He}$, alpha; $^{1}_{1}\text{H}$, proton) that bombards a nucleus (also a positive charge — the proton in the nucleus is positive) must be accelerated to overcome the repulsion.

ACCELERATORS give **charged particles** enough kinetic **energy** to penetrate the nucleus. **Neutrons** have no charge and **cannot be accelerated**.

Question: In the reaction $^{9}_{4}\text{Be} + X \rightarrow ^{12}_{6}\text{C} + ^{1}_{0}n$, the X represents
(1) an alpha particle   (3) an electron
(2) a beta particle   (4) a proton

Solution: Answer *1*. You learned that the atomic numbers on the left side of the equation must equal the atomic numbers on the right side of the equation. $^{9}_{4}\text{Be} + X \rightarrow ^{12}_{6}\text{C} + ^{1}_{0}n$

Atomic numbers on the left side = atomic numbers on the right side

$$4 + ? = 6 + 0.$$
$$? = 2.$$

Therefore, the **atomic number of X = 2**.

You also learned that the mass numbers on the left side of the equation must equal the mass numbers on the right side of the equation.

Mass numbers on the left side = mass numbers on the right side

---

$$9 + ? = 12 + 1$$
$$9 + ? = 13$$
$$? = 4$$

Therefore, the **mass number of X = 4**.

Look at Table O, at right. The particle with atomic number 2 and mass number 4 is $_2^4\text{He}$ (alpha particle).

Note: You see in Table O above, the mass of α(alpha $_2^4\text{He}$) is 4.

Table O

Symbols Used in Nuclear Chemistry		
Name	**Notation**	**Symbol**
alpha particle	$_2^4\text{He}$ or $_2^4\alpha$	α
neutron	$_0^1\text{n}$	γ
proton	$_1^1\text{H}$ or $_1^1\text{p}$	p

Question: Which nuclear equation represents artificial transmutation?

(1) $_{92}^{238}\text{U} \rightarrow _{90}^{234}\text{Th} + _2^4\text{He}$ (3) $_{88}^{226}\text{Ra} \rightarrow _2^4\text{He} + _{86}^{222}\text{Rn}$

(2) $_{13}^{27}\text{Al} + _2^4\text{He} \rightarrow _{15}^{30}\text{P} + _0^1\text{n}$ (4) $_6^{14}\text{C} \rightarrow _7^{14}\text{N} + _{-1}^0\text{e}$

Solution: Answer 2. Artificial transmutation is when the nucleus of an atom (Al, aluminum) is bombarded with a high energy particle ($_2^4\text{He}$) and transformed or changed into another element (P, phosphorus). When a question asks for artificial transmutation, eliminate any choice with only one element (and nothing else) before the arrow; therefore, choices 1, 3 and 4 are wrong. You know that artificial transmutation must have an atom and a high energy particle before the arrow. Choice 2 is correct.

Question: Given the nuclear reaction: $_1^1H + X \rightarrow _3^6Li + _2^4He$
The particle represented by X is

(1) $_4^9\text{Li}$ (2) $_4^9\text{Be}$ (3) $_5^{10}\text{Be}$ (4) $_6^{10}\text{C}$

Solution: You learned that the atomic numbers on the left side of the equation must equal the atomic numbers on the right side of the equation. $_1^1H + X \rightarrow _3^6Li + _2^4He$

Atomic numbers on the left side = atomic numbers on the right side
$$1 + ? = 3 + 2.$$
$$? = 4.$$

Therefore, the **atomic number of X = 4**.

You also learned that the mass numbers on the left side of the equation must equal the mass numbers on the right side of the equation. Mass numbers on the left side = mass numbers on the right side

$$1 + ? = 6 + 4$$
$$1 + ? = 10$$
$$? = 9$$

Therefore, the **mass number of X = 9.**

Now you have ^9_4X. X has atomic number 4. Look at the Periodic Table Atomic number 4 is Be, therefore, the particle represented by X is $^9_4$Be. Answer Choice 2.

Note: $^{226}_{288}$Ra can also be written as ^{226}Ra, Radium-226, or Ra-226. Mass number is written after the dash: Ra-226.

> **Try Sample Question #5, page 16, then do**
> **Homework Questions, #12-18, page 20, and #43, page 22.**

NUCLEAR ENERGY

NUCLEAR ENERGY: In nuclear reactions, **mass is converted into energy**. Nuclear reactions produce tremendous amounts of energy. There are two types of nuclear reactions: **fission** and **fusion**.

FISSION REACTION

In **FISSION**, one atom **splits** into two or more pieces, and **gives off** a **tremendous amount of energy**.

HINT: "**Fission**" has the letter "**i**" in it, and "**splits**" has the letter "**i**" in it. In fission, the **nucleus splits** into pieces.

In a **fission reaction**, one atom absorbs a neutron, **splits into pieces** and **gives off energy**.

When $^{235}_{92}$U (**uranium**) is **bombarded** with a $^1_0$n (**neutron**), the neutron is captured by the uranium nucleus,

$$^{235}_{92}U + ^1_0n \rightarrow [^{236}_{92}U] \rightarrow ^{141}_{56}Ba + ^{92}_{36}Kr + 3\ ^1_0n + \text{energy,}$$

Uranium Neutron   Barium Krypton Neutrons Energy

producing an unstable intermediate nucleus. This **nucleus splits** (**breaks down**) into **several pieces**, $_{56}$Ba (barium), $_{36}$Kr (krypton), and three $^1_0$n (neutrons), and **tremendous energy**. (Some mass is changed into energy.)

The three neutrons that are given off bombard other uranium nuclei, causing them to split into pieces and produce nine more neutrons; these neutrons bombard other uranium nuclei causing *them* to split and form 27 more neutrons. If fission continues, it is an uncontrolled chain reaction. The **atomic bomb** is an **uncontrolled chain reaction**.

In a fission reaction, the sum of the masses of the pieces formed (after the arrow) is less than the mass of the original heavier piece, because some mass is converted into energy. $E = mc^2$; energy = mass x (speed of light)^2.

Note: A fission chain reaction (which produces tremendous amounts of energy) can be controlled in a nuclear reactor. In some power plants, the energy from fission reactions is used to make electricity.

NUCLEAR REACTOR

In a **nuclear reactor, the (fission) chain reaction can be controlled**. A tremendous amount of energy is produced, which can be used for electric power, etc.

Parts of the Nuclear Reactor

1. FUEL: Uranium, U^{235}, and plutonium, Pu^{239}, are fissionable fuels.

2. MODERATORS: Moderators **slow down the speed of the neutrons** (so that the uranium nuclei can absorb them). Examples of moderators are **H_2O, heavy water, beryllium and graphite**.

3. CONTROL RODS: Control rods are made of **boron** and **cadmium** (can absorb neutrons) and therefore **control the number of neutrons** in the reactor.

4. COOLANTS keep the **temperature** at a reasonable level during fission reactions. **Water, heavy water, molten sodium and molten lithium** are examples of coolants.

5. SHIELDING: The shield **protects** the reactor and people **from radiation. Concrete and steel** are used for a shield.

Question: To make nuclear fission more efficient, which device is used in a nuclear reactor to slow the speed of neutrons?

   (1) internal shield   (3) control rod
   (2) external shield   (4) moderator

Solution: Answer **4**. You learned that moderators slow down the speed of neutrons so they can be absorbed by the nucleus of uranium or plutonium.

FUSION REACTION

In **FUSION**, two (light) nuclei **unite** to form a heavier nucleus. An example of a **fusion reaction** is:

$$^2_1H + ^2_1H \rightarrow ^4_2He.$$
Hydrogen Hydrogen Helium

HINT: "**Fusion**" has the letter "**u**"; "**unite**" begins with the letter "**u**." In fusion, **nuclei unite** to form one big nucleus.

Fusion of hydrogen atoms to form helium takes place in the sun and the hydrogen bomb, producing tremendous energy. **High temperature** is necessary to give the nuclei (positively charged) enough kinetic energy to overcome the repulsion and combine or **u**nite to form a heavier nucleus.

High pressure, pushing the atoms together, is also necessary in a fusion reaction. In a fusion reaction, such as $_1^2\text{H} + {}_1^2\text{H} \longrightarrow {}_2^4\text{He} +$ energy (kJ or EV), the mass of the heavier nucleus $({}_2^4\text{He})$ formed is less than the sum of the mass of the lighter nuclei $({}_1^2\text{H} + {}_1^2\text{H})$, because some mass is converted into energy. ($E = mc^2$: energy $=$ mass x speed of light^2.) The energy that is released in a fusion reaction is much more than in a fission reaction.

Question: In a fusion reaction, reacting nuclei must collide. Collisions between two nuclei are difficult to achieve because the nuclei are
  (1) both negatively charged and repel each other
  (2) both positively charged and repel each other
  (3) oppositely charged and attract each other
  (4) oppositely charged and repel each other.

Solution: Answer 2. A nucleus is made up of protons (**positively charged**) and neutrons (no charge), therefore the nucleus is positively charged. Fusion reactions are difficult because both nuclei are positively charged and repel each other.

COMPARISON OF NUCLEAR (FISSION AND FUSION) AND CHEMICAL REACTIONS

The energy released (given off) in nuclear reactions is much more than the energy released (given off) in chemical reactions. The hydrogen bomb (fusion reaction) and atomic bomb (fission reaction) are examples of nuclear reactions.

	Fission	Fusion
Similar	Reactions produce energy. Mass is converted into energy: $E = mc^2$	Same as fission
Different	nucleus splits	nuclei unite– gives much more energy than fission

BENEFITS AND RISKS OF FISSION AND FUSION

Benefits: **Fission** and **fusion** reactions **provide energy**. Fission reactions produce a great deal of energy, which is used for electric power. Fusion reactions take place in the sun and produce tremendous amounts of solar energy. In both fission and fusion, the total mass of the new nuclei formed is less than the mass of the original material. Some mass is converted into energy. A **fusion** reaction releases **much more energy than** a **fission** reaction.

Risks: **Wastes** from fission reactions in **nuclear reactors** are **very radioactive**. They must be **stored** for more than **100 thousand years** without leaking into the environment. Some of these (nuclear) wastes have long half-lives (takes a long time for half to decay). Therefore,

---

more radioactive material remains and it is more dangerous. Another risk is that accidents and fires in reactors can release dangerous levels of radioactivity (giving off rays and particles). Workers and the public can become sick from too much radiation (radioactivity) from fission reactions.

Try Sample Questions #6-10, page 17, then do Homework Questions #19-29, pages 20-21, and #44, page 22.

RADIOACTIVE ISOTOPES (RADIOISOTOPES)

A **radioactive isotope** or radioactive nuclide (nucleus of a radioactive isotope) is radioactive (gives off rays and particles).

BENEFITS OF RADIOACTIVE ISOTOPES

1. **Tracers**: To follow the course of a chemical (organic) or biological reaction (process), **carbon-14** is used as a tracer.

2. **Medical**: **Isotopes** (radioactive), with very short half-lives and which will be quickly eliminated from the body, are used in medical diagnosis and treatment of diseases.

   Radioactive iodine is used to diagnose and treat thyroid disorders:
   Iodine-131 (I-131): treatment of thyroid disorders;
   Iodine-123 (I-123): diagnosis (detection) of thyroid disorders

   Radium– treatment of a type of prostate cancer

   Cobalt-60 (Co-60): sterilizing medical equipment

3. **Food can be stored longer**. Radiation kills bacteria, yeast and molds. It permits food to be stored for a longer time.

4. **Radioactive dating**:

   a. **Geologic dating** is based on half-life. **Uranium-238** occurs naturally in rocks; it decays to **lead-206**. The ratio of **uranium-238** to **lead-206** is used to find the age of the rock or the age of the geological formation (mountain, etc.)

   b. **Dating Living Materials** (organisms that were previously alive): The **ratio of carbon-14 to carbon-12** (C-14 to C-12) can determine the age of a sample of wood, bone, animal skin, or fabric.

Some isotopes are more useful for dating objects than other isotopes. C-14 has a half-life of 5700 years (see Table N), therefore you can date objects up to 57000 years old (which is about 10 times the half-life of C-14. N-16 has a half-life of 7.2 seconds (see Table N). You can date objects up to only 72 seconds old (which is about 10 times the half-life of N-16). Most objects (bones, rocks, etc.) are older than 72 seconds, and therefore N-16 is not a useful isotope for dating objects.

5. **Nuclear Power:** Nuclear reactors are used to produce electrical energy or electricity.

---

6. **Industrial Measurement**: a beam of subatomic particles (alpha, beta, or gamma) is blocked by a certain thickness of metal. Measuring the fraction of the beam that is blocked gives a precise measurement of the thickness of the metal.

RISKS OF RADIOACTIVITY AND RADIOACTIVE ISOTOPES

1. **Biological Damage**: Exposure to radiation can **damage** or destroy the cells of an **organism, person**, etc. (examples: burns, cataracts, cancer). When **reproductive cells** are **damaged**, it is passed on to the **children** (hereditary).

2. **Long Term Storage**: Fission products from nuclear reactors are very radioactive. Gaseous radioactive wastes, such as Rn-222, Kr-85 and N-16 are stored until they are no longer radioactive. Solid and liquid wastes, such as Sr-90 and Cs-137, must be stored for a very long time (more than 100 thousand years). They are encased in special containers for permanent storage underground. It is still not known whether they can be safely stored for so long without them escaping and harming people.

3. **Accidents**: Accidents can cause fuel and wastes to escape from a nuclear reactor. In 1986, a reactor at Chernobyl in the Ukraine was destroyed by an uncontrolled chain reaction and fire; winds spread radioactivity across large parts of Europe. Thousands of fatal human cancers are expected from this exposure.

4. **Pollution**: Traces of radioactive materials are present in air, water, food, and soil either naturally or released by human activities. People can be harmed if there is too much radioactive material.

CONSTRUCTING A GRAPH OF RADIOACTIVE DECAY
(Decay of a Radioisotope)

Problem: Draw a graph from the data below, showing the amount of ^{198}Au (gold) remaining after different amounts of time. The initial amount (original amount) is 10 grams. (The half-life of ^{198}Au is 2.69 days, given on Table N on page Reference Tables 16).

Time, days	Amount remaining, grams
0	10
2.7	5
5.4	2.5
8.1	1.25

How to draw the graph:

1. On the x axis write: time, days. Space the lines along the axis equally; there must be an equal number of days between each two lines.

2. On the y axis write: amount remaining, grams. Space the lines along the axis equally; there must be an equal number of grams between each two lines.

3. Plot the experimental points. Draw a circle around each point.

4. Connect the points with a curved line. Do not continue the line past the last point.

5. Put a title on the graph which shows what the graph is about. Example:

   "Amount of ^{198}Au remaining vs. time."

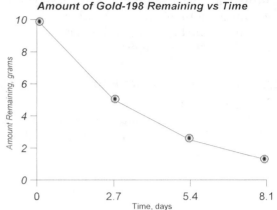

Amount of Gold-198 Remaining vs Time

Amount Remaining, grams

Time, days

Try Sample Questions #11-13, on page 17, then do Homework Questions, #30-34, pages 21-22, and #45-46, page 22.

REMEMBER: When you answer Regents questions in NUCLEAR CHEMISTRY, use Tables N, O and T and the Periodic Table.

SAMPLE REGENTS & REGENT–TYPE QUESTIONS
AND SOLUTIONS

---

1. In the equation $^{234}_{90}Th \rightarrow {}^{234}_{91}Pa + X$, the symbol X represents

   (1) $^{0}_{+1}e$ (2) $^{0}_{-1}e$ (3) $^{1}_{0}n$ (4) $^{1}_{1}H$

2. Which emanation has *no* mass and *no* charge?
   (1) alpha   (2) beta   (3) gamma   (4) neutron

3. Which kind of particle, when passed through an electric field, would be attracted to the negative electrode?
   (1) an alpha particle   (3) a neutron
   (2) a beta particle   (4) an electron

4. If one-eighth of the mass of the original sample of a radioisotope remains unchanged after 4,800 years, the isotope could be
   (1) ^{3}H (2) ^{42}K (3) ^{90}St (4) ^{226}Ra

5. An accelerator can *not* be used to speed up
   (1) alpha particles   (3) protons
   (2) beta particles   (4) neutrons

---

6. The diagram below shows a nuclear reaction in which a neutron is captured by a heavy nucleus.

Neutron   Heavy nucleus   Neutron captured in nucleus   Nuclear deformation   Products

Which type of reaction is illustrated by the diagram?
(1) an endothermic fission reaction
(2) an exothermic fission reaction
(3) an endothermic fusion reaction
(4) an exothermic fusion reaction

7. Which substance has chemical properties similar to those of radioactive ^{235}U?
(1) ^{235}Pa (2) ^{233}Pa (3) ^{233}U (4) ^{206}Pb

8. Fissionable uranium-233, uranium-235 and plutonium-239 are used in a nuclear reactor as
(1) coolants   (2) control rods   (3) moderators   (4) fuels

9. The number of neutrons available in a fission reactor is adjusted by the
(1) moderator   (2) control rods   (3) shielding   (4) coolant

10. Which substance may serve as both a moderator and coolant in some nuclear reactors?
(1) carbon dioxide   (2) boron   (3) graphite   (4) heavy water

11. Diagnostic injections of radioisotopes used in medicine normally have
(1) short half-lives and are quickly eliminated from the body
(2) short half-lives and are slowly eliminated from the body
(3) long half-lives and are quickly eliminated from the body
(4) long half-lives and are slowly eliminated from the body

12. Which radioactive isotope is used in geological dating?
(1) uranium-238   (3) cobalt-60
(2) iodine-131   (4) technetium-99

13. Which isotopic ratio needs to be determined when the age of ancient wooden objects is investigated?
(1) uranium-235 to uranium-238   (3) hydrogen-2 to hydrogen-3
(2) nitrogen-16 to nitrogen-14   (4) carbon-14 to carbon-12

SOLUTIONS

1. Answer 2. In $^{234}_{90}Th$, 90 is the atomic number and 234 is the mass number. The atomic number on the left side of the arrow must equal all the atomic numbers on the right side of the arrow:

$$_{90}Th \rightarrow _{91}Pa + x$$
left side = right side
Put -1 at the bottom of the x.

Table O	
alpha	$^{4}_{2}He$
beta	$^{0}_{-1}e$
gamma	$^{0}_{0}\gamma$
neutron	$^{1}_{0}n$

Similarly, look at the mass numbers. The mass number on the left side of the arrow must equal all the mass numbers on the right side of the arrow:

$$^{234}Th \rightarrow {}^{234}Pa + x.$$
left side = right side

Put 0 at the top of x.

The answer is $_{-1}^{0}x$. Look at Table O. It shows that an electron ("e" means electron) has the atomic number of -1 and the mass number of 0. Answer 2 is the correct answer: $_{-1}^{0}e$

2. Answer *3*. A gamma particle has no mass and no charge. Look at Table O.
The box to the right of the words "gamma radiation" shows $_{0}^{0}\gamma$. Gamma radiation has no mass and no charge.

3. Answer *1*. You learned that, in an electric field, an alpha particle, which is positive ($_{2}^{4}He$ nucleus has two protons, +), is attracted to the negative electrode.
Table O shows that beta, an electron, $_{-1}^{0}e$, has a negative charge, and a neutron, $_{0}^{1}n$, has no charge.

Table O

alpha	$_{2}^{4}He$
beta	$_{-1}^{0}e$
gamma	$_{0}^{0}\gamma$

Therefore, choices 2, 3 and 4 are wrong, because these particles are not attracted to a negative electrode.

4. Answer *4*. After 4,800 years, there was one eighth left. One half is used up in the first half-life; half of what is left is used up in the second, etc. You figured out it takes three half-lives to have ⅛ left. If three half-lives total 4,800 years, then one half-life is 1/3 x 4,800 years = 1,600 years; (or use the formula

# of Half-lives	↓ ½ remaining	Fraction Remaining
1		½
	↓ ¼ remaining	
2		¼
	↓ $^{1}/_{8}$ remaining	
3		$^{1}/_{8}$

$$\text{number of half-lives} = \frac{total\ time\ (t)}{half\text{-}life\ (T)}. \qquad 3 = \frac{4800\ years}{half\text{-}life\ (T)}.$$

Cross multiply: 3 half-lives = 4800 years. (1) half-life = 1600 years. Look at Table N, page Reference Tables 16: Answer 4 is correct; ^{226}Ra has a half-life of 1,600 years.

5. Answer *4*. Accelerators cannot speed up neutrons (no charge). Accelerators give charged particles enough kinetic energy to penetrate the nucleus.

6. Answer *2*. Fission: One big nucleus is broken into pieces. Exothermic reaction: Energy is released in a fission reaction.

7. Answer *3*. $_{92}^{233}U$ and $_{92}^{235}U$ are isotopes of the same element, have the same atomic number, same electron configuration and same number of valence electrons, and therefore have similar chemical properties.

8. Answer *4*. Uranium^{233}, uranium^{235} and plutonium^{239} are nuclear fuels.

9. Answer *2*. Control rods control the number of neutrons in a reactor.

10. Answer *4*. Heavy water is used as a moderator and coolant in nuclear reactors.

11. Answer *1*. Radioisotopes used in medicine generally have short half-lives and are quickly eliminated from the body.

12. Answer *1*. Uranium-238 (U-238) is used in geologic dating.

13. Answer *4*. The ratio of carbon-14 to carbon-12 determines the age of a sample of wood.

Now Let's Try a Few Homework Questions:

1. The ratio of stability (to see if a nucleus is stable) is
   (1) $\dfrac{proton}{neutron}$ (2) $\dfrac{neutron}{proton}$ (3) $\dfrac{proton}{positron}$ (4) $\dfrac{beta}{proton}$

2. Which nuclear reaction is classified as alpha decay?

   (1) $^{14}_{6}C \rightarrow ^{14}_{7}N + ^{0}_{-1}e$ (3) $^{226}_{88}Ra \rightarrow ^{222}_{86}Rn + ^{4}_{2}He$

   (2) $^{42}_{19}K \rightarrow ^{42}_{20}Ca + ^{0}_{-1}e$ (4) $^{3}_{1}H \rightarrow ^{0}_{-1}e + ^{3}_{2}He$

3A. Which radioactive emanations have a charge of 2+?
   (1) alpha particles (2) beta particles (3)gamma rays (4) neutrons

3B. Which group of nuclear emissions is listed in order of increasing charge?
   (1) neutron, positron, alpha particle
   (2) gamma radiation, alpha particle, beta particle
   (3) positron, alpha particle, neutron
   (4) alpha particle, beta particle, gamma radiation

4A. Given the reaction: $^{234}_{91}Pa \rightarrow X + ^{0}_{-1}e$,
   When the equation is correctly balanced, the nucleus represented by X is
   (1) $^{234}_{92}U$ (2) $^{235}_{92}U$ (3) $^{230}_{90}Th$ (4) $^{232}_{90}Th$

4B. In the reaction $^{1}_{0}n + ^{239}_{94}Pu \longrightarrow X + ^{94}_{36}Kr + 2^{1}_{0}n$, the nuclide (isotope) represented by X is
   (1) $^{144}_{58}Ce$ (2) $^{146}_{58}Ce$ (3) $^{144}_{56}Ba$ (4) $^{146}_{56}Ba$

5. Given the reaction: $^{131}_{53}I \rightarrow ^{131}_{54}Xe + X$, Which particle is represented by X?
   (1)alpha (2) beta (3) neutron (4) proton

6. A positron has
   (1) a negative charge & a mass of 0 (3) a positive charge and a mass of 0
   (2) a positive charge and a mass of 1 (4) no charge and a mass of 1

7. Gamma rays are emanations that have
   (1) mass but no charge (3) neither mass nor charge
   (2) charge but no mass (4) both mass and charge

8. In an electric field, which emanation is deflected toward the negative electrode?
   (1) beta particle (2) alpha particle (3) x rays (4) gamma rays

9. The diagram represents radiation passing through an electric field. Which type of emanation is represented by the arrow labeled 2?
   (1) alpha particle (2) beta particle
   (3) positron (4) gamma radiation

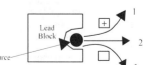

10A.Which equation represents the radioactive decay of $^{226}_{88}Ra$?
   (1) $^{226}_{88}Ra \rightarrow ^{226}_{89}Ac + ^{0}_{-1}e$ (3) $^{226}_{88}Ra \rightarrow ^{226}_{87}Fr + ^{0}_{+1}e$

(2) $^{226}_{88}$Ra $\rightarrow$ $^{225}_{88}$Ra + $^1_0$n   (4) $^{226}_{88}$Ra $\rightarrow$ $^{222}_{86}$Rn + $^4_2$He

10B. Which fraction of an original 20.00-gram sample of nitrogen-16 remains unchanged after 36.0 seconds?
    (1) 1/5   (2) 1/8   (3) 1/16   (4) 1/32

11A. Which radioactive sample would contain the greatest remaining mass of the radioactive isotope after 10 years?
    (1) 2.0 grams of ^{198}Au   (3) 4.0 grams of ^{32}P
    (2) 2.0 grams of ^{42}K   (4) 4.0 grams of ^{60}Co

11B. In how many days will a 12-gram sample of $^{131}_{53}$I decay to 1.5 grams?
    (1) 8.0   (2) 16   (3) 20   (4) 24

12. Bombarding a nucleus with high-energy particles that change it from one element into another is called
    (1) a half-reaction   (3) artificial transmutation
    (2) a breeder reaction   (4) natural transmutation

13. Given the reaction: ^9_4Be + ^1_1H $\rightarrow$ ^4_2He + X   Which species is represented by X?

    (1) $^8_3$Li   (2) $^6_3$Li   (3) $^8_5$B   (4) $^{10}_5$B

14. In the reaction

$$^{27}_{13}Al + ^4_2He \rightarrow X + ^1_0n$$

the isotope represented by X is
    1. $^{29}_{12}$Mg   (2) $^{28}_{13}$Al   (3) $^{27}_{14}$Si   (4) $^{30}_{15}$P

15. The nuclear reaction

$$^4_2He + ^{27}_{13}Al \rightarrow ^{30}_{15}P + ^1_0n$$

is an example of
    1. natural radioactivity   (3) natural transmutation
    2. nuclear fission   (4) artificial transmutation

16. Given the nuclear reaction

$$^9_4Be + X \rightarrow ^6_3Li + ^4_2He.$$

What is the identity of particle X in this equation?
    1. $^1_1$H   (2) $^2_1$H   (3) $^0_{-1}$e   (4) $^1_0$n

17. Particle accelerators can be used to increase the kinetic energy of
    1. deuterium   (2) neutrons   (3) protons   (4) tritium

18a. Particle accelerators are primarily used to
    (1) detect radioactive particles   (3) increase a particle's kinetic energy
    2. identify radioactive particles   (4) increase a particle's potential energy

  b. A particle accelerator has no effect on the velocity of
    (1) an alpha particle   (3) a neutron
    (2) a beta particle   (4) a proton

19. Which equation represents a fusion reaction?
    (1) $H_2O(g) \rightarrow H_2O(\ell)$ (2) $C(s) + O_2(g) \rightarrow CO_2(g)$

    (3) $^2_1H + ^3_1H \rightarrow ^4_2He + ^1_0n$ (4) $^{235}_{92}U + ^1_0n \rightarrow ^{142}_{56}Ba + ^{91}_{36}Kr + 3 ^1_0n$

---

20. Compared to a nuclear reaction, a chemical reaction differs in that the energy produced by a chemical reaction results primarily from
    (1) a conversion of some of the reactant's mass
    (2) a loss of potential energy by the reactants
    (3) the fusion of two nuclei
    (4) the fission of a nucleus

21. Which pair of isotopes can serve as fissionable nuclear fuels?
    (1) U-235 and Pb-208   (3) Pb-208 and Pu-239
    (2) U-235 and Pu-239   (4) Pb-206 and U-235

22. Which substance is sometimes used to slow down the neutrons in a nuclear reactor?
    (1) U-233   (2) Pu-236   (3) sulfur   (4) heavy water

23. Control rods in nuclear reactors are commonly made of boron and cadmium because these two elements have the ability to
    (1) absorb neutrons   (3) decrease the speed of neutrons
    (2) emit neutrons   (4) increase the speed of neutrons

24. The fission process in a reactor can be regulated by adjusting the number of neutrons available. This is done by the use of
    (1) moderators (2) control rods (3) coolants   (4) shielding

25. Which substance is used as a coolant in a nuclear reactor?
    (1) neutrons   (2) plutonium   (3) hydrogen   (4) heavy water

26. The temperature levels in a nuclear reactor are maintained primarily by the use of
    (1) shielding   (2) coolants   (3) moderators   (4) control rods

27. Which conditions are required to form $_{2}^{4}$He during the fusion reaction in the Sun?
    (1) high temperature & low pressure (3) high temperature & high pressure
    (2) low temperature & low pressure   (4) low temperature and high pressure

28. Which equation represents a fusion reaction?

    (1) $_{1}^{2}\text{H} + _{1}^{2}\text{H} \rightarrow _{2}^{4}\text{He}$

    (3) $_{91}^{234}\text{Pa} \rightarrow _{92}^{234}\text{U} + _{-1}^{0}\text{e}$

    (2) $_{18}^{40}\text{Ar} + _{1}^{1}\text{H} \rightarrow _{19}^{40}\text{K} + _{0}^{1}\text{n}$

    (4) $_{88}^{226}\text{Ra} \rightarrow _{86}^{222}\text{Rn} + _{2}^{4}\text{He}$

29. In a fusion reaction, a major problem related to causing the nuclei to fuse into a single nucleus is the
    (1) small mass of the nuclei   (3) attractions of the nuclei
    (2) large mass of the nuclei   (4) repulsions of the nuclei

30. Which radioactive isotope is often used as a tracer to study organic reaction mechanisms?
    (1) carbon-12   (3) uranium-235
    (2) carbon-14   (4) uranium-238

31. Radioisotopes used in medical diagnosis should have
    (1) short half-lives and be quickly eliminated from the body
    (2) short half-lives and be slowly eliminated from the body
    (3) long half-lives and be quickly eliminated from the body
    (4) long half-lives and be slowly eliminated from the body

32. A radioisotope is called a tracer when it is used to
    (1) kill bacteria in food
    (2) kill cancerous tissue
    (3) determine the age of animal skeletal remains
    (4) determine the way in which a chemical reaction occurs

33. Which radioactive isotope is used in geological dating?
    (1) uranium-238   (3) cobalt-60
    (2) iodine-131   (4) technetium-99

34. The radioisotope I-131 is used to
    (1) control nuclear reactions
    (2) determine the age of fossils
    (3) treat thyroid disorders
    (4) trigger fusion reactors.

CONSTRUCTED RESPONSE QUESTIONS: Parts B-2 and C of NYS Regents Exam

35. What ratios of particles make a nucleus stable? How can you tell whether an atom is stable?

36. A. Compare and contrast alpha, beta, gamma, and positron in terms of mass and charge
    B. Write the equation for the alpha decay that occurs in a smoke detector containing Americium-241 (Am-241).
    C. Write the equation for the beta decay of Cobalt-60 (Co-60) used in cancer treatment.
    D. How is the radioactive decay of Krypton-85 different from the radioactive decay of Americium-241?

37. ^{3}H has a half-life of 12.26 years. Find the fraction remaining after 49 years.

38. Find the fraction remaining of K-42 after 25 hours. Hint: See Table N (half-life).

39. 100 grams of a radioisotope decayed to 6¼ grams after 21 years. What is the half-life?

40. A. What is the half-life of a radioisotope if 2½ grams of a 20 gram sample remains unchanged after 3.6 seconds (20 grams decayed to 2½ grams) ?
    B. If 1/8 of an original sample of krypton-74 remains unchanged after 34.5 minutes, what is the half-life of krypton-74?

41. A. One fossil of a wooly mammoth is found to have 1/32 of the amount of C-14 found in a living organism. Determine the total time that has elapsed (age of the fossil or age of the fossilized remains) since this wooly mammoth died.
    B. A radioisotope has a half-life of 2.60 days. 5 grams remain after 10.4 days. What was the initial amount of the radioisotope?

42. A radioisotope has a half-life of 8.5 minutes. After 25.5 minutes, 10 grams remain. What was the initial amount?

43. Contrast natural and artificial transmutation.

44. A. Compare and contrast fission and fusion reactions.
    B. Some ionizing smoke detectors contain the radioisotope americium-241, which undergoes alpha decay and has a half-life of 433 years. State one scientific reason why Am-241 is a more appropriate radioactive source than Fr-220 in an ionizing smoke detector. Hint: Look at the reference table.

45. Explain the benefits and risks of radioactive isotopes.

46. Complete the data table and construct a graph showing the amount of ^{60}Co remaining after different amounts of time. The initial amount is 10 grams and the half-life is 5.25 years, as shown on *Table N*. Remember to label axes, plot and connect the points, and title the graph.

Time, years	Amount remaining, grams
0	10
5.25	
10.5	
15.75	
21	

CHAPTER 12:
SCIENCE PROCESS SKILLS

Standards 1, 2, 6, and 7,
Based on the New York State Regents Core Curriculum.

Standard 1: Analysis, Inquiry and Design
I. **Analysis.** Use representation (models, diagrams, charts) to describe and compare data.

  A. Measure and record experimental data.

   – choose correct scales and units

   – show math formulas and equations; show steps in calculations

   – estimate answers

   – use significant digits: the last digit shows uncertainty (discussed later in the chapter)

   – identify **relationships between variables** from data tables

    (Example: In Chapter 1, you had a data table and graph of Charles' Law which showed that as **one variable** (temperature) **increases**, the **other variable** (volume) also **increases.**

   – calculate percent error. Percent error compares what a student calculated, found, or measured in the lab (measured value) with the accepted value. The formula for percent error is given in Table T.

Table T

Percent Error	$\% \text{ error} = \dfrac{measured\ value\ -\ accepted\ value}{accepted\ value} \times 100$

Question: A student calculated the percent by mass of water in $BaCl_2 \cdot 2H_2O$ to be 16.4%, but the accepted value is 14.8%. What is the student's percent error?

(1) $\dfrac{1.6}{14.8} \times 100$ (2) $\dfrac{1.6}{16.4} \times 100$ (3) $\dfrac{16.4}{14.8} \times 100$ (4) $\dfrac{14.8}{1.6} \times 100$

Solution: The student calculated the percent by mass of water to be 16.4%. What the student found out, calculated, determined or measured is called the measured value. The **measured value** is **16.4%**. The **accepted value** is given as **14.8%**. Use the formula for percentage error given in Table T,

$$\% \text{ error} = \frac{\text{measured value} - \text{accepted value}}{\text{accepted value}} \times 100$$

$$\% \text{ error} = \frac{16.4 - 14.8}{14.8} \times 100 = \frac{1.6}{14.8} \times 100 \qquad \text{Answer 1}$$

You will learn about significant digits and the review of graphs at the end of Standard 1, Topic 1: Analysis, beginning on page 8.

B. Recognize and convert measurements in the metric system.

Prefixes for multiplying units are given in Table C.

Units are given in Table D.

TABLE C AND TABLE D

Table C
Selected Prefixes

Factor	Prefix	Symbol
10^3	kilo-	k
10^{-1}	deci-	d
10^{-2}	centi-	c
10^{-3}	milli-	m
10^{-6}	micro-	μ
10^{-9}	nano-	n
10^{-12}	pico-	p

Table D
Selected Units

Symbol	Name	Quantity
m	meter	length
g	gram	mass
Pa	pascal	pressure
K	kelvin	temperature
mol	mole	amount of substance
J	joule	energy, work, quantity of heat
s	second	time
min	minute	time
h	hour	time
d	day	time
y	year	time
L	liter	volume
ppm	part per million	concentration
M	molarity	solution concentration
u	atomic mass unit	atomic mass

TEMPERATURE

Celsius (°C)   Kelvin (K)

To convert Celsius to Kelvin, or Kelvin to Celsius, use the formula from Table T, K = °C + 273.

Table T

Temperature	K = °C + 273 K = kelvin
	°C = degrees Celsius

LENGTH

Length unit is the meter (m). See Table D.

1000 m = 1 km

See Table C, below, or on the previous page, or on page Reference
Tables 3.

$$1 \textbf{ kilo}\text{meter} = 10^3 \text{m} = 1000\text{m}$$

1 **meter** (m) = **100 centimeter** (cm)

$$1 \text{ cm} = \underline{\quad} \text{ m}$$

See Table C. 1 **centi**meter $= 10^{-2}$ m =
0.01 m

1 **meter** (m) = **1000 millimeter** (mm)

$$1 \text{ mm} = \underline{\quad} \text{ m}.$$

See Table C. 1 **milli**meter $= 10^{-3}$m =
.001 m

Part of Table C

Factor	Prefix	Symbol
10^3	kilo-	k
10^{-2}	centi-	c
10^{-3}	milli-	m

$$1 \textbf{ cm} = \textbf{10 mm}$$

See Table C, above. . 1 **centi**meter $= 10^{-2}$ m = .01 m

See Table C. 1 **milli**meter $= 10^{-3}$ m = .001 m

$$10 \text{ x } .001 \text{ meter } = .01 \text{ meter}$$

$$10 \text{ x } 1 \text{ mm } = \textbf{10 mm} = \textbf{1 cm}$$

Equality

SUMMARY: **1 meter** has **10 decimeters** or **100 cm** or **1000 mm**,
like **1 dollar** has **10 dimes** or **100 cents**.

See Table C on the previous page: 1 **micro**gram $= 10^{-6}$ gram (g),
1 **nano**meter $=10^{-9}$m, 1 **pico**meter $= 10^{-12}$ m.

MASS

Mass unit is the gram (g). See Table D.

$$1000 \text{ g} = 1 \text{ kg or}$$

$$1 \text{ kilogram (kg)} = 1000 \text{ grams (g)}$$

Look at Table C above or on page 2:

$$1 \textbf{ kilo}\text{gram} = \underline{\quad}\text{grams}$$

$$1 \textbf{ kilo}\text{gram} = 10^3 \text{ g} = 1000 \text{ g or}$$

$$1000 \text{ g} = 1 \text{ kg}$$

PRESSURE

Pressure unit is the pascal. See Table D.

Standard pressure (atmospheric pressure) uses the
Kilopascal $= 10^3$ pascal (Table C) or 1000 pascal.

Table A

Name	Value	Unit
Standard Pressure	101.3 kPa 1 atm	kilopascal atmosphere

$$101.3 \text{ kPa} = 1 \text{ atm}$$

TABLE C

TABLES C,D

TABLES D,A

Question: How many millimeters (mm) are in 20 meters (m)?

Solution:

Step 1: Write the **equality** 1 m = 1000 mm.

Step 2: Write the **conversion factors**:

$$\frac{1\ m}{1000\ mm} \quad \text{or} \quad \frac{1000\ mm}{1\ m}$$

(1 m has the same value as 1000 mm.)

1 meter (m) = 1000 millimeters (mm).

Step 3: **Multiply** 20 meters **by** the **conversion factor** $\frac{1000\ mm}{1\ m}$.

If the question has meters in the top number, make sure to use the conversion factor that has meters in the bottom number, so the word "meter" cancels out.

$$\frac{20\ m}{1}\ \text{x}\ \frac{1000\ mm}{1\ m} = 20000\ \text{mm}.$$

Rule:

Step 1: Write **equality** or **equation**.

Step 2: Write **conversion factor**.

Step 3: **Multiply unit** (meters, kilograms, etc.) that you **want** to **change** by the **conversion factor**. If the unit you want to change is on the top, make sure that unit is on the bottom in the conversion factor, so the units cancel out.

Question: How many grams are in 5000 mg?

Solution:

Step 1:   Write **equality** or **equation**.

1 gram(g) = 1000 milligrams (mg)

Step 2:   Write **conversion factors**:

$$\frac{1\ g}{1000\ mg} \quad \text{or} \quad \frac{1000\ mg}{1\ g}$$

Step 3:   **Multiply unit** you **want** to **change** (mg) **by** the **conversion factor**:

$$\frac{5000\ mg}{1}\ \text{x}\ \frac{1\ g}{1000\ mg} = 5\ \text{grams}$$

Do the example given on page reference tables 3.

METRIC CONVERSION-EASY METHOD

Chop a meter (m) into 10 pieces; you have 10 decimeters (dm).
Chop a decimeter into 10 pieces; you have 10 centimeters (cm).
Chop a centimeter into 10 pieces; you have 10 millimeters (mm).

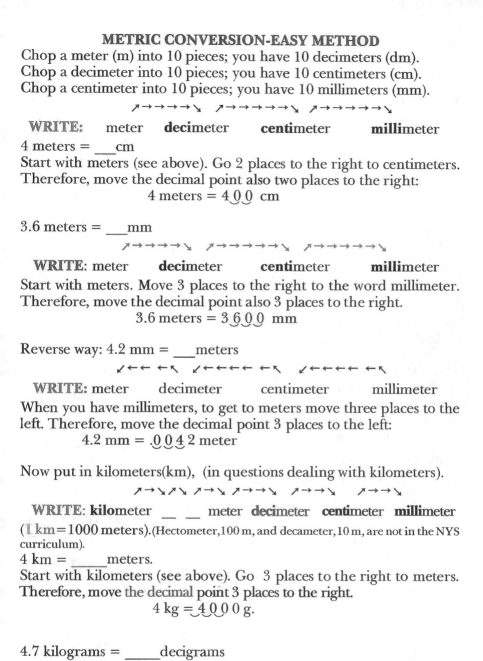

WRITE: meter   **deci**meter   **centi**meter   **milli**meter

4 meters = ___ cm
Start with meters (see above). Go 2 places to the right to centimeters.
Therefore, move the decimal point also two places to the right:

4 meters = 4 0 0 cm

3.6 meters = ___ mm

WRITE: meter   **deci**meter   **centi**meter   **milli**meter
Start with meters. Move 3 places to the right to the word millimeter.
Therefore, move the decimal point also 3 places to the right.

3.6 meters = 3 6 0 0 mm

Reverse way: 4.2 mm = ___ meters

WRITE: meter   decimeter   centimeter   millimeter
When you have millimeters, to get to meters move three places to the
left. Therefore, move the decimal point 3 places to the left:

4.2 mm = .0 0 4 2 meter

Now put in kilometers(km), (in questions dealing with kilometers).

WRITE: **kilo**meter __ __ meter decimeter **centi**meter **milli**meter
(1 km = 1000 meters).(Hectometer,100 m, and decameter, 10 m, are not in the NYS
curriculum).
4 km = ___ meters.
Start with kilometers (see above). Go 3 places to the right to meters.
Therefore, move the decimal point 3 places to the right.

4 kg = 4 0 0 0 g.

4.7 kilograms = ___ decigrams

WRITE: kilogram __ __ gram decigram centigram milligram
(These have the same prefixes as kilometer and decimeter.)
Start with kilogram (example given: 4.7 kg). Move 4 places to the right
to the word decigram. Therefore, move the decimal point also 4 places
to the right.   4.7 kg = 4 7 0 0 0 decigrams.

Reverse way: 6.4 grams = ____ kilograms

WRITE: kilogram __ __ gram decigram centigram milligram
When you have grams, to get to kilograms, move 3 places to the left.
Therefore, move the decimal point 3 places to the left .
6.4 grams = .0 0 6 4 kg.

SUMMARY:

WRITE: kilometer ____ ____ meter decimeter centimeter millimeter
or
WRITE: kilogram ____ ____ gram decigram centigram milligram

Start with what you have. The end of the arrow is the unit you want.
If you started with centimeters and want kilometers, move 5 places to
the left (see above); therefore move the decimal point 5 places to the
left.

Standard 1 (continued):
I. Analysis (continued)
  C. Use knowledge of how **particles** are **arranged** (geometrically) to
     predict properties.
     For example, if the particles are close together in a crystalline
     structure, the object is a solid.
     If the particles are far apart, the object is a gas.

  A. Organize, **graph,** and analyze data
     You learned how to draw graphs in Chapter 1:
     – Identify (choose) independent and dependent variables.
     – Draw the x axis. Space lines equally along the axis.
     – Draw the y axis. Space lines equally along the axis.
     – Label each axis with units (example: Temperature: 40, 50,
       60) and scale (example: °C).
     – Mark points. Draw a circle around each point.
     – Connect points: Lines from point to point or best fit line.
     – Do not extend line past last point.

  E. Use **Deductive** and **Inductive Reasoning** to reach conclusions
     Deductive reasoning is starting with the **general** rule and then
     going to the **specific** cases. **First**, there is a **general rule**.
     Example: In an exothermic reaction, the products have less
     potential energy than the reactants and heat is given off.
     Then, there are **specific cases** of exothermic reactions:
     $$2H_2 + O_2 \longrightarrow 2H_2O$$
     $$2Mg + O_2 \longrightarrow 2MgO$$
     In these specific examples, the products have less potential
     energy than the reactants and heat is given off.

Inductive reasoning is **beginning** with the **specific cases** and then going to the **general rule**. Examples:

First, the specific cases:
Draw a lithium atom, $_3^7 Li$.
Atomic number 3 = 3 protons = 3 electrons.
Mass number (7) - atomic number (3) =
4 neutrons.

$_3 Li$ ⊖))) with markings

Lithium

Draw a beryllium atom, $_4^9 Be$.
Atomic number 4 = 4 protons = 4 electrons.
Mass number (9) - atomic number (4) =
5 neutrons.

$_4 Be$ ⊖)) markings

Beryllium

Then there is a **general rule:**
For all elements,
Atomic number = number of protons = number of electrons
Mass number - atomic number = number of neutrons.

Use deductive reasoning (from generalization to specific) to interpret graphs made from experimental data:
Identify **relationship between variables** in the experiment.
Direct (directly proportional): when **independent variable increases, dependent variable increases** also. Example: Charles' Law: As **temperature** (independent variable) **increases, volume** (dependent variable) **increases.** In a direct relationship, both variables either increase or both variables decrease.
Inverse (inversely proportional): when **independent variable increases, dependent variable decreases.** Example: Boyle's Law: As **pressure** (independent variable) **increases, volume** (dependent variable) **decreases.** In an inverse relationship, one variable increases and the other variable decreases.
Based on the data from the graphs, you can see (interpret) **trends** or **patterns** and can **make predictions** about what will happen next.

F. Critical Thinking Skills
Mathematical equations are used to solve problems in chemistry. Example: Charles Law says that volume is directly proportional to kelvin temperature:

$$\frac{V_1}{V_2} = \frac{T_1 \ (Kelvin)}{T_2 \ (Kelvin)}$$

The equation is correct only if (we make the assumption) that pressure is kept constant.

ROUNDING

If you want to round a number, look at the digit following the digit that you are rounding to.

$$243.87$$

To round to this digit ↑↑ Look at this digit

If the digit you are looking at is **less than 5**, do **not** change the previous digit. If the digit you are looking at is **5 or more**, **increase** the **previous digit by one**. In the example 243.87, look at the 7, the digit following the 8. Since 7 is more than 5, increase the 8 by 1 to make it 9. The answer is 243.9. 243.87 is rounded to the nearest tenth, 243.9.

Question: Round 6.941 (atomic mass of lithium, Li) to the nearest tenth.

Solution: $$6.941$$

To round to this digit ↑ ↑ look at this digit

Since 4 is less than 5, do not change the previous digit 9. Answer 6.9

Now find the **formula mass** of K_2CO_3 by rounding the atomic masses to the **nearest tenth**. You found the formula mass of K_2CO_3 in whole numbers to be 138 (see page **5:9**).

Atomic mass of K = 39.0983 (See Periodic Table.)

$$39.0983$$

To round to this digit ↑ ↑ look at this digit

Since 9 is more than 5, increase the previous digit, the 0 by 1.

Atomic mass of K = 39.1 $K_2 = 39.1 \times 2 = 78.2$

Atomic mass of C = 12.0111 (See Periodic Table.)

$$12.0111$$

Round to this digit ↑ ↑ look at this digit

Since 1 is less than 5, do not change the previous digit 0.

Atomic mass of C = 12.0 $C = 12.0$

Atomic mass of O = 15.9994 (See Periodic Table.)

$$15.9994$$

Round to this digit ↑ ↑ look at this digit

Since 9 is more than 5, increase the previous digit, the 9, by 1.

Atomic mass of O = 16.0 $O_3 = 16.0 \times 3 = \underline{48.0}$

Add all atomic masses together to get the formula mass.

Formula mass to the nearest tenth =   138.2

SIGNIFICANT DIGITS

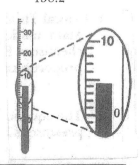

Look at the thermometer and the enlarged part from 0° to 10°. The liquid is a little above the fourth small line. Therefore, the number 4 is the **certain number** (certain digit). But the liquid is a little above the fourth small line, maybe $\frac{1}{5}$ the way up. Therefore, the **estimated number** (estimated digit) or guess is **.2** ($\frac{1}{5}$). The **correct reading** is

4. 2

↑ ↑
certain estimated

The **correct reading** (measured number) has the **cert in digits** and **one estimated digit**. The certain digits and one estimated digit are called **significant digits.**

Do Sample Question #3, page 24.
Using a Graduated Cylinder

FINDING THE NUMBER OF SIGNIFICANT FIGURES (SIGNIFICANT DIGITS)

Rule:

1. Start **counting** from the **first nonzero digit.**
    Examples: **.0025** has **2 significant digits**; start counting from the first nonzero number, which is 2.

    .0007 has **1 significant digit**; start counting from the first nonzero digit, which is 7.

2. **All digits** from **1-9** count as **significant digits.**
    Example: **8794** has **4 significant digits.**

3. **Zeros between** nonzero **(1-9)** digits are **significant.**
    Examples: **30005** has **5 significant digits.** The 3 and 5 are 2 significant digits (Rule 2); the 3 zeros between the nonzero numbers 3 and 5 are also significant, giving 5 significant digits.

    2007 has **4 significant figures.** The 2 and 7 are significant; the 2 zeros between the 2 and 7 (nonzero numbers) are also significant, giving 4 significant digits.

    407 has **3 significant figures.** The 4 and 7 are significant; the zero between the nonzero digits 4 and 7 is also significant, giving 3 significant digits.

4. In numbers with **no decimal point**, **zeros** at the **end** do **not count** as significant digits. Zeros are not significant digits. **No decimal = not count.**
    Examples: **3500** has **2 significant figures.** There is **no decimal** point; **zeros** at the **end** do **not count** (they are not significant figures)

    2470 has **3 significant figures.** It has **no decimal** point; **zeros** at the **end** do **not count**(not significant).

    But, in numbers with a **decimal point, zeros** at the **end count** as **significant figures.**
    Examples: **.3500** has **4 significant figures.** It has a decimal point; zeros at the end count as significant figures.

    .28000 has **5 significant figures.** It has a decimal point; zeros at the end count as significant figures.

    .380 has **3 significant figures.** It has a decimal point; zeros at the end count as significant figures.

Review of Rounding: You learned, when you round a number, look at the digit following the digit that you are rounding to.

$$243.87$$

To round to this digit ↑ ↑ Look at this digit

If the digit you are looking at is **less than 5**, do **not** change the previous digit. If the digit you are looking at is **5 or more, increase the previous digit by one**. Example: 243.87 Since 7 is more than 5, increase the 8 by 1 to make it 9. Answer: 243.9

SIGNIFICANT DIGITS IN MULTIPLICATION AND DIVISION

In **multiplication and division, the answer** must have the **same number** of **significant digits** as the **one** (measurement) **with** the **smallest number of significant figures.**

Question: Multiply 1.25 x 17 and give your answer to the correct number of significant figures.
Solution:

1.25	*3 significant figures*
X 17	*2 significant figures*
875	
125	
21.25	

The answer must have the same number of significant figures as the one (measurement) with the smallest number of significant figures. Therefore, the answer must have 2 significant figures. Round the number to 2 significant figures. The answer is 21. The digit following 21 is less than 5, therefore, leave 21 as your answer.

Question: Multiply 4.82 x 2.5 and give your answer to the correct number of significant digits.
Solution:

4.82	*3 significant figures*
X 2.5	*2 significant figures*
2410	
964	
12.050	

The answer must have the same number of significant figures as the one (measurement) with the smallest number of significant figures. Therefore, the answer must have 2 significant figures. Answer: 12

Question: Divide 12.05 by 2.5. Give your answer to the correct number of significant figures.
Solution: $\dfrac{12.05 \ \textit{(4 significant figures)}}{2.5 \ \textit{(2 significant figures)}} = 4.82$ The answer must have

the same number of significant figures as the number with the smallest number of significant figures = 2 significant figures. 4.8 is the answer.

SIGNIFICANT DIGITS IN ADDITION AND SUBTRACTION

When we **add** or **subtract measurements, round** the **answer** to the **same number** of **decimal places** as in the **one** (measurement) with the **smallest number** of **decimal places**.

Question: Add 2.007, 3.14, and 22.6. Give your answer to the correct number of significant figures.

Solution:

2.007	*3 decimal place*	*(3 numbers after the decimal point)*
3.14	*2 decimal places*	*(2 numbers after the decimal point)*
22.6	*1 decimal place*	*(1 number after the decimal point)*
27.747		

The answer should have the same number of decimal places as the one (measurement) with the least number of decimal places. 22.6 only has 1 decimal place (one number after the decimal point); therefore, the answer must have only one decimal place.
Answer: 27.7

Question: Add 224, 45.6, and 222.52. Give your answer to the correct number of significant figures.

Solution:

224	*No decimal places*	*(No decimal point with numbers after it)*
45.6	*1 decimal place*	*(1 number after the decimal point)*
222.52	*2 decimal places*	*(2 numbers after the decimal point)*
492.12		

The answer should have the same number of decimal places as the number with the least number of decimal places. 224 has no decimal places; therefore, the answer should have no decimal places. 492 is the answer.

Question: Subtract 392.14 − 174.2. Round the solution to the correct number of decimal places.

Solution:

392.14	*2 decimal places*	*(2 numbers after the decimal point)*
-174.2	*1 decimal place*	*(1 number after the decimal point)*
217.94		

The solution should have the same number of decimal places as the number with the smallest number of decimal places. 174.2 has only 1 decimal place (one number after the decimal point); therefore, the answer can only have 1 decimal place. Answer: 217.9

Question: Subtract

$$\begin{array}{r} 392.15 \\ -\ 174.2 \\ \hline 217.95 \end{array}$$

Round the solution to the correct number of decimal places.

Solution: The solution should have only the same number of decimal places as the number with the smallest number of decimal places. 174.2 has only 1 decimal place (one number after the decimal point); therefore, the answer can have only 1 decimal place. (The last digit of 217.95 was 5, therefore it is rounded upward). Answer: 218.0

Try Sample Questions #4-5, page 24, then do Homework Questions #8-15, pages 25-26 and #24, page 27.

REVIEW OF GRAPHS

Let's see how we can draw graphs based on experimental data. Problems 1 and 2 are on Physical Behavior of Matter (Chapter 1).

Problem 1:

A student obtained the following data showing how temperature affects the volume of a gas. Draw a graph to show the data.

Temperature, K	Volume, mL
50	100
100	200
200	400
400	800

How to draw the graph: (See graph below.)

1. On the x axis, put "Temperature, K". The **thing you change** (in this case temperature) is always put on the **x axis**. What you change is called the independent variable. Space the lines along the axis equally; there must be an equal number of degrees between each two lines. (See graph below.)

2. On the Y axis, put "Volume, mL". The **result** you get (volume) because you changed the temperature is always put on the **y axis**. It is called the dependent variable. Space the lines along the axis equally; there must be an equal number of mL between each two lines. (See graph below.)

3. Plot the experimental data on the graph. Draw a circle around each point. Connect the points with a line (or draw a straight line that is the best fit between the points). Do not continue the line past the last point.

4. On the graph, put a title which shows what the graph is about. Example: "Effect of temperature on gas volume".

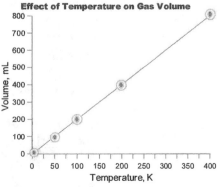

Effect of Temperature on Gas Volume

Problem 2:

A student obtained the following data for the heating curve of water, starting with ice. Draw a graph to plot the data.

Time, Minutes	Temp., °C
0	-20
1	0
3	0
5	100
16	100
17	120

How to draw the graph:

1. On the x axis, put "Time, minutes". **The thing you change** (in this case time) is always put on the **x axis**. This is the independent variable. Space the lines along the axis equally. There must be an equal number of minutes between lines. See graph below.

2. On the y axis, put "Temperature, °C". The **result** you get (temperature) is always put on the **y axis**. This is the dependent variable. Space the lines along the axis equally. There must be an equal number of degrees between lines. See graph.

Effect of Heating Time on Temperature of Water

3. Plot the experimental data on the graph. Draw a circle around each point. Connect the points with a line. Do not continue the line past the last point.

4. On the graph, put a title which shows what the graph is about. Example: "Effect of heating time on temperature of water."

Problem 3: Constructing a Solubility Curve (Chapter 6)

Draw a solubility graph from data given.

How to draw the graph:

1. On the x axis, write Temperature, °Celsius. **Always** include units of measure (in this example, °C) on the axis. The lines must be evenly spaced, and there must be an equal number of degrees between each two lines. See graph on the next page.

Solubility of KNO_3	
Temperature	**Grams Dissolved**
0	12
20	32
40	62
60	105
70	125

2. On the y axis, write Grams of solute/100 grams of water. Always include units of measure (in this example, grams). Again, the lines

---

must be evenly spaced, and there must be an equal number of grams between each two lines. See graph below.

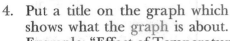

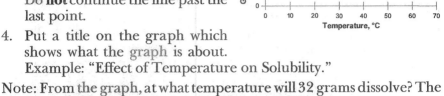

3. Plot the **experimental data for KNO₃** on the graph. Draw a circle around each point. Connect the points with a ine. Do **not** continue the line past the last point.

4. Put a title on the graph which shows what the graph is about.
   Example: "Effect of Temperature on Solubility."

Note: From the graph, at what temperature will 32 grams dissolve? The answer is 20°C. Make sure to include units (°C) in your answer.

Question: What **relationship** is shown in the **data table** and the **graph** of the Solubility Curve of KNO_3?

Solution: *Direct Relationship.* As the temperature increases, the number of grams of KNO_3 that can dissolve increases (solubility increases). You see it is a **direct relationship**: As the temperature **increases** (one variable increases), the amount that can dissolve also **increases** (the other variable increases).

Question: What do you **predict** would happen if you increase the temperature to 80°C?

Solution: You can **predict** that the number of grams of KNO_3 that can dissolve will increase (solubility increases).

Problem 4: Draw a graph of the decay of a radioisotope from the data given (Chapter 11).

Draw a graph from the data below, showing the amount of ^{198}Au (gold) remaining after different amounts of time. The starting amount (initial mass) is 10 grams. The half life of ^{198}Au is 2.69 days, (as shown on Table N on page Reference Tables 16).

Time, days	Amount remaining, grams
0	10
2.7	5
5.4	2.5
8.1	1.25

How to draw the graph:

1. On the x axis write: time, days. Space the lines along the axis equally; there must be an equal number of days between each two lines.

2. On the y axis write: amount remaining, grams. Space the lines along the axis equally; there must be an equal number of grams between each two lines.

3. Plot the experimental points. Put a circle around each point.

4. Connect the points with a curved line.

5. Put a title on the graph which shows what the graph is about. Example:

   "Amount of ^{198}Au remaining vs. time."

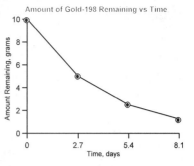

Question: What **relationship** is shown in the data table and the half-life graph of ^{198}Au (gold)?

Solution: *Inverse Relationship*. As the number of days increases, the amount remaining decreases. You see it as an **inverse** relationship: As the number of days **increases** (one variable increases), the amount remaining **decreases** (the other variable decreases).

Question: What do you **predict** will happen if more days pass?

Solution: You can **predict** that the amount remaining will decrease.

Do Homework Questions #16-17, page 26, and #25-27, pages 27-28.

SOME SAFETY MEASURES
When Doing Laboratory Experiments
Not listed in standards

1. Do not start working in the laboratory until your teacher tells you what to do.
2. If the lab equipment is not working, tell your teacher.
3. Wear safety goggles in the lab.
4. Keep sleeves and hair away from bunsen burners.
5. Do not bring food into the lab. Food can get contaminated.
6. When you heat a test tube, point the top away from you.
7. Don't heat a test tube that is stoppered.
8. Add acid to water, not water to acid. Stir while adding the acid.
9. Do not spill the chemicals.
10. Follow the teacher's instructions on how to dispose of chemicals.
11. Don't put electrical equipment near sinks (water).
12. Don't touch hot equipment.
13. Don't use glassware that is chipped or broken. If glassware is broken, tell your teacher.
14. Notify your teacher immediately of any accidents (burns, student gets hurt).
15. Know where the fire extinguisher, fire blanket and eye wash are.
16. Wash your hands before leaving the chemistry lab.
17. If a chemical (example acid or base) spills on the skin, wash with running water for 15 minutes.

Standard 1: Scientific Inquiry

II. Scientific Inquiry (investigating science):

A. Developing explanations (example: theory) to explain what is happening around us (natural phenomena).

   1. Use theories and models to explain observations and make predictions.

   2. Do research at library, have discussions with experts. Get data (information) from published sources to support, explain or defend your position.

   3. Look at competing explanations. Figure out why one theory was accepted and another was not.

B. Scientific inquiry tests explanations (example: theory):

   1. Design and carry out experiments using scientific method.

   2. Research ideas in library, get review of literature and peer feedback.

   3. Use library investigations and literature review to improve the design of experiments.

   4. Write a proposal: Write **hypothesis** to test explanation (example: Charles' Law) and what it would predict.

     Example: The **explanation** you are testing is **Charles' Law**. It states that volume of a gas is directly proportional to temperature; the higher the temperature, the larger the volume.

     Hypotheses are written as:

       If X (Charles' Law) is true, and you do test Y (doubling Kelvin temperature), then prediction Z (doubling volume) will happen.

   5. Carry out the plan for testing explanation. Select and develop techniques, getting and building apparatus, record observations. Determine safety procedures to use in your research plan.

C. After observations are made, do the following to help understand phenomena:

   1. Represent and organize observations (in data tables, diagrams, charts, graphs, and equations).

     a. Interpret organized data:

       • Look for trends or patterns .

       • Interpret the patterns in terms of scientific theories.

   2. Apply **statistical analysis** techniques to test if **chance** alone caused the results.

   3. **Compare predicted** result (Example: Charles' Law predicts that if Kelvin temperature doubles, volume doubles) with

the **actual result** (example: volume from thousands of experiments). Decide if the actual results (results from numerous experiments) show that Charles' Law is right or wrong.

4. Evaluate methods used in your experiment for error which might affect the results.

5. Compare experimental with expected (theory). Calculate percent error.

Table T

Percent Error	$\% \text{ error} = \dfrac{measured\ value\ -\ accepted\ value}{accepted\ value} \times 100$

6. Using results of the test experiment and public discussion, revise explanation and see if more research is needed.

7. Develop a written report for public scrutiny that describes the explanation, literature review, research carried out, results, and suggestions for further research.

SCIENTIFIC METHOD

You can use the scientific method to do **all types of experiments**, including experiments involving science, math, and technology, to help **solve all types of problems** on the community, regional, or global level. Here are two **experiments** using the scientific method:

Experiment 1:

1. **Problem** or **question**: Does a particular brand of detergent remove marker stains?

2. **Hypothesis**: If this detergent is used to wash clothes with a marker stain, then the detergent will remove marker stains. OR This detergent will remove marker stains from clothes.

3. **Materials**: Stained clothing, water, detergent.

4. **Procedure**: Put detergent with water on the stain. As a control, do NOT use the detergent; only put water on the stain. (A **control means** do **not use** the thing you are **testing**, the detergent.) Both stains should be the same type, the same size and in the same material.

5. **Observation:** The one with the detergent removed the stain. The one without the detergent did not remove the stain.

6. **Conclusion**: This particular brand of detergent removed marker stains.

Steps of the Scientific Method:
1. **Problem** or **Question**
2. **Hypothesis** (guess)
3. **Materials**: What materials do you need?
4. **Procedure**: How you will do the experiment?
5. **Observation**: What did you see? You can describe it. You can make a model, such as a diagram, data table or graph, to show what you saw.
6. **Conclusions**: What are the results and conclusions? Does the conclusion agree with the hypothesis (your guess)? If not, change the hypothesis and do the experiments again.

Experiment 2:

Problem: Does Antacid X decrease the acidity of orange juice?

Hypothesis: If Antacid X is added to orange juice, then the acidity should decrease. OR When you add Antacid X to orange juice, the acidity will decrease.

Materials: orange juice   indicator paper
                measuring cup   antacid tablets

Procedure: We placed ⅛ cup of orange juice in a container. We used indicator paper to measure the pH of the orange juice. We added one crushed antacid tablet and used indicator paper again to measure the pH. We then added one more crushed tablet (2 tablets total) and measured the pH. We continued to add one more tablet at a time and measured the pH after each one. As a control, do **not** add the antacid. Put salt (NaCl) (about the same volume as the crushed antacid) into the orange juice. (As you know, the control should **not** have the thing you are testing, the antacid.)

Observations: The pH of the orange juice was 3.7. The more antacid we added, the more the pH increased. (As you know, an increase in pH means less acid.) See the data table to the right. In the control reaction, the salt had no effect on the pH of the orange juice.

# Antacids	pH
0	3.7
1	4.5
2	5.5
3	6.0
4	7.0

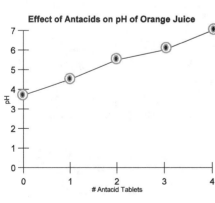

Effect of Antacids on pH of Orange Juice

Conclusion: This antacid decreases the acidity (raises the pH) of orange juice.

Rules for Experiments:
1. Experiments should have a control.
2. Experiments should be able to be repeated many times and give the same results; then the results are valid and reliable.
   Large sample makes the results of the experiment valid.
3. You can calculate percent error to see how your results differ from the accepted value.

Sometimes there is a slight error in the results of an experiment because of heat loss to the environment, or heat absorbed by equipment.

Standard 1.

III. Engineering Design uses modeling to find the **best solution** to a problem, considering **time, money, space, equipment**, etc.

Steps:

A. Study the problem–identify needs and ways to meet the needs.

B. Use information sources. Document how they help the solution.

C. **Generate solutions; improve** each part of the **solution; pick** the **best solution**-document **how it is best** and how **cost, ergonomics** (how people interact with machinery; example: do your arms hurt after using the machinery), and **environmental issues** were considered.

D. Develop a work schedule and plans for efficiency and cost control; build a model of the solution.

E. Develop and do a test of the solution-record results and predict new problems and their solutions.

Standard 2: Information Systems

I. **Use of the Internet**: Use the **Internet** to **get information** on any topic, such as Periodic Table, acid rain, etc. If you cannot locate the topic on the Internet, **ask the school librarian** or the **librarian** in the **public library** for assistance.

II. **Evaluating Information Sources**: Evaluate information on the Internet to **make sure** it is scientifically **correct**. Anyone can post material on the internet, whether it is accurate or not. **Make sure** there is **data** to **support** the **scientists' claims**.

Peer Review: Any good scientific reporting has peer review. **Peer review** means that the work has been **looked at** by **other experts** in the field to make sure the **conclusions** are **correct (valid)**.

Standard 6: Interconnectedness, Common Themes

I. **Systems Thinking**: One part of a system interrelates with another part of a system and combine to perform specific functions. Example: **Dissolving** a **solid** in **water**:

When **NaOH** dissolves in **water**, **heat goes from** the **NaOH** to the **water**, which **causes** the **water** to get **hotter**. **NaOH** and **water interrelate (interact)** together to dissolve.

When NH_4Cl (ammonium chloride) **dissolves** in **water, heat goes from** the **water** (surroundings) **to** the **solid** (system), **causing** the **water to get colder**. The solid and water interrelate together to dissolve.

II. **Models** are a simplified way to show objects, processes, etc.
   A. A model can be:
      1. An **object** (see Chapter 1, ball models of elements, compounds, and mixtures).
      2. A **drawing** (see Chapter 10, structural formulas of organic compounds). The limitation of a drawing is that it is two-dimensional (length and width). It does not show depth (not 3-D.) With a drawing, it is difficult to show shapes of compounds, bond angles, and positioning of atoms.
      3. A **theory** (atomic theory).
      4. A **mathematical model (equation)**. Equations describe things. Equations of the combined gas law show how gases behave when temperature, pressure, and volume are changed.
   B. Models are made from information (about a system). Experimental data about the gas laws was used to develop kinetic molecular theory.
   C. A model (theory) must be revised when there is evidence showing that the model (theory) is wrong. Example: the theory of the atom was revised by many scientists (Rutherford, Bohr, etc.)
   D. Use percent error to test whether a model is correct (valid). Charles' Law states, if Kelvin temperature doubles, volume doubles. The **prediction** from Charles' Law is **volume doubles. Charles' Law** can be called the **measured** value and the **results** of very many **experiments** the **accepted value**.

   You can use the equation in Table T for percent error. You must know what to substitute in the equation to prove the theory. In Table T,

$$\% \ error = \frac{\overset{\text{measured value}}{(\text{volume from Charles' Law theory})} - \overset{\text{accepted value}}{(\text{volume from many experiments})}}{\text{accepted value}} \times 100$$

   As you saw, to find out whether a model (example: Charles' Law) is valid (correct), **compare** the **prediction** from a model (Charles' Law) (if temperature doubles) **volume doubles, with** the **result (volume) of many experiments** and find the percent error.

III. **Scales and Powers of Ten**
   A. Effect of **Change in Scale**
      For chemical experiments, you can use very small amounts (microscale) or large amounts (macroscale) of reactant.
      Many high schools today use **microscale laboratory** experiments. For example, in titration experiments, instead of

using burettes made of glass, which are expensive and fragile, many high schools use **smaller** and **less expensive equipment**, which also provides **easier cleanup** and produces **less waste**.

Similarly, in the **real world**, **microscale** is used for **laboratory research** in industry. However, **production** (not laboratory research) in industry is on a **large scale**, **not microscale**. For example, industry produces **large amounts** of gasoline. Industry is concerned about **cost-benefit**. It is more cost efficient (costs less per gallon or pound) to make a product on a large scale.

B. **Powers of 10** are used to :
1. Show pH. pH indicates the strength of an acid or base (how strong the acid or base is).
If H^+ concentration is 1 x 10^{-4}, pH is 4 (see Chapter 8). The solution is acidic.
If H^+ concentration is 1 x 10^{-8}, pH is 8. The solution is basic.

2. Multiply very large and small numbers. When you multiply exponents with the same base, you add the exponents:
$$X^3 \bullet X^2 = X^{3+2} = X^5.$$
When you divide exponents with the same base, you subtract the exponents: $X^6 / X^2 = X^{6-2} = X^4$.

For example, in a solution, $[H^+]$ x $[OH^-] = 10^{-14}$.([ ] means concentration). If $[H^+] = 10^{-3}$, then $[OH^-]$ is 10^{-11}, because 10^{-3} x $10^{-11} = 10^{-14}$ (add the exponents).

3. Write numbers in **Scientific Notation**: Use scientific notation to write very large or very small numbers.

SCIENTIFIC NOTATION

495.27 = 4.9527 X 10^2		**Example 1:** Change 495.27 into scientific notation. A. **Move** the **decimal point 2 places to the left** so you only have one digit (4 in this example) before the decimal point. In general, move the decimal point so you only have 1 digit (a number from 1 to 9) before the decimal point.
A	B	B. Since you moved the decimal point 2 places to the left (making the number smaller by 10 x 10 = 100), you need to **multiply** the **number** by 10^2 (**positive exponent**) which means 10 x 10 = 100 times larger, to have the same value.
move decimal point 2 places to the left.	multiply by 10^2 (exponent is 2, positive exponent)	

36740 = 3.6740 x 10^4		**Example 2:** Change 36740 into scientific notation. A. If there is no decimal point, put a decimal point at the end of the number. **Move** the **decimal point 4 places to the left,** so you only have one digit, 3, before the decimal point. B. **Multiply** the **number by** 10^4 (**positive exponent**), to have the same value.
A	B	

.0032 =
003.2 x 10⁻³
A   B

Example 3. Change .0032 into scientific notation.
A. **Move** the **decimal point 3 places to the right** so you only have 1 digit, 3 in this example, before the decimal point. If a number **starts** with a decimal point (example .0032, or 0.0032, which is the same), move the decimal point to the right.

B. Since you moved the decimal point 3 places to the right (making the number bigger by 10 x 10 x 10 = 1000), you need to **multiply** the **number by** 10^{-3} (**negative exponent**) which means 10 x 10 x 10 = 1000 times smaller, to have the same value.

IV. Equilibrium

A **system** at **equilibrium** is **stable** (the amount of reactants and products does not change-it remains constant).

A system at equilibrium is stable because of:

- **Static Equilibrium**: Nothing happens; there is no change. Or
- **Dynamic Equilibrium**: Rates of forward and reverse reactions are equal.

Static Equilibrium: A mixture of hydrogen and oxygen does not change. If a small amount of energy is put in (warming it, a small disturbance) **nothing happens**.

But, if a flame or spark is added to the mixture of hydrogen and oxygen, there is now enough energy (activation energy-energy reaches the threshold level) to make the hydrogen and oxygen react to form water. $2H_2 + O_2 \longrightarrow 2H_2O$. (See activation energy, beginning of Chapter 7).

Dynamic Equilibrium: Sometimes, the amounts of reactants and products stay the same because the reaction is going forward and backward at the same rate. (Example: In a saturated solution of sugar in water, some sugar on the bottom dissolves at the same rate as some dissolved sugar goes to the bottom. (See *Equilibrium*, Chapter 7).

But, when something is added to a mixture (change in concentration, pressure, or heat), it may shift the equilibrium. (See *LeChatelier*, Chapter 7). In the reaction

$$3H_2 + N_2 \longrightarrow 2NH_3 + energy,$$

adding H_2 or N_2 shifts equilibrium to the right, producing more NH_3. But, adding heat shifts equilibrium to the left, and would give less NH_3.

Do Homework Questions #18-20, pages 26-27, and #28-35, page 28

V. Patterns of Change

In Chapter 1, we drew graphs for heating curves and Charles' Law. In

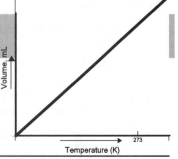

Effect of Temperature on Gas Volume

Volume, mL

Temperature (K)

273

Chapter 6, we drew a solubility curve and in Chapter 11 a half life curve. Here is a graph of Charles' Law, as we did in Chapter 1. **By looking** at the **graph**, you can **predict** that, as **temperature increases, volume increases**.

Chapter 1 has a cooling curve. By looking at and interpreting the graph, you can see that when liquid changes to solid, there was no change in temperature, therefore no change in kinetic energy.

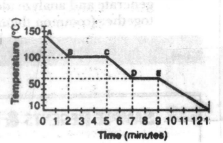

Standard 7: Interdisciplinary Problem Solving

I. **Connections: Use knowledge** from **many subject areas** (disciplines) to solve problems. Skills of **math, science, and technology** are **used together** to make informed decisions and **solve problems** dealing with **science, technology, society, consumer decision making** , designing **experiments** and science inquiry.

A. Analyze science, technology, society problems and issues on a community, national, or global scale and plan and carry out (by communicating it in a letter to the editor) a **remedial course** of action.

B. Analyze and measure consumer product data (information about consumer products), understand their **effect on** the **environment,** judge the **efficacy** of **competing products,** analyze for **cost-benefit** (is the benefit worth the cost) and risk-benefit (is the benefit worth the risks) **trade-offs.** Compare and analyze consumer products, such as antacids and vitamins.

C. Design solutions to **real-world problems** on a **community, national, or global scale.** Use **math** for **measurement, science** to **help get solutions,** and **technology** for **machinery** to apply the solution. Make (design) a potential solution to a problem which affects a large area. Example: Suggest a plan to adjust the acidity of a lake in the Adirondacks. Use a technological design process:

Use science: knowledge of pH.

math: measurements

technology: machinery

D. **Do** a **scientific experiment** following the **scientific method** (science inquiry) that also involves **math, science**, and **technology** (technological tools) and communicate the results.

Example: Design an experiment that uses math to solve a scientific problem. Example: an experiment to compare the density of different brands of soda pop.

E. Students solving a problem using math, science, and technology should work effectively, gather and process information, generate and analyze ideas, use math, science, and technology together (common themes), realize ideas, and present results.

Do Homework Questions #36-37, page 28.

SAMPLE REGENTS & REGENT-TYPE QUESTIONS AND SOLUTIONS

1. A student determined the percentage of water of hydration in $BaCl_2 \bullet 2H_2O$ by using the data in the table at right.

Quantity Measured	Value Obtained
mass of $BaCl_2 \bullet 2H_2O$	3.80 grams
mass of $BaCl_2$	3.20 grams
% of water calculated	15.79%

The accepted percentage value for the water of hydration is 14.75%. What is the student's percent error?
(1) 1.04% (2) 6.00% (3) 6.59% (4) 7.05%

2. How many cm are in 7 meters?

3. What is the reading of the meniscus at A and B? (What is the correct reading of the volume of water in the graduated cylinder at A and B?)

4. Which measurement contains 4 significant figures?
1. 0.0004 (2) 0.040 (3) 4000 (4) 3007

5. What is the product of (2.324 cm x 1.11cm) expressed to the correct number of significant figures?
1. 2.58 cm^2 (2) 2.5780 cm^2 (3) 2.5796 cm^2 (4) 2.57964 cm^2

Solutions

1. Answer *4*. Use the formula for percent error given in Table T:

Table T

Percent Error	% error = $\dfrac{\text{measured value - accepted value}}{\text{accepted value}}$ x 100

Percent error = $\dfrac{15.79 - 14.75}{14.75}$ x 100 = $\dfrac{1.04}{14.75}$ x 100 = 7.05%

2. 1 meter = 100 cm

$$7 \text{ m} \times \frac{100 \text{ cm}}{1 \text{ m}} = 700 \text{ cm}$$

3. A. 30 mL and 40 mL are marked on the graduated cylinder. Each line is 1 mL. Read the bottom of the curve of the water (called the bottom of the meniscus). It is 32 mL. 32 mL is the certain digits–you read them on the graduated cylinder. The water level is half way up between 32 and 33; therefore, the estimated number is .5 (½). The correct reading (with significant digits) is 32.5. The correct reading has the certain digits and one estimated digit.
   B. The certain digits are 37. The bottom of the meniscus (bottom of the curve) is on the line for 37; therefore, the estimated digit is .0. The correct reading is 37.0.

4. Answer *4*. 3007 has 4 significant digits. The 3 and the 7 are two significant digits because all digits from 1 to 9 are significant. The 2 zeros between the nonzero digits 3 and 7 are also significant, giving 4 significant digits.
   Wrong choices:
   (1) 0.0004 has only 1 significant figure. To find the number of significant figures, start counting from the first nonzero digit, which is 4.
   (2) 0.040 has 2 significant figures. Start counting from the first nonzero digit, which is 4. In numbers with a decimal point, zeros at the end count as significant figures. Therefore, 40 are the 2 significant figures.
   (3) 4000 has 1 significant figure. In numbers with **no decimal** point, zeros at the end do not count.

5. In multiplication and division, the answer must have the same number of significant digits as the one (measurement) with the smallest number of significant figures. The answer must have 3 significant figures. Answer *1* has 3 significant figures.

<hr>

Now let's try a few Homework Questions:

<hr>

1. 250 mm is equal to
   (1) 25 meters (2) 25 kilometers (3) 25 cm (4) 2500

2. 1 milliliter equals
   (1) .01 liter (2) .1 liter (3) .001 liter (4) .000001 liter

3. In an experiment, a student found 18.6% by mass of water in a sample of $BaCl_2 \bullet 2H_2O$. The accepted value is 14.8%. What was the student's experimental percent error?

   (1) $\frac{3.8}{18.6} \times 100$ (2) $\frac{3.8}{14.8} \times 100$ (3) $\frac{14.8}{18.6} \times 100$ (4) $\frac{18.6}{14.8} \times 100$

4. a. A student determined the heat of fusion of water to be 367 joules per gram. If the accepted value is 333.6 joules per gram, what is the student's percent error?
   (1) 8.0% (2) 10.% (3) 11% (4) 90.%

   b. A student measures the mass and volume of a piece of aluminum as 25.6 grams and 9.1 cubic centimeters. The student calculates the density of the aluminum. What is the percent error of the student's calculated density?
   (1) 1% (2) 4% (3) 3% (4) 2%
   Hint: Review density on pages 5:12-5:13 and Table S.

5. A student calculated the percent by mass of water in a hydrate to be 37.2%. If the accepted value is 36.0%, the percent error in the student's calculation is equal to

   (1) $\frac{1.2}{37.2} \times 100$ (2) $\frac{37.2}{36.0} \times 100$ (3) $\frac{1.2}{36.0} \times 100$ (4) $\frac{36.0}{37.2} \times 100$

<hr>

6. In a laboratory exercise to determine the volume of a mole of gas at STP, a student determines the volume to be 2.25 liters greater than the accepted value of 22.4 liters. The percentage error in the student's value is
   (1) 2.25%   (2) 10.0%   (3) 20.2%   (4) 24.7%

7. A laboratory experiment was performed to determine the percent by mass of water in a hydrate. The accepted value is 36.0% water. Which observed value has an error of 5.00%?
   (1) 31.0% water (2) 37.8% water (3) 36.0% water (4) 41.0% water

8. A. Which measurement below contains 3 significant figures?
   B. Which measurement below contains 2 significant figures?
   C. Which two measurements below contain one significant figure?
   (1) .08 cm   (2) .080 cm   (3) 8.08 cm   (4) 800 cm

9. Given: (52.6 cm)(1.214 cm). What is the product expressed to the correct number of significant figures?
   (1) 64 cm^2 (2) 63.8564 cm^2 (3) 63.86 cm^2 (4) 63.9 cm^2

10. What is the product of (2.324 cm x 1.11 cm) expressed to the correct number of significant figures?
    (1) 2.5780 cm^2 (2) 2.58 cm^2 (3) 2.5796 cm^2 (4) 2.57964 cm^2

11. A. Divide 38.57 by 5.2. The answer, to the correct number of significant figures, is
    (1) 7.4   (2) 7.4173   (3) 7.42   (4) 7.417

    B. The density of hydrogen at STP is 0.0899 gram per liter. This density, to two significant figures, is
    (1) 0.090 g/l   (2) 0.09 g/l   (3) 0.089 g/l   (4) 0.0899 g/l

12. Which quantity expresses the sum of the given masses to the correct number of significant figures?
    (1) 5800 g (2) 5798 g (3) 797.9 g (4) 5797.892 g

    22.1   g
    375.66   g
    5400.132 g

13. What is the sum of 6.6412 g + 12.85 g + 0.046 g + 3.48 g, expressed to the correct number of significant figures?
    (1) 23 g   (2) 23.0 g   (3) 23.017 g   (4) 23.02 g

14. Which measurement below has the greatest number of significant figures?
    (1) 6.060 mg   (2) 60.6 mg   (3) 606 mg   (4) 60600 mg

15. A. Which measurement below contains a total of 3 significant figures?
    B. Which measurements below have 4 significant figures?
    (1) 0.012 g   (2) 0.125 g   (3) 1205 g   (4) 12050 g

16. Solubility data for salt X is shown in the table below.
    Which graph most closely represents the data shown in the table?

Temperature °C	Solubility $\frac{g\ salt\ X}{100\ g\ H_2O}$
10	5
20	10
30	15
40	20
50	25
60	30

17. The graph represents the cooling curve of a substance starting at a temperature below the boiling point of the substance.

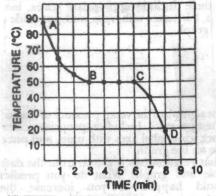

During which interval was the substance completely in the solid phase?
(1) A to B   (2) A to C
(3) B to C   (4) C to D

18. A student observed that when sodium hydroxide is dissolved in water, the temperature of the water increased. The student should conclude that the dissolving of sodium hydroxide
(1) is endothermic   (3) produces an acid solution
(2) is exothermic   (4) produces a salt solution

19. What occurs as potassium nitrate is dissolved in a beaker of water, indicating that the process is endothermic?
(1) The temperature of the solution decreases.
(2) The temperature of the solution increases.
(3) The solution changes color.
(4) The solution gives off a gas.

20. A solid is dissolved in a beaker of water. Which observation suggests that the process is endothermic?
(1) The solution gives off a gas.
(2) The solution changes color.
(3) The temperature of the solution decreases.
(4) The temperature of the solution increases.

CONSTRUCTED RESPONSE QUESTIONS: Parts B-2 and C of NYS Regents Exam

21. Convert:
5 meters = ___mm   240 gram = ___kg
24 kilograms = ___grams   4000 mm = ___meter
43 meters = ___cm   675 cm = ___mm
8 kilopascal = ___pascal   5000 mm = ___cm
5 cm = ___mm   200 °C = ___K

22. Compare deductive and inductive reasoning.

23. A. Based on knowing how the parts are arranged (geometrically), how can you predict properties?
   B. From looking at and interpreting a graph, how can you tell if it is directly proportional (a direct relationship) or inversely proportional (an inverse relationship)?

24. A. Find the gram formula masses of $MgCl_2$ and Na_2SO_4. Round atomic masses from the Periodic Table to the nearest tenth. Show all work.
   B. During a laboratory experiment, a sample of zinc was found to have a mass of 63.20 grams and a volume of 8.9 milliliters. What is the density, expressed to the correct number of significant figures? (Hint: See density, **Chap. 5:**13.)

25. Plot a graph (data at right) of the volume of a gas vs temperature. Include appropriate scales, mark axes, label axes with units, and put a title on the graph.

Temperature, K	Volume, mL
100	300
200	600
300	900
400	1200
500	1500

26. Plot a graph of the amount of ^{42}K remaining vs time. Include appropriate scales, mark axes, label axes with units, and put a title on the graph.

Time, hours	Amount of ^{42}K remaining, grams
0	10
12.5	5
25	2.5
37.5	1.25
50	0.625

27. Plot a graph of the vapor pressure of a liquid vs temperature. Include appropriate scales, mark axes, label axes with units, and put a title on the graph.
What **relationship** is shown from the data table and graph? What do you **predict** would happen if you increase the temperature to 150°C?

Temperature, °C	Pressure, kPa
0	2
10	3
25	4
50	7
75	23
100	57
125	124

28. Devise an experiment using the scientific method to show how a type of antacid affected the pH of orange pineapple juice, vinegar, etc. Ask your teacher for indicator or pH paper.

29. Devise an experiment using different detergents and see which removes an ink stain, crayon stain, or marker stain most. Use the scientific method, a control, and data table to show results.

30. Devise any experiment using the scientific method. Have a control. Draw a data table and a graph to show your observations.

31. How can you get information on a science topic, and how can you tell whether the information is correct?

32. Draw a model showing compound formation.

33. A. How do you multiply and divide exponents with the same base?
B. Change into scientific notation: 3627.5   .00053   .752

34. How can you explain why equilibrium (static and dynamic) is stable?

35. Explain how equilibrium is disturbed (spark, adding something) and becomes unstable (changes).

36. Design a solution to a real world problem on a community, national, or global scale. Use math, science, and technology to solve the problem.

37. Do a scientific experiment following the scientific method (science inquiry) that involves math, science, and technology, and communicate the results.

CHAPTER QUESTION: Parts B-2 and C of NYS Regents Exam

38. Read Point B on the graduated cylinder.
Hint: Look at Point A. 5.2 is the certain digit.
.05 is the estimated digit. Therefore, the reading is 5.25.

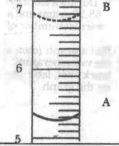

CHAPTER 13:
APPLIED PRINCIPLES

THIS MATERIAL IS NOT IN THE
NEW YORK STATE REGENTS CURRICULUM,
BUT THERE CAN BE APPLICATIONS ON IT

HABER PROCESS: PRODUCES AMMONIA

HABER PROCESS:

Nitrogen + Hydrogen → Ammonia + Heat
$$N_2 + 3 H_2 \quad 2 NH_3 + 91.8 \text{ kJ}$$

Let's apply **LeChatelier's Principle** that you learned in Chapter 6 to the Haber Process.

An **increase** in **concentration** of N_2 or H_2 causes the **reaction** to **go to** the **right** (displaces or shifts the equilibrium to the right), and more ammonia (NH_3) is formed (produced).

An **increase** in **pressure** causes reaction to go in the **direction** of **less gas molecules.**

Left Side		Right Side
$N_2(g) + 3H_2(g)$	⇌	$2NH_3(g)$

1 molecule N_2 + 3 molecules H_2 → **2 molecules NH_3**

4 molecules on left side   2 molecules on right side

Smaller number of gas molecules is on right side (2 is less than 4)

Therefore, when there is **more pressure**, the **reaction goes to the right;** the right side has a **smaller number of gas molecules (2 gas molecules).** This favors the production of NH_3.

In the Haber Process, the manufacturers look for the best temperature to get the most production of NH_3. (The best temperature is based on LeChatelier's Principle and effective collisions.)

A **catalyst** is used in the production of NH_3, to **increase the rate of reaction.**

CONTACT PROCESS PRODUCES SULFURIC ACID

1. Sulfur (S) is burned to produce sulfur dioxide (SO_2).
$$S + O_2 \rightarrow SO_2 + heat$$

2. SO_2 is then oxidized to produce sulfur trioxide (SO_3).
$$2\,SO_2 + O_2 \rightarrow 2\,SO_3 + heat$$

3. Sulfur trioxide (SO_3) is then absorbed by sulfuric acid (H_2SO_4).
$$H_2SO_4 + SO_3 \rightarrow H_2S_2O_7$$
sulfuric acid + sulfur trioxide → pyrosulfuric acid

4. Pyrosulfuric acid and water produce a large amount of H_2SO_4.
$$H_2S_2O_7 + H_2O \rightarrow 2\,H_2SO_4$$

OBTAINING METALS FROM COMPOUNDS (ORES)

To obtain metals from compounds (ores):

In the earth, the metal is in a compound. The manufacturers want to get the metal **alone** out of the ore (compound).

1. You learned (Chapter 3, Section A) that **metals in Group 1 and Group 2** (the most active metals) are obtained by **electrolysis** of their fused compounds (salts). Examples are sodium (Na) and potassium (K) (both in Group 1), obtained by electrolysis of their fused salts. **Aluminum** metal (from aluminum oxide, Al_2O_3, **in** the mineral **bauxite**) is also obtained by **electrolysis**.

2. Moderately active metals like **zinc** (Zn) and **iron** (Fe) are obtained by **reduction** of their oxides **by carbon (coke)** or **carbon monoxide (CO)**.
$$\mathbf{ZnO} + \mathbf{C} + heat \rightarrow \mathbf{Zn} + CO$$
(zinc oxide + carbon → zinc + carbon monoxide)
$$\mathbf{Fe_2O_3} + 3\,\mathbf{CO} + heat \rightarrow 3\,CO_2 + \mathbf{2\,Fe}$$
(iron oxide + carbon monoxide → carbon dioxide + iron)

3. **Chromium** can be obtained from its oxide, a relatively stable compound, by **reduction with aluminum**.
$$2\,Al + Cr_2O_3 \rightarrow Al_2O_3 + 2\,Cr$$
(aluminum + chromium oxide → aluminum oxide + chromium)

Now you have chromium alone, not in a compound.

Question: Given the reaction:
$$2Al + Cr_2O_3 \rightarrow Al_2O_3 + 2Cr.$$
When this reaction is used to produce chromium, the aluminum is acting as

    (1) a catalyst   (3) an oxidizing agent
    (2) an alloy   (4) a reducing agent

Solution: Answer *4*. Look at the equation:
$$2Al^0 + Cr_2O_3 \rightarrow Al_2^{3+}O_3 + 2Cr.$$
You learned that aluminum, or any element alone, has an oxidation number of zero. The oxidation number of $_{13}Al$ (look at the Periodic Table) is $+3$ in compounds.

$Al^0 - 3e \rightarrow Al^{3+}$. Al loses 3 electrons and becomes Al^{3+}. Al is oxidized. Al is a reducing agent.

CORROSION

CORROSION is the gradual attack on a metal by its surroundings (moisture, some gases in the air), with the result that the usefulness of the metal is destroyed. Corrosion is an oxidation-reduction reaction. **Iron (Fe) corrodes** (rusts) because of moisture and air, and loses its strength.

Metals like ALUMINUM **and** ZINC form a SELF-PROTECTIVE COVERING TO PREVENT FURTHER CORROSION. Example: Aluminum corrodes and forms aluminum oxide, a self-protecting covering which adheres to the aluminum and prevents further corrosion. **But iron corrodes** and KEEPS ON CORRODING.

Ways to prevent corrosion:

1. **PLATE** FE WITH ALUMINUM AND ZINC, because, as you learned, ALUMINUM AND ZINC FORM A SELF-PROTECTIVE COATING to prevent corrosion. Cover the iron with paint, oil or porcelain to prevent corrosion.

2. ALLOY: MIX IRON WITH NICKEL OR CHROMIUM TO PRODUCE STAINLESS STEEL to prevent corrosion.

3. Corrosion is a **redox reaction.** Connect a more active metal to the iron. The more active metal will oxidize or corrode instead of the iron.

Question: Iron corrodes more easily than aluminum and zinc because aluminum and zinc both
- (1) are reduced
- (2) are oxidizing agents
- (3) form oxides that are self-protective
- (4) form oxides that are very reactive

Solution: Answer *3*. Aluminum and zinc form oxides that form a self-protective covering that adheres to the metal and prevents further corrosion.

BATTERIES / VOLTAIC CELLS

Batteries are voltaic cells. You learned (in Chapter 9) that, in voltaic cells, chemical reactions produce electricity. A battery **discharges** (produces electricity) when the reaction goes from left to right (just like you read the equation).

LEAD-ACID BATTERY (used in cars, boats, etc.)
$$Pb^0 + Pb^{4+}O_2 + 2H_2SO_4 \rightarrow 2Pb^{2+}SO_4 + 2H_2O$$
(lead + lead dioxide + sulfuric acid → lead sulfate + water)

$Pb - 2e \rightarrow Pb^{2+}$	$Pb^{4+} + 2e \rightarrow Pb^{2+}$
oxidation	reduction

In a **voltaic cell, oxidation** takes place at the **anode**. Pb is **oxidized (loses electrons)**. Pb is the anode. In the voltaic cell, the anode is **negative**.

In a **voltaic cell, reduction** takes place at the **cathode**. Pb^{4+} is reduced. Pb^{4+} is the cathode. In a voltaic cell, the cathode is **positive**.

In a voltaic cell, the redox reaction produces electricity, and the **electrolyte, sulfuric acid, H_2SO_4**, conducts electricity.

Question: Given the lead-acid battery reaction:

$$Pb + PbO_2 + 2H_2SO_4 \underset{charge}{\overset{discharge}{\rightleftharpoons}} 2PbSO_4 + 2H_2O$$

Which species is oxidized during battery discharge?
   (1) Pb   (2) PbO_2 (3) SO_4^{2-} (4) H_2O

Solution: Answer *1*:

$$Pb^0 + PbO_2 + 2H_2SO_4 \rightarrow Pb^{2+}SO_4 + 2H_2O$$

Pb, or any element alone, has an oxidation number of zero.

In $PbSO_4$, and all other compounds, the sum of the oxidation numbers must be zero. Look at Table E, on page Reference Tables-5. SO_4 is written as SO_4^{2-}; therefore, Pb must be Pb^{2+}.

  Pb is oxidized: $Pb^0 - 2e \rightarrow Pb^{2+}$ **- OR -** $Pb^0 \rightarrow Pb^{2+} + 2e$.

NICKEL OXIDE-CADMIUM BATTERY
$$2Ni^{3+}OOH + Cd^0 + 2H_2O \rightarrow 2Ni^{2+}(OH)_2 + Cd^{2+}(OH)_2$$

$Ni^{3+} + e \rightarrow Ni^{2+}$	$Cd^0 - .2e \rightarrow Cd^{2+}$
reduction	oxidation

In a **voltaic cell**, the **cathode** is where **reduction** takes place. Ni^{3+} gains 1 electron; Ni^{3+} is reduced; Ni^{3+} is the cathode. In a voltaic cell, the cathode is **positive**.

In a **voltaic cell**, the **anode** is where **oxidation** takes place. Cd^0 loses 2 electrons; Cd^0 is oxidized; Cd^0 is the anode. In a voltaic cell, the anode is **negative**.

Question: A battery consists of which type of cells?
   (1) electrolytic   (3) electroplating
   (2) voltaic   (4) electromagnetic

Solution: Answer *2*. Batteries are voltaic cells.

Question: Given the nickel oxide-cadmium reaction:
$$2NiOOH + Cd + 2H_2O \rightarrow 2Ni(OH)_2 + Cd(OH)_2.$$
During discharge, the Cd electrode
   (1) is oxidized   (3) gains electrons
   (2) is reduced   (4) gains mass
Solution: Answer *1*.
$$2NiOOH + Cd^0 + 2H_2O \rightarrow 2Ni(OH)_2 + Cd^{2+}(OH)_2$$
Cd, or any free element alone, has an oxidation number of zero.
Look at the Periodic Table: The oxidation number of $_{48}Cd$ is $+2$ in
a compound, such as $Cd(OH)_2$. Cd loses electrons:
   Cd is oxidized: $Cd^0 - 2e \rightarrow Cd^{2+}$ $-$ **OR** $-$ $Cd \rightarrow Cd^{2+} + 2e.$

PETROLEUM

PETROLEUM is a **mixture** of **hydrocarbons** (examples: **gasoline, kerosene, fuel oil**), used as fuels and in making **plastics, textiles,** rubber and detergents. **Natural gas** (a fuel found with petroleum) is **mostly methane**, with some ethane. **Bottled gas**, also a fuel, is **propane and butane.**

Two methods are used to make petroleum more useful. As you know, petroleum is a mixture of hydrocarbons:

FRACTIONAL DISTILLATION SEPARATES petroleum BY BOILING POINT into mixtures of hydrocarbons, such as gasoline, kerosene, and fuel oil.

CRACKING breaks large molecules of hydrocarbons into smaller, useful molecules of gasoline and fuel oil.
$$\text{Example: } C_{14}H_{30} \rightarrow C_7H_{16} + C_7H_{14}$$
You can recognize cracking when C_{14} (14 C atoms) is broken into smaller pieces of C_7 (7 C atoms) and C_7 (7 C atoms).

Try Sample Questions #5-9, on page 6, and then do Homework Questions, #8-22, pages 7-8.

REMEMBER: When you answer Regents questions in APPLIED PRINCIPLES, use the Periodic Table and Table E.

SAMPLE REGENTS & REGENT-TYPE QUESTIONS
AND SOLUTIONS

1. The Haber process is used in the commercial preparation of
    (1) hydrochloric acid   (3) ammonia
    (2) sulfuric acid   (4) sulfur

2. What is produced when sulfur is burned during the first step of the contact process?
    (1) sulfuric acid   (3) sulfur trioxide
    (2) sulfur dioxide   (4) pyrosulfuric acid

3. Which element is commercially obtained by the electrolysis of its fused salt?
    (1) silver   (2) copper   (3) helium   (4) sodium

4. Given the reaction:
    $$ZnO + X + heat \rightarrow Zn + XO$$
    Which element, represented by X, is used industrially to reduce the ZnO to Zn?
    (1) Cu   (2) C   (3) Sn   (4) Pb

5. Which type of reaction is occurring when a metal undergoes corrosion?
    (1) oxidation-reduction   (3) polymerization
    (2) neutralization   (4) saponification

6. Which substance functions as the electrolyte in an automobile battery?
    (1) PbO_2 (2) $PbSO_4$ (3) H_2SO_4 (4) H_2O

7. Which commercial products are derived primarily from petroleum?
    (1) mineral acids and plastics   (3) plastics and textiles
    (2) fertilizers and rubber   (4) textiles and fertilizers

8. Natur l gas is composed mostly of
    (1) butane   (2) octane   (3) methane   (4) propane

9. During fractional distillation, hydrocarbons are separated according to their
    (1) boiling points   (3) triple points
    (2) melting points   (4) saturation points

SOLUTIONS

1. Answer *3*. The Haber process produces ammonia (NH_3).

2. Answer *2*. In the first step, sulfur (S) is burned to produce sulfur dioxide (SO_2).
    $$S \quad + \quad O_2 \quad \rightarrow \quad SO_2 \quad + \quad heat$$
    sulfur   oxygen   sulfur dioxide

3. Answer *4*. Elements of Group 1 and Group 2 are obtained by electrolysis of their fused salts. Na is an element in Group 1, and therefore is obtained by electrolysis. All the other choices are not in Group 1 or Group 2.

4. Answer *2*. ZnO reacts with C to produce Zn and CO.

5. Answer *1*. Corrosion is a redox reaction (oxidation-reduction reaction).

6. Answer *3*. Sulfuric acid, H_2SO_4 is the electrolyte in the lead-acid battery. Lead-acid batteries are used in cars.

7. Answer *3*. Petroleum is used in making plastics and textiles.

8. Answer *3*. Natural gas is mostly methane (CH_4), with ethane (C_2H_6).

9. Answer *1*. In fractional distillation, hydrocarbons are separated according to their boiling points.

NOW LET'S TRY A FEW HOMEWORK QUESTIONS:

1. Which substance is produced by the Haber process?
    (1) aluminum   (2) ammonia   (3) nitric acid   (4) sulfuric acid

2. Given the reaction at equilibrium:
    $$2SO_2(g) + O_2(g) \rightleftharpoons 2SO_3(g) + heat$$
    Which change will shift the equilibrium to the right?
    (1) adding a catalyst   (3) decreasing the pressure
    (2) adding more $O_2(g)$ (4) increasing the temperature

3. Which acid is formed during the contact process?
    (1) HNO_2 (2) HNO_3 (3) H_2SO_4 (4) H_2S

4. Group 1 and Group 2 metals are obtained commercially from their fused compounds by
    (1) reduction with CO   (3) reduction with Al
    (2) reduction by heat   (4) electrolytic reduction

5. Which metal is obtained commercially by the electrolysis of its salt?
    (1) Zn   (2) K   (3) Fe   (4) Ag

6. Which metal oxide is most easily reduced by carbon?
    (1) aluminum   (2) iron   (3) magnesium   (4) sodium

7. Which element can be found in nature in the free (uncombined) state?
    (1) Ca   (2) Ba   (3) Au   (4) Al

8. The corrosion of aluminum (Al) is a less serious problem than the corrosion of iron (Fe) because
    (1) Al does not oxidize   (3) Al oxidizes to form a protective layer
    (2) Fe does not oxidize   (4) Fe oxidizes to form a protective layer

9. Which metal forms a self-protective coating against corrosion?
    (1) Fe   (2) Cu   (3) Zn   (4) Mg

10. Given the reaction in a lead storage battery:
    $$Pb + PbO_2 + 2H_2SO_4 \rightarrow 2PbSO_4 + 2H_2O$$

    When the battery is being discharged, which change in the oxidation state of lead occurs?
    (1) Pb is oxidized to Pb^{2+}.   (3) Pb^{2+} is reduced to Pb.
    (2) Pb is oxidized to Pb^{4+} (4) Pb^{4+} is reduced to Pb.

11. Which type of chemical reaction generates the electrical energy produced by a battery?
    (1) oxidation-reduction   (3) neutralization
    (2) substitution   (4) addition

12. Given the overall reaction for the lead-acid battery:

$$Pb + PbO_2 + 2H_2SO_4 \overset{discharge}{\underset{charge}{\rightleftharpoons}} 2PbSO_4 + 2H_2O$$

    Which element changes oxidation state when electric energy is produced?
    (1) hydrogen   (2) oxygen   (3) sulfur   (4) lead

13. Given the reaction for the nickel-cadmium battery:
    $$2Ni^{3+}OOH + Cd^0 + 2H_2O \rightarrow 2Ni^{2+}(OH)_2 + Cd^{2+}(OH)_2$$
    Which species is oxidized during the discharge of the battery?
    (1) Ni^{3+} (2) Ni^{2+} (3) Cd   (4) Cd^{2+}

14. When a battery is in use, stored chemical energy is first changed to
    (1) electrical energy   (3) light energy
    (2) heat energy   (4) mechanical energy

15. A common gaseous fuel that is often found with petroleum is
    (1) carbon monoxide   (3) methane
    (2) carbon dioxide   (4) ethene

16. Petroleum is classifed chemically as
    (1) a substance   (2) a compound   (3) an element   (4) a mixture

17. Which products are obtained from the fractional distillation of petroleum?
    (1) esters and acids   (3) soaps and starches
    (2) alcohols and aldehydes   (4) kerosene and gasoline

18. Which balanced equation represents a cracking reaction?
    (1) $C_4H_{10} \rightarrow C_2H_6 + C_2H_4$ (3) $C_4H_{10} + Br_2 \rightarrow C_4H_9Br + HBr$
    (2) $C_4H_8 + 6O_2 \rightarrow 4CO_2 + 4H_2O$ (4) $C_4H_8 + Br_2 \rightarrow C_4H_8Br_2$

19. Given the equation for the overall reaction in a lead-acid storage battery,

$$Pb(s) + PbO_2(s) + 2H_2SO_4(aq) \overset{discharge}{\underset{charge}{\rightleftharpoons}} 2PbSO_4(s) + 2H_2O(\ell)$$

    which occurs during the charging of the battery?
    (1) concentration of H_2SO_4 decreases, number of moles of Pb(s) inc eases.
    (2) concentration of H_2SO_4 decreases, number of moles of $H_2O(\ell)$ increases.
    (3) concentration of H_2SO_4 increases, number of moles of Pb(s) decreases.
    (4) concentration of H_2SO_4 increases, number of moles of $H_2O(\ell)$ decreases.

20. The corrosion of iron is an example of
    (1) an oxidation-reduction reaction (3) a substitution reaction
    (2) an addition reaction   (4) a neutralization reaction

21. The separation of petroleum into components based on their boiling points is accomplished by
    (1) cracking   (3) fractional distillation
    (2) melting   (4) addition polymerization

22. Petroleum is a complex mixture of many
    (1) hydrocarbons   (3) organic halides
    (2) aldehydes   (4) ketones

APPENDIX I: REFERENCE TABLES

You will be given Reference Tables A-T and the Periodic Table with the Regents exam. Use the Reference Tables the way you learned in the book. If you don't understand how to use a Reference Table, go back to the page that explains the Table.

Table A
Standard Temperature and Pressure

Name	Value	Unit
Standard Pressure	101.3 kPa 1 atm	kilopascal atmosphere
Standard Temperature	273 K 0°C	kelvin degree Celsius

TABLE A has standard **temperature** and **pressure**. The boiling point of a liquid is when vapor pressure (of a liquid) = **standard pressure** (atmospheric pressure), **101.3 kPa**, given in Table A.

If you have a combined gas law problem, and a **gas** is at **STP** (standard temperature and pressure), use **temperature 273 K** and **pressure 101.3kPa** from Table A in the combined gas law.

Review pages 1:21, 1:26.

Table B
Physical Constants for Water

Heat of Fusion	334 J/g
Heat of Vaporization	2260 J/g
Specific Heat Capacity of H_2O (ℓ)	4.18 J/g•°C

TABLE B has physical constants for water. Use the **heat of fusion H_f** (heat needed to change 1 gram of ice to water) and **heat of vaporization H_v** (heat needed to change 1 gram of water to water vapor) given in this table to solve heat energy problems:

q (heat) =   mH_f and
q (heat) =   mH_v } Formulas given in Table T.

Use the **specific heat capacity C** of H_2O(l) to solve heat (energy) problems.

q (heat) = mCΔT   Formula given in Table T

Review pages 1:6, 1:8, 1:9, 1:11, 1:12.

Table C
Selected Prefixes

Factor	Prefix	Symbol
10^3	kilo-	k
10^{-1}	deci-	d
10^{-2}	centi-	c
10^{-3}	milli-	m
10^{-6}	micro-	μ
10^{-9}	nano-	n
10^{-12}	pico-	p

TABLE C gives prefixes that are used in front of meters, grams, pascals, moles, joules, liters, parts per million, etc.

Examples:
- **kilo**gram = 10^3 = 1000 grams.
- **centi**meter= 10^{-2}= 0.01 meter
- **milli**liter = 10^{-3}= 0.001 liter
- **micro**gram = 10^{-9} = 0.000001 gram
- **nano**meter = 10^{-9} = 0.000000001 meter
- **pico**meter = 10^{-12} = 0.000000000001 meter

Example: Convert 102 picometers into meters.

Look at Table C. pico = 10^{-12}

Step 1: Write the equality 1 picometer = 10^{-12} meter

Step 2: Write conversion factors:

$$\frac{10^{-12} \; meter}{1 \; picometer} \quad or \quad \frac{1 \; picometer}{10^{-12} \; meter}$$

Step 3: Multiply unit you want to change (picometers) by the conversion factor:

$$102 \; picometer \; x \; \frac{10^{-12} \; meter}{1 \; picometer} = 102 \; x \; 10^{-12} \; meter = 1.02 \; x \; 10^{-10} \; m$$

Review pages 12:2-12:4.

Table D
Selected Units

Symbol	Name	Quantity
m	meter	length
g	gram	mass
Pa	pascal	pressure
K	kelvin	temperature
mol	mole	amount of substance
J	joule	energy, work, quantity of heat
s	second	time
min	minute	time
h	hour	time
d	day	time
y	year	time
L	liter	volume
ppm	parts per million	concentration
M	molarity	solution concentration
u	atomic mass unit	atomic mass

TABLE D gives the name of the unit and symbol for different quantities.
Review pages 12:2-12:4.

Table E
Selected Polyatomic Ions

H_3O^+	hydronium	CrO_4^{2-}	chromate
Hg_2^{2+}	dimercury (I)	$Cr_2O_7^{2-}$	dichromate
NH_4^+	ammonium	MnO_4^-	permanganate
$C_2H_3O_2^-$ CH_3COO^- } acetate		NO_2^-	nitrite
		NO_3^-	nitrate
CN^-	cyanide	O_2^{2-}	peroxide
CO_3^{2-}	carbonate	OH^-	hydroxide
HCO_3^-	hydrogen carbonate	PO_4^{3-}	phosphate
$C_2O_4^{2-}$	oxalate	SCN^-	thiocyanate
ClO^-	hypochlorite	SO_3^{2-}	sulfite
ClO_2^-	chlorite	SO_4^{2-}	sulfate
ClO_3^-	chlorate	HSO_4^-	hydrogen sulfate
ClO_4^-	perchlorate	$S_2O_3^{2-}$	thiosulfate

TABLE E gives the names, the formulas and the charges of many polyatomic ions. What is the formula of magnesium hydroxide? In a compound, the sum of the oxidation numbers must equal zero. Look at the upper right hand corner of the box on the Periodic Table for magnesium (Mg). **Mg** has an oxidation number of **+2**. Look at the oxidation number of hydroxide (OH^-) on Table F. OH^- has an oxidation number of -1.

$$2\ OH^- = 2 \text{ times } OH^- = (2 \text{ times } -1) = -2$$
$$+2 \text{ and } -2 = 0$$

The formula is **Mg** $(OH)_2$.

Review the criss-cross method to write formulas, page 4:25.

Review page 4:25.

Table F
Solubility Guidelines for Aqueous Solutions

Ions That Form Soluble Compounds	Exceptions
Group 1 ions (Li$^+$, Na$^+$, etc.)	
ammonium (NH$_4^+$)	
nitrate (NO$_3^-$)	
acetate (C$_2$H$_3$O$_2^-$ or CH$_3$COO$^-$)	
hydrogen carbonate (HCO$_3^-$)	
chlorate (ClO$_3^-$)	
perchlorate (ClO$_4^-$)	
halides (Cl$^-$, Br$^-$, I$^-$)	when combined with Ag$^+$, Pb^{2+}, and Hg$_2^{2+}$
sulfates (SO$_4^{2-}$)	when combined with Ag$^+$, Ca^{2+}, Sr^{2+}, Ba^{2+}, and Pb^{2+}

TABLE F is used to find out if a compound is soluble.

The left box of the table lists ions that form soluble compounds and the exceptions. Group I ions (Li$^+$, Na$^+$) form soluble compounds: LiBr and Na$_2$CO$_3$ are soluble. Sulfates are soluble, except when combined with Ag$^+$, Ca^{2+}, Sr^{2+}, Ba^{2+}, and Pb^{2+}: Ag$_2$SO$_4$, CaSO$_4$, SrSO$_4$, BaSO$_4$, and PbSO$_4$ are insoluble.

Table F
Solubility Guidelines for Aqueous Solutions

Ions That Form Insoluble Compounds*	Exceptions
carbonate (CO_3^{2-})	when combined with Group 1 ions or ammonium (NH_4^+)
chromate (CrO_4^{2-})	when combined with Group 1 ions, Ca^{2+}, Mg^{2+}, or ammonium (NH_4^+)
phosphate (PO_4^{3-})	when combined with Group 1 ions or ammonium (NH_4^+)
sulfide (S^{2-})	when combined with Group 1 ions or ammonium (NH_4^+)
hydroxide (OH^-)	when combined with Group 1 ions, Ca^{2+}, Ba^{2+}, Sr^{2+}, or ammonium (NH_4^+)

*compounds having very low solubility in H_2O

TABLE F(continued)

The right box of the table lists ions that form insoluble compounds and the exceptions. Carbonates form insoluble compounds: $CaCO_3$ is insoluble. Carbonates are insoluble, except when combined with Group 1 ions or ammonium (NH_4^+): $(NH_4)_2CO_3$ is soluble.

Review pages 3:10, 6:5, 6:6, 7:14

Table G
Solubility Curves at Standard Pressure

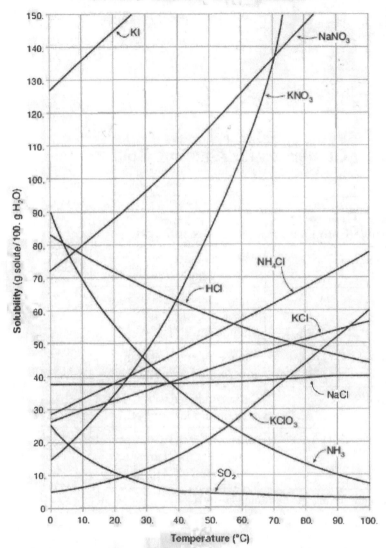

Solubility (g solute/100. g H₂O)

Temperature (°C)

Labeled curves: KI, NaNO₃, KNO₃, NH₄Cl, HCl, KCl, NaCl, KClO₃, NH₃, SO₂

TABLE G: The curved line for each compound shows how many grams of solute is the most that can dissolve in 100 grams of water at a given temperature. Look at the line labeled SO_2 on the Reference Table. At 0°C, 23 grams is the most that can dissolve (in 100 grams of water).. At 60°C, about 4 grams is the most that can dissolve (in 100 grams of water). This shows that, for SO_2, as temperature increases, solubility decreases.

At 60°C, about 4 grams of SO_2 is the most that can dissolve in 100 grams of water. Therefore, you know, at 60°C, about 8 grams is the most that can dissolve in **200 grams** of water.

Review pages 6:1-6:3

Table H
Vapor Pressure of Four Liquids

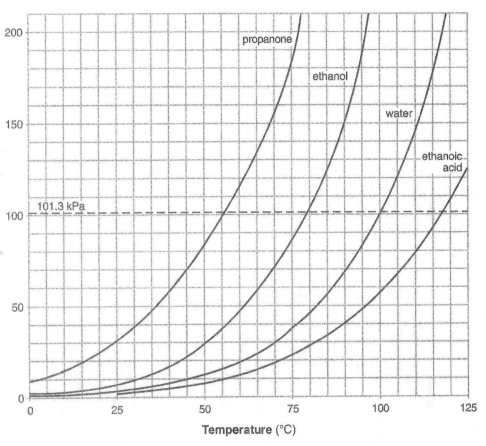

propanone

ethanol

water

ethanoic acid

101.3 kPa

Temperature (°C)

TABLE H shows the vapor pressure of 4 liquids. The dashed line at **101.3 kPa** is standard pressure (atmospheric pressure), given on Table A. The boiling point is when vapor pressure equals atmospheric pressure. The **boiling point** of **propanone** is when the **vapor pressure** of propanone **equals atmospheric pressure** = 101.3 kPa (the two lines cross) = **55°C.** The **boiling point** of **water** is when the **vapor pressure** of water **equals atmospheric pressure** = 101.3 kPa (the two lines cross) = **100°C.**

Review pages 1:20-1:21, 4:23.

Table I
Heats of Reaction at 101.3 kPa and 298 K

Reaction	ΔH (kJ)*
$CH_4(g) + 2O_2(g) \longrightarrow CO_2(g) + 2H_2O(\ell)$	−890.4
$C_3H_8(g) + 5O_2(g) \longrightarrow 3CO_2(g) + 4H_2O(\ell)$	−2219.2
$2C_8H_{18}(\ell) + 25O_2(g) \longrightarrow 16CO_2(g) + 18H_2O(\ell)$	−10943
$2CH_3OH(\ell) + 3O_2(g) \longrightarrow 2CO_2(g) + 4H_2O(\ell)$	−1452
$C_2H_5OH(\ell) + 3O_2(g) \longrightarrow 2CO_2(g) + 3H_2O(\ell)$	−1367
$C_6H_{12}O_6(s) + 6O_2(g) \longrightarrow 6CO_2(g) + 6H_2O(\ell)$	−2804
$2CO(g) + O_2(g) \longrightarrow 2CO_2(g)$	−566.0
$C(s) + O_2(g) \longrightarrow CO_2(g)$	−393.5
$4Al(s) + 3O_2(g) \longrightarrow 2Al_2O_3(s)$	−3351
$N_2(g) + O_2(g) \longrightarrow 2NO(g)$	+182.6
$N_2(g) + 2O_2(g) \longrightarrow 2NO_2(g)$	+66.4
$2H_2(g) + O_2(g) \longrightarrow 2H_2O(g)$	−483.6
$2H_2(g) + O_2(g) \longrightarrow 2H_2O(\ell)$	−571.6
$N_2(g) + 3H_2(g) \longrightarrow 2NH_3(g)$	−91.8
$2C(s) + 3H_2(g) \longrightarrow C_2H_6(g)$	−84.0
$2C(s) + 2H_2(g) \longrightarrow C_2H_4(g)$	+52.4
$2C(s) + H_2(g) \longrightarrow C_2H_2(g)$	+227.4
$H_2(g) + I_2(g) \longrightarrow 2HI(g)$	+53.0
$KNO_3(s) \xrightarrow{H_2O} K^+(aq) + NO_3^-(aq)$	+34.89
$NaOH(s) \xrightarrow{H_2O} Na^+(aq) + OH^-(aq)$	−44.51
$NH_4Cl(s) \xrightarrow{H_2O} NH_4^+(aq) + Cl^-(aq)$	+14.78
$NH_4NO_3(s) \xrightarrow{H_2O} NH_4^+(aq) + NO_3^-(aq)$	+25.69
$NaCl(s) \xrightarrow{H_2O} Na^+(aq) + Cl^-(aq)$	+3.88
$LiBr(s) \xrightarrow{H_2O} Li^+(aq) + Br^-(aq)$	−48.83
$H^+(aq) + OH^-(aq) \longrightarrow H_2O(\ell)$	−55.8

*The ΔH values are based on molar quantities represented in the equations.
A minus sign indicates an exothermic reaction.

TABLE I gives some reactions and tells how much heat is absorbed or released during the reaction.

ΔH = negative (**ΔH = −**) means exothermic, heat is given off.

ΔH = positive (**ΔH = +**) means endothermic, heat is absorbed.

Review pages 7:7-7:8.

TABLE J: Table J shows how active each metal and nonmetal is.

Most active metal: Top of Table J. Most easily loses an electron. **Most easily oxidized.**

As you go down Table J, metals are less active and less easily oxidized, as shown by the arrow going down, from most to least.

Least active metal: Bottom of Table J. Hardest to lose electrons, **least easily oxidized** (hardest to be oxidized).

Most active nonmetal: Top of Table J: Most easily gains an electron, **most easily reduced.**

As you go down Table J, nonmetals are less active and less easily reduced, as shown by the arrow going down, from most to least.

Least active nonmetal: Bottom of Table J: Hardest to gain electrons, **least easily reduced.**

To find the **properties** of the **ion** (least or most easily oxidized, least or most easily reduced), take the properties of the element and **reverse** them:

1. **Reverse** most to least or least to most.
2. **Reverse** reduced to oxidized or oxidized to reduced.

F is **most** easily **reduced.** For F⁻, change most to least and reduced to oxidized: F⁻ (ion) is **least** easily **oxidized.**

Metals higher up are **more active** and **replace metals below** them **from compounds.** This **reaction** is **spontaneous.**

Li is a **metal above K** and **replaces K** from its compounds.

$$Li + KCl \longrightarrow LiCl + K$$

Metals below are less active and do **not** replace metal above them from compounds.

K is below Li and does not replace Li from its compounds.

$$K + LiCl \longrightarrow No\ reaction$$

Nonmetals higher up in Table J are more active and **replace nonmetals below from compounds.** The **reaction** is **spontaneous** (reaction takes place).

F_2 is a **nonmetal above Cl** and **replaces Cl** from its compounds.

$$F_2 + 2NaCl \longrightarrow 2NaF + Cl_2$$

Nonmetals below are less active and do **not** replace nonmetals above them from compounds.

$$Cl_2 + 2NaF \longrightarrow No\ reaction$$
Review pages 8:1,9:14-19

Table J
Activity Series**

Most Active	Metals	Nonmetals	Most Active
	Li	F_2	
	Rb	Cl_2	
	K	Br_2	
	Cs	I_2	
	Ba		
	Sr		
	Ca		
	Na		
	Mg		
	Al		
	Ti		
	Mn		
	Zn		
	Cr		
	Fe		
	Co		
	Ni		
	Sn		
	Pb		
	H_2		
	Cu		
	Ag		
Least Active	Au		Least Active

**Activity Series is based on the hydrogen
standard. H_2 is *not* a metal.

Table K
Common Acids

Formula	Name
HCl(aq)	hydrochloric acid
HNO_2(aq)	nitrous acid
HNO_3(aq)	nitric acid
H_2SO_3(aq)	sulfurous acid
H_2SO_4(aq)	sulfuric acid
H_3PO_4(aq)	phosphoric acid
H_2CO_3(aq) or CO_2(aq)	carbonic acid
CH_3COOH(aq) or $HC_2H_3O_2$(aq)	ethanoic acid (acetic acid)

TABLE K lists common acids. Acids contain hydrogen and yield H^+ in solution.

Look at Table K above. (aq) means in water. For example: HCl(aq) means HCl in water, HNO_3(aq) means HNO_3 in water, etc.

Review pages 8:1-8:4

Table L
Common Bases

Formula	Name
$NaOH(aq)$	sodium hydroxide
$KOH(aq)$	potassium hydroxide
$Ca(OH)_2(aq)$	calcium hydroxide
$NH_3(aq)$	aqueous ammonia

TABLE L lists common bases. Bases end in OH and produce OH⁻ ions in solution. You can combine an acid from Table K with a base from Table L to form salt and water.

$$\text{Acid} + \text{Base} \longrightarrow \text{Salt} + \text{Water}$$
$$HCl + NaOH \longrightarrow NaCl + H_2O$$
$$H_2SO_4 + Ca(OH)_2 \longrightarrow CaSO_4 + H_2O$$

Review pages 8:2, 8:6.

Table M
Common Acid–Base Indicators

Indicator	Approximate pH Range for Color Change	Color Change
methyl orange	3.1–4.4	red to yellow
bromthymol blue	6.0–7.6	yellow to blue
phenolphthalein	8–9	colorless to pink
litmus	4.5–8.3	red to blue
bromcresol green	3.8–5.4	yellow to blue
thymol blue	8.0–9.6	yellow to blue

Source: *The Merck Index*, 14th ed., 2006, Merck Publishing Group

TABLE M gives common acid-base indicators. Use an indicator to determine the pH of a solution (how strong the acid or base is). For methyl orange, the pH range for color change is **3.1** – **4.4**, and the color change is **red** to **yellow**. This means that if the solution has a pH of **3.1 or less**, the color of the solution is **red**. If the solution has a pH of 4.4 or more, the color of the solution is yellow.

Two indicators can be used to find a narrower range of pH. If a solution turns methyl orange **yellow,** you know the pH is **4.4 or above.** If the same solution turns bromthymol blue **yellow,** you know the pH is **6.0 or less.** From both indicators, you know the pH is between 4.4 and 6.0. (Using methyl orange alone, you know pH is 4.4 or more. Using bromthymol blue, you know the pH is 6.0 or less.

Review page 8:9.

Table N
Selected Radioisotopes

Nuclide	Half-Life	Decay Mode	Nuclide Name
^{198}Au	2.695 d	β^-	gold-198
^{14}C	5715 y	β^-	carbon-14
^{37}Ca	182 ms	β^+	calcium-37
^{60}Co	5.271 y	β^-	cobalt-60
^{137}Cs	30.2 y	β^-	cesium-137
^{53}Fe	8.51 min	β^+	iron-53
^{220}Fr	27.4 s	α	francium-220
^{3}H	12.31 y	β^-	hydrogen-3
^{131}I	8.021 d	β^-	iodine-131
^{37}K	1.23 s	β^+	potassium-37
^{42}K	12.36 h	β^-	potassium-42
^{85}Kr	10.73 y	β^-	krypton-85
^{16}N	7.13 s	β^-	nitrogen-16
^{19}Ne	17.22 s	β^+	neon-19
^{32}P	14.28 d	β^-	phosphorus-32
^{239}Pu	2.410×10^4 y	α	plutonium-239
^{226}Ra	1599 y	α	radium-226
^{222}Rn	3.823 d	α	radon-222
^{90}Sr	29.1 y	β^-	strontium-90
^{99}Tc	2.13×10^5 y	β^-	technetium-99
^{232}Th	1.40×10^{10} y	α	thorium-232
^{233}U	1.592×10^5 y	α	uranium-233
^{235}U	7.04×10^8 y	α	uranium-235
^{238}U	4.47×10^9 y	α	uranium-238

Source: CRC Handbook of Chemistry and Physics, 91st ed., 2010–2011, CRC Press

TABLE N gives some radioactive nuclides, their half-lives and their decay mode. You need to know the half-life (time it takes for half of the substance to decay) to solve problems on Radioactive Decay.

Review pages 11:5-11:8.

Table O
Symbols Used in Nuclear Chemistry

Name	Notation	Symbol
alpha particle	^4_2He or $^4_2\alpha$	α
beta particle	$^0_{-1}\text{e}$ or $^0_{-1}\beta$	β^-
gamma radiation	$^0_0\gamma$	γ
neutron	^1_0n	n
proton	^1_1H or ^1_1p	p
positron	$^0_{+1}\text{e}$ or $^0_{+1}\beta$	β^+

TABLE O is used in nuclear chemistry. It lists the name of the particle, notation (how it is written) and the symbol. Table O gives the atomic number, mass number and charge of the particle. In a nuclear reaction, when it says alpha is given off, write ^4_2He (from Table O). When the equation says beta is given off, write $^0_{-1}\text{e}$ (from Table O).

Review pages 11:2-11:4, 11:8-11:10.

Table P
Organic Prefixes

Prefix	Number of Carbon Atoms
meth-	1
eth-	2
prop-	3
but-	4
pent-	5
hex-	6
hept-	7
oct-	8
non-	9
dec-	10

TABLE P is used in organic chemistry. Table P gives prefixes. By looking at the beginning of the word (prefix) you know how many carbon atoms there are. Example: propane begins with prop-. Look at Table P and you see prop-means 3 carbon atoms; propane has 3 carbon atoms.

When you draw propane, draw it with three carbon atoms.

Review pages 10:2-10:17, 10:31-10:33, 10:35-10:38.

Table Q
Homologous Series of Hydrocarbons

Name	General Formula	Examples	
		Name	Structural Formula
alkanes	C_nH_{2n+2}	ethane	H H $\mid$ $\mid$ H—C—C—H $\mid$ $\mid$ H H
alkenes	C_nH_{2n}	ethene	H $\quad$ H \ $\quad$ / C=C / $\quad$ \ H $\quad$ H
alkynes	C_nH_{2n-2}	ethyne	H—C≡C—H

Note: n = number of carbon atoms

TABLE Q is also used in organic chemistry. It shows alkanes, alkenes, and alkynes; it gives a general formula and the name and structural formula of an example. Propane ends in -ane. Look at Table Q. Alk**anes** end in -ane; they have the formula C_nH_{2n+2}. Apply the formula to propane. (You know the number of carbon atoms from Table P.) $C_3H_{6+2} = C_3H_8$. You also see the structural formula of ethane. Ethane and all alkanes have single carbon-carbon bonds. Therefore, you know propane has single carbon-carbon bonds.

Review pages 10:2-10:17, 10:31-10:33, 10:35-10:37.

Periodic Table

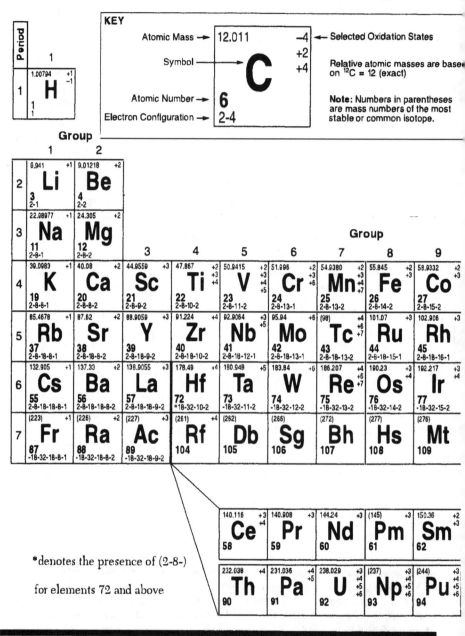

KEY

Atomic Mass → 12.011

Symbol → **C**

Atomic Number → 6

Electron Configuration → 2-4

Selected Oxidation States → -4, +2, +4

Relative atomic masses are based on $^{12}C = 12$ (exact)

Note: Numbers in parentheses are mass numbers of the most stable or common isotope.

Period

Group 1 2

Group 1 2 3 4 5 6 7 8 9

*denotes the presence of (2-8-)

for elements 72 and above

PERIODIC TABLE OF THE ELEMENTS: The Periodic Table gives the atomic numbers, atomic mass, oxidation states, electron configuration, and symbols of the elements.

The elements are divided into vertical columns called groups. Group 1 has 1 valence electron; Group 2 has 2 valence electrons. Periods go across the

---

High Marks: Regents Chemistry Made Easy

of the Elements

18

4.00260	0
He	
2	
2	

Group

13	14	15	16	17	18
10.81 +3 **B** 5 2-3	12.011 −4 +2 +2 **C** 6 2-4	14.0067 −3 −2 −1 +1 +2 +3 +4 +5 **N** 7 2-5	15.9994 −2 **O** 8 2-6	18.9984 −1 **F** 9 2-7	20.180 0 **Ne** 10 2-8
26.98154 +3 **Al** 13 2-8-3	28.0855 −4 +2 +4 **Si** 14 2-8-4	30.97376 −3 +3 +5 **P** 15 2-8-5	32.065 −2 +4 +6 **S** 16 2-8-6	35.453 −1 +1 +5 +7 **Cl** 17 2-8-7	39.948 0 **Ar** 18 2-8-8

10	11	12						
58.693 +2 +3 **Ni** 28 2-8-16-2	63.546 +1 +2 **Cu** 29 2-8-18-1	65.409 +2 **Zn** 30 2-8-18-2	69.723 +3 **Ga** 31 2-8-18-3	72.64 +2 +4 **Ge** 32 2-8-18-4	74.9216 −3 +3 +5 **As** 33 2-8-18-5	78.96 −2 +4 +6 **Se** 34 2-8-18-6	79.904 −1 +1 +5 **Br** 35 2-8-18-7	83.798 0 +2 **Kr** 36 2-8-18-8
106.42 +2 +4 **Pd** 46 2-8-18-18	107.868 +1 **Ag** 47 2-8-18-18-1	112.41 +2 **Cd** 48 2-8-18-18-2	114.818 +3 **In** 49 2-8-18-18-3	118.71 +2 +4 **Sn** 50 2-8-18-18-4	121.760 −3 +3 +5 **Sb** 51 2-8-18-18-5	127.60 −2 +4 +6 **Te** 52 2-8-18-18-6	126.904 −1 +1 +5 +7 **I** 53 2-8-18-18-7	131.29 0 +2 +4 +6 **Xe** 54 2-8-18-18-8
195.08 +2 +4 **Pt** 78 -18-32-17-1	196.967 +1 +3 **Au** 79 -18-32-18-1	200.59 +1 +2 **Hg** 80 -18-32-18-2	204.383 +1 +3 **Tl** 81 -18-32-18-3	207.2 +2 +4 **Pb** 82 -18-32-18-4	208.980 +3 +5 **Bi** 83 -18-32-18-5	(209) +2 +4 **Po** 84 -18-32-18-6	(210) **At** 85 -18-32-18-7	(222) 0 **Rn** 86 -18-32-18-8
(281) **Ds** 110	(280) **Rg** 111	(285) **Cn** 112	(284) **Uut** 113**	(289) **Uuq** 114	(288) **Uup** 115	(292) **Uuh** 116	(?) **Uus** 117	(294) **Uuo** 118

151.964 +2 +3 **Eu** 63	157.25 +3 **Gd** 64	158.925 +3 **Tb** 65	162.500 +3 **Dy** 66	164.930 +3 **Ho** 67	167.259 +3 **Er** 68	168.934 +3 **Tm** 69	173.04 +2 +3 **Yb** 70	174.9668 +3 **Lu** 71
(243) +3 +4 +5 +6 **Am** 95	(247) +3 **Cm** 96	(247) +3 +4 **Bk** 97	(251) +3 **Cf** 98	(252) +3 **Es** 99	(257) +3 **Fm** 100	(258) +2 +3 **Md** 101	(259) +2 +3 **No** 102	(262) +3 **Lr** 103

**The systematic names and symbols for elements of atomic numbers 113 and above will be used until the approval of trivial names by IUPAC.

Periodic Table. Period 1 has $_1$H and $_2$He, which have 1 shell. Period 2 has 2 shells.
Review pages 1:1, 2:7-2:8, 2:10, 2:11, 3:1-3:13, 4:3, 4:6-4:9, 4:24, 4:25, 5:1, 5:3, 5:9-5:13, 5:19-5:27, 9:1, 9:3, 9:5-9:6, 11:2, 13:2, 13.5.

Table R
Organic Functional Groups

Class of Compound	Functional Group	General Formula	Example
halide (halocarbon)	—F (fluoro-) —Cl (chloro-) —Br (bromo-) —I (iodo-)	R—X (X represents any halogen)	$CH_3CHClCH_3$ 2-chloropropane
alcohol	—OH	R—OH	$CH_3CH_2CH_2OH$ 1-propanol
ether	—O—	R—O—R'	$CH_3OCH_2CH_3$ methyl ethyl ether
aldehyde	$\overset{\overset{\text{O}}{\|\|}}{-\text{C}-\text{H}}$	$\overset{\overset{\text{O}}{\|\|}}{R-\text{C}-\text{H}}$	$\overset{\overset{\text{O}}{\|\|}}{CH_3CH_2\text{C}-\text{H}}$ propanal
ketone	$\overset{\overset{\text{O}}{\|\|}}{-\text{C}-}$	$\overset{\overset{\text{O}}{\|\|}}{R-\text{C}-R'}$	$\overset{\overset{\text{O}}{\|\|}}{CH_3\text{C}CH_2CH_2CH_3}$ 2-pentanone
organic acid	$\overset{\overset{\text{O}}{\|\|}}{-\text{C}-\text{OH}}$	$\overset{\overset{\text{O}}{\|\|}}{R-\text{C}-\text{OH}}$	$\overset{\overset{\text{O}}{\|\|}}{CH_3CH_2\text{C}-\text{OH}}$ propanoic acid
ester	$\overset{\overset{\text{O}}{\|\|}}{-\text{C}-\text{O}-}$	$\overset{\overset{\text{O}}{\|\|}}{R-\text{C}-\text{O}-R'}$	$\overset{\overset{\text{O}}{\|\|}}{CH_3CH_2\text{C}OCH_3}$ methyl propanoate
amine	$\overset{\overset{\|}{}}{-\text{N}-}$	$\overset{\overset{R'}{\|}}{R-\text{N}-R''}$	$CH_3CH_2CH_2NH_2$ 1-propanamine
amide	$\overset{\overset{\text{O}}{\|\|}\;\|}{-\text{C}-\text{NH}}$	$\overset{\text{O}\;\;R'}{R-\text{C}-\text{NH}}$	$\overset{\overset{\text{O}}{\|\|}}{CH_3CH_2\text{C}-NH_2}$ propanamide

Note: R represents a bonded atom or group of atoms.

TABLE R gives you type of compound, functional group, general formula (formula of compound) and example.

This table helps you to draw structural formulas. If you need to draw an alcohol, look at column 2 in the table and write the functional group OH.

If you need to draw a ketone, write the functional group $\overset{\overset{\text{O}}{\|\|}}{-\text{C}-}$.

When you draw these compounds, look at which functional group to put in. See the example in the last column. See where the functional group is placed.

Review pages 10:23, 10:26-10:38.

TABLE S. Look at Reference Table S on the next two pages. It shows you the atomic number, symbol, and name of the elements, with the properties of the elements: ionization energy, electronegativity, melting point, boiling point, density, and atomic radius. You can compare these properties when you go down a group or across a period.

Compare ionization energy, electronegativity, atomic readius as you go down a group:

Example: **What happens** to **ionization energy** as you go **down a group**? Look at Group 1.

Step 1: List the **Group 1 elements** with their **atomic numbers** in a vertical column (just **like the Periodic Table** has the elements in Group 1):

Step 2: Look at Reference **Table S.** The elements in Table S are listed in order of atomic number. **Copy** the ionization energy for elements listed in Step 1 ($_3$Li, $_{11}$Na, $_{19}$K, and $_{37}$Rb) from Table S on the next page and write the **ionization energy** next to the element.

(If you were comparing electronegativity or atomic radius as you go down a group, you would copy electronegativity or atomic radius instead of ionization energy). You can see, as you go **down** Group 1 or any other **group, ionization energy decreases.**

Step 1		Step 2
Atomic #	Symbol	Ionization energy
3	Li	520
11	Na	496
19	K	419
37	Rb	403

Compare ionization energy, electronegativity, atomic readius as you go across a period:

Example: **What happens** to **electronegativity** as you go **across a period**? Look at period 2 on the Periodic Table on page Reference Tables 20-21 or below. Remember: periods are horizontal rows on the Periodic Table.

Period	2	Li	Be	B	C	N	O	F	Ne

Step 1: Look at **Table S.** Start with $_3$Li and end with $_{10}$Ne (Period 2). These elements in Table S are listed in the same order as Period 2 on the Periodic Table. (All elements in Table S are listed in the same order as periods on the Periodic Table).

Step 2: Look at the **electronegativities.**

(If you were comparing ionization energy or atomic radius as you go across a period, instead look at ionization energy or atomic radius).

Step 3: If it makes it easier, put a piece of **paper** across Table S **above atomic #3** (lithium) and another one **below atomic #10** (neon).

Step 4: You can see that, as you go **across Period 2** from $_3$Li to $_9$F, or any other period, **electronegativity increases.**

Review pages 3:24-3:30, 4:1-4:4.

Table S
Properties of Selected Elements

Atomic Number	Symbol	Name	First Ionization Energy (kJ/mol)	Electro-negativity	Melting Point (K)	Boiling Point (K)	Density*** (g/cm³)	Atomic Radius (pm)
1	H	hydrogen	1312	2.2	14	20.	0.000082	32
2	He	helium	2372	—	—	4	0.000164	37
3	Li	lithium	520.	1.0	454	1615	0.534	130.
4	Be	beryllium	900.	1.6	1560.	2744	1.85	99
5	B	boron	801	2.0	2348	4273	2.34	84
6	C	carbon	1086	2.6	—	—	—	75
7	N	nitrogen	1402	3.0	63	77	0.001145	71
8	O	oxygen	1314	3.4	54	90.	0.001308	64
9	F	fluorine	1681	4.0	53	85	0.001553	60.
10	Ne	neon	2081	—	24	27	0.000825	62
11	Na	sodium	496	0.9	371	1156	0.97	160.
12	Mg	magnesium	738	1.3	923	1363	1.74	140.
13	Al	aluminum	578	1.6	933	2792	2.70	124
14	Si	silicon	787	1.9	1687	3538	2.3296	114
15	P	phosphorus (white)	1012	2.2	317	554	1.823	109
16	S	sulfur (monoclinic)	1000.	2.6	388	718	2.00	104
17	Cl	chlorine	1251	3.2	172	239	0.002898	100.
18	Ar	argon	1521	—	84	87	0.001633	101
19	K	potassium	419	0.8	337	1032	0.89	203.
20	Ca	calcium	590.	1.0	1115	1757	1.54	174
21	Sc	scandium	633	1.4	1814	3109	2.99	159
22	Ti	titanium	659	1.5	1941	3560.	4.506	148
23	V	vanadium	651	1.6	2183	3680.	6.0	144
24	Cr	chromium	653	1.7	2180.	2944	7.15	130.
25	Mn	manganese	717	1.6	1519	2334	7.3	139
26	Fe	iron	762	1.8	1811	3134	7.87	124
27	Co	cobalt	760.	1.9	1768	3200.	8.86	118
28	Ni	nickel	737	1.9	1728	3186	8.90	117
29	Cu	copper	745	1.9	1358	2835	8.96	122
30	Zn	zinc	906	1.7	693	1180.	7.134	120.
31	Ga	gallium	579	1.8	303	2477	5.91	123
32	Ge	germanium	762	2.0	1211	3106	5.3234	120.
33	As	arsenic (grey)	944	2.2	1090.	—	5.75	120.
34	Se	selenium (grey)	941	2.6	494	958	4.809	118
35	Br	bromine	1140	3.0	266	332	3.1028	117
36	Kr	krypton	1351	—	116	120.	0.003425	116
37	Rb	rubidium	403	0.8	312	961	1.83	215
38	Sr	strontium	549	1.0	1050.	1655	2.64	190.
39	Y	yttrium	600.	1.2	1795	3618	4.47	176
40	Zr	zirconium	640.	1.3	2128	4682	6.52	164

High Marks: Regents Chemistry Made Easy

Atomic Number	Symbol	Name	First Ionization Energy (kJ/mol)	Electro-negativity	Melting Point (K)	Boiling* Point (K)	Density** (g/cm³)	Atomic Radius (pm)
41	Nb	niobium	652	1.6	2750.	5017	8.57	156
42	Mo	molybdenum	684	2.2	2896	4912	10.2	146
43	Tc	technetium	702	2.1	2430.	4538	11	138
44	Ru	ruthenium	710.	2.2	2606	4423	12.1	135
45	Rh	rhodium	720.	2.3	2237	3968	12.4	134
46	Pd	palladium	804	2.2	1828	3236	12.0	130.
47	Ag	silver	731	1.9	1235	2435	10.5	136
48	Cd	cadmium	868	1.7	594	1040.	8.69	140.
49	In	indium	558	1.8	430.	2345	7.31	142
50	Sn	tin (white)	709	2.0	505	2875	7.287	140.
51	Sb	antimony (gray)	831	2.1	904	1860.	6.68	140.
52	Te	tellurium	869	2.1	723	1261	6.232	137
53	I	iodine	1008	2.7	387	457	4.933	136
54	Xe	xenon	1170.	2.6	161	165	0.005366	136
55	Cs	cesium	376	0.8	302	944	1.873	238
56	Ba	barium	503	0.9	1000.	2170.	3.62	206
57	La	lanthanum	538	1.1	1193	3737	6.15	194

Elements 58-71 have been omitted.

Atomic Number	Symbol	Name	First Ionization Energy (kJ/mol)	Electro-negativity	Melting Point (K)	Boiling* Point (K)	Density** (g/cm³)	Atomic Radius (pm)
72	Hf	hafnium	659	1.3	2506	4876	13.3	164
73	Ta	tantalum	798	1.5	3290	5731	16.4	158
74	W	tungsten	759	1.7	3695	5828	19.3	150.
75	Re	rhenium	756	1.9	3458	5869	20.8	141
76	Os	osmium	814	2.2	3306	5285	22.587	136
77	Ir	iridium	865	2.2	2719	4701	22.562	132
78	Pt	platinum	864	2.2	2041	4098	21.5	130.
79	Au	gold	890.	2.4	1337	3129	18.3	130.
80	Hg	mercury	1007	1.9	234	630.	13.5336	132
81	Tl	thallium	589	1.8	577	1746	11.8	144
82	Pb	lead	716	1.8	600.	2022	11.3	145
83	Bi	bismuth	703	1.9	544	1837	9.79	150.
84	Po	polonium	812	2.0	527	1235	9.20	142
85	At	astatine	—	2.2	575	—	—	148
86	Rn	radon	1037	—	202	211	0.009074	146
87	Fr	francium	383	0.7	300.	—	—	242
88	Ra	radium	509	0.9	969	—	5	211
89	Ac	actinium	499	1.1	1323	3471	10.	201

Elements 90 and above have been omitted.

* boiling point at standard pressure
** density of solids and liquids at room temperature and density of gases at 298 K and 101.3 kPa
— no data available

Source: CRC Handbook of Chemistry and Physics, 91st ed., 2010-2011, CRC Press

TABLE T lists formulas and equations. Density, mole calculations, and percent composition are used in math. Percent error is used in experiments. Concentration is used in solutions. Combined gas law, heat, and temperature are used in physical behavior of matter. Titration is used in acids.

Review pages:
Density: 5:12-5:13:
Mole calculations: 5:20:
Percent error: 12:1-12:2;
Percent composition: 5:19;
Concentration: 6:6-6:9;
Combined gas law: 1:22-1:26;
Titration: 8:6-8:8;
Heat: 1:5-1:6, 1:11-1:12;
Temperature: 1:7, 12:2;

Table T
Important Formulas and Equations

Density	$d = \dfrac{m}{V}$	d = density m = mass V = volume
Mole Calculations	number of moles $= \dfrac{\text{given mass}}{\text{gram-formula mass}}$	
Percent Error	% error $= \dfrac{\text{measured value} - \text{accepted value}}{\text{accepted value}} \times 100$	
Percent Composition	% composition by mass $= \dfrac{\text{mass of part}}{\text{mass of whole}} \times 100$	
Concentration	parts per million $= \dfrac{\text{mass of solute}}{\text{mass of solution}} \times 1\,000\,000$ molarity $= \dfrac{\text{moles of solute}}{\text{liter of solution}}$	
Combined Gas Law	$\dfrac{P_1 V_1}{T_1} = \dfrac{P_2 V_2}{T_2}$	P = pressure V = volume T = temperature
Titration	$M_A V_A = M_B V_B$	M_A = molarity of H^+ M_B = molarity of OH^- V_A = volume of acid V_B = volume of base
Heat	$q = mC\Delta T$ $q = mH_f$ $q = mH_v$	q = heat H_f = heat of fusion m = mass H_v = heat of vaporization C = specific heat capacity ΔT = change in temperature
Temperature	$K = {}^\circ C + 273$	K = kelvin ${}^\circ C$ = degree Celsius

These tables were used for the NY State Regents Exam under the v
previous curriculum before September, 2001.

TABLE OF STANDARD ENERGIES OF
FORMATION OF COMPOUNDS

STANDARD ENERGIES OF FORMATION
OF COMPOUNDS AT 1 atm AND 298 K

Compound	Heat (Enthalpy) of Formation* kcal/mol (ΔH_f^o)	Free Energy of Formation kcal/mol (ΔG_f^o)
Aluminum oxide $Al_2O_3(s)$	−400.5	−378.2
Ammonia $NH_3(g)$	−11.0	−3.9
Barium sulfate $BaSO_4(s)$	−352.1	−325.6
Calcium hydroxide $Ca(OH)_2(s)$	−235.7	−214.8
Carbon dioxide $CO_2(g)$	−94.1	−94.3
Carbon monoxide $CO(g)$	−26.4	−32.8
Copper (II) sulfate $CuSO_4(s)$	−184.4	−158.2
Ethane $C_2H_6(g)$	−20.2	−7.9
Ethene (ethylene) $C_2H_4(g)$	12.5	16.3
Ethyne (acetylene) $C_2H_2(g)$	54.2	50.0
Hydrogen fluoride $HF(g)$	−64.8	−65.3
Hydrogen iodide $HI(g)$	6.3	0.4
Iodine chloride $ICl(g)$	4.3	−1.3
Lead (II) oxide $PbO(s)$	−51.5	−45.0
Magnesium oxide $MgO(s)$	−143.8	−136.1
Nitrogen (II) oxide $NO(g)$	21.6	20.7
Nitrogen (IV) oxide $NO_2(g)$	7.9	12.3
Potassium chloride $KCl(s)$	−104.4	−97.8
Sodium chloride $NaCl(s)$	−98.3	−91.8
Sulfur dioxide $SO_2(g)$	−70.9	−71.7
Water $H_2O(g)$	−57.8	−54.6
Water $H_2O(\ell)$	−68.3	−56.7

* Minus sign indicates an exothermic reaction.

Sample equations:

$$2Al(s) + \frac{3}{2}O_2(g) \rightarrow Al_2O_3(s) + 400.5 \text{ kcal}$$

$$2Al(s) + \frac{3}{2}O_2(g) \rightarrow Al_2O_3(s) \quad \Delta H \approx -400.5 \text{ kcal/mol}$$

Table of Standard Energies of Formation of Compounds (Table G) gives the
heat of formation (how much heat is absorbed or released when the
compound is formed) for many compounds.

ΔH = negative ($\Delta H = -$) means exothermic reaction
ΔH = positive ($\Delta H = +$) means endothermic reaction
ΔG = negative means spontaneous
Review pages 7:7, 7:10

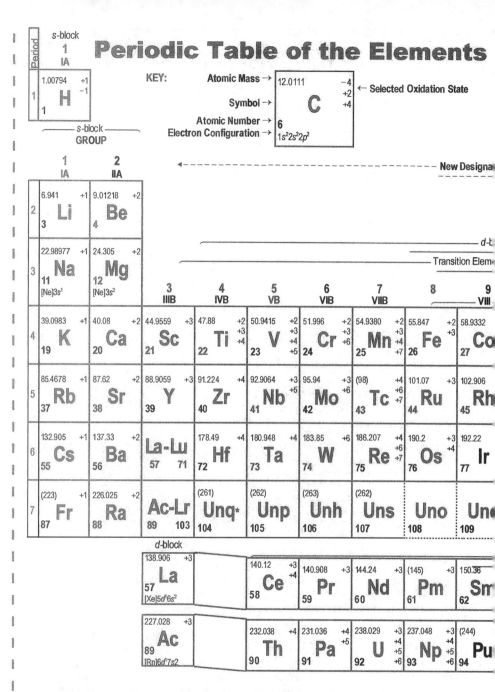

Periodic Table of the Elements

KEY:

Atomic Mass → 12.0111 − 4, +2, +4 ← Selected Oxidation State

Symbol → C

Atomic Number → 6

Electron Configuration → $1s^2 2s^2 2p^2$

s-block

GROUP

This table is NOT part of the New York State Regents Curriculum

PERIODIC TABLE OF THE ELEMENTS: The Periodic Table gives the atomic numbers, atomic mass, oxidation states, electron configuration, and symbols of the elements.

The elements are divided into vertical columns called groups. Group 1 has 1 valence electron; Group 2 has 2 valence electrons.

Relative atomic masses are
based on $_{12}C = 12.00000$

4.00260	0
He	

p-block

13	14	15 GROUP 16	17	18
IIIA	IVA	VA VIA	VIIA	O

10.81 +3 **B** 5	12.0111 −4 +2 +4 **C**	14.00673 +1 −2 +2 −1 +3 +4 **O** 8	15.9994 −2 **O**	18.998403 1 **F** 9	20.179 0 **Ne** 10

| | | | 10 | 11 | 12 | 26.98154 +3 **Al** 13 [Ne]3s²3p¹ | 28.0855 −4 +2 +4 **Si** 14 | 30.97376 −3 +3 +5 **P** 15 [Ne]3s²3p³ | 32.06 −2 +4 +6 **S** 16 [Ne]3s²3p⁴ | 35.453 −1 +1 +3 +5 +7 **Cl** 17 [Ne]3s²3p⁵ | 39.948 0 **Ar** 18 [Ne]3s²3p⁶ |

10	11	12

58.69 +2 +3 **Ni**	63.546 +1 +2 **Cu**	65.39 +2 **Zn**	69.72 +3 **Ga**	72.59 −4 +2 **Ge⁴**	74.9216 −3 +3 +5 **As⁺⁵**	78.96 −2 +4 +6 **Se⁺⁶**	79.904 −1 +1 +5 **Br⁺⁵**	83.80 0 +2 **Kr**

106.42 +2 +4 **Pd**	107.868 +1 **Ag**	112.41 +2 **Cd**	114.82 +3 **In**	118.71 +2 +4 **Sn**	121.75 −3 +3 +5 **Sb⁵**	127.60 −2 +4 +6 **Te⁺⁶**	126.905 −1 +1 +5 +7 **I**	131.29 0 +2 +4 +6 **Xe⁺⁴⁺⁶**

195.08 +2 +4 **Pt**	196.967 +1 +3 **Au**	200.59 +1 +2 **Hg⁺²**	204.383 +1 +3 **Tl**	207.2 +2 +4 **Pb**	208.980 +2 +5 **Bi⁺⁵**	(209) +2 +4 **Po**	(210) **At**	(222) 0 **Rn**

*The systematic names and symbols for elements of atomic numbers greater than 103 will be used until the approval of trivial

MASS NUMBERS IN PARENTHESES ARE THE MASS NUMBERS OF THE MOST STABLE OR COMMON ISOTOPES

f block

151.96 +2 +3 **Eu⁺³** 63	157.25 +3 **Gd** 64	158.925 +3 **Tb** 65	162.50 +3 **Dy** 66	164.930 +3 **Ho** 67	167.26 +3 **Er** 68	168.934 +3 **Tm** 69	173.04 +2 +3 **Yb** 70	174.967 +3 **Lu** 71	Lanthanoid Series

(243) +3 +4 +5 +6 **Am** 96	(247) +3 **Cm** 97	(247) +3 +4 **Bk** 98	(251) +3 **Cf** 99	(252) **Es** 100	(257) **Fm** 101	(258) **Md** 102	(259) **No** 103	(260) **Lr** 103	Actinoid Series

Periods go across the Periodic Table. Period 1 has $_1$H and $_2$He, which have 1 shell. Period 2 has 2 shells.
The Periodic Table shows s-block, p-block, d-block and transition elements.
The bottom of the Table shows the lanthanide and actinide series.

Review: 2:18-2:25

TABLE OF STRENGTHS OF ACIDS

RELATIVE STRENGTHS OF ACIDS IN AQUEOUS SOLUTION AT 1 atm AND 298 K	
Conjugate Pairs *ACID* *BASE*	K_a
$HI = H^+ + I^-$	very large
$HBr = H^+ + Br^-$	very large
$HCl = H^+ + Cl^-$	very large
$HNO_3 = H^+ + NO_3^-$	very large
$H_2SO_4 = H^+ + HSO_4^-$	large
$H_2O + SO_2 = H^+ + HSO_3^-$	1.5×10^{-2}
$HSO_4^- = H^+ + SO_4^{2-}$	1.2×10^{-2}
$H_3PO_4 = H^+ + H_2PO_4^-$	7.5×10^{-3}
$Fe(H_2O)_6^{3+} = H^+ + Fe(H_2O)_5(OH)^{2+}$	8.9×10^{-4}
$HNO_2 = H^+ + NO_2^-$	4.6×10^{-4}
$HF = H^+ + F^-$	3.5×10^{-4}
$Cr(H_2O)_6^{3+} = H^+ + Cr(H_2O)_5(OH)^{2+}$	1.0×10^{-4}
$CH_3COOH = H^+ + CH_3COO^-$	1.8×10^{-5}
$Al(H_2O)_6^{3+} = H^+ + Al(H_2O)_5(OH)^{2+}$	1.1×10^{-5}
$H_2O + CO_2 = H^+ + HCO_3^-$	4.3×10^{-7}
$HSO_3^- = H^+ + SO_3^{2-}$	1.1×10^{-7}
$H_2S = H^+ + HS^-$	9.5×10^{-8}
$H_2PO_4^- = H^+ + HPO_4^{2-}$	6.2×10^{-8}
$NH_4^+ = H^+ + NH_3$	5.7×10^{-10}
$HCO_3^- = H^+ + CO_3^{2-}$	5.6×10^{-11}
$HPO_4^{2-} = H^+ + PO_4^{3-}$	2.2×10^{-13}
$HS^- = H^+ + S^{2-}$	1.3×10^{-14}
$H_2O = H^+ + OH^-$	1.0×10^{-14}
$OH^- = H^+ + O^{2-}$	$< 10^{-36}$
$NH_3 = H^+ + NH_2^-$	very small

Note: $H^+(aq) = H_3O^+$

Sample equation: $HI + H_2O = H_3O^+ + I^-$

Table of Strengths of Acids (Table L):

1. Strong acids are on the top of Table L, very large K_a (ionization constant). Examples: HI, HBr.

   As you go down Table L, the acids get weaker and the value of K_a (ionization constant) gets smaller. Remember, a bigger negative exponent is a smaller K_a.

2. Acids on the top of Table L are good conductors of electricity.

3. The same compound or ion on both sides of Table L is amphoteric (examples: NH_3 and HS^-). Amphoteric means it can act like an acid or a base.

Review pages 8:19, 8:20, 8:21.

TABLE OF EQUILIBRIA

<table>
<tr><th colspan="3">CONSTANTS FOR VARIOUS EQUILIBRIA
AT 1 atm AND 298 K</th></tr>
</table>

$H_2O(\ell) = H^+(aq) + OH^-(aq)$	$K_w = 1.0 \times 10^{-14}$
$H_2O(\ell) + H_2O(\ell) = H_3O^+(aq) + OH^-(aq)$	$K_w = 1.0 \times 10^{-14}$
$CH_3COO^-(aq) + H_2O(\ell) = CH_3COOH(aq) + OH^-(aq)$	$K_b = 5.6 \times 10^{-10}$
$NaF(aq) + H_2O(\ell) = Na^+(aq) + OH^-(aq) + HF(aq)$	$K_b = 1.5 \times 10^{-11}$
$NH_3(aq) + H_2O(\ell) = NH_4^+(aq) + OH^-(aq)$	$K_b = 1.8 \times 10^{-5}$
$CO_3^{2-}(aq) + H_2O(\ell) = HCO_3^-(aq) + OH^-(aq)$	$K_b = 1.8 \times 10^{-4}$
$Ag(NH_3)_2^+(aq) = Ag^+(aq) + 2NH_3(aq)$	$K_{eq} = 8.9 \times 10^{-8}$
$N_2(g) + 3H_2(g) = 2NH_3(g)$	$K_{eq} = 6.7 \times 10^{5}$
$H_2(g) + I_2(g) = 2HI(g)$	$K_{eq} = 3.5 \times 10^{-1}$

Compound	K_{sp}	Compound	K_{sp}
AgBr	5.0×10^{-13}	Li_2CO_3	2.5×10^{-2}
AgCl	1.8×10^{-10}	$PbCl_2$	1.6×10^{-5}
Ag_2CrO_4	1.1×10^{-12}	$PbCO_3$	7.4×10^{-14}
AgI	8.3×10^{-17}	$PbCrO_4$	2.8×10^{-13}
$BaSO_4$	1.1×10^{-10}	PbI_2	7.1×10^{-9}
$CaSO_4$	9.1×10^{-6}	$ZnCO_3$	1.4×10^{-11}

Table of Equilibria (Table M): At the top of Table M, you are given the equilibrium constant for water: $K_w = 1.0 \times 10^{-14}$, which you will need for the Regents. You are also given other equilibrium constants on the top half of this table.

The bottom part of Table M gives K_{sp} (**solubility product**) of various compounds. The larger the K_{sp}, the more soluble the compound is. $LiCO_3$ has $K_{sp} = 2.5 \times 10^{-2}$, the biggest K_{sp}, most soluble. AgI has $K_{sp} = 8.3 \times 10^{-17}$, smallest K_{sp}, least soluble. Remember, the biggest negative exponent has the smallest K_{sp}.

Review pages 7:27, 7:28, 8:21.

Table of Reduction Potentials: This table lists the reduction half-reaction with its half-cell potential, E° (measured by reacting it with the hydrogen half-cell.)

HOW TO USE TABLE OF REDUCTION POTENTIALS
TABLE N

1. Calculate E°:

$$Al(s) + 2Ag^+ \rightarrow Al^{3+} + 3Ag(s)$$

This example starts with Al. This Table starts with Al^{3+} (reverse or opposite reaction). The Table has E° = -1.66; therefore, in this example, change the sign of E°, making E° = $+1.66$.

This example and the Table start with Ag^+. Leave the sign of this E° the way it is: E° = $+0.80$

Add the two E°'s:

$$Al \rightarrow Al^{3+} \quad E° = +1.66$$

$$Ag^+ \rightarrow Ag \quad E° = \underline{+\ 0.80}$$

$$E° = +2.46$$

E° is positive (E° = +); the reaction is spontaneous.

2. Top of Table of Reduction Potentials: Most easily reduced. Bottom of Table (right side): Most easily oxidized.

3. Anything on the left side reacts spontaneously with anything below it to the right. F_2 reacts spontaneously with Cl^-. Anything on the right side reacts spontaneously with anything to the left and above it. Cl^- reacts spontaneously with F_2.

Table of Reduction Potentials	
Half Reaction	E° *(volts)*
Left Right	
$F_2 + 2e \rightarrow 2F^-$	
$Au^{+3} + 3e \rightarrow Au$	
$Cl_2 + 2e \rightarrow 2Cl^-$	

4. Any metal below H_2 reacts with acid to liberate hydrogen

Review pages 9:19-9:25.

STANDARD ELECTRODE POTENTIALS

Ionic Concentrations 1 M Water At 298 K, 1 atm

Half-Reaction	E^0 (volts)
$F_2(g) + 2e^- \rightarrow 2F^-$	+2.87
$8H^+ + MnO_4^- + 5e^- \rightarrow Mn^{2+} + 4H_2O$	+1.51
$Au^{3+} + 3e^- \rightarrow Au(s)$	+1.50
$Cl_2(g) + 2e^- \rightarrow 2Cl^-$	+1.36
$14H^+ + Cr_2O_7^{2-} + 6e^- \rightarrow 2Cr^{3+} + 7H_2O$	+1.23
$4H^+ + O_2(g) + 4e^- \rightarrow 2H_2O$	+1.23
$4H^+ + MnO_2(s) + 2e^- \rightarrow Mn^{2+} + 2H_2O$	+1.22
$Br_2(\ell) + 2e^- \rightarrow 2Br^-$	+1.09
$Hg^{2+} + 2e^- \rightarrow Hg(\ell)$	+0.85
$Ag^+ + e^- \rightarrow Ag(s)$	+0.80
$Hg_2^{2+} + 2e^- \rightarrow 2Hg(\ell)$	+0.80
$Fe^{3+} + e^- \rightarrow Fe^{2+}$	+0.77
$I_2(s) + 2e^- \rightarrow 2I^-$	+0.54
$Cu^+ + e^- \rightarrow Cu(s)$	+0.52
$Cu^{2+} + 2e^- \rightarrow Cu(s)$	+0.34
$4H^+ + SO_4^{2-} + 2e^- \rightarrow SO_2(aq) + 2H_2O$	+0.17
$Sn^{4+} + 2e^- \rightarrow Sn^{2+}$	+0.15
$2H^+ + 2e^- \rightarrow H_2(g)$	0.00
$Pb^{2+} + 2e^- \rightarrow Pb(s)$	-0.13
$Sn^{2+} + 2e^- \rightarrow Sn(s)$	-0.14
$Ni^{2+} + 2e^- \rightarrow Ni(s)$	-0.26
$Co^{2+} + 2e^- \rightarrow Co(s)$	-0.28
$Fe^{2+} + 2e^- \rightarrow Fe(s)$	-0.45
$Cr^{3+} + 3e^- \rightarrow Cr(s)$	-0.74
$Zn^{2+} + 2e^- \rightarrow Zn(s)$	-0.76
$2H_2O + 2e^- \rightarrow 2OH^- + H_2(g)$	-0.83
$Mn^{2+} + 2e^- \rightarrow Mn(s)$	-1.19
$Al^{3+} + 3e^- \rightarrow Al(s)$	-1.66
$Mg^{2+} + 2e^- \rightarrow Mg(s)$	-2.37
$Na^+ + e^- \rightarrow Na(s)$	-2.71
$Ca^{2+} + 2e^- \rightarrow Ca(s)$	-2.87
$Sr^{2+} + 2e^- \rightarrow Sr(s)$	-2.89
$Ba^{2+} + 2e^- \rightarrow Ba(s)$	-2.91
$Cs^+ + e^- \rightarrow Cs(s)$	-2.92
$K^+ + e^- \rightarrow K(s)$	-2.93
$Rb^+ + e^- \rightarrow Rb(s)$	-2.98
$Li^+ + e^- \rightarrow Li(s)$	-3.04

INTERNET RESOURCES

1. The following web sites are useful resources for chemistry:
    www.encarta.msn.homework
    www.chem4kids.com
    www.howstuffworks.com
    www.jpl.org/teen
    www.thinkquest.org

2. For additional sites, you can try the following search engines:
    askjeeves.com
    about.com
    altavista.com
    yahoo.com
    google.com
    webcrawler.com

3. The author can take no responsibility for errors, omissions, or misinformation on any of these sites. Students undertaking an experiment suggested in these sites must become fully aware of safe procedures and risks.

APPENDIX III: GLOSSARY

A

absolute zero: $0K = -273\,^{\circ}C$

accelerator: gives charged particles enough kinetic energy to penetrate the nucleus.

acid, Arrhenius: has H and yields (releases or gives off) H^+ in aqueous solutions.

acid, Bronsted-Lowry: a proton donor (gives away a proton, H^+).

activation energy: smallest amount of energy needed to start a reaction.

addition: in organic reactions, adding one or more atoms at a double or triple bond.

addition polymerization: monomers joining together by breaking a double or triple bond (adding together the monomers) to form a polymer. Example: Ethene (ethylene) + ethene (ethylene) + ethene (ethylene) join together by breaking the double bond to form polyethylene.

alcohol: an organic compound in which one or more hydrogens are replaced by an OH (hydroxyl) group.

aldehyde: functional group, $\overset{\overset{H}{|}}{C{=}O}$, with a single bond between carbon and hydrogen and a double bond between C and O.

alkali metals: Elements in Group 1 are called alkali metals.

alkaline earth metals: Elements in Group 2 are called alkaline earth metals.

alkane: a group of hydrocarbons with the general formula C_nH_{2n+2}. An alkane has single bonds and is saturated.

alkene: a group of hydrocarbons also known as the *ethylene series*, with the formula C_nH_{2n}. An alkene has one double bond and is unsaturated.

alkyl: This group has one less hydrogen than the corresponding alkane.

alkyne: also known as the *acetylene series*, C_nH_{2n-2}. An alkyne has a triple bond and is unsaturated.

allotropes: forms of the same element that have different molecular formulas or crystal structures.

alloy: a mixture of two metals. Example: Iron with nickel or chromium to produce stainless steel to prevent corrosion.

alpha decay: a nucleus decays and gives off alpha particles, ^4_2He.

alternate acid-base theory: theory that describes acids as H^+ donors and bases as H^+ acceptors; formerly known as Bronsted-Lowry theory.

amide: organic compound made by reacting an acid with an amine.

amine: organic compound that has nitrogen attached to carbon.

amino acid: organic compound that has both an acid and an amine group.

amphoteric: sometimes acts like an acid and sometimes acts like a base. Examples: Water and HSO_4^-.

anode: the name of the electrode where oxidation takes place. An anode is *negative* in an *electrochemical* cell. An anode is *positive* in an *electrolytic cell*.

aqueous solution: Water is the solvent in the solution.

Arrhenius acid: an acid that has H and releases (yields or gives off) H^+ in an aqueous solution.

Arrhenius base: has OH and yields (releases or gives off) OH^- in an aqueous solution.

artificial radioactivity: Elements can be made radioactive by bombarding their nuclei with high energy particles such as protons, neutrons and alpha particles.

artificial transmutation: transforming one element into another element by bombarding the nucleus with high energy particles.

atomic mass of an element: the weighted average mass of the naturally occurring isotopes of that element.

atomic mass unit: $1/12^{th}$ the mass of ^{12}C.

atomic number: equals the number of protons in the nucleus of an atom.

atomic radius: half the distance between adjacent nuclei.

Avogadro's Hypothesis or Law: equal volumes of all gases under the same conditions of temperature and pressure have an equal number of molecules.

B

base, Arrhenius: has OH and yields (releases or gives away) OH^- in aqueous solutions.

base, Bronsted-Lowry: a proton acceptor (accepts protons).

battery: electrochemical cell.

benzene family: cyclic aromatic hydrocarbons having the formula C_nH_{2n-6}. Benzene is the first and simplest member of the benzene series.

beta decay: Nucleus decays and gives off beta particles.

Boyle's Law: At constant temperature, the volume of a gas is inversely proportional to pressure.

binary compound: consists of only two elements, e.g., NaCl.

Bohr model of atom: electrons revolve (go around) the nucleus in concentric circular orbits.

boiling point of water: 100°C. Also called water-steam equilibrium temperature. Water will boil at 100°C when vapor pressure equals atmospheric pressure (pressure of the atmosphere).

bottled gas: a fuel made of propane and butane.

bright line spectrum: spectrum of color given off by an element when it goes back from an excited state to the ground state.

Bronsted-Lowry acid: a proton donor (gives away a proton, H^+).

Bronsted-Lowry base: a proton acceptor (accepts a proton, H^+)

C

calorie: the amount of heat necessary to raise 1 gram of water 1°C.

catalyst: lowers the activation energy and gives a faster rate of reaction.

cathode: the name of the electrode where reduction takes place. Cathode is positive in an electro-

chemical cell; cathode is negative in an electrolytic cell.

centrifuge: an instrument that spins very fast and separates mixtures by density.

Charles's Law: At constant pressure, volume is directly proportional to Kelvin (absolute) temperature.

High Marks: Regents Chemistry Made Easy

chemical bond: the attachment between atoms caused by a sharing or transferring of electrons.

chemical change: produces a new substance with different properties.

chemical equation: shows which bonds are broken and which bonds are built.

chemical equilibrium: when the concentration of the reactants and products of a reaction remain constant. Equilibrium means the forward and reverse reactions take place at the same rate.

chemical formula: describes the composition of an element or compound.

chemical property: how a substance reacts to form new substances.

Chemistry: the study of matter.

chromatography: a way to separate mixtures. The mixture is put near one end of a filter paper. The paper is put into a solvent. As the solvent moves up, different substances move different distances up the paper.

collision theory: explanation of rates of reaction. Rate depends on how often collisions occur and how many of them are effective.

Combined Gas Law: Use this law when **both** temperature and pressure **change** at the same time. Temperature must be in Kelvin.

common ion effects: In general, when an ionic compound like AgCl breaks up into ions (here, $AgCl \rightarrow Ag^+ + Cl^-$), an increase in the concentration of one of the products (such as Cl^-), causes the other product (this example, Ag^+) to decrease, and the reactant (this example, AgCl) to increase.

compound: a substance which can be broken down into two or more

elements by chemical means. A compound consists of two or more elements, chemically united, and has a fixed composition.

concentrated solution: contains a large amount of dissolved substance.

condensation polymerization: monomers joining together by dehydration synthesis (removing water) to form a polymer.

conjugate acid: substance formed when a base takes on a proton. Add H^+. H_2O and H_3O^+ are a conjugate acid-base pair; they differ by H^+.

conjugate base: substance that remains after the acid gives up a proton. If HSO_4^- is the acid, the conjugate base is SO_4^{2-}.

contact process: produces H_2SO_4, sulfuric acid.

control rods: In a nuclear reactor, control rods control the number of neutrons in the reactor. Control rods are made of boron and cadmium to absorb neutrons, therefore control the supply of neutrons in the reactor.

coolants: In a nuclear reactor, coolants keep the temperature at a reasonable level during a fission reaction. Examples: H_2O, heavy water, molten sodium, and molten lithium.

coordinate covalent: a bond formed when one atom donates both electrons that are shared.

corrosion: gradual attack on a metal by its surroundings (moisture, some gases in the air), with the result that the usefulness of the metal is destroyed.

covalent: a bond formed when two atoms share electrons. In covalent bonds, the electronegativity difference must be less than 1.7.

cracking: breaking large molecules of hydrocarbons into smaller

---

molecules. It is used because it breaks large molecules into gaso-line and fuel oil.

D

dating wood: a ratio of C-14 to C-12 determines the age of a sample of wood.

deductive reasoning: logic that starts with a general rule, then goes to individual cases.

$$\text{density} = \frac{mass}{volume}$$

dienes: unsaturated hydrocarbons that have two double bonds. They are not alkenes.

dihydroxy alcohol: a compound containing 2 OH groups is a dihydroxy alcohol or glycol.

dilute solution: contains very little dissolved substance (example: salt).

dipole: molecule with charge distributed unevenly, one end is more positive and one end is more negative.

dispersion forces: weak attractive forces between nonpolar molecules.

distillation: a process that separates a mixture of liquids or liquids and solids by boiling points.

dynamic equilibrium: equilibrium where the rate of forward and reverse reactions are equal. The same amount of material dissolves as comes out of solution.

E

electrochemical cell: a cell that uses 2 metals in conducting liquid, either to generate electricity or for electrolysis.

electrode: metal strip in a half cell where oxidation or reduction takes place.

electrolysis of water: Electricity breaks down water into hydrogen and oxygen, $2 H_2O \rightarrow 2 H_2 + O_2$.

electrolyte: a substance that dissolves in water and forms a solution that conducts an electric current (electricity).

electrolytic cell: a cell in which electric current (electricity) produces a chemical reaction.

electron: goes around the nucleus. An electron has a negative charge and very little mass (1/1836 amu).

electron configuration: shows how many electrons are in each principal energy level.

electron dot method: shows how many electrons are in the last shell (valence electrons). Dots around the symbol of the element represent valence electrons.

electronegativity: the attraction of an atom for electrons in a bond.

electroplating: another example of an electrolytic cell. A layer of a metal such as silver or copper coats any object (spoon or fork) to be plated.

element: a substance which cannot be decomposed (broken down) into anything simpler by a chemical change (chemical means). Examples: hydrogen, oxygen and nitrogen.

empirical formula: represents the simplest ratio in which atoms combine to form a compound.

endothermic: chemical reaction in which energy is absorbed. ΔH is positive in an endothermic reaction.

energy: ability to do work.

equilibrium: when the forward and reverse reactions occur at the same rate.

equilibrium constant: equilibrium constant K_{eq} = product of the molar concentrations on the right side of the equation divided by the product of the molar concentrations on the left side of the equation. The coefficient of any substance becomes the exponent (or power) for that substance.

equilibrium vapor pressure: the pressure caused by the water vapor at equilibrium.

esterification: acid and alcohol producing ester and H_2O.

ethene: C_2H_4, first member of the alkene group, commonly called ethylene. C_2H_4

$$\begin{array}{cc} H & H \\ | & | \\ H-C{=}C-H \end{array}$$

, has 1

double bond, and is unsaturated.

ethers: -O- functional group. R_1-O-R_2 R_1 and R_2 are hydrocarbon groups which contain carbon.

ethylene glycol: a dihydroxy alcohol (2 OH).

ethyne: first member of the alkyne group, commonly called acetylene. C_2H_2: H-C≡C-H: has a triple bond and is unsaturated.

evaporation: when water changes into a gas (water vapor).

excited: an atom is excited when electrons absorb energy and jump ahead to a higher energy level, leaving one of the previous orbitals half empty.

exothermic: a chemical reaction that gives off energy. ΔH is negative in an exothermic reaction.

F

fats: esters derived from glycerol (trihydroxy alcohol) and long-chain fatty acids.

fermentation:
$$C_6H_{12}O_6 \rightarrow 2C_2H_5OH + 2\ CO_2$$
$$\text{ethanol} \quad \text{carbon dioxide}$$
Yeast produces an enzyme (zymase) that breaks down glucose into ethanol and carbon dioxide.

filtration: a process that separates the solid and liquid parts of a mixture by pouring through tiny openings.

fission reaction: In a fission reaction, an atom absorbs a neutron, splits into pieces, and gives off energy.

fixed points: the boiling point (100°C) and freezing point (0°C) of water are the two fixed points on the thermometer.

fractional distillation: separates the different (mixtures of) hydrocarbons in petroleum by their boiling points.

free-energy change:
$$\Delta G = \Delta H - T\Delta S.$$
To determine if a reaction is spontaneous, use this formula.
ΔG = free energy change.
ΔH = heat of reaction.
ΔS = change of entropy.
T = Kelvin temperature.
If ΔG is negative, the reaction is spontaneous. If ΔG = zero, it is at equilibrium.

freezing point of water: 0°C = ice-water equilibrium temperature.

fuel: In a nuclear reactor: uranium, ^{235}U, and plutonium, ^{239}Pu, are fissionable fuels used in a nuclear reactor.

fusion reaction: In a fusion reaction, light nuclei combine or unite to form a heavier nucleus.

G

gamma rays: rays given off when a nucleus decays. Gamma rays are similar to high-energy X-rays. Gamma rays are not particles. They have no charge and no mass.

gas: no definite volume, no definite shape of its own, spreads all over. A gas takes the shape and volume of the container you put it in.

geologic dating: ratio of uranium-238 to lead-206 in a mineral determines the age of the mineral.

glycerol: a trihydroxy alcohol.

Graham's Law: Under the same conditions of temperature and pressure, gases diffuse (spread out) at a rate inversely (opposite) proportional to the square roots of their molecular masses.

groups: vertical colums of the Periodic Table are called groups.

H

Haber Process: produces ammonia.
$$N_2 + 3 H_2 \rightarrow 2 NH_3$$
nitrogen + hydrogen → ammonia

half life: time required for half of the sample to decay.

halide (halocarbon): organic compound that contains halogen (fluorine, chlorine, bromine, iodine) instead of 1 or more hydrogens in the molecule.

halogen group: Group 17 of the Periodic Table.

Heat of Fusion: the amount of heat needed to change solid to liquid at a constant temperature.

Heat of Reaction: the amount of heat given off or absorbed in a chemical reaction.
$$\Delta H = H_{products} - H_{reactants}.$$

Heat of Vaporization: the amount of heat needed to change liquid to gas at a constant temperature.

heterogeneous: made up of two different things not evenly spread out.

homogeneous: a substance or mixture with particles spread out evenly. Solutions are homogeneous mixtures.

hydrocarbons: organic compounds that contain only carbon and hydrogen atoms.

hydrogen bomb: releases tremendous energy from a fusion reaction using a fission reaction as a trigger.

hydrogen bonding: connects one water molecule with another water molecule.

hydrogen bonds: formed between molecules when hydrogen is covalently bonded to a small, highly electronegative atom; therefore, hydrogen bonds occur in compounds with nitrogen (N), oxygen (O), or fluorine (F).

hydrolysis: salts react with water to form solutions that are acidic or basic.

indicator: dye that changes color in acid or base.

inductive reasoning: logic that starts with individual cases, then derives a general rule.

insoluble: the compound hardly dissolves; very, very, very little.

iodine-131: radioisotope that diagnoses thyroid disorders.

ionic bonds: formed by the transfer of 1 or more electrons from one atom to another to form ions (charged particles).

ionic radius: the size of an ion. In metals, the ionic radius is smaller than the atomic radius; in nonmetals, the ionic radius is bigger than the atomic radius.

ionization constant: the equilibrium constant for ionic substances is called K_{ion}. The bigger the ionization constant, the more ions there are and the greater the degree of ionization.

ionization constant of water: K_w (H^+ times OH^-) = $[H^+][OH^-]$ = 1 x 10^{-14}

ionic substances: have high melting points, conduct electricity in solution and in liquid form but not in solid form, and dissolve in polar substances such as water.

ionization energy: the amount of energy needed to remove the most loosely bound electron from an atom.

ions: charged particles.

isomers: have the same molecular formula but different structural formulas.

n-pentane and 2-methyl butane are isomers. Both have the same molecular formula C_5H_{12}.

isotopes: have the same atomic number but different mass numbers.

joule: unit for measuring energy.

Kelvin temperature: Kelvin = Celsius + 273°

Ketone: The functional group is C=O. In a ketone, C=O is in the middle, and carbon atoms are on both sides of the C=O.

kilocalorie: 1 kilocalorie equals 1000 calories.

kinetic energy: energy having to do with motion (things moving).

kinetics: concerned with rates of reaction (how fast, how many moles are consumed or produced in a unit of time) and mechanism (steps in a chemical reaction).

kinetic theory: a model that tells how gases should behave: also called ideal gas law.

Law of Conservation of Energy: Energy may be changed from one form to another, but the total amount is the same.

Le Chatelier's Principle: If a system is subjected to a stress, the equilibrium will be displaced in the direction that relieves the stress.

Lead-acid battery: A battery is an electrochemical cell. Pb is the anode where oxidation takes place. PbO_2 is the cathode where reduction takes place. Lead-acid batteries are used in cars.

Lewis electron dot structures: a way to show valence electrons by drawing x's or dots around the symbol of the element.

liquid: definite volume (4 ounces or 8 ounces, etc.), takes the shape of the container.

M

mass number: equal to the total number of protons and neutrons in the nucleus.

matter: anything that takes up space and has mass: Examples: desk, book, salt, sugar, etc.

mechanism: steps in a reaction.

melting point: temperature when a solid changes to a liquid at 1 atmosphere pressure.

metals: good conductors of heat and electricity. They have metallic luster and are malleable and ductile. Metals are to the left of the zigzag line on the Periodic Table.

metal hydride: a compound of an active metal and hydrogen (example: Na and H).

metallic bonding: occurs in metals. A metal consists of positive ions surrounded by a "sea" of mobile electrons.

metalloids: have characteristics of both metals and nonmetals. Metalloids are on the zigzag line of the Periodic Table.

methanal: $\begin{matrix} H \\ | \\ H-C=O \end{matrix}$: Common name is formaldehyde. First member of aldehydes.

methane: CH_4

methanoic acid: first member of the organic acid series. Common name is formic acid.

mixture: two or more substances mixed together. Example: salt and sugar mixed together.

model: a picture, drawing, chart, equation, etc., used to represent something.

moderator: In a nuclear reactor, a moderator slows down the speed of neutrons (so that the uranium nuclei can absorb them.) Examples: water, heavy water, beryllium, graphite.

molal boiling point elevation (for water): $0.52°C$. Boiling point elevates (rises, goes up) by $0.52°C$ for each mole of particles in a kilogram of water.

molal freezing point depression (for water): $1.86°C$. One mole of particles in a kilogram of water lowers the freezing point by $1.86°C$.

molality: number of moles of solute dissolved in a kilogram of solvent.

molarity of a solution: the number of moles of solute in 1 liter of solution.

mole (of molecules): 1 gram molecular mass = 6.02×10^{23} molecules = 22.4 liters for gases at STP.

molecular formula: indicates the total number of atoms of each element needed to form the molecule.

molecular substance: element or compound made of molecules.

molecule-ion attraction: The positive ions of the salt go to the negative part of the solvent (example: water), and the negative ions of the salt go to the positive part of the solvent (example: water).

molecular mass: the sum of the masses of the atoms in a molecule. The molecular mass of a gas can be calculated by using this formula: molecular mass = density at STP x 22.4 liters.

molecule: A molecule is

the smallest discrete particle of an element or compound that has covalent bonds.

monohydroxy alcohol: has only one OH (hydroxyl group).

monomer: a small molecule that can be joined together with other small molecules to form a polymer.

N

natural gas: A fuel, this is mostly methane and ethane.

network solid: A solid that has covalently bonded atoms (covalent bonds) linked in 1 big network or one big macromolecule.

neutralization: Equivalent amounts of acid + base → salt + water. H^+ + OH^- → H_2O.

nickel oxide-cadmium battery: A battery is a electrochemical cell. Ni^{3+} is the cathode, and Cd is the anode.

noble gases: The elements of Group 18 are called noble gases. The atoms have a complete outer shell: electron configuration is stable.

nonmetals: poor conductors of heat and electricity. They lack luster and are brittle. Nonmetals are to the right of the zigzag on the Periodic Table.

nonpolar covalent: When electrons are shared between atoms of the same element (therefore, same electronegativity) or different elements with the same electronegativity, the electrons are shared equally and the bond is nonpolar. Electrons are in the middle between the atoms.

nucleus: The nucleus has protons and neutrons and is in the center of the atom.

neutron: in the nucelus, has no charge, and a mass of one atomic mass unit.

nuclear energy: In nuclear reactions, mass is converted into energy.

nuclear reactor: device that contains a fuel that undergoes a controlled fission reaction to produce tremendous amounts of energy.

O

orbital: the place where the electron may be found.

orbital model of an atom: tells us the place (orbitals) where we will probably find the electrons.

organic acids: a double bond between C and O and a single bond between C and OH is the functional group of organic acids.

organic chemistry: the study of carbon and carbon compounds.

oxidation: uniting with oxygen. Hydrocarbons react with oxygen, producing carbon dioxide and water.

oxidation: in electrochemistry, loss of electrons.

oxidized: Na loses an electron. Sodium is oxidized. Sodium is a reducing agent.

oxidizing agent: is reduced = gain of electrons.

P

partial pressure: pressure produced by 1 gas in a mixture of gases. N_2 and O_2 gases are mixedtogether. The pressure produced by N_2

alone is called the partial pressure of N_2.

parts per million: amount of dissolved substance, expressed as $\dfrac{grams\ of\ solute}{grams\ of\ solution} \times 1000000$.

peer review: having scientific work checked by experts in the field before it is published.

percent by mass: amount of dissolved substance, expressed as $\dfrac{grams\ of\ solute}{100\ grams\ of\ solution}$.

percent by volume: amount of dissolved substance, expressed as $\dfrac{grams\ of\ solute\ (salt)}{100\ ml\ of\ solution}$ or $\dfrac{milliliters\ of\ solute\ (liquid)}{100\ ml\ of\ solution}$

percent error: how much a measurement is different from the accepted value.

periods: Horizontal rows of the Periodic Table are called periods.

Periodic Law: states that the properties of the elements are a periodic function of their atomic number.

peroxide: compound in which oxygen has an oxidation state of -1. Example: H_2O_2: Hydrogen Peroxide.

petroleum: a mixture of hydrocarbons (example: gasoline, kerosene, and fuel oil).

pH: indicates the strength of the acid or base (how strong the acid or base is). pH less than 7 is acidic. The lower the number, the more acidic. pH 7 is neutral, pH more than 7 is basic; the higher the number the more basic.

phase equilibrium (dynamic equilibrium): rate of change from liquid to gas equals rate of change from gas to liquid. Phase equilibrium is also when rate of change from solid to liquid equals rate of change from liquid to solid.

physical change: change in appearance, but the substance itself is not changed.

physical equilibrium: rates of forward and reverse reactions are equal for a physical change, such as change of state or dissolving in water.

physical property: a characteristic that can be recognized without changing the substance to anything else.

polar covalent: When electrons are shared between atoms of different elements (different electronegativities), they are shared unequally, and the bond between the atoms is polar.

polymer: A large molecule is called a polymer.

polymerization: smaller molecules joining together to form one big molecule.

positron decay: breakdown of a nucleus that gives off positrons, particles with the same mass as an electron but a positive charge.

precipitate: an insoluble compound which does not dissolve.

pressure: amount of force exerted per unit-of-area.

primary alcohol: carbon that is attached to the OH group is attached to only 1 carbon or no carbons.

principal energy level: shows how far the electron is from the nucleus. First energy level (shell #1) is closest to the nucleus, while other shells are further away from the nucleus.

propanone: has 3 C atoms and is a ketone. All ketones have $\overset{O}{\underset{C}{\|}}$ between 2 carbon atoms.

propene: C_3H_6

proton: in the nucleus, has a positive charge, and a mass of 1 atomic mass unit.

Q

quanta: when the excited electron (the electron that jumped ahead) goes back to lower energy levels, it gives off specific amounts of energy called quanta.

R

radioactive isotope (radioisotope): isotope that breaks down into smaller atoms, giving off particles and rays.

radioactivity: nucleus of certain atoms decays (breaks down) spontaneously and gives off rays and particles.

radioactive wastes: Fission products from nuclear reactors are very radioactive and these wastes must be stored for a very long time. Examples of radioactive wastes: strontium-90, cesium-137, radon-222, krypton-85, and nitrogen-16.

radium and cobalt-60: Radioisotopes used for cancer therapy.

rate of reaction: How fast, how many moles are produced or consumed in a unit of time

redox: Reactions involving oxidation and reduction are called redox reactions.

reduced: Fluorine gains one electron. F is reduced. F is an oxidizing agent.

reducing agent: Oxidized: loss of electrons.

reduction: Gain of electrons.

S

salt bridge: permits ions to flow (to migrate) between the two half cells of an electrochemical cell.

saponification (hydrolysis): an ester breaks up into acid and alcohol (reverse of esterification). Saponification produces soap.

saturated organic compound: compound that has only single bonds between carbon atoms.

saturated solution: A solution that has the most solute (salt) that it can hold. In a saturated solution, rate of crystallization equals rate of dissolving.

secondary alcohol: carbon attached to OH is attached to 2 other carbon (C) atoms.

shielding: In a nuclear reactor, the shield protects the reactor and people from radioactivity. Concrete and steel are used for a shield.

significant digits: a way to tell how precise a number is. Significant digits are the certain ones and one estimated digit.

solar energy: from fusion of hydrogen atoms to form helium.

solid: has definite shape, definite volume and crystalline structure. A crystal has particles arranged in a regular geometric pattern.

solubility of a solute: mass of solute that dissolves in a given volume of solvent at equilibrium. Or, solubility is the concentration of solute in a saturated solution (most that it can hold).

solubility product constant: the product of the molar concentrations of the compound's ions (in a saturated solution), each raised to the appropriate power.

solute: substance like salt or sugar dissolved in the solvent (example: water).

solution: a homogeneous mixture (made up of 2 or more substances).

solution equilibrium: rate at which solute (sugar) crystallizes equal rate at which solute (sugar) dissolves.

solvent: usually a liquid (example: water) that the solute (salt or sugar) is dissolved in.

specific heat capacity: the amount of heat needed to raise 1 gram of a substance $1\,°C$.

spectrum: lines of energy that are produced when electrons go from higher to lower energy.

spontaneous reaction: a reaction that, under a specific set of conditions, will take place. Spontaneous reactions occur in the direction of less energy and greater entropy (randomness, disorder).

stable: a nucleus that does not break up (decay) into anything else.

static equilibrium: equilibrium where nothing happens, e.g., a mixture of hydrogen and oxygen at room temperature.

STP: Standard Temperature and Pressure (of a gas) is $0°C$ (273K) (temperature) and 760 mm Hg or 760 torr. Or 1 atmosphere (pressure).

supersaturated solution: contains more dissolved substance than should be able to dissolve at that temperature.

sublevels: principal energy levels are divided into sublevels.

sublimation: change from solid to gas without passing through liquid phase (no liquid phase). Examples of solid directly to gas: dry ice (carbon dioxide), $CO_2(s)$ → $CO_2(g)$; iodine, $I_2(s)$ → $I_2(g)$

substance: any variety of matter that has the same properties and composition throughout.

substitution: in organic reactions, replacement of one kind of atom or group.

superposition: the bond between the carbon atoms in benzene, C_6H_6, is the average of single and double bonds.

T

technetium-99: radioisotope that pinpoints brain tumors.

temperature: a measure of the average kinetic energy of the molecules.

tertiary alcohol: C attached to OH is attached to three other carbon atoms.

titration: adding measured volumes of an acid or base of known molarity to a base or acid of unknown molarity until neutralization occurs.

toluene: C_7H_8: second member of benzene series.

tracer: follows the course of an organic reaction. Carbon-14 is used as a tracer.

transition elements: found in Groups 3-11 of the Periodic Table. These are the elements in which d-orbitals of the next to the outermost principal energy level are being filled. They have colored ions and multiple positive oxidation states.

transmutation: when the nucleus of an atom decays and one element is changed into another element.

trihydroxy alcohol: Compound with 3 OH groups is a trihydroxy (trihydric) alcohol. An example is glycerol.

U

unsaturated organic compound: compound that has at least one double or triple bond.

unsaturated solution: When you can still add solute (salt) to a solution and it will dissolve, it is called an unsaturated solution.

V

Van der Waals Forces: weak attractive forces between nonpolar molecules.

vapor pressure: Some water is changed into vapor (gas) and vapor exerts a pressure on the sides of the container. This pressure is called vapor pressure.

voltaic cell: cell that uses redox reactions to produce electricity.

APPENDIX IV: INDEX

A

Absolute zero 1:23;
Accelerator 11:9;
Acid, Arrhenius 8:3
Acid, Bronsted-Lowry 8:4
Activation energy......... 7:2,7:4-7:6
Addition polymerization 10:43;
Alcohol   1:4; 6:5, 10:23, 10:26, 10:30,
    10:31, 10:39, 10:40, 10:42, 10:45,
Aldehyde 10:26, 10:31, 10:32;
Alkane   10:2-9,10:23,10:24, 10:26,
    10:27, 10:39, 10:40,
Alkene 10:10-13;
Alkyl 10:7, 10:17;
Alkyne 10:14-16;
Alloy 13:2, 13:3;
Alpha decay ... 11:2, 11:3, 11:5, 11:18;
Alternate acid-base theories 8:4
Amide 10:30, 10:37, 10:38;
Amine 10:29, 10:30, 10:36, 10:37;
Amino Acid 10:29, 10:43;
Amphoteric 8:5, 8:20;
Anode 9:13, 9:14, 9:25-27, 13:4;
Aqueous solution. . . .4:28; 6:1, 6:12; 8:3,
    8:11, 8:13;
Arrhenius acid 8:3;
Arrhenius base 8:3;
Artificial radioactivity 11:8;
Artificial transmutation. . 11:8-10, 11:19,
    11:20, 11:22;
Asymmetrical molecule 4:20
Atomic mass. 1:35; 2:1, 2:4, 2:5, 2:12-14;
    3:1; 5:4, 5:9, 5:11-13, 5:16, 5:19,
    5:20, 5:23, 5:25, 5:26;
Atomic mass unit .. 2:1, 2:4, 2:12, 2:13;
Atomic number..... 2:2-4, 2:6, 2:7, 2:9,
    2:12-14, 2:16; 3:1, 3:4, 3:8, 3:9; 4:2,
    4:17; 11:1-4, 11:9, 11:10, 11:17,
    11:18; 12:6, 12:7;
Atomic radius 3:28;
Attractive force 1:29, 1:30
Avogadro's hypothesis 1:29; 5:16;
Avogadro's number 1:29

B

Base, Arrhenius 8:3
Base, Bronsted-Lowry 8:4
Battery 9:26; 13:3, 13:4, 13:6-8;
Benzene 10:1, 10:18;

Beta decay 11:2, 11:3;
Binary compound .... 1:1, 1:13 ;

Bohr model 2:5;
Boiling point.....1:2, 1:3, 1:7, 1:8,
    1:11, 1:12, 1:21, 3:7, 3:9, 3:10; 4:21,
    4:22, 6:9-12, 6:14;   10:4, 10:21,
    10:22, 10:39, 10:40; 13:5;
Boiling point elevation 6:9;
Bond . 4:1-4, 4:7-18, 4:20, 4:21, 4:25-28;
    7:1; 10:1, 10:2, 10:4, 10:7-18, 10:20,
    10:22, 10:26, 10:30, 10:32-36,
    10:39-43, 10:48;
Bottled gas 13:5;
Boyle's Law. .1:21, 1:22, 1:25, 1:28, 1:34;
    12:7;
Bright line spectrum 2:8, 2:9;
Bromcresol green 8:9
Bromthymol blue 8:9, 8:10;
Bronsted-Lowry Acid 8:4, 8:17;
Bronsted-Lowry Base 8:4;

C

Catalyst. . .7:2, 7:4-6, 7:13, 7:15, 7:17-19,
    7:23; 13:1, 13:2;
Cathode 9:13, 9:14, 9:25-27, 13:4;
Centrifuge 1:4;
Charles's Law 1:23-25, 1:28, 1:34;
Chemical bond 4:1, 4:3, 4:17;
Chemical change. . .1:1, 1:13, 7:11; 9:25,
    9:26;
Chemical compound 5:1
Chemical equation 5:4, 5:6;
Chemical equilibrium 7:11;
Chemical formula 4:23, 5:1;
Chemical property 1:2; 3:10;
Chemistry. . . 1:1, 1:2, 1:4; 2:1; 3:13; 5:3;
    10:1; 11:8, 11:10, 11:15; 12:7;
Chromatography 1:3, 1:4;
Collision theory 7:1;
Common ion effect 7:13
Compound1:1, 1:2, 1:4, 1:13, 1:15, 1:16,
    1:19; 4:9, 4:10, 4:14, 4:17, 4:19,
    4:22-25, 4:28; 5:1, 5:2, 5:4, 5:5, 5:8,
    5:10, 5:11, 5:15-17, 5:21-25; 6:11,
    6:12; 7:7, 7:10, 7:14-17, 7:21,
    7:27-29, 7:31; 8:3, 8:5, 8:13, 8:23,
    8:25; 9:1, 9:3, 9:5, 9:6, 9:8, 9:9,9:11,
    9:15, 9:16; 10:2-7, 10:10, 10:11,

T

Technetium-99 . . . . 11:13, 11:14,

Temperature 1:5-26, 1:28-31; 3:6; 4:22;
    5:25-26; 6:1-5, 7:1-3, 7:9, 7:11, 7:13,
    7:26, 11:11, 11:12,; 12:1, 12:2, 12:6,
    12:7, 12:11-13, 12:16, 12:20, 12:22,
    13:1,
Tertiary alcohol 10:26,
Titration 8:6, 8:7, 12:20;
Toluene 10:20,;
Tracers 11:13
Transition elements . . . 2:18; 3:5, 3:10;
Transmutation 11:3, 11:8-10,
Trihydroxy alcohols 10:26

U

Unsaturated organic compound. . . 4:12,
    10:11,10:15
Unsaturated solution 6:3;

V

Vapor pressure 1:20, 1:21; 7:11;
Voltaic cell 9:12-14, 9:19-21, 9:25, 9:26,

W

Wave mechanical model 2:5

APPENDIX V: NYS REGENTS EXAMS

DESCRIPTION OF CHEMISTRY PHYSICAL SETTING REGENTS, STARTING JUNE 2002

The Regents consists of Parts A, B, and C.

Part A has 30 multiple choice questions: 30 points.

Part B has multiple choice and short response questions.

Part C has extended constructed response questions-a series of interrelated questions based on topics and applications. Some of these questions may have several parts, which can include examples, similarities and differences, and applications.

Parts B and C together total 55 points.

The 85 points for Parts A, B, and C are scaled up to 100 points.

All questions on the Chemistry: The Physical Setting Regents must be answered.

STRATEGIES FOR TEST TAKING

1 **Use the Reference Tables.** In the book, you learned which tables to use for physical behavior of matter, atomic concepts, periodic table, bonding, etc. You can get a much higher mark on the Regents by using the Tables. Many of the answers to the questions are right there!

2 When you guess **eliminate the wrong choices**. This increases the chance of getting the right answer.

3 Read each question carefully.
Read each reading passage carefully to help you answer the questions correctly.

4 For a **lengthy question**, pick out the **main words** of the question to make it easier to find the answer.

5 If you do not understand the choices (which choice is correct), look at the question, think of how you would answer it, then see which choice is closest to your answer.

6 You must answer all questions. If you do not know the answer, take a guess. You don't lose points for wrong answers.

7 Stay with your first guess if you don't have a better one.

8 For questions which have models, graphs, data tables, or pictures:

Read every word in the model, graph, data table, or picture very carefully.

    a. For graphs, look at the title and the labels on the x and y axes. See what the relationship is between the two variables.

    b. For data tables, see the title and the headings on the top and sides of the columns. Look for the relationships shown in the table. You can be asked to draw a graph based on the data table.

    c. For models, diagrams, and pictures, look carefully at all details in the model, picture, diagram, etc, and use **all** the information given to answer the question.

9 For longer constructed responses, each part of the question should be answered in a separate sentence or paragraph, so the one marking it knows where each answer is.

If a question has 3 or 4 parts, make sure you answer **all** the parts.

See what the question asks. Does the question ask for similarities, differences, or examples? Does the question ask why, explain why (give a reason) or how (by what means or what method)? Make sure you answer what the question asks for.

10 Go to the easy questions first. If there is a hard question, skip it and come back to it. Make a mark near the question to remember to come back for it.

11 Pace yourself. Make sure you have enough time to complete the exam.

12 Review the test.

Make sure you are given the Reference Tables with the Chemistry Regents.

ENTS HIGH SCHOOL EXAMINATION

Part A

Date: Tuesday June 20, 2017
Time: 9:15 a.m. to 12:15 p.m., only

Answer all questions in this part.

Directions (1–30): For *each* statement or question, record on your separate answer sheet the *number* of the word or expression that, of those given, best completes the statement or answers the question. Some questions may require the use of the *2011 Edition Reference Tables for Physical Setting/Chemistry*.

1 Which statement describes the structure of an atom?
 (1) The nucleus contains positively charged electrons.
 (2) The nucleus contains negatively charged protons.
 (3) The nucleus has a positive charge and is surrounded by negatively charged electrons.
 (4) The nucleus has a negative charge and is surrounded by positively charged electrons.

2 Which term is defined as the region in an atom where an electron is most likely to be located?
 (1) nucleus   (3) quanta
 (2) orbital   (4) spectra

3 What is the number of electrons in an atom of scandium?
 (1) 21   (3) 45
 (2) 24   (4) 66

4 Which particle has the *least* mass?
 (1) a proton   (3) a helium atom
 (2) an electron   (4) a hydrogen atom

5 Which electron transition in an excited atom results in a release of energy?
 (1) first shell to the third shell
 (2) second shell to the fourth shell
 (3) third shell to the fourth shell
 (4) fourth shell to the second shell

6 On the Periodic Table, the number of protons in an atom of an element is indicated by its
 (1) atomic mass
 (2) atomic number
 (3) selected oxidation states
 (4) number of valence electrons

7 Which type of formula shows an element symbol for each atom and a line for each bond between atoms?
 (1) ionic   (3) empirical
 (2) structural   (4) molecular

8 What is conserved during all chemical reactions?
 (1) charge   (3) vapor pressure
 (2) density   (4) melting point

9 In which type of reaction can two compounds exchange ions to form two different compounds?
 (1) synthesis
 (2) decomposition
 (3) single replacement
 (4) double replacement

10 At STP, two 5.0-gram solid samples of different ionic compounds have the same density. These solid samples could be differentiated by their
 (1) mass   (3) temperature
 (2) volume   (4) solubility in water

11 What is the number of electrons shared between the atoms in an I_2 molecule?
 (1) 7   (3) 8
 (2) 2   (4) 4

12 Which substance has nonpolar covalent bonds?
 (1) Cl_2 (3) SiO_2
 (2) SO_3 (4) CCl_4

13 Compared to a potassium atom, a potassium ion has
 (1) a smaller radius   (3) fewer protons
 (2) a larger radius   (4) more protons

P.S./Chem.–June '17   [2]

14 Which form of energy is associated with the random motion of particles in a gas?

(1) chemical   (3) nuclear
(2) electrical   (4) thermal

15 The average kinetic energy of water molecules *decreases* when

(1) $H_2O(\ell)$ at 337 K changes to $H_2O(\ell)$ at 300. K
(2) $H_2O(\ell)$ at 373 K changes to $H_2O(g)$ at 373 K
(3) $H_2O(s)$ at 200. K changes to $H_2O(s)$ at 237 K
(4) $H_2O(s)$ at 273 K changes to $H_2O(\ell)$ at 273 K

16 The joule is a unit of

(1) concentration   (3) pressure
(2) energy   (4) volume

17 Compared to a sample of helium at STP, the same sample of helium at a higher temperature and a lower pressure

(1) condenses to a liquid
(2) is more soluble in water
(3) forms diatomic molecules
(4) behaves more like an ideal gas

18 A sample of a gas is in a sealed, rigid container that maintains a constant volume. Which changes occur between the gas particles when the sample is heated?

(1) The frequency of collisions increases, and the force of collisions decreases.
(2) The frequency of collisions increases, and the force of collisions increases.
(3) The frequency of collisions decreases, and the force of collisions decreases.
(4) The frequency of collisions decreases, and the force of collisions increases.

19 At STP, which gaseous sample has the same number of molecules as 3.0 liters of $N_2(g)$?

(1) 6.0 L of $F_2(g)$ (3) 3.0 L of $H_2(g)$
(2) 4.5 L of $N_2(g)$ (4) 1.5 L of $Cl_2(g)$

20 Distillation of crude oil from various parts of the world yields different percentages of hydrocarbons. Which statement explains these different percentages?

(1) Each component in a mixture has a different solubility in water.
(2) Hydrocarbons are organic compounds.
(3) The carbons in hydrocarbons may be bonded in chains or rings.
(4) The proportions of components in a mixture can vary.

21 In which 1.0-gram sample are the particles arranged in a crystal structure?

(1) $CaCl_2(s)$ (3) $CH_3OH(\ell)$
(2) $C_2H_6(g)$ (4) $CaI_2(aq)$

22 When a reversible reaction is at equilibrium, the concentration of products and the concentration of reactants must be

(1) decreasing   (3) constant
(2) increasing   (4) equal

23 In chemical reactions, the difference between the potential energy of the products and the potential energy of the reactants is equal to the

(1) activation energy
(2) ionization energy
(3) heat of reaction
(4) heat of vaporization

24 What occurs when a catalyst is added to a chemical reaction?

(1) an alternate reaction pathway with a lower activation energy
(2) an alternate reaction pathway with a higher activation energy
(3) the same reaction pathway with a lower activation energy
(4) the same reaction pathway with a higher activation energy

25 What is the name of the compound with the formula $CH_3CH_2CH_2NH_2$?

(1) 1-propanol   (3) propanal
(2) 1-propanamine   (4) propanamide

High Marks: Regents Chemistry Made Easy

26 Which compound is an isomer of $C_2H_5OC_2H_5$?

(1) CH_3COOH (3) $C_3H_7COCH_3$

(2) $C_2H_5COOCH_3$ (4) C_4H_9OH

27 Ethanoic acid and 1-butanol can react to produce water and a compound classified as an

(1) aldehyde   (3) ester

(2) amide   (4) ether

28 During an oxidation-reduction reaction, the number of electrons gained is

(1) equal to the number of electrons lost
(2) equal to the number of protons gained
(3) less than the number of electrons lost
(4) less than the number of protons gained

29 Which process requires energy for a nonspontaneous redox reaction to occur?

(1) deposition   (3) alpha decay

(2) electrolysis   (4) chromatography

30 Which pair of compounds represents one Arrhenius acid and one Arrhenius base?

(1) CH_3OH and NaOH (3) HNO_3 and NaOH

(2) CH_3OH and HCl   (4) HNO_3 and HCl

Part B–1

Answer all questions in this part.

Directions (31–50): For *each* statement or question, record on your separate answer sheet the *number* of th
word or expression that, of those given, best completes the statement or answers the question. Some question
may require the use of the 2011 *Edition Reference Tables for Physical Setting/Chemistry*.

31 Which electron configuration represents the
electrons of an atom of neon in an excited state?

(1) 2-7   (3) 2-7-1
(2) 2-8   (4) 2-8-1

32 Some information about the two naturally
occurring isotopes of gallium is given in the table
below.

**Natural Abundance of
Two Gallium Isotopes**

Isotope	Natural Abundance (%)	Atomic Mass (u)
Ga-69	60.11	68.926
Ga-71	39.89	70.925

Which numerical setup can be used to calculate
the atomic mass of gallium?

(1) (0.6011)(68.926 u) + (0.3989)(70.925 u)
(2) (60.11)(68.926 u) + (39.89)(70.925 u)
(3) (0.6011)(70.925 u) + (0.3989)(68.926 u)
(4) (60.11)(70.925 u) + (39.89)(68.926 u)

33 A student measures the mass and volume of
a sample of copper at room temperature and
101.3 kPa. The mass is 48.9 grams and the volume
is 5.00 cubic centimeters. The student calculates
the density of the sample. What is the percent
error of the student's calculated density?

(1) 7.4%   (3) 9.2%
(2) 8.4%   (4) 10.2%

34 What is the chemical formula for sodium sulfate?

(1) Na_2SO_4 (3) $NaSO_4$
(2) Na_2SO_3 (4) $NaSO_3$

35 Given the balanced equation representing
reaction:

$$2Na(s) + Cl_2(g) \rightarrow 2NaCl(s) + energy$$

If 46 grams of Na and 71 grams of Cl_2 rea
completely, what is the total mass of NaC
produced?

(1) 58.5 g   (3) 163 g
(2) 117 g   (4) 234 g

36 Given the balanced equation representing
reaction:

$$2NO + O_2 \rightarrow 2NO_2 + energy$$

The mole ratio of NO to NO_2 is

(1) 1 to 1   (3) 3 to 2
(2) 2 to 1   (4) 5 to 2

37 The particle diagram below represents a sol
sample of silver.

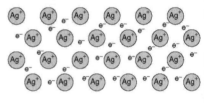

Which type of bonding is present when valenc
electrons move within the sample?

(1) metallic bonding   (3) covalent bonding
(2) hydrogen bonding   (4) ionic bonding

High Marks: Regents Chemistry Made Easy

38 Given the formula representing a molecule:

Which statement explains why the molecule is nonpolar?

(1) Electrons are shared between the carbon atoms and the hydrogen atoms.
(2) Electrons are transferred from the carbon atoms to the hydrogen atoms.
(3) The distribution of charge in the molecule is symmetrical.
(4) The distribution of charge in the molecule is asymmetrical.

39 A solid sample of a compound and a liquid sample of the same compound are each tested for electrical conductivity. Which test conclusion indicates that the compound is ionic?

(1) Both the solid and the liquid are good conductors.
(2) Both the solid and the liquid are poor conductors.
(3) The solid is a good conductor, and the liquid is a poor conductor.
(4) The solid is a poor conductor, and the liquid is a good conductor.

40 Which statement explains why 10.0 mL of a 0.50 M H_2SO_4(aq) solution exactly neutralizes 5.0 mL of a 2.0 M NaOH(aq) solution?

(1) The moles of H^+(aq) equal the moles of OH^-(aq).
(2) The moles of H_2SO_4(aq) equal the moles of NaOH(aq).
(3) The moles of H_2SO_4(aq) are greater than the moles of NaOH(aq).
(4) The moles of H^+(aq) are greater than the moles of OH^-(aq).

41 Which particle diagram represents *one* substance in the gas phase?

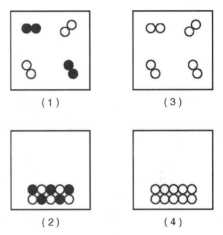

42 Given the equation representing a chemical reaction at equilibrium in a sealed, rigid container:

$$H_2(g) + I_2(g) + energy \rightleftharpoons 2HI(g)$$

When the concentration of $H_2(g)$ is increased by adding more hydrogen gas to the container at constant temperature, the equilibrium shifts

(1) to the right, and the concentration of HI(g) decreases
(2) to the right, and the concentration of HI(g) increases
(3) to the left, and the concentration of HI(g) decreases
(4) to the left, and the concentration of HI(g) increases

43 Which diagram represents the potential energy changes during an exothermic reaction?

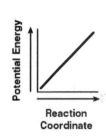

Reaction Coordinate

(1)

Reaction Coordinate

(3)

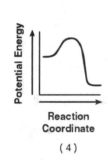

Reaction Coordinate

(2)

Reaction Coordinate

(4)

44 Which compound is classified as an ether?
(1) CH_3CHO
(2) CH_3OCH_3
(3) CH_3COCH_3
(4) CH_3COOCH_3

45 Given the equation representing a reversibl reaction:

$$HCO_3^-(aq) + H_2O(\ell) \rightleftharpoons H_2CO_3(aq) + OH^-(aq$$

Which formula represents the H^+ acceptor i the forward reaction?
(1) $HCO_3^-(aq)$
(2) $H_2O(\ell)$
(3) $H_2CO_3(aq)$
(4) $OH^-(aq)$

46 What is the mass of an original 5.60-gram sampl of iron-53 that remains unchanged after 25.5 minutes?
(1) 0.35 g
(2) 0.70 g
(3) 1.40 g
(4) 2.80 g

47 Given the equation representing a nuclea reaction:

$$^1_1H + X \rightarrow ^6_3Li + ^4_2He$$

The particle represented by X is
(1) ^9_4Li
(2) ^9_4Be
(3) $^{10}_5Be$
(4) $^{10}_6C$

48 Fission and fusion reactions both release energy However, only fusion reactions
(1) require elements with large atomic number
(2) create radioactive products
(3) use radioactive reactants
(4) combine light nuclei

49 The chart below shows the crystal shapes and melting points of two forms of solid phosphorus.

Two Forms of Phosphorus

Form of Phosphorus	Crystal Shape	Melting Point (°C)
white	cubic	44
black	orthorhombic	610

Which phrase describes the two forms of phosphorus?

(1) same crystal structure and same properties
(2) same crystal structure and different properties
(3) different crystal structures and different properties
(4) different crystal structures and same properties

50 Which graph shows the relationship between pressure and Kelvin temperature for an ideal gas at constant volume?

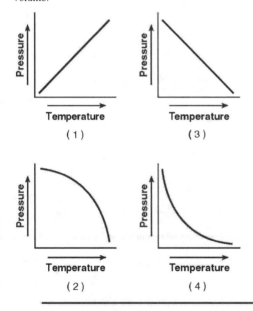

Part B–2

Answer all questions in this part.

Directions (51–65): Record your answers in the spaces provided in your answer booklet. Some questions may require the use of the 2011 *Edition Reference Tables for Physical Setting/Chemistry.*

Base your answers to questions 51 through 53 on the information below and on your knowledge of chemistry.

The elements in Group 17 are called halogens. The word "halogen" is derived from Greek and means "salt former."

51 State the trend in electronegativity for the halogens as these elements are considered in order of increasing atomic number. [1]

52 Identify the type of chemical bond that forms when potassium reacts with bromine to form a salt. [1]

53 Based on Table *F*, identify *one* ion that reacts with iodide ions in an aqueous solution to form an insoluble compound. [1]

Base your answers to questions 54 through 57 on the information below and on your knowledge of chemistry.

The diagrams below represent four different atomic nuclei.

Four Atomic Nuclei

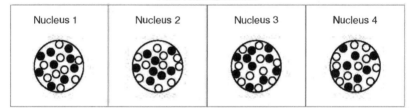

| Nucleus 1 | Nucleus 2 | Nucleus 3 | Nucleus 4 |

Key
● = proton
○ = neutron

54 Identify the element that has atomic nuclei represented by nucleus 1. [1]

55 Determine the mass number of the nuclide represented by nucleus 2. [1]

56 Explain why nucleus 2 and nucleus 4 represent the nuclei of two different isotopes of the same element. [1]

57 Identify the nucleus above that is found in an atom that has a stable valence electron configuration. [1]

Base your answers to questions 58 through 60 on the information below and on your knowledge of chemistry.

The equation below represents a chemical reaction at 1 atm and 298 K.

$$2H_2(g) + O_2(g) \rightarrow 2H_2O(g)$$

58 State the change in energy that occurs in order to break the bonds in the hydrogen molecules. [1]

59 In the space *in your answer booklet*, draw a Lewis electron-dot diagram for a water molecule. [1]

60 Compare the strength of attraction for electrons by a hydrogen atom to the strength of attraction for electrons by an oxygen atom within a water molecule. [1]

Base your answers to questions 61 through 63 on the information below and on your knowledge of chemistry.

- A test tube contains a sample of solid stearic acid, an organic acid.
- Both the sample and the test tube have a temperature of 22.0°C.
- The stearic acid melts after the test tube is placed in a beaker with 320. grams of water at 98.0°C.
- The temperature of the liquid stearic acid and water in the beaker reaches 74.0°C.

61 Identify the element in stearic acid that makes it an organic compound. [1]

62 State the direction of heat transfer between the test tube and the water when the test tube was placed in the water. [1]

63 Show a numerical setup for calculating the amount of thermal energy change for the water in the beaker. [1]

Base your answers to questions 64 and 65 on the information below and on your knowledge of chemistry.

A nuclear reaction is represented by the equation below.

$$^3_1H \rightarrow {}^3_2He + {}^0_{-1}e$$

64 Identify the decay mode of hydrogen-3. [1]

65 Explain why the equation represents a transmutation. [1]

Part C

Answer all questions in this part.

Directions (66–85): Record your answers in the spaces provided in your answer booklet. Some question may require the use of the *2011 Edition Reference Tables for Physical Setting/Chemistry.*

Base your answers to questions 66 through 68 on the information below and on your knowledge of chemistr

A technician recorded data for two properties of Period 3 elements. The data are shown in the table below.

Two Properties of Period 3 Elements

Element	Na	Mg	Al	Si	P	S	Cl	Ar
Ionic Radius (pm)	95	66	51	41	212	184	181	—
Reaction with Cold Water	reacts vigorously	reacts very slowly	no observable reaction	no observable reaction	no observable reaction	no observable reaction	reacts slowly	no observable reaction

66 Identify the element in this table that is classified as a metalloid. [1]

67 State the phase of chlorine at 281 K and 101.3 kPa. [1]

68 State evidence from the technician's data which indicates that sodium is more active than aluminum. [1]

Base your answers to questions 69 through 71 on the information below and on your knowledge of chemistr

Ammonia, $NH_3(g)$, can be used as a substitute for fossil fuels in some internal combustion engines. The reaction between ammonia and oxygen in an engine is represented by the unbalanced equation below.

$$NH_3(g) + O_2(g) \rightarrow N_2(g) + H_2O(g) + energy$$

69 Balance the equation *in your answer booklet* for the reaction of ammonia and oxygen, using the smallest whole-number coefficients. [1]

70 Show a numerical setup for calculating the mass, in grams, of a 4.2-mole sample of O_2. Use 32 g/mol as the gram-formula mass of O_2. [1]

71 Determine the new pressure of a 6.40-L sample of oxygen gas at 300. K and 100. kPa after the gas is compressed to 2.40 L at 900. K. [1]

Base your answers to questions 72 through 76 on the information below and on your knowledge of chemistry.

 Fruit growers in Florida protect oranges when the temperature is near freezing by spraying water on them. It is the freezing of the water that protects the oranges from frost damage. When $H_2O(\ell)$ at 0°C changes to $H_2O(s)$ at 0°C, heat energy is released. This energy helps to prevent the temperature inside the orange from dropping below freezing, which could damage the fruit. After harvesting, oranges can be exposed to ethene gas, C_2H_4, to improve their color.

72 Write the empirical formula for ethene. [1]

73 Explain, in terms of bonding, why the hydrocarbon ethene is classified as unsaturated. [1]

74 Determine the gram-formula mass of ethene. [1]

75 Explain, in terms of particle arrangement, why the entropy of the water *decreases* when the water freezes. [1]

76 Determine the quantity of heat released when 2.00 grams of $H_2O(\ell)$ freezes at 0°C. [1]

Base your answers to questions 77 through 80 on the information below and on your knowledge of chemistry.

A student constructs an electrochemical cell during a laboratory investigation. When the switch is closed, electrons flow through the external circuit. The diagram and ionic equation below represent this cell and the reaction that occurs.

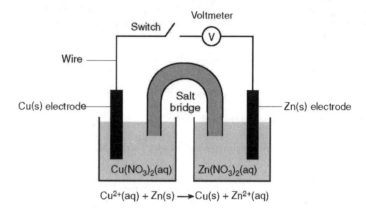

$$Cu^{2+}(aq) + Zn(s) \longrightarrow Cu(s) + Zn^{2+}(aq)$$

77 State the form of energy that is converted to electrical energy in the operating cell. [1]

78 State, in terms of the Cu(s) electrode and the Zn(s) electrode, the direction of electron flow in the external circuit when the cell operates. [1]

79 Write a balanced equation for the half-reaction that occurs in the Cu half-cell when the cell operates. [1]

80 State what happens to the mass of the Cu electrode and the mass of the Zn electrode in the operating cell. [1]

Base your answers to questions 81 and 82 on the information below and on your knowledge of chemistry.

A solution is made by dissolving 70.0 grams of $KNO_3(s)$ in 100. grams of water at 50.°C and standard pressure.

81 Show a numerical setup for calculating the percent by mass of KNO_3 in the solution. [1]

82 Determine the number of additional grams of KNO_3 that must dissolve to make this solution saturated. [1]

Base your answers to questions 83 through 85 on the information below and on your knowledge of chemistry.

Vinegar is a commercial form of acetic acid, $HC_2H_3O_2(aq)$. One sample of vinegar has a pH value of 2.4.

83 Explain, in terms of particles, why $HC_2H_3O_2(aq)$ can conduct an electric current. [1]

84 State the color of bromthymol blue indicator in a sample of the commercial vinegar. [1]

85 State the pH value of a sample that has ten times *fewer* hydronium ions than an equal volume of a vinegar sample with a pH value of 2.4. [1]

PHYSICAL SETTING
CHEMISTRY

Tuesday, June 20, 2017 — 9:15 a.m. to 12:15 p.m., only

ANSWER BOOKLET

☐ Male

Student .... Sex: ☐ Female

Teacher ....

School .... Grade ....

Record your answers for Part B–2 and Part C in this booklet.

Part B–2

51 _____

_____

_____

52 _____

53 _____

54 _____

55 _____

56 _____

_____

_____

57 _____

58 _____

_____

_____

59

60 _____

_____

_____

High Marks: Regents Chemistry Made Easy   Page Regents-17

61 _____

62 _____

_____

_____

63

64 _____

65 _____

_____

_____

66 _____

67 _____

68 _____

_____

_____

69 _____ $NH_3(g)$ + _____ $O_2(g)$ → _____ $N_2(g)$ + _____ $H_2O(g)$ + energy

70

71 _____ kPa

72 _____

73 _____

_____

_____

74 _____ g/mol

75 _____

_____

_____

76 _____ J

77 _____

78 _____

_____

_____

79 _____

80 Cu electrode: _____

Zn electrode: _____

P.S./Chem. Answer Booklet–June '17   [5]   [●VER

Page Regents-20   **High Marks: Regents Chemistry Made Easy**

81

82 _____ g

83 _____

_____

_____

84 _____

85 _____

REGENTS HIGH SCHOOL EXAMINATION

Part A

Date: Thursday, January 25, 2018
Time: 9:15 a.m. to 12:15 p.m., only

Answer all questions in this part.

Directions (1–30): For *each* statement or question, record on your separate answer sheet the *number* of the word or expression that, of those given, best completes the statement or answers the question. Some questions may require the use of the *2011 Edition Reference Tables for Physical Setting/Chemistry.*

1 Which statement describes the location of protons and neutrons in an atom of helium?

(1) Protons and neutrons are in the nucleus.

(2) Protons and neutrons are outside the nucleus.

(3) Protons are outside the nucleus, and neutrons are in the nucleus.

(4) Protons are in the nucleus, and neutrons are outside the nucleus.

2 Given a list of atomic model descriptions:

A: electron shells outside a central nucleus
B: hard, indivisible sphere
C: mostly empty space

Which list of atomic model descriptions represents the order of historical development from the earliest to most recent?

(1) A, B, C   (3) B, C, A
(2) A, C, B   (4) B, A, C

3 Which list represents the classification of the elements nitrogen, neon, magnesium, and silicon, respectively?

(1) metal, metalloid, nonmetal, noble gas

(2) nonmetal, noble gas, metal, metalloid

(3) nonmetal, metalloid, noble gas, metal

(4) noble gas, metal, metalloid, nonmetal

4 In the ground state, all atoms of Group 15 elements have the same number of

(1) valence electrons

(2) electron shells

(3) neutrons

(4) protons

5 What is the chemical formula for ammonium sulfide?

(1) $(NH_4)_2S$ (3) $(NH_4)_2SO_4$
(2) $(NH_4)_2SO_3$ (4) $(NH_4)_2S_2O_3$

6 Which formula is an empirical formula?

(1) N_2O_4 (3) C_3H_6
(2) NH_3 (4) P_4O_{10}

7 Chemical properties can be used to

(1) determine the temperature of a substance

(2) determine the density of a substance

(3) differentiate between two compounds

(4) differentiate between two neutrons

8 Ice, $H_2O(s)$, is classified as

(1) an ionic compound

(2) a molecular compound

(3) a homogeneous mixture

(4) a heterogeneous mixture

9 Which phrase describes the molecular polarity and distribution of charge in a molecule of carbon dioxide, CO_2?

(1) polar and symmetrical

(2) polar and asymmetrical

(3) nonpolar and symmetrical

(4) nonpolar and asymmetrical

10 Which element tends *not* to react with other elements?

(1) helium   (3) phosphorus
(2) hydrogen   (4) potassium

11 Given the equation representing a reaction:

$$O + O \rightarrow O_2$$

Which statement describes the changes that occur as the oxygen molecule is produced?

(1) Energy is absorbed as bonds are broken.
(2) Energy is absorbed as bonds are formed.
(3) Energy is released as bonds are broken.
(4) Energy is released as bonds are formed.

12 Which term represents the strength of the attraction an atom has for the electrons in a chemical bond?

(1) electrical conductivity
(2) electronegativity
(3) first ionization energy
(4) specific heat capacity

13 Compared to a 15-gram sample of $Cu(s)$ at $25°C$, a 25-gram sample of $Cu(s)$ at $25°C$ has

(1) the same density and the same chemical properties
(2) the same density and different chemical properties
(3) a different density and the same chemical properties
(4) a different density and different chemical properties

14 Which substance can *not* be broken down by a chemical change?

(1) ammonia   (3) tungsten
(2) ethanol   (4) water

15 The kinetic molecular theory states that all particles of an ideal gas are

(1) colliding without transferring energy
(2) in random, constant, straight-line motion
(3) arranged in a regular geometric pattern
(4) separated by small distances relative to their size

16 Which sample of gas at STP has the same number of molecules as 6 liters of $Cl_2(g)$ at STP?

(1) 3 liters of $O_2(g)$ (3) 3 moles of $O_2(g)$
(2) 6 liters of $N_2(g)$ (4) 6 moles of $N_2(g)$

17 A chemical reaction is most likely to occur when the colliding particles have the proper

(1) energy and orientation
(2) solubility and density
(3) ionic radii and mass
(4) atomic radii and volume

18 The energy absorbed and the energy released during a chemical reaction are best represented by a

(1) cooling curve
(2) heating curve
(3) kinetic energy diagram
(4) potential energy diagram

19 A catalyst increases the rate of a chemical reaction by

(1) providing an alternate reaction pathway
(2) providing the required heat of reaction
(3) increasing the potential energy of the products
(4) increasing the activation energy of the reaction

20 Which formula represents an alkyne?

(1) C_nH_n (3) C_nH_{2n+2}
(2) $C_{2n}H_n$ (4) C_nH_{2n-2}

21 Which process involves the transfer of electrons?

(1) double replacement
(2) neutralization
(3) oxidation-reduction
(4) sublimation

22 Which change occurs at the anode in an operating electrochemical cell?

(1) gain of protons   (3) loss of protons
(2) gain of electrons   (4) loss of electrons

23 Which device requires electrical energy to produce a chemical change?

(1) electrolytic cell   (3) voltaic cell
(2) salt bridge   (4) voltmeter

24 Which substance is an Arrhenius acid?

(1) HBr   (3) NaOH
(2) NaBr   (4) NH_3

25 Which laboratory process is used to determine the concentration of one solution by using a volume of another solution of known concentration?

(1) crystallization   (3) filtration
(2) distillation   (4) titration

26 Which type of reaction occurs when $H^+(aq)$ reacts with $OH^-(aq)$?

(1) combustion   (3) fermentation
(2) decomposition   (4) neutralization

27 According to one acid-base theory, a molecule acts as an acid when the molecule

(1) accepts an H^+ (3) donates an H^+
(2) accepts an OH^- (4) donates an OH^-

28 In which type of reaction can an atom of one element be converted to an atom of another element?

(1) addition   (3) substitution
(2) reduction   (4) transmutation

29 An unstable nucleus spontaneously releases a positron. This is an example of

(1) radioactive decay
(2) nuclear fusion
(3) chemical decomposition
(4) thermal conductivity

30 Which phrase describes a risk associated with producing energy in a nuclear power plant?

(1) depletion of atmospheric hydrogen (H_2)
(2) depletion of atmospheric carbon dioxide (CO_2)
(3) production of wastes needing long-term storage
(4) production of wastes that cool surrounding water supplies

Part B–1

Answer all questions in this part.

Directions (31–50): For *each* statement or question, record on your separate answer sheet the *number* of the word or expression that, of those given, best completes the statement or answers the question. Some questions may require the use of the *2011 Edition Reference Tables for Physical Setting/Chemistry.*

31 An ion that consists of 7 protons, 9 neutrons, and 10 electrons has a net charge of

(1) 2–   (3) 3+

(2) 2+   (4) 3–

32 Which electron configuration represents the electrons of an atom in an excited state?

(1) 2–2   (3) 2–8

(2) 2–2–1   (4) 2–8–1

33 The table below gives the atomic mass and the abundance of the two naturally occurring isotopes of boron.

Naturally Occurring Isotopes of Boron

Isotope	Atomic Mass (u)	Natural Abundance (%)
B–10	10.01	19.9
B–11	11.01	80.1

Which numerical setup can be used to determine the atomic mass of the element boron?

(1) $\dfrac{(10.01\ u)(19.9) + (11.01\ u)(80.1)}{100}$

(2) $\dfrac{(10.01\ u)(0.199) + (11.01\ u)(0.801)}{100}$

(3) $\dfrac{10.01\ u + 11.01\ u}{2}$

(4) $\dfrac{19.9\% + 80.1\%}{2}$

34 In which group on the Periodic Table would a nonmetallic element belong if atoms of this element tend to gain two electrons to complete their valence shell?

(1) 14   (3) 16

(2) 15   (4) 17

35 Which trend is observed as the first four elements in Group 17 on the Periodic Table are considered in order of increasing atomic number?

(1) Electronegativity increases.

(2) First ionization energy decreases.

(3) The number of valence electrons increases.

(4) The number of electron shells decreases.

36 What is the number of moles of KF in a 29-gram sample of the compound?

(1) 1.0 mol   (3) 0.50 mol

(2) 2.0 mol   (4) 5.0 mol

37 Which bond is most polar?

(1) C–O   (3) N–O

(2) H–O   (4) S–O

38 Based on Table *F*, which equation represents a saturated solution having the *lowest* concentration of Cl^- ions?

(1) $NaCl(s) \rightleftharpoons Na^+(aq) + Cl^-(aq)$

(2) $AgCl(s) \rightleftharpoons Ag^+(aq) + Cl^-(aq)$

(3) $NH_4Cl(s) \rightleftharpoons NH_4^+(aq) + Cl^-(aq)$

(4) $KCl(s) \rightleftharpoons K^+(aq) + Cl^-(aq)$

39 What is the molarity of a solution that contains 0.500 mole of KNO_3 dissolved in 0.500-liter of solution?

(1) 1.00 M   (3) 0.500 M
(2) 2.00 M   (4) 4.00 M

40 Given samples of water:

    Sample 1: 100. grams of water at 10.°C
    Sample 2: 100. grams of water at 20.°C

Compared to sample 1, sample 2 contains

(1) molecules with a lower average kinetic energy
(2) molecules with a lower average velocity
(3) less heat energy
(4) more heat energy

41 Given the key:

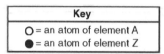

Key
O = an atom of element A
● = an atom of element Z

Which particle model diagram represents a chemical change?

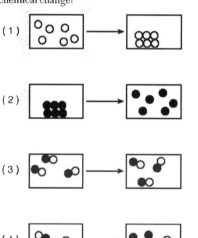

42 Based on Table H, what is the vapor pressure of CH_3COOH at 90.°C?

(1) 40. kPa   (3) 114 kPa
(2) 48 kPa   (4) 150. kPa

43 The arrangement of particles is most ordered in a sample of

(1) NaCl(aq)   (3) NaCl(g)
(2) NaCl(ℓ)   (4) NaCl(s)

44 What is the net amount of heat released when two moles of $C_2H_6(g)$ are formed from its elements at 101.3 kPa and 298 K?

(1) 42.0 kJ   (3) 126.0 kJ
(2) 84.0 kJ   (4) 168.0 kJ

45 Which compounds are isomers of each other?

(1) methanol and methanal
(2) propanoic acid and pentanoic acid
(3) 1-propanol and 2-propanol
(4) 1-chloropropane and 2-bromopropane

46 A reaction between an alcohol and an organic acid is classified as

(1) esterification   (3) saponification
(2) fermentation   (4) substitution

47 Why is potassium nitrate classified as an electrolyte?

(1) It is a molecular compound.
(2) It contains a metal.
(3) It can conduct electricity as a solid.
(4) It releases ions in an aqueous solution.

48 When the concentration of hydrogen ions in a solution is *decreased* by a factor of ten, the pH of the solution

(1) increases by 1   (3) decreases by 1
(2) increases by 10   (4) decreases by 10

49 The cooling curve below represents the uniform cooling of a substance, starting at a temperature above its boiling point.

Cooling Curve

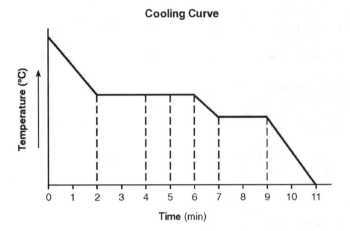

During which time interval does the substance exist as both a liquid and a solid?

(1) min 2 to min 4   (3) min 5 to min 7

(2) min 4 to min 5   (4) min 7 to min 9

50 Given the balanced equation representing a reaction:

$$CH_4(g) + 2O_2(g) \rightarrow CO_2(g) + 2H_2O(g) + energy$$

Which change in reaction conditions will increase the frequency of effective collisions between reactant molecules?

(1) decreasing the pressure of the reactants

(2) decreasing the temperature of the reactants

(3) increasing the concentration of the reactants

(4) increasing the volume of the reactants

Part B–2

Answer all questions in this part.

Directions (51–65): Record your answers in the spaces provided in your answer booklet. Some questions may require the use of the *2011 Edition Reference Tables for Physical Setting/Chemistry.*

51 Convert the melting point of mercury to degrees Celsius. [1]

52 Draw a Lewis electron-dot diagram for a molecule of hydrogen fluoride, HF. [1]

53 Show a numerical setup for calculating the quantity of heat in joules required to completely vaporize 102.3 grams of $H_2O(\ell)$ at 100.°C and 1.0 atm. [1]

54 State the color of methyl orange indicator after the indicator is placed in a solution of 0.10 M $NH_3(aq)$. [1]

Base your answers to questions 55 and 56 on the information below and on your knowledge of chemistry.

The bright-line spectra for four elements and a mixture of elements are shown in the diagram below.

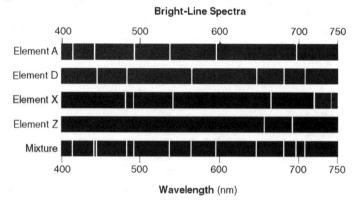

Bright-Line Spectra

Wavelength (nm)

55 Write the letter of each element present in the mixture. [1]

56 Explain, in terms of electrons and energy states, how the light emitted by excited atoms is produced. [1]

Base your answers to questions 57 through 59 on the information below and on your knowledge of chemistry.

Rubidium and iodine have different chemical and physical properties. Some of these properties are shown in the table below.

Some Physical and Chemical Properties of Rubidium and Iodine

Rubidium	Iodine
silvery-white solid	bluish-black lustrous solid
forms ionic compounds with nonmetals	forms ionic bonds with active metals
reacts with oxygen in the air	sublimes at room temperature
specific heat = 0.363 J/g•K	specific heat = 0.214 J/g•K

57 State the chemical property of iodine listed in this table. [1]

58 Compare the atomic radius of an atom of iodine to the atomic radius of an atom of rubidium when both atoms are in the ground state. [1]

59 Compare the electrical conductivity of these two elements at STP. [1]

Base your answers to questions 60 through 62 on the information below and on your knowledge of chemistry.

Given the unbalanced equation showing the reactants and product of a reaction occurring at 298 K and 100. kPa:

$$P_4(s) + Cl_2(g) \rightarrow PCl_3(\ell) + energy$$

60 Balance the equation *in your answer booklet* for the reaction, using the smallest whole-number coefficients. [1]

61 State why this reaction is a synthesis reaction. [1]

62 Show a numerical setup for calculating the percent composition by mass of chlorine in $PCl_3(\ell)$ (gram-formula mass = 137 g/mol). [1]

Base your answers to questions 63 through 65 on the information below and on your knowledge of chemistry.

The diagram below shows the first three steps in the uranium-238 radioactive decay series.

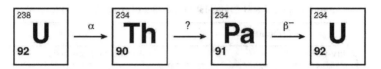

The decay mode for the first and third steps is shown above the arrows. The decay mode for the second step is not shown in the diagram. Thorium-234 has a half-life of 24.10 days.

63 Explain, in terms of neutrons and protons, why U-238 and U-234 are different isotopes of uranium. [1]

64 Identify the decay mode particle emitted from the Th-234. [1]

65 Determine the total time that must elapse until only $\frac{1}{16}$ of an original sample of Th-234 remains unchanged. [1]

Part C

Answer all questions in this part.

Directions (66–85): Record your answers in the spaces provided in your answer booklet. Some questions may require the use of the *2011 Edition Reference Tables for Physical Setting/Chemistry*.

Base your answer to question 66 on the information below and on your knowledge of chemistry.

Tetrachloroethene, C_2Cl_4, is a solvent used in many dry cleaning processes.

66 Write the empirical formula for tetrachloroethene. [1]

_____

Base your answers to questions 67 through 69 on the information below and on your knowledge of chemistry.

Thermal energy is absorbed as chemical reactions occur during the process of baking muffins. The batter for muffins often contains baking soda, $NaHCO_3(s)$, which decomposes as the muffins are baked in an oven at 200.°C. The balanced equation below represents this reaction, which releases $CO_2(g)$ and causes the muffins to rise as they bake. The $H_2O(\ell)$ is released into the air of the oven as it becomes a vapor.

$$2NaHCO_3(s) + heat \rightarrow Na_2CO_3(s) + H_2O(\ell) + CO_2(g)$$

67 Based on Table *E*, identify the polyatomic ion in the solid product of the reaction. [1]

68 State the direction of heat flow between the air in the oven and the muffin batter when the muffin batter is first placed in the preheated oven at 200.°C. [1]

69 Compare the potential energy of the liquid water molecules to the potential energy of the water vapor molecules. [1]

_____

Base your answers to questions 70 through 72 on the information below and on your knowledge of chemistry.

A bubble of air at the bottom of a lake rises to the surface of the lake. Data for the air inside the bubble at the bottom of the lake and at the surface of the lake are listed in the table below.

Data for the Air Inside the Bubble

Location in Lake	Temperature (K)	Pressure (kPa)	Volume (mL)	Density (g/mL)
surface	293	104.0	2.5	0.0012
bottom	282	618.3	?	———

70 State the number of significant figures used to express the pressure at the surface of the lake. [1]

71 Show a numerical setup for calculating the volume of the bubble at the bottom of the lake. [1]

72 Determine the mass of the air in the bubble at the surface of the lake. [1]

Base your answers to questions 73 through 77 on the information below and on your knowledge of chemistry.

Nitrogen dioxide, NO_2, is a dark brown gas that is used to make nitric acid and to bleach flour. Nitrogen dioxide has a boiling point of 294 K at 101.3 kPa. In a rigid cylinder with a movable piston, nitrogen dioxide can be in equilibrium with colorless dinitrogen tetroxide, N_2O_4. This equilibrium is represented by the equation below.

$$2NO_2(g) \rightleftharpoons N_2O_4(g) + 58 \text{ kJ}$$

73 State evidence from the equation that the forward reaction is exothermic. [1]

74 Compare the rate of the forward reaction to the rate of the reverse reaction when the system has reached equilibrium. [1]

75 State one stress, other than adding or removing $NO_2(g)$ or $N_2O_4(g)$, that would increase the amount of the dark brown gas. [1]

76 At standard pressure, compare the strength of intermolecular forces in $NO_2(g)$ to the strength of intermolecular forces in $N_2(g)$. [1]

77 Determine the oxidation state of nitrogen in nitrogen dioxide. [1]

Base your answers to questions 78 through 81 on the information below and on your knowledge of chemistry.

A student sets up a voltaic cell using magnesium and zinc electrodes. The porous barrier in the cell has the same purpose as a salt bridge. The diagram and the ionic equation below represent this operating cell.

Voltaic Cell

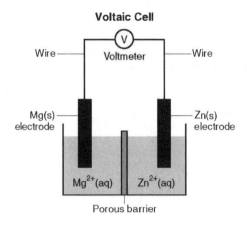

$$Mg(s) + Zn^{2+}(aq) \rightarrow Zn(s) + Mg^{2+}(aq)$$

78 Determine the number of moles of $Mg^{2+}(aq)$ ions produced when 2.5 moles of $Zn^{2+}(aq)$ react completely in this cell. [1]

79 State, in terms of ions, how the porous barrier functions as a salt bridge in this cell. [1]

80 State, in terms of the relative activity of metals, why the reaction in this cell occurs. [1]

81 Write a balanced half-reaction equation for the oxidation that occurs in this operating cell. [1]

Base your answers to questions 82 through 85 on the information below and on your knowledge of chemistry.

Polyvinyl chloride (PVC) is a polymer used to make drain pipes, flooring, electric wire insulation, and some plastic bottles. Making PVC requires several reactions. The first step is represented by the equation below.

$$\text{Equation 1:} \quad C_2H_4 \quad + \quad Cl_2 \quad \rightarrow \quad C_2H_4Cl_2$$

$$\text{ethene} \qquad \text{chlorine} \qquad \text{1,2-dichloroethane}$$

The 1,2-dichloroethane is converted to vinyl chloride. To produce PVC, the vinyl chloride monomer is polymerized, as represented by the equation below.

Equation 2:

Vinyl chloride (monomer)

PVC (polymer)

Note: ⊓ and n represent the same large number in the equation.

82 Explain, in terms of chemical bonds, why the hydrocarbon in equation 1 is unsaturated. [1]

83 Identify the class of organic compounds to which the product of equation 1 belongs. [1]

84 Draw a structural formula for the product of equation 1. [1]

85 State the number of electrons shared between the carbon atoms in a molecule of vinyl chloride. [1]

The University of the State of New York

REGENTS HIGH SCHOOL EXAMINATION

PHYSICAL SETTING
CHEMISTRY

Thursday, January 25, 2018 — 9:15 a.m. to 12:15 p.m., only

ANSWER BOOKLET

Student . . . .

Teacher . . . .

School . . . . Grade . . . .

Record your answers for Part B–2 and Part C in this booklet.

Part B–2

51 _____ °C

52

53

54 _____

55 _____

56 _____

_____

_____

57 _____

58 _____

_____

_____

59 _____

_____

_____

P.S./Chem. Answer Booklet–Jan. '18   [2]

Page Regents-36   **High Marks: Regents Chemistry Made Easy**

60 _____ $P_4(s)$ + _____ $Cl_2(g)$ → _____ $PCl_3(\ell)$ + energy

61 _____

_____

_____

62

63 _____

_____

_____

64 _____

65 _____ d

Part C

66 _____

67 _____

68 From _____ to _____

69 _____

_____

_____

70 _____

71

72 _____ g

P.S./Chem. Answer Booklet–Jan. '18   [4]

Page Regents-38   **High Marks: Regents Chemistry Made Easy**

73 _____

_____

_____

74 _____

_____

_____

75 _____

76 _____

_____

_____

77 _____

78 _____ mol

79 _____

_____

_____

80 _____

_____

_____

81 _____

82 _____

_____

_____

83 _____

84

85 _____

REGENTS HIGH SCHOOL EXAMINATION

Part A

Date: Wednesday, June 18, 2018
Time: 9:15 a.m. to 12:15 p.m., only

Answer all questions in this part.

Directions (1–30): For *each* statement or question, record on your separate answer sheet the *number* of the word or expression that, of those given, best completes the statement or answers the question. Some questions may require the use of the *2011 Edition Reference Tables for Physical Setting/Chemistry.*

1 Which statement describes the charge and location of an electron in an atom?

(1) An electron has a positive charge and is located outside the nucleus.
(2) An electron has a positive charge and is located in the nucleus.
(3) An electron has a negative charge and is located outside the nucleus.
(4) An electron has a negative charge and is located in the nucleus.

2 Which statement explains why a xenon atom is electrically neutral?

(1) The atom has fewer neutrons than electrons.
(2) The atom has more protons than electrons.
(3) The atom has the same number of neutrons and electrons.
(4) The atom has the same number of protons and electrons.

3 If two atoms are isotopes of the same element, the atoms must have

(1) the same number of protons and the same number of neutrons
(2) the same number of protons and a different number of neutrons
(3) a different number of protons and the same number of neutrons
(4) a different number of protons and a different number of neutrons

4 Which electrons in a calcium atom in the ground state have the greatest effect on the chemical properties of calcium?

(1) the two electrons in the first shell
(2) the two electrons in the fourth shell
(3) the eight electrons in the second shell
(4) the eight electrons in the third shell

5 The weighted average of the atomic masses of the naturally occuring isotopes of an element is the

(1) atomic mass of the element
(2) atomic number of the element
(3) mass number of each isotope
(4) formula mass of each isotope

6 Which element is classified as a metalloid?

(1) Cr   (3) Sc
(2) Cs   (4) Si

7 Which statement describes a chemical property of iron?

(1) Iron oxidizes.
(2) Iron is a solid at STP.
(3) Iron melts.
(4) Iron is attracted to a magnet.

8 Graphite and diamond are two forms of the same element in the solid phase that differ in their

(1) atomic numbers
(2) crystal structures
(3) electronegativities
(4) empirical formulas

9 Which ion has the largest radius?

(1) Br^- (3) F^-
(2) Cl^- (4) I^-

10 Carbon monoxide and carbon dioxide have

(1) the same chemical properties and the same physical properties
(2) the same chemical properties and different physical properties
(3) different chemical properties and the same physical properties
(4) different chemical properties and different physical properties

11 Based on Table S, which group on the Periodic Table has the element with the highest electronegativity?

(1) Group 1   (3) Group 17
(2) Group 2   (4) Group 18

12 What is represented by the chemical formula $PbCl_2(s)$?

(1) a substance
(2) a solution
(3) a homogeneous mixture
(4) a heterogeneous mixture

13 What is the vapor pressure of propanone at 50.°C?

(1) 37 kPa   (3) 83 kPa
(2) 50. kPa   (4) 101 kPa

14 Which statement describes the charge distribution and the polarity of a CH_4 molecule?

(1) The charge distribution is symmetrical and the molecule is nonpolar.
(2) The charge distribution is asymmetrical and the molecule is nonpolar.
(3) The charge distribution is symmetrical and the molecule is polar.
(4) The charge distribution is asymmetrical and the molecule is polar.

15 In a laboratory investigation, a student separates colored compounds obtained from a mixture of crushed spinach leaves and water by using paper chromatography. The colored compounds separate because of differences in

(1) molecular polarity
(2) malleability
(3) boiling point
(4) electrical conductivity

16 Which phrase describes the motion and attractive forces of ideal gas particles?

(1) random straight-line motion and no attractive forces
(2) random straight-line motion and strong attractive forces
(3) random curved-line motion and no attractive forces
(4) random curved-line motion and strong attractive forces

17 At which temperature will $Hg(\ell)$ and $Hg(s)$ reach equilibrium in a closed system at 1.0 atmosphere?

(1) 234 K   (3) 373 K
(2) 273 K   (4) 630. K

18 A molecule of any organic compound has at least one

(1) ionic bond   (3) oxygen atom
(2) double bond   (4) carbon atom

19 A chemical reaction occurs when reactant particles

(1) are separated by great distances
(2) have no attractive forces between them
(3) collide with proper energy and proper orientation
(4) convert chemical energy into nuclear energy

20 Systems in nature tend to undergo changes toward

(1) lower energy and lower entropy
(2) lower energy and higher entropy
(3) higher energy and lower entropy
(4) higher energy and higher entropy

21 Which formula can represent an alkyne?

(1) C_2H_4 (3) C_3H_4
(2) C_2H_6 (4) C_3H_6

22 Given the formula representing a compound:

$$H-C-H$$

with structure:

```
        H
        |
     H–C–H
      H | H
      | | |
   H–C–C–C–H
      | | |
      H H H
```

Which formula represents an isomer of this compound?

```
  H   H H H
  \   | | |
   C=C–C–C–H
  /     | |
  H     H H

   (1)
```

```
          H H
          | |
   H–C≡C–C–C–H
          | |
          H H

   (3)
```

```
   H H H H
   | | | |
 H–C–C–C–C–H
   | | | |
   H H H H

   (2)
```

```
    H H H
    | | |
 H–C–C–C–H
    | | |
    H | H
      |
   H–C–H
      |
      H

   (4)
```

23 Which energy conversion occurs in an operating voltaic cell?

(1) chemical energy to electrical energy
(2) chemical energy to nuclear energy
(3) electrical energy to chemical energy
(4) electrical energy to nuclear energy

24 Which process requires energy to decompose a substance?

(1) electrolysis   (3) sublimation
(2) neutralization   (4) synthesis

25 The concentration of which ion is increased when LiOH is dissolved in water?

(1) hydroxide ion   (3) hydronium ion
(2) hydrogen ion   (4) halide ion

26 Which equation represents neutralization?

(1) $6Li(s) + N_2(g) \rightarrow 2Li_3N(s)$

(2) $2Mg(s) + O_2(g) \rightarrow 2MgO(s)$

(3) $2KOH(aq) + H_2SO_4(aq) \rightarrow$
     $K_2SO_4(aq) + 2H_2O(\ell)$

(4) $Pb(NO_3)_2(aq) + K_2CrO_4(aq) \rightarrow$
     $2KNO_3(aq) + PbCrO_4(s)$

27 The stability of an isotope is related to its ratio of

(1) neutrons to positrons
(2) neutrons to protons
(3) electrons to positrons
(4) electrons to protons

28 Which particle has the *least* mass?

(1) alpha particle   (3) neutron
(2) beta particle   (4) proton

29 The energy released during a nuclear reaction is a result of

(1) breaking chemical bonds
(2) forming chemical bonds
(3) mass being converted to energy
(4) energy being converted to mass

30 The use of uranium-238 to determine the age of a geological formation is a beneficial use of

(1) nuclear fusion
(2) nuclear fission
(3) radioactive isomers
(4) radioactive isotopes

Part B–1

Answer all questions in this part.

Directions (31–50): For *each* statement or question, record on your separate answer sheet the *number* of the word or expression that, of those given, best completes the statement or answers the question. Some questions may require the use of the *2011 Edition Reference Tables for Physical Setting/Chemistry*.

Base your answers to questions 31 and 32 on your knowledge of chemistry and the bright-line spectra produced by four elements and the spectrum of a mixture of elements represented in the diagram below.

Bright-Line Spectra

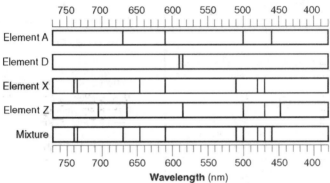

31 Which elements are present in this mixture?

(1) D and A   (3) X and A

(2) D and Z   (4) X and Z

32 Each line in the spectra represents the energy

(1) absorbed as an atom loses an electron

(2) absorbed as an atom gains an electron

(3) released as an electron moves from a lower energy state to a higher energy state

(4) released as an electron moves from a higher energy state to a lower energy state

33 The table below shows the number of protons, neutrons, and electrons in four ions.

Four Ions

Ion	Number of Protons	Number of Neutrons	Number of Electrons
A	8	10	10
E	9	10	10
G	11	12	10
J	12	12	10

Which ion has a charge of 2–?

(1) A   (3) G
(2) E   (4) J

34 What is the approximate mass of an atom that contains 26 protons, 26 electrons and 19 neutrons?

(1) 26 u   (3) 52 u
(2) 45 u   (4) 71 u

35 Which electron configuration represents a potassium atom in an excited state?

(1) 2-7-6   (3) 2-8-8-1
(2) 2-8-5   (4) 2-8-7-2

36 What is the total number of neutrons in an atom of K-42?

(1) 19   (3) 23
(2) 20   (4) 42

37 Given the equation representing a reaction:

$$2C + 3H_2 \rightarrow C_2H_6$$

What is the number of moles of C that must completely react to produce 2.0 moles of C_2H_6?

(1) 1.0 mol   (3) 3.0 mol
(2) 2.0 mol   (4) 4.0 mol

38 Given the equation representing a reaction:

$$Mg(s) + 2HCl(aq) \rightarrow MgCl_2(aq) + H_2(g)$$

Which type of chemical reaction is represented by the equation?

(1) synthesis
(2) decomposition
(3) single replacement
(4) double replacement

39 The table below lists properties of selected elements at room temperature.

Properties of Selected Elements at Room Temperature

Element	Density (g/cm³)	Malleability	Conductivity
sodium	0.97	yes	good
gold	19.3	yes	good
iodine	4.933	no	poor
tungsten	19.3	yes	good

Based on this table, which statement describes how two of these elements can be differentiated from each other?

(1) Gold can be differentiated from tungsten based on density.
(2) Gold can be differentiated from sodium based on malleability.
(3) Sodium can be differentiated from tungsten based on conductivity.
(4) Sodium can be differentiated from iodine based on malleability.

40 Which particle diagram represents a mixture?

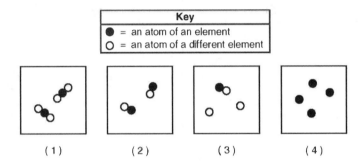

Key
● = an atom of an element
○ = an atom of a different element

(1)   (2)   (3)   (4)

41 An atom of which element reacts with an atom of hydrogen to form a bond with the greatest degree of polarity?
(1) carbon
(3) nitrogen
(2) fluorine
(4) oxygen

42 What is the concentration of an aqueous solution that contains 1.5 moles of NaCl in 500. milliliters of this solution?
(1) 0.30 M
(3) 3.0 M
(2) 0.75 M
(4) 7.5 M

43 The table below shows data for the temperature, pressure, and volume of four gas samples.

Data for Four Gases

Gas Sample	Temperature (K)	Pressure (atm)	Volume (L)
I	600.	2.0	5.0
II	300.	1.0	10.0
III	600.	3.0	5.0
IV	300.	1.0	10.0

Which two gas samples contain the same number of molecules?
(1) I and II
(3) II and III
(2) I and III
(4) II and IV

44 Based on Table I, what is the ΔH value for the production of 1.00 mole of $NO_2(g)$ from its elements at 101.3 kPa and 298 K?
(1) +33.2 kJ
(3) +132.8 kJ
(2) −33.2 kJ
(4) −132.8 kJ

45 Which equation represents an addition reaction?
(1) $C_3H_8 + Cl_2 \rightarrow C_3H_7Cl + HCl$
(2) $C_3H_6 + Cl_2 \rightarrow C_3H_6Cl_2$
(3) $CaCl_2 + Na_2CO_3 \rightarrow CaCO_3 + 2NaCl$
(4) $CaCO_3 \rightarrow CaO + CO_2$

46 Given the balanced equation representing a reaction:

$$Ni(s) + 2HCl(aq) \rightarrow NiCl_2(aq) + H_2(g)$$

In this reaction, each Ni atom
(1) loses 1 electron
(3) gains 1 electron
(2) loses 2 electrons
(4) gains 2 electrons

47 Which equation represents a reduction half-reaction?
(1) $Fe \rightarrow Fe^{3+} + 3e^-$
(3) $Fe^{3+} \rightarrow Fe + 3e^-$
(2) $Fe + 3e^- \rightarrow Fe^{3+}$
(4) $Fe^{3+} + 3e^- \rightarrow Fe$

48 Given the balanced ionic equation representing a reaction:

$$Cu(s) + 2Ag^+(aq) \rightarrow Cu^{2+}(aq) + 2Ag(s)$$

During this reaction, electrons are transferred from
(1) $Cu(s)$ to $Ag^+(aq)$
(2) $Cu^{2+}(aq)$ to $Ag(s)$
(3) $Ag(s)$ to $Cu^{2+}(aq)$
(4) $Ag^+(aq)$ to $Cu(s)$

49 Which metal reacts spontaneously with Sr^{2+} ions?
(1) $Ca(s)$
(3) $Cs(s)$
(2) $Co(s)$
(4) $Cu(s)$

50 Given the balanced equation representing a reaction:

$$HCl + H_2O \rightarrow H_3O^+ + Cl^-$$

The water molecule acts as a base because it
(1) donates an H^+
(3) donates an OH^-
(2) accepts an H^+
(4) accepts an OH^-

Part B–2

Answer all questions in this part.

Directions (51–65): Record your answers in the spaces provided in your answer booklet. Some questions may require the use of the *2011 Edition Reference Tables for Physical Setting/Chemistry.*

51 State the general trend in first ionization energy as the elements in Period 3 are considered from left to right. [1]

52 Identify a type of strong intermolecular force that exists between water molecules, but does *not* exist between carbon dioxide molecules. [1]

53 Draw a structural formula for 2-butanol. [1]

Base your answers to questions 54 through 56 on the information below and on your knowledge of chemistry.

Some compounds of silver are listed with their chemical formulas in the table below.

Silver Compounds

Name	Chemical Formula
silver carbonate	Ag_2CO_3
silver chlorate	$AgClO_3$
silver chloride	$AgCl$
silver sulfate	Ag_2SO_4

54 Explain, in terms of element classification, why silver chloride is an ionic compound. [1]

55 Show a numerical setup for calculating the percent composition by mass of silver in silver carbonate (gram-formula mass = 276 g/mol). [1]

56 Identify the silver compound in the table that is most soluble in water. [1]

Base your answers to questions 57 through 59 on the information below and on your knowledge of chemistry.

When a cobalt-59 atom is bombarded by a subatomic particle, a radioactive cobalt-60 atom is produced. After 21.084 years, 1.20 grams of an original sample of cobalt-60 produced remains unchanged.

57 Complete the nuclear equation by writing a notation for the missing particle. [1]

58 Based on Table *N*, identify the decay mode of cobalt-60. [1]

59 Determine the mass of the original sample of cobalt-60 produced. [1]

High Marks: Regents Chemistry Made Easy   Page Regents-49

Base your answers to questions 60 through 62 on the information below and on your knowledge of chemistry.

A sample of a molecular substance starting as a gas at 206°C and 1 atm is allowed to cool for 16 minutes. This process is represented by the cooling curve below.

Cooling Curve for a Substance

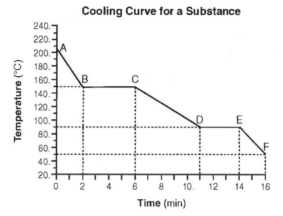

60 Determine the number of minutes that the substance was in the liquid phase, only. [1]

61 Compare the strength of the intermolecular forces within this substance at 180.°C to the strength of the intermolecular forces within this substance at 120.°C. [1]

62 Describe what happens to the potential energy and the average kinetic energy of the molecules in the sample during interval *DE*. [1]

Base your answers to questions 63 through 65 on the information below and on your knowledge of chemistry.

The diagram below represents a cylinder with a moveable piston containing 16.0 g of $O_2(g)$. At 298 K and 0.500 atm, the $O_2(g)$ has a volume of 24.5 liters.

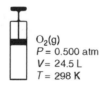

$O_2(g)$
$P = 0.500$ atm
$V = 24.5$ L
$T = 298$ K

63 Determine the number of moles of $O_2(g)$ in the cylinder. The gram-formula mass of $O_2(g)$ is 32.0 g/mol. [1]

64 State the changes in *both* pressure and temperature of the gas in the cylinder that would increase the frequency of collisions between the $O_2(g)$ molecules. [1]

65 Show a numerical setup for calculating the volume of $O_2(g)$ in the cylinder at 265 K and 1.00 atm. [1]

[10]

Answer all questions in this part.

Directions (66–85): Record your answers in the spaces provided in your answer booklet. Some questions may require the use of the *2011 Edition Reference Tables for Physical Setting/Chemistry.*

Base your answers to questions 66 through 69 on the information below and on your knowledge of chemistry.

In the late 1800s, Dmitri Mendeleev developed a periodic table of the elements known at that time. Based on the pattern in his periodic table, he was able to predict properties of some elements that had not yet been discovered. Information about two of these elements is shown in the table below.

Some Element Properties Predicted by Mendeleev

Predicted Elements	Property	Predicted Value	Actual Value
eka-aluminum (Ea)	density at STP	5.9 g/cm^3	5.91 g/cm^3
	melting point	low	30.°C
	oxide formula	Ea$_2$O$_3$	
	approximate molar mass	68 g/mol	
eka-silicon (Es)	density at STP	5.5 g/cm^3	5.3234 g/cm^3
	melting point	high	938°C
	oxide formula	EsO$_2$	
	approximate molar mass	72 g/mol	

66 Identify the phase of Ea at 310. K. [1]

67 Write a chemical formula for the compound formed between Ea and Cl. [1]

68 Identify the element that Mendeleev called eka-silicon, Es. [1]

69 Show a numerical setup for calculating the percent error of Mendeleev's predicted density of Es. [1]

Base your answers to questions 70 through 73 on the information below and your knowledge of chemistry.

Methanol can be manufactured by a reaction that is reversible. In the reaction, carbon monoxide gas and hydrogen gas react using a catalyst. The equation below represents this system at equilibrium.

$$CO(g) + 2H_2(g) \rightleftharpoons CH_3OH(g) + energy$$

70 State the class of organic compounds to which the product of the forward reaction belongs. [1]

71 Compare the rate of the forward reaction to the rate of the reverse reaction in this equilibrium system. [1]

72 Explain, in terms of collision theory, why increasing the concentration of $H_2(g)$ in this system will increase the concentration of $CH_3OH(g)$. [1]

73 State the effect on the rates of both the forward and reverse reactions if no catalyst is used in the system. [1]

Base your answers to questions 74 through 76 on the information below and on your knowledge of chemistry.

Fatty acids, a class of compounds found in living things, are organic acids with long hydrocarbon chains. Linoleic acid, an unsaturated fatty acid, is essential for human skin flexibility and smoothness. The formula below represents a molecule of linoleic acid.

74 Write the molecular formula of linoleic acid. [1]

75 Identify the type of chemical bond between the oxygen atom and the hydrogen atom in the linoleic acid molecule. [1]

76 On the diagram *in your answer booklet*, circle the organic acid functional group. [1]

Base your answers to questions 77 through 79 on the information below and on your knowledge of chemistry.

Fuel cells are voltaic cells. In one type of fuel cell, oxygen gas, $O_2(g)$, reacts with hydrogen gas, $H_2(g)$, producing water vapor, $H_2O(g)$, and electrical energy. The unbalanced equation for this redox reaction is shown below.

$$H_2(g) + O_2(g) \rightarrow H_2O(g) + energy$$

A diagram of the fuel cell is shown below. During operation of the fuel cell, hydrogen gas is pumped into one compartment and oxygen gas is pumped into the other compartment. Each compartment has an inner wall that is a porous carbon electrode through which ions flow. Aqueous potassium hydroxide, KOH(aq), and the porous electrodes serve as the salt bridge.

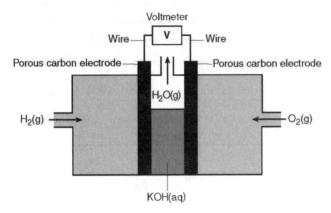

77 Balance the equation *in your answer booklet* for the reaction in this fuel cell, using the smallest whole-number coefficients. [1]

78 Determine the change in oxidation number for oxygen in this operating fuel cell. [1]

79 State the number of moles of electrons that are gained when 5.0 moles of electrons are lost in this reaction. [1]

Base your answers to questions 80 through 82 on the information below and on your knowledge of chemistry.

In a laboratory investigation, a student compares the concentration and pH value of each of four different solutions of hydrochloric acid, HCl(aq), as shown in the table below.

Data for HCl(aq) Solutions

Solution	Concentration of HCl(aq) (M)	pH Value
W	1.0	0
X	0.10	1
Y	0.010	2
Z	0.0010	3

80 State the number of significant figures used to express the concentration of solution Z. [1]

81 Determine the concentration of an HCl(aq) solution that has a pH value of 4. [1]

82 Determine the volume of 0.25 M NaOH(aq) that would exactly neutralize 75.0 milliliters of solution X. [1]

Base your answers to questions 83 through 85 on the information below and on your knowledge of chemistry.

Carbon dioxide is slightly soluble in seawater. As carbon dioxide levels in the atmosphere increase, more CO_2 dissolves in seawater, making the seawater more acidic because carbonic acid, $H_2CO_3(aq)$, is formed.

Seawater also contains aqueous calcium carbonate, $CaCO_3(aq)$, which is used by some marine organisms to make their hard exoskeletons. As the acidity of the sea water changes, the solubility of $CaCO_3$ also changes, as shown in the graph below.

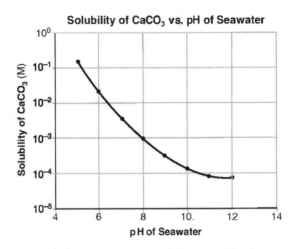

83 State the trend in the solubility of $CaCO_3$ as seawater becomes more acidic. [1]

84 State the color of bromcresol green in a sample of seawater in which the $CaCO_3$ solubility is 10^{-2} M. [1]

85 A sample of seawater has a pH of 8. Determine the new pH of the sample if the hydrogen ion concentration is increased by a factor of 100. [1]

PHYSICAL SETTING
CHEMISTRY

Wednesday, June 20, 2018 — 9:15 a.m. to 12:15 p.m., only

ANSWER BOOKLET

Student . . . .

Teacher . . . .

School . . . . Grade . . . .

Record your answers for Part B–2 and Part C in this booklet.

Part B–2

51 _____

52 _____

53

54 _____

_____

_____

55

56 _____

57 $^{59}_{27}\text{Co} + $ _____ $\rightarrow {}^{60}_{27}\text{Co}$

58 _____

59 _____ g

60 _____ min

61 _____

_____

_____

62 Potential energy: _____

Average kinetic energy: _____

63 _____ mol

64 Change in pressure: _____

Change in temperature: _____

65

P.S./Chem. Answer Booklet–June '18   [3]   [OVER]

Page Regents-58   **High Marks: Regents Chemistry Made Easy**

Part C

66 _____

67 _____

68 _____

69

70 _____

71 _____

_____

_____

72 _____

_____

_____

73 Rate of forward reaction: _____

Rate of reverse reaction: _____

74 _____

75 _____

76

77 _____ $H_2(g)$ + _____ $O_2(g) \rightarrow$ _____ $H_2O(g)$ + energy

78 From _____ to _____

79 _____ **mol**

P.S./Chem. Answer Booklet–June '18   [5]   [●VER]

Page Regents-60   **High Marks: Regents Chemistry Made Easy**

80 _____

81 _____ M

82 _____ mL

83

_____

_____

84 _____

85 _____

REGENTS HIGH SCHOOL EXAMINATION

Part A

Date: Thursday, August 16, 2018
Time: 8:30 to 11:30 a.m., only

Answer all questions in this part.

Directions (1–30): For *each* statement or question, record on your separate answer sheet the *number* of the word or expression that, of those given, best completes the statement or answers the question. Some questions may require the use of the *2011 Edition Reference Tables for Physical Setting/Chemistry.*

1 According to the wave-mechanical model, an orbital is defined as the most probable location of

(1) a proton   (3) a positron
(2) a neutron   (4) an electron

2 The part of an atom that has an overall positive charge is called

(1) an electron   (3) the first shell
(2) the nucleus   (4) the valence shell

3 Which subatomic particles each have a mass of approximately 1 u?

(1) proton and electron
(2) proton and neutron
(3) neutron and electron
(4) neutron and positron

4 The discovery of the electron as a subatomic particle was a result of

(1) collision theory
(2) kinetic molecular theory
(3) the gold-foil experiment
(4) experiments with cathode ray tubes

5 The elements on the Periodic Table of the Elements are arranged in order of increasing

(1) atomic mass   (3) atomic number
(2) formula mass   (4) oxidation number

6 Which element is classified as a metalloid?

(1) Te   (3) Hg
(2) S   (4) I

7 At STP, $O_2(g)$ and $O_3(g)$ are two forms of the same element that have

(1) the same molecular structure and the same properties
(2) the same molecular structure and different properties
(3) different molecular structures and the same properties
(4) different molecular structures and different properties

8 Which substance can be broken down by chemical means?

(1) ammonia   (3) antimony
(2) aluminum   (4) argon

9 Which statement describes $H_2O(\ell)$ and $H_2O_2(\ell)$?

(1) Both are compounds that have the same properties.
(2) Both are compounds that have different properties.
(3) Both are mixtures that have the same properties.
(4) Both are mixtures that have different properties.

10 Which two terms represent major categories of compounds?

(1) ionic and nuclear
(2) ionic and molecular
(3) empirical and nuclear
(4) empirical and molecular

11 Which formula represents an asymmetrical molecule?

(1) CH_4 (3) N_2
(2) CO_2 (4) NH_3

12 Which statement describes the energy changes that occur as bonds are broken and formed during a chemical reaction?

(1) Energy is absorbed when bonds are both broken and formed.
(2) Energy is released when bonds are both broken and formed.
(3) Energy is absorbed when bonds are broken, and energy is released when bonds are formed.
(4) Energy is released when bonds are broken, and energy is absorbed when bonds are formed.

13 A solid sample of copper is an excellent conductor of electric current. Which type of chemical bonds are in the sample?

(1) ionic bonds
(2) metallic bonds
(3) nonpolar covalent bonds
(4) polar covalent bonds

14 Which list includes three forms of energy?

(1) thermal, nuclear, electronegativity
(2) thermal, chemical, electromagnetic
(3) temperature, nuclear, electromagnetic
(4) temperature, chemical, electronegativity

15 Based on Table S, an atom of which element has the strongest attraction for electrons in a chemical bond?

(1) chlorine   (3) oxygen
(2) nitrogen   (4) selenium

16 At which temperature and pressure would a sample of helium behave most like an ideal gas?

(1) 75 K and 500. kPa
(2) 150. K and 500. kPa
(3) 300. K and 50. kPa
(4) 600. K and 50. kPa

17 A cube of iron at 20.°C is placed in contact with a cube of copper at 60.°C. Which statement describes the initial flow of heat between the cubes?

(1) Heat flows from the copper cube to the iron cube.
(2) Heat flows from the iron cube to the copper cube.
(3) Heat flows in both directions between the cubes.
(4) Heat does not flow between the cubes.

18 Which sample at STP has the same number of atoms as 18 liters of $Ne(g)$ at STP?

(1) 18 moles of $Ar(g)$
(2) 18 liters of $Ar(g)$
(3) 18 grams of $H_2O(g)$
(4) 18 milliliters of $H_2O(g)$

19 Compared to H_2S, the higher boiling point of H_2O is due to the

(1) greater molecular size of water
(2) stronger hydrogen bonding in water
(3) higher molarity of water
(4) larger gram-formula mass of water

20 In terms of entropy and energy, systems in nature tend to undergo changes toward

(1) lower entropy and lower energy
(2) lower entropy and higher energy
(3) higher entropy and lower energy
(4) higher entropy and higher energy

21 Amines, amides, and amino acids are categories of

(1) isomers
(2) isotopes
(3) organic compounds
(4) inorganic compounds

22 A molecule of which compound has a multiple covalent bond?

(1) CH_4 (3) C_3H_8
(2) C_2H_4 (4) C_4H_{10}

23 Which type of reaction produces soap?

(1) polymerization   (3) fermentation

(2) combustion   (4) saponification

24 For a reaction system at equilibrium, LeChatelier's principle can be used to predict the

(1) activation energy for the system

(2) type of bonds in the reactants

(3) effect of a stress on the system

(4) polarity of the product molecules

25 Which value changes when a Cu atom becomes a Cu^{2+} ion?

(1) mass number

(2) oxidation number

(3) number of protons

(4) number of neutrons

26 Which reaction occurs at the anode in an electrochemical cell?

(1) oxidation   (3) combustion

(2) reduction   (4) substitution

27 What evidence indicates that the nuclei of strontium-90 atoms are unstable?

(1) Strontium-90 electrons are in the excited state.

(2) Strontium-90 electrons are in the ground state.

(3) Strontium-90 atoms spontaneously absorb beta particles.

(4) Strontium-90 atoms spontaneously emit beta particles.

28 Which nuclear emission is listed with its notation?

(1) gamma radiation, $^{0}_{0}\gamma$

(2) proton, $^{4}_{2}He$

(3) neutron, $^{0}_{-1}\beta$

(4) alpha particle, $^{1}_{1}H$

29 The energy released by a nuclear fusion reaction is produced when

(1) energy is converted to mass

(2) mass is converted to energy

(3) heat is converted to temperature

(4) temperature is converted to heat

30 Dating once-living organisms is an example of a beneficial use of

(1) redox reactions

(2) organic isomers

(3) radioactive isotopes

(4) neutralization reactions

Part B–1

Answer all questions in this part.

Directions (31–50): For *each* statement or question, record on your separate answer sheet the *number* of the word or expression that, of those given, best completes the statement or answers the question. Some questions may require the use of the *2011 Edition Reference Tables for Physical Setting/Chemistry.*

31 What is the net charge of an ion that has 11 protons, 10 electrons, and 12 neutrons?

(1) 1+   (3) 1−
(2) 2+   (4) 2−

32 Which electron configuration represents the electrons of an atom in an excited state?

(1) 2-5   (3) 2-5-1
(2) 2-8-5   (4) 2-6

33 Which element is a liquid at 1000. K?

(1) Ag   (3) Ca
(2) Al   (4) Ni

34 Which formula represents ammonium nitrate?

(1) NH_4NO_3 (3) $NH_4(NO_3)_2$
(2) NH_4NO_2 (4) $NH_4(NO_2)_2$

35 The empirical formula for butene is

(1) CH_2 (3) C_4H_6
(2) C_2H_4 (4) C_4H_8

36 Which equation represents a conservation of charge?

(1) $2Fe^{3+} + Al \rightarrow 2Fe^{2+} + Al^{3+}$
(2) $2Fe^{3+} + 2Al \rightarrow 3Fe^{2+} + 2Al^{3+}$
(3) $3Fe^{3+} + 2Al \rightarrow 2Fe^{2+} + 2Al^{3+}$
(4) $3Fe^{3+} + Al \rightarrow 3Fe^{2+} + Al^{3+}$

37 Given the balanced equation representing a reaction:

$$2H_2 + O_2 \rightarrow 2H_2O + energy$$

Which type of reaction is represented by this equation?

(1) decomposition
(2) double replacement
(3) single replacement
(4) synthesis

38 When a Mg^{2+} ion becomes a Mg atom, the radius increases because the Mg^{2+} ion

(1) gains 2 protons   (3) loses 2 protons
(2) gains 2 electrons   (4) loses 2 electrons

39 The *least* polar bond is found in a molecule of

(1) HI   (3) HCl
(2) HF   (4) HBr

40 A solution is prepared using 0.125 g of glucose, $C_6H_{12}O_6$, in enough water to make 250. g of total solution. The concentration of this solution, expressed in parts per million, is

(1) 5.00×10^1 ppm   (3) 5.00×10^3 ppm
(2) 5.00×10^2 ppm   (4) 5.00×10^4 ppm

41 What is the amount of heat, in joules, required to increase the temperature of a 49.5-gram sample of water from 22°C to 66°C?

(1) 2.2×10^3 J   (3) 9.1×10^3 J
(2) 4.6×10^3 J   (4) 1.4×10^4 J

42 A sample of a gas in a rigid cylinder with a movable piston has a volume of 11.2 liters at STP. What is the volume of this gas at 202.6 kPa and 300. K?

(1) 5.10 L   (3) 22.4 L
(2) 6.15 L   (4) 24.6 L

43 The equation below represents a reaction between two molecules, X_2 and Z_2. These molecules form an "activated complex," which then forms molecules of the product.

Key
●● = a molecule of X_2
◯◯ = a molecule of Z_2

Reactants   Activated   Product
          complex

Which diagram represents the most likely orientation of X_2 and Z_2 when the molecules collide with proper energy, producing an activated complex?

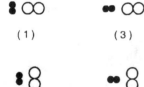

(1)   (3)

(2)   (4)

44 What is the chemical name for the compound $CH_3CH_2CH_2CH_3$?

(1) butane   (3) decane
(2) butene   (4) decene

45 In a laboratory activity, the density of a sample of vanadium is determined to be 6.9 g/cm^3 at room temperature. What is the percent error for the determined value?

(1) 0.15%   (3) 13%
(2) 0.87%   (4) 15%

46 Given the equation representing a reaction:

$$Cd + NiO_2 + 2H_2O \rightarrow Cd(OH)_2 + Ni(OH)_2$$

Which half-reaction equation represents the oxidation in the reaction?

(1) $Ni^{4+} + 2e^- \rightarrow Ni^{2+}$
(2) $Ni^{4+} \rightarrow Ni^{2+} + 2e^-$
(3) $Cd \rightarrow Cd^{2+} + 2e^-$
(4) $Cd + 2e^- \rightarrow Cd^{2+}$

47 Which metal reacts spontaneously with $NiCl_2(aq)$?

(1) Au(s)   (3) Sn(s)
(2) Cu(s)   (4) Zn(s)

48 Which solution is the best conductor of an electric current?

(1) 0.001 mole of NaCl dissolved in 1000. mL of water
(2) 0.005 mole of NaCl dissolved in 1000. mL of water
(3) 0.1 mole of NaCl dissolved in 1000. mL of water
(4) 0.05 mole of NaCl dissolved in 1000. mL of water

49 Compared to a 1.0-liter aqueous solution with a pH of 7.0, a 1.0-liter aqueous solution with a pH of 5.0 contains

(1) 10 times more hydronium ions
(2) 100 times more hydronium ions
(3) 10 times more hydroxide ions
(4) 100 times more hydroxide ions

50 Given the equation representing a reaction:

$$^{208}_{82}Pb + ^{70}_{30}Zn \rightarrow ^{277}_{112}Cn + ^{1}_{0}n$$

Which type of reaction is represented by this equation?

(1) neutralization   (3) substitution
(2) polymerization   (4) transmutation

Part B-2

Answer all questions in this part.

Directions (51–65): Record your answers in the spaces provided in your answer booklet. Some questions may require the use of the *2011 Edition Reference Tables for Physical Setting/Chemistry.*

51 Show a numerical setup for calculating the percent composition by mass of oxygen in Al_2O_3 (gram-formula mass = 102 g/mol). [1]

52 Identify a laboratory process that can be used to separate a liquid mixture of methanol and water, based on the differences in their boiling points. [1]

Base your answers to questions 53 through 55 on the information below and on your knowledge of chemistry.

The table below shows data for three isotopes of the same element.

Data for Three Isotopes of an Element

Isotopes	Number of Protons	Number of Neutrons	Atomic Mass (u)	Natural Abundance (%)
Atom D	12	12	23.99	78.99
Atom E	12	13	24.99	10.00
Atom G	12	14	25.98	11.01

53 Explain, in terms of subatomic particles, why these three isotopes represent the same element. [1]

54 State the number of valence electrons in an atom of isotope D in the ground state. [1]

55 Compare the energy of an electron in the first electron shell to the energy of an electron in the second electron shell in an atom of isotope E. [1]

Base your answers to questions 56 through 58 on the information below and on your knowledge of chemistry.

The elements in Group 2 on the Periodic Table can be compared in terms of first ionization energy, electronegativity, and other general properties.

56 Describe the general trend in electronegativity as the metals in Group 2 on the Periodic Table are considered in order of increasing atomic number. [1]

57 Explain, in terms of electron configuration, why the elements in Group 2 have similar chemical properties. [1]

58 Explain, in terms of atomic structure, why barium has a lower first ionization energy than magnesium. [1]

Base your answers to questions 59 through 61 on the information below and on your knowledge of chemistry.

A saturated solution of sulfur dioxide is prepared by dissolving $SO_2(g)$ in 100. g of water at 10.°C and standard pressure.

59 Determine the mass of SO_2 in this solution. [1]

60 Based on Table G, state the general relationship between solubility and temperature of an aqueous SO_2 solution at standard pressure. [1]

61 Describe what happens to the solubility of $SO_2(g)$ when the pressure is increased at constant temperature. [1]

Base your answers to questions 62 through 65 on the information below and on your knowledge of chemistry.

Starting as a solid, a sample of a molecular substance is heated, until the entire sample of the substance is a gas. The graph below represents the relationship between the temperature of the sample and the elapsed time.

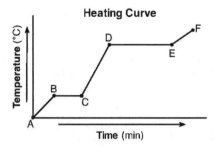

Heating Curve

62 Using the key *in your answer booklet*, draw a particle diagram to represent the sample during interval *AB*. Your response must include *at least six* molecules. [1]

63 Compare the average kinetic energy of the molecules of the sample during interval *BC* to the average kinetic energy of the molecules of the sample during interval *DE*. [1]

64 On the graph *in your answer booklet*, mark an **X** on the axis labeled "Temperature (°C)" to indicate the boiling point of the substance. [1]

65 State evidence that indicates the sample undergoes only physical changes during this heating. [1]

Part C

Answer all questions in this part.

Directions (66–85): Record your answers in the spaces provided in your answer booklet. Some question may require the use of the *2011 Edition Reference Tables for Physical Setting/Chemistry.*

Base your anwers to questions 66 through 68 on the information below and on your knowledge chemistry.

"Water gas," a mixture of hydrogen and carbon monoxide, is an industrial fuel and source of commercial hydrogen. Water gas is produced by passing steam over hot carbon obtained from coal. The equation below represents this system at equilibrium.

$$C(s) + H_2O(g) + heat \rightleftharpoons CO(g) + H_2(g)$$

66 State, in terms of the rates of the forward and reverse reactions, what occurs when dynamic equilibrium is reached in this system. [1]

67 In the space *in your answer booklet,* draw a Lewis electron-dot diagram for a molecule of H_2O. [1]

68 Explain, in terms of collisions, why increasing the surface area of the hot carbon increases the rate of the forward reaction. [1]

Base your answers to questions 69 through 71 on the information below and on your knowledge of chemistry.

In a laboratory activity, each of four different masses of $KNO_3(s)$ is placed in a separate test tube that contains 10.0 grams of H_2O at 25°C.

When each sample is first placed in the water, the temperature of the mixture decreases. The mixture in each test tube is then stirred while it is heated in a hot water bath until all of the $KNO_3(s)$ is dissolved. The contents of each test tube are then cooled to the temperature at which KNO_3 crystals first reappear. The procedure is repeated until the recrystallization temperatures for each mixture are consistent, as shown in the table below.

Data Table for the Laboratory Activity

Mixture	Mass of KNO₃ (g)	Mass of H₂O (g)	Temperature of Recrystallization (°C)
1	4.0	10.0	24
2	5.0	10.0	32
3	7.5	10.0	45
4	10.0	10.0	58

69 Based on Table I, explain why there is a *decrease* in temperature when the $KNO_3(s)$ was first dissolved in the water. [1]

70 Determine the percent by mass concentration of KNO_3 in mixture 2 after heating. [1]

71 Compare the freezing point of mixture 4 at 1.0 atm to the freezing point of water at 1.0 atm. [1]

Base your answers to questions 72 through 74 on the information below and on your knowledge o chemistry.

The balanced equation below represents the reaction between carbon monoxide and oxygen to produce carbon dioxide.

$$2CO(g) + O_2(g) \rightarrow 2CO_2(g) + energy$$

72 On the potential energy diagram *in your answer booklet*, draw a double-headed arrow ($\updownarrow$) to indicate the interval that represents the heat of reaction. [1]

73 Determine the number of moles of $O_2(g)$ needed to completely react with 8.0 moles of $CO(g)$. [1]

74 On the potential energy diagram *in your answer booklet*, draw a dashed line to show how the potential energy diagram changes when the reaction is catalyzed. [1]

Base your answers to questions 75 through 77 on the information below and on your knowledge o chemistry.

The equation below represents an industrial preparation of diethyl ether.

Compound A   **Compound B**

75 Write the name of the class of organic compounds to which compound A belongs. [1]

76 Identify the element in compound B that makes it an organic compound. [1]

77 Explain, in terms of elements, why compound B is *not* a hydrocarbon. [1]

Base your answers to questions 78 through 81 on the information below and on your knowledge of chemistry.

A student is to determine the concentration of an NaOH(aq) solution by performing two different titrations. In a first titration, the student titrates 25.0 mL of 0.100 M H_2SO_4(aq) with NaOH(aq) of unknown concentration.

In a second titration, the student titrates 25.0 mL of 0.100 M HCl(aq) with a sample of the NaOH(aq). During this second titration, the volume of the NaOH(aq) added and the corresponding pH value of the reaction mixture is measured. The graph below represents the relationship between pH and the volume of the NaOH(aq) added for this second titration.

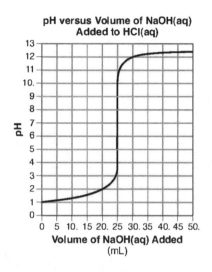

pH versus Volume of NaOH(aq)
Added to HCl(aq)

Volume of NaOH(aq) Added
(mL)

78 Identify the positive ion present in the H_2SO_4(aq) solution before the titration. [1]

79 Complete the equation *in your answer booklet* for the neutralization that occurs in the first titration by writing a formula of the missing product. [1]

80 Based on the graph, determine the volume of NaOH(aq) used to exactly neutralize the HCl(aq). [1]

81 State the color of phenolphthalein indicator if it were added after the HCl(aq) was titrated with 50. mL of NaOH(aq). [1]

Base your answers to questions 82 through 85 on the information below and on your knowledge of chemistry.

When uranium-235 nuclei are bombarded with neutrons, many different combinations of smaller nuclei can be produced. The production of neodymium-150 and germanium-81 in one of these reactions is represented by the equation below.

$$\ _0^1n + \ _{92}^{235}U \rightarrow \ _{60}^{150}Nd + \ _{32}^{81}Ge + 5\ _0^1n$$

Germanium-81 and uranium-235 have different decay modes. Ge-81 emits beta particles and has a half-life of 7.6 seconds.

82 Explain, in terms of nuclides, why the reaction represented by the nuclear equation is a fission reaction. [1]

83 State the number of protons and number of neutrons in a neodymium-150 atom. [1]

84 Complete the equation *in your answer booklet* for the decay of Ge-81 by writing a notation for the missing nuclide. [1]

85 Determine the time required for a 16.00-gram sample of Ge-81 to decay until only 1.00 gram of the sample remains unchanged. [1]

The University of the State of New York

REGENTS HIGH SCHOOL EXAMINATION

PHYSICAL SETTING
CHEMISTRY

Thursday, August 16, 2018 — 8:30 to 11:30 a.m., only

ANSWER BOOKLET

Student...

Teacher...

School... Grade ....

Record your answers for Part B–2 and Part C in this booklet.

Part B–2

51

52 _____

53 _____

_____

_____

54 _____

55 _____

_____

_____

56 _____

_____

_____

57 _____

_____

_____

58 _____

_____

_____

59 _____ g

60 _____

_____

_____

61 _____

_____

_____

62

Key
◯ = a molecule of the substance

High Marks: Regents Chemistry Made Easy   Page Regents-77

63 _____

_____

_____

64

Heating Curve

65 _____

_____

_____

Part C

66 _____

_____

_____

67

68 _____

_____

_____

69 _____

_____

_____

70 _____ %

71 _____

_____

_____

72

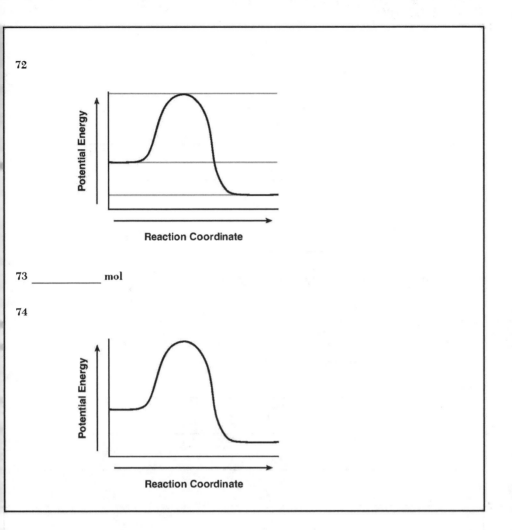

73 _____ mol

74

High Marks: Regents Chemistry Made Easy   Page Regents-81

75 _____

76 _____

77 _____

_____

_____

78 _____

79 $2NaOH(aq) + H_2SO_4(aq) \rightarrow 2H_2O(\ell) +$ _____ (aq)

80 _____ mL

81 _____

82 _____

_____

_____

83 Protons: _____

Neutrons: _____

84 $^{81}_{32}Ge \rightarrow ^{\;\;0}_{-1}e +$ _____

85 _____ s

    High Marks: Regents Chemistry Made Easy

ENTS HIGH SCHOOL EXAMINATION

Part A

Date: Friday, January 25, 2019
Time: 9:15 a.m. to 12:15 p.m., only

Answer all questions in this part.

Directions (1–30): For *each* statement or question, record on your separate answer sheet the *number* of the word or expression that, of those given, best completes the statement or answers the question. Some questions may require the use of the *2011 Edition Reference Tables for Physical Setting/Chemistry.*

1 The results of the gold foil experiment led to the conclusion that an atom is

(1) mostly empty space and has a small, negatively charged nucleus
(2) mostly empty space and has a small, positively charged nucleus
(3) a hard sphere and has a large, negatively charged nucleus
(4) a hard sphere and has a large, positively charged nucleus

2 Atoms are neutral because the number of

(1) protons equals the number of neutrons
(2) protons equals the number of electrons
(3) neutrons is greater than the number of protons
(4) neutrons is greater than the number of electrons

3 In the ground state, valence electrons of a krypton atom are found in

(1) the first shell
(2) the outermost shell
(3) both the nucleus and the first shell
(4) both the first shell and the outermost shell

4 According to the wave-mechanical model of the atom, electrons are located in

(1) orbitals
(2) circular paths
(3) a small, dense nucleus
(4) a hard, indivisible sphere

5 Which electron configuration represents the electrons in an atom of sodium in the ground state at STP?

(1) 2-8-1   (3) 2-8-6
(2) 2-7-2   (4) 2-7-7

6 The elements on the Periodic Table of the Elements are arranged in order of increasing

(1) atomic number
(2) mass number
(3) number of neutrons
(4) number of valence electrons

7 Which element is malleable at STP?

(1) chlorine   (3) helium
(2) copper   (4) sulfur

8 At 298 K and 1 atm, which noble gas has the lowest density?

(1) Ne   (3) Xe
(2) Kr   (4) Rn

9 Which two terms represent types of chemical formulas?

(1) empirical and molecular
(2) polar and nonpolar
(3) synthesis and decomposition
(4) saturated and concentrated

10 Which quantities are conserved in all chemical reactions?

(1) charge, pressure, and energy
(2) charge, mass, and energy
(3) volume, pressure, and energy
(4) volume, mass, and pressure

11 Which term represents the sum of the atomic masses of the atoms in a molecule?

(1) atomic number
(2) mass number
(3) formula mass
(4) percent composition by mass

12 Which equation represents energy being absorbed as a bond is broken?

(1) $H + H \rightarrow H_2$ + energy
(2) $H + H$ + energy $\rightarrow H_2$
(3) $H_2 \rightarrow H + H$ + energy
(4) H_2 + energy $\rightarrow H + H$

13 Which term is used to describe the attraction that an oxygen atom has for the electrons in a chemical bond?

(1) alkalinity
(2) electronegativity
(3) electron configuration
(4) first ionization energy

14 Which substance can *not* be decomposed by chemical means?

(1) C   (3) CO_2
(2) CO   (4) C_3O_2

15 A beaker contains a dilute sodium chloride solution at 1 atmosphere. What happens to the number of solute particles in the solution and the boiling point of the solution, as more sodium chloride is dissolved?

(1) The number of solute particles increases, and the boiling point increases.
(2) The number of solute particles increases, and the boiling point decreases.
(3) The number of solute particles decreases, and the boiling point increases.
(4) The number of solute particles decreases, and the boiling point decreases.

16 Which form of energy is transferred when an ice cube at 0°C is placed in a beaker of water at 50°C?

(1) chemical   (3) nuclear
(2) electrical   (4) thermal

17 The average kinetic energy of the particles in a sample of matter is expressed as

(1) density   (3) pressure
(2) volume   (4) temperature

18 At STP, which gas sample has the same number of molecules as 2.0 liters of $CH_4(g)$ at STP?

(1) 1.0 liter of $C_2H_6(g)$
(2) 2.0 liters of $O_2(g)$
(3) 5.0 liters of $N_2(g)$
(4) 6.0 liters of $CO_2(g)$

19 Given the equation:

$$I_2(s) \rightarrow I_2(g)$$

Which phrase describes this change?

(1) endothermic chemical change
(2) endothermic physical change
(3) exothermic chemical change
(4) exothermic physical change

20 Which term identifies a factor that will shift chemical equilibrium?

(1) atomic radius   (3) decay mode
(2) catalyst   (4) temperature

21 According to which theory or law is a chemical reaction most likely to occur when two particles with the proper energy and orientation interact with each other?

(1) atomic theory
(2) collision theory
(3) combined gas law
(4) law of conservation of matter

22 Addition of a catalyst can speed up a reaction by providing an alternate reaction pathway that has a

(1) lower activation energy
(2) higher activation energy
(3) lower heat of reaction
(4) higher heat of reaction

23 Which compound is saturated?

(1) butane   (3) heptene
(2) ethene   (4) pentyne

High Marks: Regents Chemistry Made Easy

24 An alcohol and an ether have the same molecular formula, C_2H_6O. These two compounds have

(1) the same functional group and the same physical and chemical properties
(2) the same functional group and different physical and chemical properties
(3) different functional groups and the same physical and chemical properties
(4) different functional groups and different physical and chemical properties

25 Which metal is most easily oxidized?

(1) Ag   (3) Cu
(2) Co   (4) Mg

26 Which substance is an Arrhenius acid?

(1) H_2 (3) KCl
(2) HCl   (4) NH_3

27 Which statement describes an electrolyte?

(1) An electrolyte conducts an electric current as a solid and dissolves in water.
(2) An electrolyte conducts an electric current as a solid and does not dissolve in water.
(3) When an electrolyte dissolves in water, the resulting solution conducts an electric current.
(4) When an electrolyte dissolves in water, the resulting solution does not conduct an electric current.

28 Which type of reaction occurs when an Arrhenius acid reacts with an Arrhenius base to form a salt and water?

(1) combustion   (3) neutralization
(2) decomposition   (4) saponification

29 Compared to the energy released per mole of reactant during chemical reactions, the energy released per mole of reactant during nuclear reactions is

(1) much less   (3) slightly less
(2) much greater   (4) slightly greater

30 Which phrase describes a risk of using the radioisotope Co-60 in treating cancer?

(1) production of acid rain
(2) production of greenhouse gases
(3) increased biological exposure
(4) increased ozone depletion

Answer all questions in this part.

Directions (31–50): For *each* statement or question, record on your separate answer sheet the *number* of the word or expression that, of those given, best completes the statement or answers the question. Some questions may require the use of the *2011 Edition Reference Tables for Physical Setting/Chemistry*.

31 The three nuclides, U-233, U-235, and U-238, are isotopes of uranium because they have the same number of protons per atom and

(1) the same number of electrons per atom
(2) the same number of neutrons per atom
(3) a different number of electrons per atom
(4) a different number of neutrons per atom

32 Given the information in the table below:

Two Forms of Carbon

Form	Bonding	Hardness	Electrical Conductivity
diamond	Each carbon atom bonds to four other carbon atoms in a three-dimensional network.	very hard	no
graphite	Each carbon atom bonds to three other carbon atoms in two-dimensional sheets.	soft	yes

Diamond and graphite have different properties because they have different

(1) crystal structures
(2) electronegativities
(3) numbers of protons per atom
(4) numbers of valence electrons per atom

33 Given the equation representing a chemical reaction:

$$NaCl(aq) + AgNO_3(aq) \rightarrow NaNO_3(aq) + AgCl(s)$$

This reaction is classified as a

(1) synthesis reaction
(2) decomposition reaction
(3) single replacement reaction
(4) double replacement reaction

High Marks: Regents Chemistry Made Easy

34 What is the formula for iron(II) oxide?
(1) FeO   (3) Fe_2O
(2) FeO_2 (4) Fe_2O_3

35 Given the reaction:

$$2KClO_3(s) \rightarrow 2KCl(s) + 3O_2(g)$$

How many moles of $KClO_3$ must completely react to produce 6 moles of O_2?

(1) 1 mole   (3) 6 moles
(2) 2 moles   (4) 4 moles

36 What is the number of moles of CO_2 in a 220.-gram sample of CO_2 (gram-formula mass = 44 g/mol)?

(1) 0.20 mol   (3) 15 mol
(2) 5.0 mol   (4) 44 mol

37 A solution contains 25 grams of KNO_3 dissolved in 200. grams of H_2O. Which numerical setup can be used to calculate the percent by mass of KNO_3 in this solution?

(1) $\dfrac{25\,g}{175\,g} \times 100$ (3) $\dfrac{25\,g}{225\,g} \times 100$

(2) $\dfrac{25\,g}{200.\,g} \times 100$ (4) $\dfrac{200.\,g}{225\,g} \times 100$

38 What is the molarity of 0.50 liter of an aqueous solution that contains 0.20 mole of NaOH (gram-formula mass = 40. g/mol)?

(1) 0.10 M   (3) 2.5 M
(2) 0.20 M   (4) 0.40 M

39 A mixture consists of ethanol and water. Some properties of ethanol and water are given in the table below.

Some Properties of Ethanol and Water

Property	Ethanol	Water
boiling point at standard pressure	78°C	100.°C
density at STP	0.80 g/cm³	1.00 g/cm³
flammability	flammable	nonflammable
melting point	−114°C	0.°C

Which statement describes a property of ethanol after being separated from the mixture?

(1) Ethanol is nonflammable.
(2) Ethanol has a melting point of 0.°C.
(3) Ethanol has a density of 0.80 g/cm³ at STP.
(4) Ethanol has a boiling point of 89°C at standard pressure.

40 A rigid cylinder with a movable piston contains a sample of hydrogen gas. At 330. K, this sample has a pressure of 150. kPa and a volume of 3.50 L. What is the volume of this sample at STP?

(1) 0.233 L   (3) 4.29 L
(2) 1.96 L   (4) 6.26 L

41 Which numerical setup can be used to calculate the heat energy required to completely melt 100. grams of $H_2O(s)$ at 0°C?

(1) (100. g)(334 J/g)
(2) (100. g)(2260 J/g)
(3) (100. g)(4.18 J/g•K)(0°C)
(4) (100. g)(4.18 J/g•K)(273 K)

42 During which phase change does the entropy of a sample of H_2O increase?

(1) $H_2O(g) \rightarrow H_2O(\ell)$
(2) $H_2O(g) \rightarrow H_2O(s)$
(3) $H_2O(\ell) \rightarrow H_2O(g)$
(4) $H_2O(\ell) \rightarrow H_2O(s)$

43 Given the formula for a compound:

$$H-\overset{\overset{\displaystyle H}{|}}{\underset{\underset{\displaystyle H}{|}}{C}}-\overset{\overset{\displaystyle H}{|}}{\underset{\underset{\displaystyle H}{|}}{C}}-\overset{\overset{\displaystyle H}{|}}{\underset{\underset{\displaystyle H}{|}}{C}}-\overset{\overset{\displaystyle H}{|}}{\underset{\underset{\displaystyle H}{|}}{C}}-N\overset{\displaystyle H}{\underset{\displaystyle H}{}}$$

What is a chemical name for this compound?

(1) 1-butanamide   (3) 1-butanamine
(2) 4-butanamide   (4) 4-butanamine

44 Given the equation for a reaction:

$$C_4H_{10} + Cl_2 \rightarrow C_4H_9Cl + HCl$$

Which type of reaction is represented by the equation?

(1) addition   (3) fermentation
(2) substitution   (4) polymerization

45 Which half-reaction equation represents reduction?

(1) $Cu \rightarrow Cu^{2+} + 2e^-$
(2) $Cu^{2+} + 2e^- \rightarrow Cu$
(3) $Ag + e^- \rightarrow Ag^+$
(4) $Ag^+ \rightarrow Ag + e^-$

46 Given the balanced ionic equation representing a reaction:

$$Zn(s) + Co^{2+}(aq) \rightarrow Zn^{2+}(aq) + Co(s)$$

Which statement describes the electrons involved in this reaction?

(1) Each Zn atom loses 2 electrons, and each Co^{2+} ion gains 2 electrons.
(2) Each Zn atom loses 2 electrons, and each Co^{2+} ion loses 2 electrons.
(3) Each Zn atom gains 2 electrons, and each Co^{2+} ion loses 2 electrons.
(4) Each Zn atom gains 2 electrons, and each Co^{2+} ion gains 2 electrons.

47 What are the two oxidation states of nitrogen in NH_4NO_2?

(1) +3 and +5   (3) −3 and +3
(2) +3 and −5   (4) −3 and −3

48 The table below shows the molar concentrations of hydronium ion, H_3O^+, in four different solutions.

Molar Concentration of H_3O^+ Ions in Four Solutions

Solution	Molar Concentration of H_3O^+ Ion (M)
A	0.1
B	0.01
C	0.001
D	0.0001

Which solution has the highest pH?

(1) A   (3) C
(2) B   (4) D

49 Given the equation:

$$^{235}_{92}U + ^{1}_{0}n \rightarrow ^{140}_{56}Ba + ^{93}_{36}Kr + 3^{1}_{0}n + energy$$

Which type of nuclear reaction is represented by the equation?

(1) fission   (3) beta decay
(2) fusion   (4) alpha decay

50 Which nuclear emission has the *least* penetrating power and the greatest ionizing ability?

(1) alpha particle   (3) gamma ray
(2) beta particle   (4) positron

Part B–2

Answer all questions in this part.

Directions (51–65): Record your answers in the spaces provided in your answer booklet. Some questions may require the use of the *2011 Edition Reference Tables for Physical Setting/Chemistry.*

Base your answers to questions 51 through 54 on the information below and on your knowledge of chemistry.

The formulas and names of four chloride compounds are shown in the table below.

Formula	Name
CCl_4	carbon tetrachloride
RbCl	rubidium chloride
CsCl	cesium chloride
HCl	hydrogen chloride

51 Identify the noble gas that has atoms with the same electron configuration as the metal ions in rubidium chloride, when both the atoms and the ions are in the ground state. [1]

52 Explain, in terms of atomic structure, why the radius of a cesium ion in cesium chloride is smaller than the radius of a cesium atom when both are in the ground state. [1]

53 In the space *in your answer booklet*, draw a Lewis electron-dot diagram for a molecule of HCl. [1]

54 Explain, in terms of charge distribution, why a molecule of carbon tetrachloride is a nonpolar molecule. [1]

Base your answers to questions 55 through 57 on the information below and on your knowledge of chemistry.

Some isotopes of neon are Ne-19, Ne-20, Ne-21, Ne-22, and Ne-24. The neon-24 decays by beta emission. The atomic mass and natural abundance for the naturally occurring isotopes of neon are shown in the table below.

Naturally Occurring Isotopes of Neon

Isotope Notation	Atomic Mass (u)	Natural Abundance (%)
Ne-20	19.99	90.48
Ne-21	20.99	0.27
Ne-22	21.99	9.25

55 Identify the decay mode of Ne-19. [1]

56 State the number of neutrons in an atom of Ne-20 and the number of neutrons in an atom of Ne-22. [1]

57 Show a numerical setup for calculating the atomic mass of neon. [1]

Base your answers to questions 58 through 60 on the information below and on your knowledge of chemistry.

Periodic trends are observed in the properties of the elements in Period 3 on the Periodic Table. These elements vary in physical properties, such as phase, and in chemical properties, such as their ability to lose or gain electrons during a chemical reaction.

58 Identify the metals in Period 3 on the Periodic Table. [1]

59 Identify the element in Period 3 that requires the *least* amount of energy to remove the most loosely held electrons from a mole of gaseous atoms of the element in the ground state. [1]

60 State the general trend in atomic radius as the elements in Period 3 are considered in order of increasing atomic number. [1]

Base your answers to questions 61 through 63 on the information below and on your knowledge of chemistry.

A thiol is very similar to an alcohol, but a thiol has a sulfur atom instead of an oxygen atom in the functional group. The equation below represents a reaction of methanethiol and iodine, producing dimethyl disulfide and hydrogen iodide.

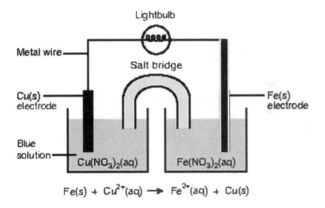

Methanethiol   Dimethyl disulfide

61 State the number of electrons shared between the sulfur atoms in the dimethyl disulfide. [1]

62 Identify the polarity of an H–I bond and the polarity of an S–S bond. [1]

63 Explain, in terms of electron configuration, why sulfur atoms and oxygen atoms form compounds with similar molecular structures. [1]

Base your answers to questions 64 and 65 on the information below and on your knowledge of chemistry.

A student constructs an electrochemical cell. A diagram of the operating cell and the unbalanced ionic equation representing the reaction occurring in the cell are shown below. The blue color of the solution in the copper half-cell indicates the presence of Cu^{2+} ions. The student observes that the blue color becomes less intense as the cell operates.

Lightbulb

Metal wire

Salt bridge

Cu(s) electrode   Fe(s) electrode

Blue solution

$Cu(NO_3)_2(aq)$ $Fe(NO_3)_2(aq)$

$$Fe(s) + Cu^{2+}(aq) \longrightarrow Fe^{2+}(aq) + Cu(s)$$

64 Identify the type of electrochemical cell represented by the diagram. [1]

65 State one inference that the student can make about the concentration of the Cu^{2+} ions based on the change in intensity of the color of the $Cu(NO_3)_2(aq)$ solution as the cell operates. [1]

Part C

Answer all questions in this part.

Directions (66–85): Record your answers in the spaces provided in your answer booklet. Some questions may require the use of the *2011 Edition Reference Tables for Physical Setting/Chemistry*.

Base your answers to questions 66 through 69 on the information below and on your knowledge of chemistry.

> In a laboratory investigation, a student is given a sample that is a mixture of 3.0 grams of NaCl(s) and 4.0 grams of sand, which is mostly SiO$_2$(s). The purpose of the investigation is to separate and recover the compounds in the sample. In the first step, the student places the sample in a 250-mL flask. Then, 50. grams of distilled water are added to the flask, and the contents are thoroughly stirred. The mixture in the flask is then filtered, using the equipment represented by the diagram below.

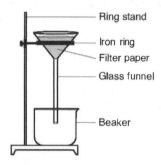

66 Explain, in terms of solubility, why the mixture in the flask remains heterogeneous even after thorough stirring. [1]

67 Based on Table G, state evidence that all of the NaCl(s) in the flask would dissolve in the distilled water at 20.°C. [1]

68 Describe a procedure to remove the water from the mixture that passes through the filter and collects in the beaker. [1]

69 The student reports that 3.4 grams of NaCl(s) were recovered from the mixture. Show a numerical setup for calculating the student's percent error. [1]

Base your answers to questions 70 through 73 on the information below and on your knowledge of chemistry.

In a laboratory activity, the volume of helium gas in a rigid cylinder with a movable piston is varied by changing the temperature of the gas. The activity is done at a constant pressure of 100. kPa. Data from the activity are plotted on the graph below.

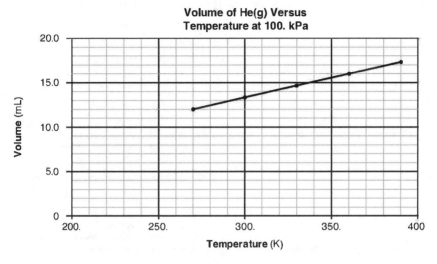

Volume of He(g) Versus Temperature at 100. kPa

70 Determine the temperature of the He(g) at a volume of 15.0 mL. [1]

71 Explain, in terms of particle volume, why the sample of helium can *not* be compressed by the piston to zero volume. [1]

72 State what happens to the average distance between the He atoms as the gas is heated. [1]

73 State a change in pressure that will cause the helium in the cylinder to behave more like an ideal gas. [1]

Base your answers to questions 74 through 76 on the information below and on your knowledge of chemistry.

The balanced equation below represents the reaction between a 5.0-gram sample of zinc metal and a 0.5 M solution of hydrochloric acid. The reaction takes place in an open test tube at 298 K and 1 atm in a laboratory activity.

$$Zn(s) + 2HCl(aq) \rightarrow H_2(g) + ZnCl_2(aq) + energy$$

74 State *one* change in reaction conditions, other than adding a catalyst, that will increase the rate of the reaction. [1]

75 On the labeled axes *in your answer booklet*, draw a potential energy diagram for this reaction. [1]

76 Explain why this reaction will *not* reach equilibrium. [1]

___

Base your answers to questions 77 through 79 on the information below and on your knowledge of chemistry.

Crude oil, primarily a mixture of hydrocarbons, is separated into useful components in a fractionating tower. At the bottom of the tower, the crude oil is heated to about 400°C. The gases formed rise and cool. Most of the gases condense and are collected as liquid fractions. The table below shows the temperature ranges for collecting various hydrocarbon fractions.

Hydrocarbon Fractions Collected

Number of Carbon Atoms per Molecule	Temperature Range (°C)
1-4	below 40
5-12	40-200
12-16	200-300
16-20	300-370
>20	above 370

77 Determine the number of carbon atoms in one molecule of an alkane that has 22 hydrogen atoms in the molecule. [1]

78 State the temperature range for the fraction collected that contains octane molecules. [1]

79 Draw a structural formula for 3-ethylhexane. [1]

___

Base your answers to questions 80 through 82 on the information below and on your knowledge of chemistry.

In a laboratory activity, a student titrates a 20.0-milliliter sample of HCl(aq) using 0.025 M NaOH(aq). In one of the titration trials, 17.6 milliliters of the base solution exactly neutralizes the acid sample.

80 Identify the positive ion in the sample of HCl(aq). [1]

81 Show a numerical setup for calculating the concentration of the hydrochloric acid using the titration data. [1]

82 The concentration of the base is expressed to what number of significant figures? [1]

Base your answers to questions 83 through 85 on the information below and on your knowledge of chemistry.

In the past, some paints that glowed in the dark contained zinc sulfide and salts of Ra-226. As the radioisotope Ra-226 decayed, the energy released caused the zinc sulfide in these paints to emit light. The half-lives for Ra-226 and two other radioisotopes used in these paints are listed on the table below.

Radioisotopes in the Paints

Radioisotope	Half-Life (y)
Pm-147	2.6
Ra-226	1599
Ra-228	5.8

83 Explain, in terms of half-lives, why Ra-226 may have been used more often than the other isotopes in these paints. [1]

84 Complete the nuclear equation *in your answer booklet* for the beta decay of Pm-147 by writing an isotopic notation for the missing product. [1]

85 What fraction of an original Ra-228 sample remains unchanged after 17.4 years? [1]

The University of the State of New York

REGENTS HIGH SCHOOL EXAMINATION

PHYSICAL SETTING
CHEMISTRY

Friday, January 25, 2019 — 9:15 a.m to 12:15 p.m., only

ANSWER BOOKLET

Student...

Teacher...

School .... Grade ....

Record your answers for Part B–2 and Part C in this booklet.

Part B–2
51
52
53

54 _____

_____

_____

55 _____

56 Ne-20: _____

Ne-22: _____

57

58 _____

59 _____

60 _____

_____

_____

61 _____

62 H–I bond: _____

S–S bond: _____

63 _____

_____

_____

64 _____

65 _____

_____

_____

Part C

66 _____

_____

_____

67 _____

_____

_____

68 _____

_____

_____

69

P.S./Chem. Answer Booklet–Jan. '19   [4]

Page Regents-100   **High Marks: Regents Chemistry Made Easy**

70 _____ K

71 _____

_____

_____

72 _____

_____

_____

73 _____

74 _____

_____

_____

75

Potential Energy

Reaction Coordinate

76 _____

_____

_____

77 _____

78 _____ °C to _____ °C

79

80 _____

81

82 _____

83 _____

_____

_____

84 $^{147}_{61}Pm \rightarrow ^{0}_{-1}e +$ _____

85 _____

P.S./Chem. Answer Booklet–Jan. '19   [8]   Printed on Recycled Paper

P.S./CHEMISTRY

    High Marks: Regents Chemistry Made Easy

GENTS HIGH SCHOOL EXAMINATION

Part A

Date: Tuesday June 25, 2019
Time: 9:15 a.m. to 12:15 p.m., only

Answer all questions in this part.

Directions (1–30): For *each* statement or question, record on your separate answer sheet the *number* of the word or expression that, of those given, best completes the statement or answers the question. Some questions may require the use of the *2011 Edition Reference Tables for Physical Setting/Chemistry*.

1 Which particles are found in the nucleus of an argon atom?

(1) protons and electrons
(2) positrons and neutrons
(3) protons and neutrons
(4) positrons and electrons

2 The diagram below represents a particle traveling through an electric field.

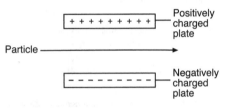

Particle ⟶

An electric field exists between the two plates.

Which particle remains undeflected when passing through this electric field?

(1) proton   (3) neutron
(2) electron   (4) positron

3 The mass of an electron is

(1) equal to the mass of a proton
(2) equal to the mass of a neutron
(3) greater than the mass of a proton
(4) less than the mass of a neutron

4 Compared to the energy of an electron in the second shell of an atom of sulfur, the energy of an electron in the

(1) first shell is lower
(2) first shell is the same
(3) third shell is lower
(4) third shell is the same

5 In the ground state, an atom of which element has seven valence electrons?

(1) sodium   (3) nitrogen
(2) phosphorus   (4) fluorine

6 Which information is sufficient to differentiate a sample of sodium from a sample of silver?

(1) the mass of each sample
(2) the volume of each sample
(3) the reactivity of each sample with water
(4) the phase of each sample at room temperature

7 Graphite and diamond are two forms of solid carbon at STP. These forms have

(1) different molecular structures and different properties
(2) different molecular structures and the same properties
(3) the same molecular structures and different properties
(4) the same molecular structures and the same properties

8 As the first five elements in Group 14 are considered in order from top to bottom, there are changes in both the

(1) number of valence shell electrons and number of first shell electrons
(2) electronegativity values and number of first shell electrons
(3) number of valence shell electrons and atomic radii
(4) electronegativity values and atomic radii

9 Which statement explains why NaBr is classified as a compound?

(1) Na and Br are chemically combined in a fixed proportion.
(2) Na and Br are both nonmetals.
(3) NaBr is a solid at 298 K and standard pressure.
(4) NaBr dissolves in H_2O at 298 K.

10 Which two terms represent types of chemical formulas?

(1) fission and fusion
(2) oxidation and reduction
(3) empirical and structural
(4) endothermic and exothermic

11 During all chemical reactions, charge, mass and energy are

(1) condensed   (3) decayed
(2) conserved   (4) decomposed

12 The degree of polarity of a covalent bond between two atoms is determined by calculating the difference in their

(1) atomic radii   (3) electronegativities
(2) melting points   (4) ionization energies

13 Which substance can *not* be broken down by a chemical change?

(1) ammonia   (3) methane
(2) magnesium   (4) water

14 Which statement describes the components of a mixture?

(1) Each component gains new properties.
(2) Each component loses its original properties.
(3) The proportions of components can vary.
(4) The proportions of components cannot vary.

15 Table sugar can be separated from a mixture of table sugar and sand at STP by adding

(1) sand, stirring, and distilling at 100.°C
(2) sand, stirring, and filtering
(3) water, stirring, and distilling at 100.°C
(4) water, stirring, and filtering

16 Which statement describes the particles of an ideal gas, based on the kinetic molecular theory?

(1) The volume of the particles is considered negligible.
(2) The force of attraction between the particles is strong.
(3) The particles are closely packed in a regular, repeating pattern.
(4) The particles are separated by small distances, relative to their size.

17 During which two processes does a substance release energy?

(1) freezing and condensation
(2) freezing and melting
(3) evaporation and condensation
(4) evaporation and melting

18 Based on Table *I*, which compound dissolves in water by an exothermic process?

(1) NaCl   (3) NH_4Cl
(2) NaOH   (4) NH_4NO_3

19 At STP, which property of a molecular substance is determined by the arrangement of its molecules?

(1) half-life
(2) molar mass
(3) physical state
(4) percent composition

20 Equilibrium can be reached by

(1) physical changes, only
(2) nuclear changes, only
(3) both physical changes and chemical changes
(4) both nuclear changes and chemical changes

21 Which value is defined as the difference between the potential energy of the products and the potential energy of the reactants during a chemical change?

(1) heat of fusion
(2) heat of reaction
(3) heat of deposition
(4) heat of vaporization

22 The effect of a catalyst on a chemical reaction is to provide a new reaction pathway that results in a different

(1) potential energy of the products
(2) heat of reaction
(3) potential energy of the reactants
(4) activation energy

23 Chemical systems in nature tend to undergo changes toward

(1) lower energy and lower entropy
(2) lower energy and higher entropy
(3) higher energy and lower entropy
(4) higher energy and higher entropy

24 The atoms of which element bond to one another in chains, rings, and networks?

(1) barium   (3) iodine
(2) carbon   (4) mercury

25 What is the general formula for the homologous series that includes ethene?

(1) C_nH_{2n} (3) C_nH_{2n-2}
(2) C_nH_{2n-6} (4) C_nH_{2n+2}

26 When an F atom becomes an F^- ion, the F atom

(1) gains a proton   (3) gains an electron
(2) loses a proton   (4) loses an electron

27 Which substance is an Arrhenius base?

(1) HNO_3 (3) $Ca(OH)_2$
(2) H_2SO_3 (4) CH_3COOH

28 In which type of nuclear reaction do two light nuclei combine to produce a heavier nucleus?

(1) positron emission   (3) fission
(2) gamma emission   (4) fusion

29 Using equal masses of reactants, which statement describes the relative amounts of energy released during a chemical reaction and a nuclear reaction?

(1) The chemical and nuclear reactions release equal amounts of energy.
(2) The nuclear reaction releases half the amount of energy of the chemical reaction.
(3) The chemical reaction releases more energy than the nuclear reaction.
(4) The nuclear reaction releases more energy than the chemical reaction.

30 The ratio of the mass of U-238 to the mass of Pb-206 can be used to

(1) diagnose thyroid disorders
(2) diagnose kidney function
(3) date geological formations
(4) date once-living things

Answer all questions in this part.

Directions (31–50): For *each* statement or question, record on your separate answer sheet the *number* of the word or expression that, of those given, best completes the statement or answers the question. Some questions may require the use of the *2011 Edition Reference Tables for Physical Setting/Chemistry*.

31 The bright-line spectra of four elements, *G, J, L,* and *M,* and a mixture of *at least two* of these elements is given below.

Bright-Line Spectra

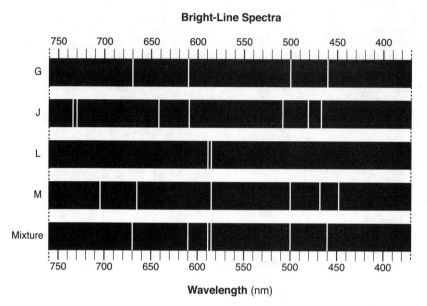

Wavelength (nm)

Which elements are present in the mixture?

(1) *G* and *J*   (3) *M, J,* and *G*
(2) *G* and *L*   (4) *M, J,* and *L*

32 Which electron configuration represents an atom of chlorine in an excited state?

(1) 2-8-7-2   (3) 2-8-8
(2) 2-8-7   (4) 2-7-8

33 A student measures the mass and volume of a sample of aluminum at room temperature, and calculates the density of Al to be 2.85 grams per cubic centimeter. Based on Table *S*, what is the percent error for the student's calculated density of Al?

(1) 2.7%   (3) 5.6%
(2) 5.3%   (4) 95%

34 Magnesium and calcium have similar chemical properties because their atoms in the ground state have
(1) equal numbers of protons and electrons
(2) equal numbers of protons and neutrons
(3) two electrons in the first shell
(4) two electrons in the outermost shell

35 As the elements in Period 2 of the Periodic Table are considered in order from left to right, which property generally *decreases*?
(1) atomic radius   (3) ionization energy
(2) electronegativity   (4) nuclear charge

36 Given the balanced equation for the reaction of butane and oxygen:

$$2C_4H_{10} + 13O_2 \rightarrow 8CO_2 + 10H_2O + energy$$

How many moles of carbon dioxide are produced when 5.0 moles of butane react completely?
(1) 5.0 mol   (3) 20. mol
(2) 10. mol   (4) 40. mol

37 What is the percent composition by mass of nitrogen in the compound N_2H_4 (gram-formula mass = 32 g/mol)?
(1) 13%   (3) 88%
(2) 44%   (4) 93%

38 Which ion in the ground state has the same electron configuration as an atom of neon in the ground state?
(1) Ca^{2+} (3) Li^+
(2) Cl^- (4) O^{2-}

39 The molar masses and boiling points at standard pressure for four compounds are given in the table below.

Compound	Molar Mass (g/mol)	Boiling Point (K)
HF	20.01	293
HCl	36.46	188
HBr	80.91	207
HI	127.91	237

Which compound has the strongest intermolecular forces?
(1) HF   (3) HBr
(2) HCl   (4) HI

40 Which particle model diagram represents xenon at STP?

Key
O = an atom of xenon

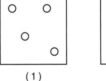

(1)

(3)

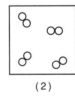

(2)

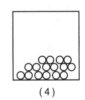

(4)

41 What is the amount of heat absorbed when the temperature of 75 grams of water increases from 20.°C to 35°C?
(1) 1100 J   (3) 6300 J
(2) 4700 J   (4) 11 000 J

42 Which sample of HCl(aq) reacts at the fastest rate with a 1.0-gram sample of iron filings?

(1) 10. mL of 1 M HCl(aq) at 10.°C
(2) 10. mL of 1 M HCl(aq) at 25°C
(3) 10. mL of 3 M HCl(aq) at 10.°C
(4) 10. mL of 3 M HCl(aq) at 25°C

43 Given the equation representing a system at equilibrium:

$$N_2O_4(g) \rightleftharpoons 2NO_2(g)$$

Which statement describes the concentration of the two gases in this system?

(1) The concentration of $N_2O_4(g)$ must be less than the concentration of $NO_2(g)$.

(2) The concentration of $N_2O_4(g)$ must be greater than the concentration of $NO_2(g)$.

(3) The concentration of $N_2O_4(g)$ and the concentration of $NO_2(g)$ must be equal.

(4) The concentration of $N_2O_4(g)$ and the concentration of $NO_2(g)$ must be constant.

44 Given the equation representing a system at equilibrium:

$$PCl_5(g) + energy \rightleftharpoons PCl_3(g) + Cl_2(g)$$

Which change will cause the equilibrium to shift to the right?

(1) adding a catalyst
(2) adding more $PCl_3(g)$
(3) increasing the pressure
(4) increasing the temperature

45 Given the formula representing a molecule:

$$H-\overset{\overset{\displaystyle H}{|}}{\underset{\underset{\displaystyle H}{|}}{C}}-\overset{\overset{\displaystyle H}{|}}{\underset{\underset{\displaystyle H}{|}}{C}}-\overset{\overset{\displaystyle H}{|}}{\underset{\underset{\displaystyle H}{|}}{C}}-\overset{\overset{\displaystyle H}{|}}{\underset{\underset{\displaystyle H}{|}}{C}}-\overset{\overset{\displaystyle H}{|}}{\underset{\underset{\displaystyle H}{|}}{C}}-N\overset{\diagup H}{\diagdown_H}$$

A chemical name for this compound is

(1) pentanone   (3) 1-pentanamine
(2) 1-pentanol   (4) pentanamide

46 Given the formula of a compound:

$$H-C\equiv C-\overset{\overset{\displaystyle H}{|}}{\underset{\underset{\displaystyle H}{|}}{C}}-\overset{\overset{\displaystyle H}{|}}{\underset{\underset{\displaystyle H}{|}}{C}}-H$$

This compound is classified as an

(1) aldehyde   (3) alkyne
(2) alkene   (4) alcohol

47 Which equation represents fermentation?

(1) $C_2H_4 + H_2O \rightarrow CH_3CH_2OH$
(2) $C_2H_4 + HCl \rightarrow CH_3CH_2Cl$
(3) $C_6H_{12}O_6 \rightarrow 2CH_3CH_2OH + 2CO_2$
(4) $2CH_3CHO \rightarrow C_3H_5CHO + H_2O$

48 Given the equation representing a reaction:

$$3CuCl_2(aq) + 2Al(s) \rightarrow 3Cu(s) + 2AlCl_3(aq)$$

The oxidation number of copper changes from

(1) +1 to 0   (3) +2 to +1
(2) +2 to 0   (4) +6 to +3

49 Given the equation representing a reversible reaction:

$$CH_3COOH(aq) + H_2O(\ell) \rightleftharpoons$$
$$CH_3COO^-(aq) + H_3O^+(aq)$$

According to one acid-base theory, the two H^+ donors in the equation are

(1) CH_3COOH and H_2O
(2) CH_3COOH and H_3O^+
(3) CH_3COO^- and H_2O
(4) CH_3COO^- and H_3O^+

50 Which nuclear equation represents a spontaneous decay?

(1) $^{222}_{86}Rn \rightarrow ^{218}_{84}Po + ^{4}_{2}He$

(2) $^{27}_{13}Al + ^{4}_{2}He \rightarrow ^{30}_{15}P + ^{1}_{0}n$

(3) $^{235}_{92}U + ^{1}_{0}n \rightarrow ^{139}_{56}Ba + ^{94}_{36}Kr + 3^{1}_{0}n$

(4) $^{7}_{3}Li + ^{1}_{1}H \rightarrow ^{4}_{2}He + ^{4}_{2}He$

Part B–2

Answer all questions in this part.

Directions (51–65): Record your answers in the spaces provided in your answer booklet. Some questions may require the use of the *2011 Edition Reference Tables for Physical Setting/Chemistry.*

51 Draw a structural formula for methanal. [1]

Base your answers to questions 52 through 54 on the information below and on your knowledge of chemistry.

The atomic mass and natural abundance of the naturally occuring isotopes of hydrogen are shown in the table below.

Naturally Occuring Isotopes of Hydrogen

Isotope	Common Name of Isotope	Atomic Mass (u)	Natural Abundance (%)
H-1	protium	1.0078	99.9885
H-2	deuterium	2.0141	0.0115
H-3	tritium	3.0160	negligible

The isotope H-2, also called deuterium, is usually represented by the symbol "D." Heavy water forms when deuterium reacts with oxygen, producing molecules of D_2O.

52 Explain, in terms of subatomic particles, why atoms of H-1, H-2, and H-3 are each electrically neutral. [1]

53 Determine the formula mass of heavy water, D_2O. [1]

54 Based on Table *N*, identify the decay mode of tritium. [1]

Base your answers to questions 55 through 57 on the information below and on your knowledge of chemistry.

At 23°C, 85.0 grams of $NaNO_3(s)$ are dissolved in 100. grams of $H_2O(\ell)$.

55 Convert the temperature of the $NaNO_3(s)$ to kelvins. [1]

56 Based on Table *G*, determine the additional mass of $NaNO_3(s)$ that must be dissolved to saturate the solution at 23°C. [1]

57 State what happens to the boiling point and freezing point of the solution when the solution is diluted with an additional 100. grams of $H_2O(\ell)$. [1]

Base your answers to questions 58 through 61 on the information below and on your knowledge of chemistry.

A 200.-milliliter sample of $CO_2(g)$ is placed in a sealed, rigid cylinder with a movable piston at 296 K and 101.3 kPa.

58 State a change in temperature and a change in pressure of the $CO_2(g)$ that would cause it to behave more like an ideal gas. [1]

59 Determine the volume of the sample of $CO_2(g)$ if the temperature and pressure are changed to 336 K and 152.0 kPa. [1]

60 State, in terms of *both* the frequency and force of collisions, what would result from decreasing the temperature of the original sample of $CO_2(g)$, at constant volume. [1]

61 Compare the mass of the original 200.-milliliter sample of $CO_2(g)$ to the mass of the $CO_2(g)$ sample when the cylinder is adjusted to a volume of 100. milliliters. [1]

High Marks: Regents Chemistry Made Easy

Base your answers to questions 62 through 65 on the information below and on your knowledge of chemistry.

Cobalt-60 is an artificial isotope of Co-59. The incomplete equation for the decay of cobalt-60, including beta and gamma emissions, is shown below.

$$^{60}_{27}\text{Co} \rightarrow X + {}^{0}_{-1}\text{e} + {}^{0}_{0}\gamma$$

62 Explain, in terms of *both* protons and neutrons, why Co-59 and Co-60 are isotopes of cobalt. [1]

63 Compare the penetrating power of the beta and gamma emissions. [1]

64 Complete the nuclear equation, *in your answer booklet*, for the decay of cobalt-60 by writing a notation for the missing product. [1]

65 Based on Table N, determine the total time required for an 80.00-gram sample of cobalt-60 to decay until only 10.00 grams of the sample remain unchanged. [1]

Part C

Answer all questions in this part.

Directions (66-85): Record your answers in the spaces provided in your answer booklet. Some questions may require the use of the *2011 Edition Reference Tables for Physical Setting/Chemistry.*

Base your answers to questions 66 through 69 on the information below and on your knowledge of chemistry.

During a laboratory activity, appropriate safety equipment was used and safety procedures were followed. A laboratory technician heated a sample of solid $KClO_3$ in a crucible to determine the percent composition by mass of oxygen in the compound. The unbalanced equation and the data for the decomposition of solid $KClO_3$ are shown below.

$$KClO_3(s) \rightarrow KCl(s) + O_2(g)$$

Lab Data and Calculated Results

Object or Material	Mass (g)
empty crucible and cover	22.14
empty crucible, cover, and $KClO_3$	24.21
$KClO_3$	2.07
crucible, cover, and KCl after heating	23.41
KCl	?
O_2	0.80

66 Write a chemical name for the compound that decomposed. [1]

67 Based on the lab data, show a numerical setup to determine the number of moles of O_2 produced. Use 32 g/mol as the gram-formula mass of O_2. [1]

68 Based on the lab data, determine the mass of KCl produced in the reaction. [1]

69 Balance the equation *in your answer booklet* for the decomposition of $KClO_3$, using the smallest whole-number coefficients. [1]

Base your answers to questions 70 through 73 on the information below and on your knowledge of chemistry.

A bottled water label lists the ions dissolved in the water. The table below lists the mass of some ions dissolved in a 500.-gram sample of the bottled water.

Ions in 500. g of Bottled Water

Ion Formula	Mass (g)
Ca^{2+}	0.040
Mg^{2+}	0.013
Na^+	0.0033
SO_4^{2-}	0.0063
HCO_3^-	0.180

70 State the number of significant figures used to express the mass of hydrogen carbonate ions in the table above. [1]

71 Based on Table F, write the formula of the ion in the bottled water table that would form the *least* soluble compound when combined with the sulfate ion. [1]

72 Show a numerical setup for calculating the parts per million of the Na^+ ions in the 500.-gram sample of the bottled water. [1]

73 Compare the radius of a Mg^{2+} ion to the radius of a Mg atom. [1]

Base your answers to questions 74 through 77 on the information below and on your knowledge of chemistry.

Ethyl ethanoate is used as a solvent for varnishes and in the manufacture of artificial leather. The formula below represents a molecule of ethyl ethanoate.

$$
\begin{array}{ccccc}
& H & O & & H & H \\
& | & || & & | & | \\
H- & C- & C- & O- & C- & C-H \\
& | & & & | & | \\
& H & & & H & H
\end{array}
$$

74 Identify the element in ethyl ethanoate that makes it an organic compound. [1]

75 Write the empirical formula for this compound. [1]

76 Write the name of the class of organic compounds to which this compound belongs. [1]

77 Determine the number of electrons shared in the bond between a hydrogen atom and a carbon atom in the molecule. [1]

Base your answers to questions 78 through 80 on the information below and on your knowledge of chemistry.

An operating voltaic cell has magnesium and silver electrodes. The cell and the ionic equation representing the reaction that occurs in the cell are shown below.

Voltaic Cell

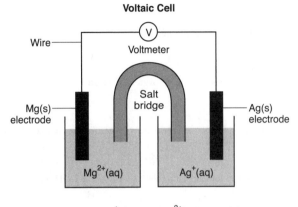

$$Mg(s) + 2Ag^+(aq) \longrightarrow Mg^{2+}(aq) + 2Ag(s)$$

78 State the purpose of the salt bridge in this cell. [1]

79 Write a balanced equation for the half-reaction that occurs at the magnesium electrode in this cell. [1]

80 Explain, in terms of electrical energy, how electrolysis reactions differ from voltaic cell reactions. [1]

Base your answers to questions 81 through 85 on the information below and on your knowledge of chemistry.

In a laboratory investigation, an HCl(aq) solution with a pH value of 2 is used to determine the molarity of a KOH(aq) solution. A 7.5-milliliter sample of the KOH(aq) is exactly neutralized by 15.0 milliliters of the 0.010 M HCl(aq). During this laboratory activity, appropriate safety equipment is used and safety procedures are followed.

81 Determine the pH value of a solution that is ten times *less* acidic than the HCl(aq) solution. [1]

82 State the color of the indicator bromcresol green if it is added to a sample of the KOH(aq) solution. [1]

83 Complete the equation *in your answer booklet* by writing the chemical formula for *each* product. [1]

84 Show a numerical setup for calculating the molarity of the KOH solution. [1]

85 Explain, in terms of aqueous ions, why 15.0 mL of a 1.0 M HCl(aq) solution is a better conductor of electricity than 15.0 mL of a 0.010 M HCl(aq) solution. [1]

The University of the State of New York

REGENTS HIGH SCHOOL EXAMINATION

PHYSICAL SETTING
CHEMISTRY

Tuesday, June 25, 2019 — 9:15 a.m to 12:15 p.m., only

ANSWER BOOKLET

Student. . . . .

Teacher. . . . .

School . . . .   Grade . . . .

Record your answers for Part B–2 and Part C in this booklet.

Part B–2

51

52 _____

_____

_____

53 _____ u

54 _____

---

55 _____ K

56 _____ g

57 Boiling point: _____

Freezing point: _____

58 Temperature: _____

Pressure: _____

59 _____ mL

60 _____

_____

_____

61 _____

_____

_____

P.S./Chem. Answer Booklet–June '19   [2]

Page Regents-120   **High Marks: Regents Chemistry Made Easy**

62 _____

_____

_____

63 _____

_____

_____

64 $^{60}_{27}\text{Co} \rightarrow$ _____ $+ ^{0}_{-1}\text{e} + ^{0}_{0}\gamma$

65 _____ y

Part C

66 _____

67

68 _____ g

69 _____ $KClO_3(s) \rightarrow$ _____ $KCl(s) +$ _____ $O_2(g)$

70 _____

71 _____

72

73 _____

_____

_____

P.S./Chem. Answer Booklet–June '19   [4]

Page Regents-122   **High Marks: Regents Chemistry Made Easy**

74 _____

75 _____

76 _____

77 _____

78 _____

_____

_____

79 _____

80 _____

_____

_____

81 _____

82 _____

83 HCl(aq) + KOH(aq) → _____ + _____

84

85 _____

_____

_____